SPECTRA FOR THE IDENTIFICATION OF ADDITIVES IN FOOD PACKAGING

EUROPEAN COMMISSION
Edith CRESSON, Member of the Commission responsible
for Research, Innovation, Education, Training
and Youth
DG XII-C-05 - SMT Programme

European Commission
Directorate-General for Science, Research and Development
Standards Measurement and Testing Programme

Spectra for the Identification of Additives in Food Packaging

by

Ben van Lierop
Food Inspection Service, Utrecht, The Netherlands

Laurence Castle
Ministry of Agriculture, Fisheries and Food,
CSL Food Science Laboratory,
Norwich, United Kingdom

Alexandre Feigenbaum
Institut National de la Recherche Agronomique, Reims, France

and

Achim Boenke
European Commission, SMT Programme, DG XII/C5, Brussels, Belgium

KLUWER ACADEMIC PUBLISHERS
DORDRECHT / BOSTON / LONDON

A C.I.P. Catalogue record for this book is available from the Library of Congress.

ISBN 0-7923-4986-5

Published for the
European Commission
Directorate-General Science, Research and Development.

EUR 18027 EN

Coverdesign by Merel van Lierop

Published by Kluwer Academic Publishers,
P.O. Box 17, 3300 AA Dordrecht, The Netherlands.

Sold and distributed in the U.S.A. and Canada
by Kluwer Academic Publishers,
101 Philip Drive, Norwell, MA 02061, U.S.A.

In all other countries, sold and distributed
by Kluwer Academic Publishers,
P.O. Box 322, 3300 AH Dordrecht, The Netherlands.

Printed on acid-free paper

CONTENTS

CHAPTER 1

Introduction

This new book on additives used in plastics for food contact, can be seen as a companion to the 1993 publication *Spectra for the Identification of Monomers in Food Packaging* [1, 2]. That earlier book presented information on monomeric substances listed in Directive 90/128/EEC [3], which restricts the range of monomers and other starting substances that can be used for the production of plastics materials and articles intended for food contact applications.

As a logical supplement to the collection of monomers and other starting substances, the preparation of a reference collection and a Handbook of analytical data of additives was undertaken with funding from The European Commission under the Standards Measurements and Testing programme. We then give a collection of spectra for the identification of 100 of the most important additives used in plastics packaging and coatings. Infra-red (FT-IR) and mass spectra (MS) are presented, as in the monomers book, but we have extended the scope to include proton nuclear magnetic resonance (^{1}H-NMR) spectra and gas-chromatographic (GC) data.

Legal Framework

The Commission of the European Communities provides in Synoptic Document N. 7 [4] a provisional list of additives used for the production of food contact plastics. This Synoptic Document anticipates a Directive on additives for food contact plastics. One hundred of the most important additives were selected from this provisional list after extensive consultation with researchers in the field and with representatives from European industry (Food Contact Additives Panel (FCA) sector group of the European Chemical Industry Council (CEFIC).

Although the additives are listed in [4], it is still yet undecided exactly how control on their use would be best exercised. The two most likely forms of control are a restriction on the migration of the specific substances to food or food simulants (SMLs, specific migration limits) or controls on the composition of the plastic (QMs, maximum quantity of the substance in the material). Several studies have been carried out to inform this debate, to relate the maximum level of residual substance in the finished plastic (QM limit) with the amount of substance migrating to foods or food simulants under defined conditions (SML limit). Most recently, in the EU project AIR-941025 *Safety and Quality Control of Plastic Materials for Food Contact* [5] coordinated by one of us (Alexandre Feigenbaum, Institut National de la Recherche Agronomique), the substances in the polymer are first identified with the aid of a reference collection of additives. In subsequent parts of the project, a mathematical model was developed to relate SM values to QM values. It will be necessary to validate the migration model by generating the necessary experimental data.

These developments notwithstanding, in fact whichever form of control, SML or QM, is finally adopted for additives, the first step in any investigation must be to identify the substances that may migrate and therefore need to be quantified. This book delivers data for this identification.

Implementation of legislation

Although the principles intended to govern the control of materials and articles are clear from the relevant Directives, the practical problems of implementation and the development of an approach that should be adopted by enforcement authorities in real situations, have not yet been addressed. The most systematic approach to control has been elaborated in the Netherlands to meet Dutch Regulations [6-9] in existence before Directive 90/128/EEC. The approach used has been to initially identify polymeric materials by infra-red spectra and then to identify the substances in solvent extracts by gas-chromatography - mass spectrometry (GC-MS) and by liquid chromatography (LC). Practical application of this approach over more than ten years has shown that considerable experience is required in knowing what type of polymer is used in what food contact situation, as well as the likely additives and other constituents that might be present. The Dutch test method has been discussed in the CEN working group WG2 of Technical Committee TC 194 and a CEN standard is in preparation [10]. The Dutch method is applied in its original or slightly modified form in e.g. government laboratories in Denmark, Greece, Norway, Sweden and Switzerland and in industrial test labs (especially by converters). Spectroscopic and chromatographic information as well as access to authentic reference substances are essential to carry out this work.

This Handbook will be of practical assistance to enforcement laboratories in the European Union and to industry laboratories needing to demonstrate compliance. Also, to Universities and other research establishments who undertake more fundamental research into migration from plastics. If the models for predicting migration will be generally applied then identification of additives is essential. In addition to the spectroscopic and other information contained in this Handbook, there exists a physical collection of these reference substances of known provenance and from which the spectra were obtained. Reasonable requests for these reference substances themselves will be met on application to the address below. These substances will be supplied neat or as a reference solution where appropriate from:

Plastics Reference Collection of Additives
Food Inspection Service Utrecht
Specialisatie Verpakkingen
Nijenoord 6, 3552 AS Utrecht
Postbox 10123, 3505 AB Utrecht
The Netherlands
tel +31-30-2461-611
fax +31-30-2422-566
E-mail igbut@wxs.nl and/or lieropb@worldonline.nl

Scope of the Handbook

EC Directive 90/128/EEC controls the safety of plastics food packaging materials by limiting through a positive list the substances permitted to be used as monomers and other starting substances for the production of plastics. After monomers the European Commission will regulate plastics additives. The additives used in plastics intended for food contact are listed in Commission working documents ('Synoptic Documents'). This project was started on the basis of Synoptic Document N. 5 and for the final selection of additives the Synoptic Document N. 7 [4] was used.

The basis for selection of the one hundred most important additives were the usage, the level of the tolerable daily intake figure (which indicates if any control e.g. by QM or SML is needed) and, finally, the intrinsic likelihood of migration.

Entries for each substance provide the structural formula, the CAS number and the PM reference number, which is the number by which substances are indexed by the Commission services e.g. in Directives and Synoptic Documents. Alternative names whether systematic IUPAC nomenclature or trivial names, are given as a further aid to identification.

Physical characteristics such as solubility in common organic solvents are described although these are indicative only and not definitive nor exhaustive. These have been taken from datasheets from the suppliers of each substance. Stability of the additive under test is important and the user of this handbook is recommended to check for stability before conducting quantitative extraction or migration tests. Information on the stability of each additive can only be obtained by extensive investigation. The stability of additives in food simulants has been investigated in EU project AIR3-CT94-2360 *Stability of Plastic Additives in Food Simulants* and this is a useful reference source [11].

Some indications for handling are included, but safety requirements for the handling of chemicals vary from country to country. Therefore the safety precautions are given in very general terms and are not intended to have any legal status.

Information on current use of the additives is given. This information was obtained after consultation with industry (FCA, sector group of CEFIC). Also the personal experience of the participants was taken into account.

The retention times of the additives which could be analysed by GC under the conditions described in Chapter II, are given in minutes. GC was performed using the normal hydrocarbons C_{12}, C_{20} and C_{24} as retention time markers. This allows the calculation of retention indices which will facilitate the identification of additives. This calculation can be carried out by relating the measured retention time of the reference substances to the retention times of the hydrocarbons C_{12}, C_{20} and C_{24}. A possible formula is given by Guiochon and Guillemin [12]. All the additives were analysed under the same gas chromatographic conditions.

The Fourier transform infra-red spectra were recorded in two laboratories as a check of the data. Comparison with literature spectra could seldom be carried out because these were not usually available. The FT-IR spectra can be used either to identify the substance as such, or to identify additives in subtracted spectra. The spectra can be added to existing data bases.

Mass spectra are similarly meant to complement existing data bases. Mass spectra are very suitable to identify additives in solvent extracts of food contact polymers. For substances that can be analysed by GC the mass spectra were measured on two GC-MS systems to generate classical electron-impact spectra. However many additives cannot be analysed by GC for reasons of e.g. low volatility or thermal instability. For these substances the rapidly developing technique of LC-MS is suitable. A limitation of current LC-MS instrumentation is that most ionisation interfaces do not give consistent and library-searchable spectra. One exception to this is the particle-beam interface which gives classical electron-impact spectra. So for additives that could not be analysed by GC, then direct source injection or flow-injection particle beam MS analysis was applied. These spectra were recorded twice in the same laboratory (separated by several weeks) to check for consistency.

Proton magnetic resonance (^{1}H-NMR) spectra displayed in this Handbook were recorded in two different laboratories as a double check. The spectra can be used to identify additives either pure, or as constituents of polymer extracts [5]. In the research project AIR-9411025 [5], it has been shown that ^{1}H-NMR is a powerful tool for the quality control of most food packaging plastics, specially in an industrial framework. The materials are extracted by a suitable solvent. After evaporation to dryness, the ^{1}H-NMR spectrum of the extract is recorded. This spectrum can be used as a fingerprint of the material. Using the data base of this Handbook and relying on experience for interpretation it is possible to decide very quickly about the presence or the absence of an additive in the material. ^{1}H-NMR requires quite expensive equipment, but the essential information can often be obtained in a very short time from the spectra [13-15] This approach therefore efficiently complements methods previously developed.

It should be noted that the substances supplied by industry were representative of current usage and were of good technical quality but were not necessarily fine chemicals of analytical purity (e.g. $>99.5\%$). They could contain therefore the normal and inevitable minor feedstock and reaction impurities and especially (where applicable) isomeric 'impurities'. For FT-IR, ^{1}H-NMR and flow-injection MS analysis, these impurities will be included in the spectra and may or may not be discernible. In contrast, for GC-MS analysis where acquisition of a spectrum is preceded by some chromatographic separation, these impurities will have been removed to a greater or lesser extent from the resulting spectrum. This is also true for sample introduction into the MS by direct probe where preferential evaporation from the heated probe can occur. The user of this handbook should bear this in mind, the nature of spectra acquisition, when comparing these reference spectra to any spectrum generated in his/her own investigations.

Acknowledgments

Special thanks go to Karl Ehlert from FISU, who undertook the large task of preparing the camera-ready datasheets. We are very grateful to our colleagues who provided the analytical data. ^{1}H-NMR spectra were provided by prof James Bouquant (Université de Reims) and Denise Scholler (INRA) and by Chris Honeybone and Katrina Mountfort (MAFF). Special thanks to prof Bouquant for interpretation of the ^{1}H-NMR spectra and his scientific contribution to the project. The infra-red spectra were provided by Marjan van der Beek (FISU) and Janine Le Sech (INRA). The mass spectra were provided by Teuni Volp (FISU) and by Phil Clarke, Simon Hird and Andy Damant (MAFF). The gas-chromatographic retention times were measured by Karl Ehlert (FISU).

The success in producing this Handbook has also been due to assistance in supplying samples and to the information made freely available by European industrial and Government sources too numerous to mention. We are specially grateful to R. Ashby (ICI), J. Autin (CEFIC-FCA), J. Gonzalez (CICC), K. Hinrichs (Henkel), S.E.P.G. Merckx (General Electric), U. Schönhausen (Ciba Geigy) and H. Vergallen (Premark Resources).

References

[1] Bush, J., Gilbert, J. and Goenaga, X. Spectra for the identification of monomers in food packaging. Kluwer Academic Publishers, Dordrecht, The Netherlands, 1993.

[2] Gilbert, J., Bush, J., Lopez de Sa, A., Lierop, J.B.H. van and Goenaga, X. Establishment of a reference collection of substances and an analytical handbook of reference data to support enforcement of EC regulations on food contact materials. *Food Additives and Contaminants,* 1994, **11,** 71-74.

[3] Commission Directive 90/128/EEC of 23 February 1990 relating to plastics materials and articles intended to come into contact with foodstuffs. *Official Journal of the European Communities* **L349**, 26-47.

[4] Synoptic document N. 7. Draft of provisional list of monomers and additives used in the manufacture of plastics and coatings intended to come into contact with foodstuffs. *Commission document CS/PM/2356*. Brussels, 15 May 1994.

[5] Project AIR-941025. Safety and Quality Control of Plastic Materials for Food Contact Materials. *Final report* 1997. In preparation.

[6] WARENWET, 1988, Dutch Food Law, Packaging and Utensils Regulation (VGB).

[7] Battum, D. van and Lierop, J.B.H. van. Testing of food contact materials in the Netherlands. *Food Additives and Contaminants,* 1988, **5**, 381-395.

[8] Lierop, J.B.H. van. Enforcement of European Community legislation at the national level. *Food Additives and Contaminants,* 1994, **11**, 131-139.

[9] Lierop, J.B.H. van. Enforcement of food packaging legislation. *Food Additives and Contaminants*, 1997, **14,** 555-561.

[10] Battum, D. van and Lierop, J.B.H. van, 1997 Materials and articles in contact with foodstuffs, Guide for examination of plastic food contact materials. CEN TC 194/SC1/WG2 document N118.

[11] Project AIR3-CT94-2360. Stability of plastic additives in food simulants. Final report 1997.

[12] Guiochon, G. and Guillemin, C.L., *Quantitative Gas Chromatography, Elsevier,* 1988.

[13] Castle, L., Nichol, J. and Gilbert, J. Migration of polyisobutylene additive from polyethylene films into foods during microwave and domestic oven use. *Food Additives and Contaminants*, 1992, **9**, 315-330.

[14] Feigenbaum, A., Bouquant, J., Hamadi, M., Métois, P., Riquet, A. M. and Scholler, D. Quick methods to control compliance of plastic materials with food packaging regulations. *Food Additives and Contaminants,* 1997, **14**, 571-582.

[15] Métois, P., Scholler, D., Bouquant, J. and Feigenbaum, A. Alternative test methods to control the compliance of food packaging materials with the European Union regulation: the case of aromatic antioxidants and of bis(ethanolamine) antistatics based on ^{1}H-NMR and UV-visible spectrophotometry. *Food Additives and Contaminants,* 1997, in press.

CHAPTER 2

GC retention times

It is described in the Dutch test method that a first identification of the additives can be performed by determination of the gas-chromatographic retention times. Therefore we provide the retention times of the additives which could be analysed by GC. The GC analysis was carried out on a Perkin Elmer autosystem XL with the following conditions. The column was a Hewlett Packard Ultra 1 column, 50 m x 0.32 mm with a film thickness of 0.52 μm of crosslinked methylsiloxane. The temperature program was 5 min at 50°C then raised at 5°C/min to 320°C and held for 26 min. The carrier gas was helium at 3.5 mL/min. The temperature of the injector was 260°C and injections of 1 μL of substance dissolved in diethylether were injected splitless. The flame ionisation detector used was held at 325°C. Under these conditions the retention times of the hydrocarbon markers were 20.1, 38.5 and 45.2 min for C_{12}, C_{20} and C_{24} respectively. Retention indexes of the reference substances can be calculated on basis of these retention times [12].

Infra red spectra

The manner in which Fourier Transform Infra-Red (FT-IR) spectra were obtained depended on the physical nature of the substance. The techniques used were formulation into KBr pellets and deposition onto KBr, KRS-5 or NaCl windows. Thus, ca. 1 mg of substance was mixed with KBr (50 mg) in a mortar. When transmission of the major peaks was low a larger amount of substance was used until all peaks had a satisfactory intensity. Spectra were acquired at 4 cm^{-1} resolution using a MIDAC spectrometer. The sample and reference cell were scanned 60 times at 1.4 s/scan and the mean difference spectrum was plotted. FT-IR spectra were checked in a second laboratory using different instrumentation and conditions.

Mass spectra

Mass spectra were recorded under standardised conditions and a common format for data presentation has been adopted for each entry.

For GC-MS, spectra were obtained using an updated Finnigan Mat SSQ 70 mass spectrometer coupled to a Varian 3400 gas chromatograph. Ionisation was by electron impact at 70 eV with a source temperature of 180°C. The scan range was 40 - 700 daltons. The GC was equipped with a J&W DB5ms capillary column, 30 m x 0.32 mm with a film thickness of 0.52 μm 5% phenyl / 95% dimethylpolysiloxane. The GC temperature programme was 3 min at 40°C then raised at 20°C/min to 320°C and held at that temperature for 20 min. A solution of 100 μg/mL of the reference substance was prepared for the measurement in the GC-MS. As a check and also for substances that could not be analysed by GC-MS, mass spectra were also obtained in a second laboratory using different instrumentation and conditions.

Direct insertion probe MS spectra were acquired on a VG 7070 mass spectrometer. The substance dissolved in a suitable solvent was introduced into the mass spectrometer by the direct insertion probe heated at 350-400°C. In a small number of cases it was necessary to introduce the solid substance (for reasons of poor solubility) or to heat the probe to a higher temperature (for reasons of involatility). Electron impact spectra at 70 eV were recorded. Spectra were averaged and background-subtracted.

For particle - beam MS analysis, a VG Autospec magnetic sector mass spectrometer fitted with a VG LINC-Particle Beam Interface was used. Electron impact ionisation was used with 70 eV electron energy, a trap current of 500 μA and a source temperature of 250°C. The temperature of the expansion/desolvation chamber was set at 60°C. An HPLC pump was used to deliver the mobile phase (methanol) at a flow rate of 0.5 mL/min to the PB interface. The samples were dissolved in a suitable solvent at 1 mg/mL then diluted immediately prior to analysis to 100 μg/mL. Portions (10 μL) were injected to give a spectrum of appropriate intensity. The scan range was 40 to 1000 daltons with a 1.5 s scan time and an inter-scan delay of 0.5 s. The spectrometer was tuned using PFK to a resolution of better than 1000 and then calibrated. The nebuliser gas pressure and nebuliser position were optimised using 50 μL injections of 1.0 μg/mL caffeine solution.

NMR spectra

^{1}H-NMR spectra were recorded using a Bruker AC 250 spectrometer (250 MHz) to average 16 scans of a 30 mg/ml solution of each additive at 25°C in $CDCl_3$ containing tetramethylsilane (TMS). Where solubility in $CDCl_3$ was poor, an alternative solvent was used and this is noted on the spectrum. For substances with a complex spectrum, a greater number of scans were averaged and this is also noted.

Acids, fatty (C_{14}-C_{18}), alkyl (C_{14}-C_{18}) ester

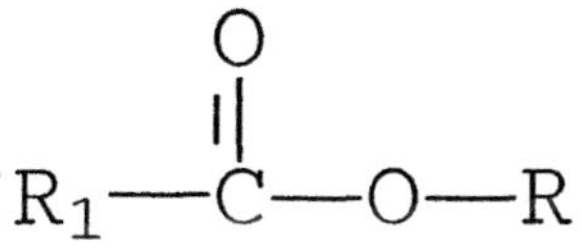

CAS No.	- 85566-24-1
PM Ref. No.	- 31304
Restrictions	- None
Formula	-
Molecular weight	-
Alternative names	-
Physical Characteristics	- White flakes, mp 46 - 51°C. - Soluble in ether, hexane, toluene, chloroform, dichloromethane.
Handling	- Store at room temperature.
Safety	-
Availability	- Standard sample was supplied.
Current uses	- Lubricant.
Applications	- General use.
GC Retention time	- 53.3, 55.9, 58.2, 60.6 min.

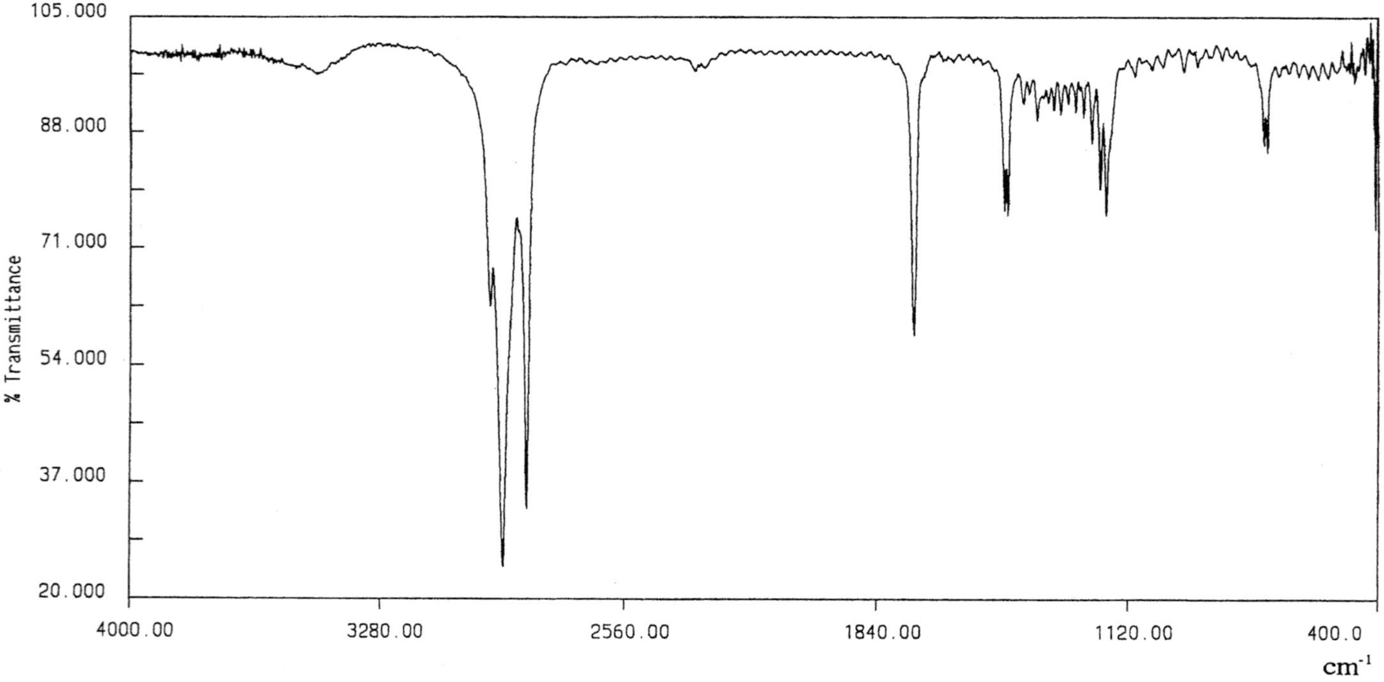

Name : ACIDS, FATTY (C14-C18), ALKYL (C14-C18) ESTERS

PM Nr : 31304 CAS Nr : 85566-24-1

KBr pellet

ACIDS, FATTY (C_{14}-C_{18}), ALKYL (C_{14}-C_{18}) ESTERS

Peak wavenumber	% T	Relative intensity
407.0	74.0	34.6
719.5	85.1	19.7
729.2	86.4	18.1
1182.5	75.8	32.1
1199.9	79.6	27.1
1221.1	86.5	18.0
1311.8	90.9	12.1
1377.3	89.9	13.4
1464.2	75.8	32.1
1473.8	76.6	31.1
1732.3	58.5	55.1
2849.2	33.0	88.9
2916.7	24.7	100.0
2957.2	62.5	49.8

Acids, fatty (C_{14}-C_{18}), alkyl (C_{14}-C_{18}) ester

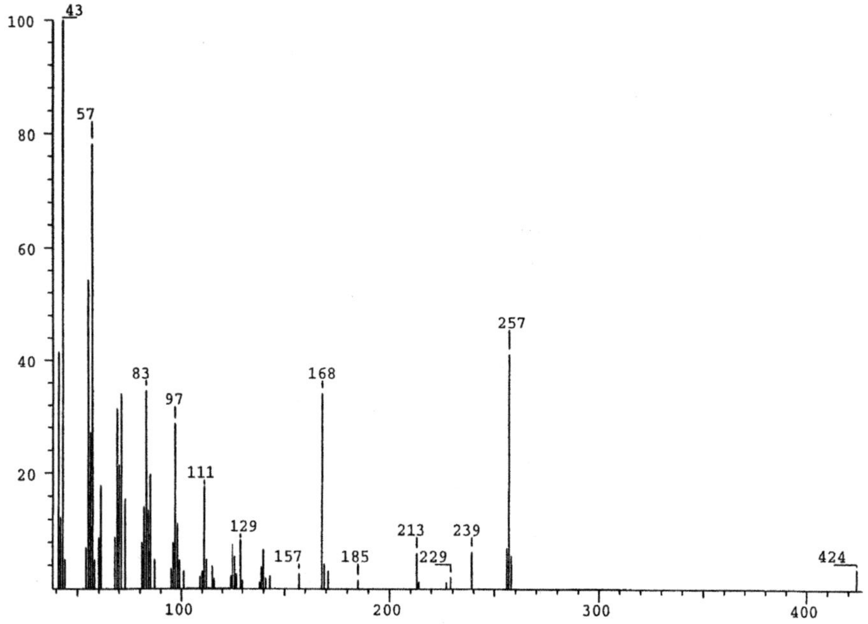

M/Z	Ion Intensity (%)	M/Z	Ion Intensity (%)
43	100.0	97	28.8
57	78.2	111	17.7
69	31.4	168	34.1
71	34.0	213	6.2
73	15.5	239	6.4
83	34.6	257	41.2
84	13.7	424	3.7
85	19.8		

Spectrometer	: Finnigan Mat SSQ 700
Inletsystem	: Capillary GC/MS
Source Temperature	: 180°C
Electron Energy	: 70 eV
Scan Range	: 40-500

Acids, fatty (C_{14}-C_{18}), alkyl (C_{14}-C_{18}) ester

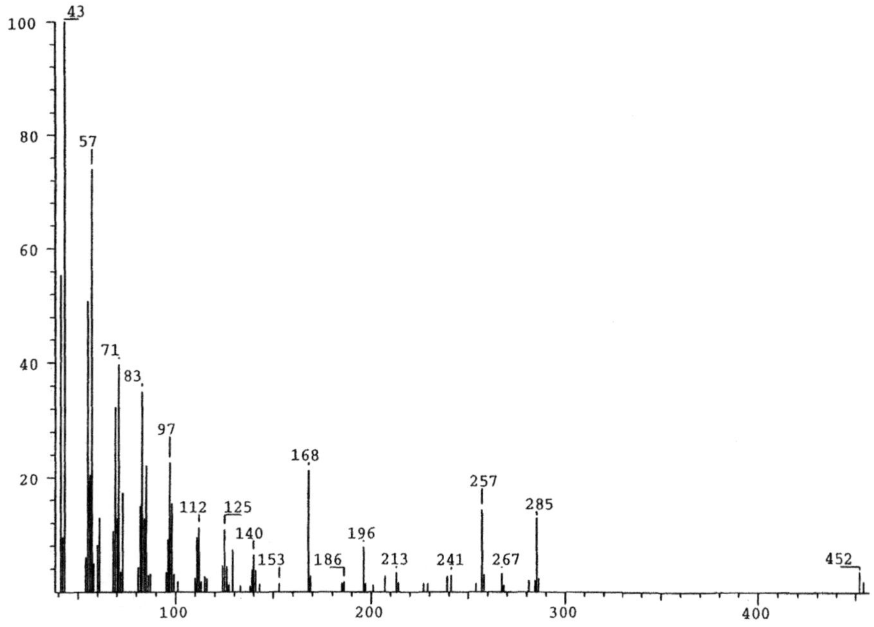

M/Z	Ion Intensity (%)	M/Z	Ion Intensity (%)
41	55.4	97	22.7
43	100.0	111	9.5
57	73.9	112	11.3
69	32.3	125	10.8
71	39.7	168	21.2
73	17.2	257	14.3
83	34.9	285	12.9

Spectrometer	: Finnigan Mat SSQ 700
Inletsystem	: Capillary GC/MS
Source Temperature	: 180°C
Electron Energy	: 70 eV
Scan Range	: 40-500

Acids, fatty (C_{14}-C_{18}), alkyl (C_{14}-C_{18}) ester

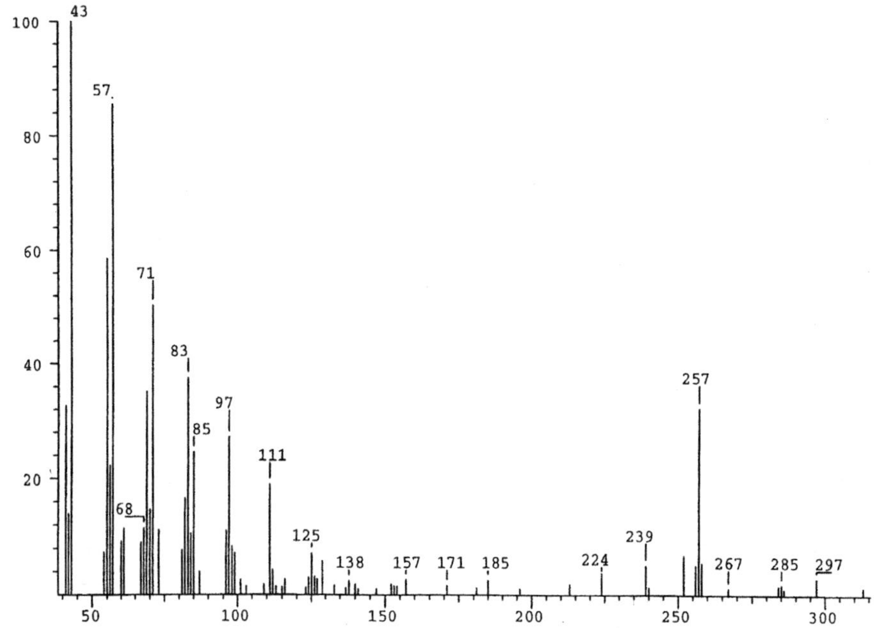

M/Z	Ion Intensity (%)	M/Z	Ion Intensity (%)
41	32.7	83	37.6
43	100.0	85	24.6
57	85.6	97	27.2
69	35.2	111	19.0
71	50.4	125	7.2
73	11.2	257	32.4

Spectrometer : Finnigan Mat SSQ 700
Inletsystem : Capillary GC/MS
Source Temperature : 180°C
Electron Energy : 70 eV
Scan Range : 40-500

Acids, fatty (C_{14}-C_{18}), alkyl (C_{14}-C_{18}) ester

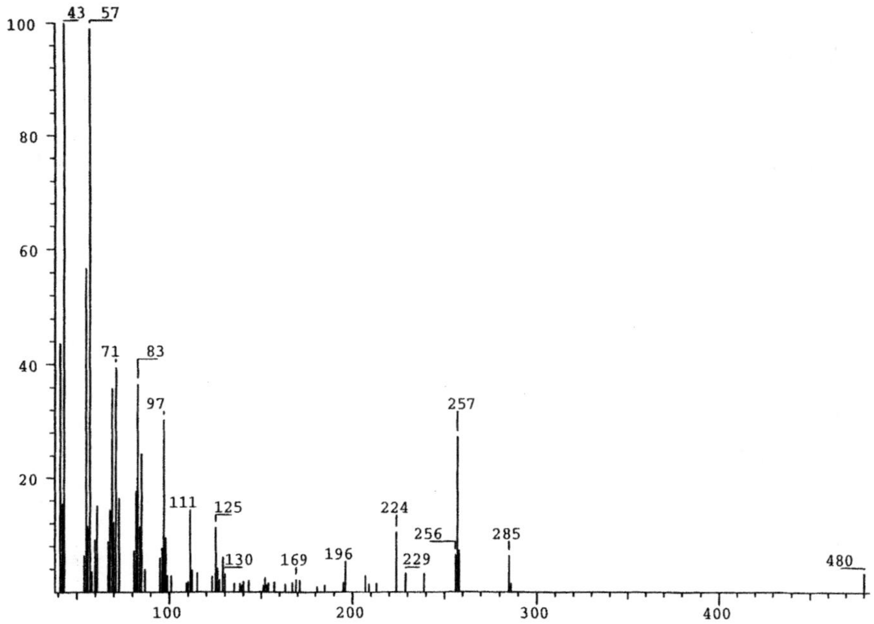

M/Z	Ion Intensity (%)	M/Z	Ion Intensity (%)
41	43.7	85	24.3
43	100.0	97	30.2
57	99.0	111	13.3
69	35.9	125	11.2
71	39.4	224	10.5
73	16.4	257	32.4
83	36.6		

Spectrometer	: Finnigan Mat SSQ 700
Inletsystem	: Capillary GC/MS
Source Temperature	: 180°C
Electron Energy	: 70 eV
Scan Range	: 40-500

PM n°	CAS n°	Name
31304	85566-24-1	ACIDS, FATTY (C_{14}-C_{18}), ALKYL (C_{14}-C_{18}) ESTERS

Solvent	$CDCL_3$
SF	250.13
SI	16384
SW	3125
NS	32

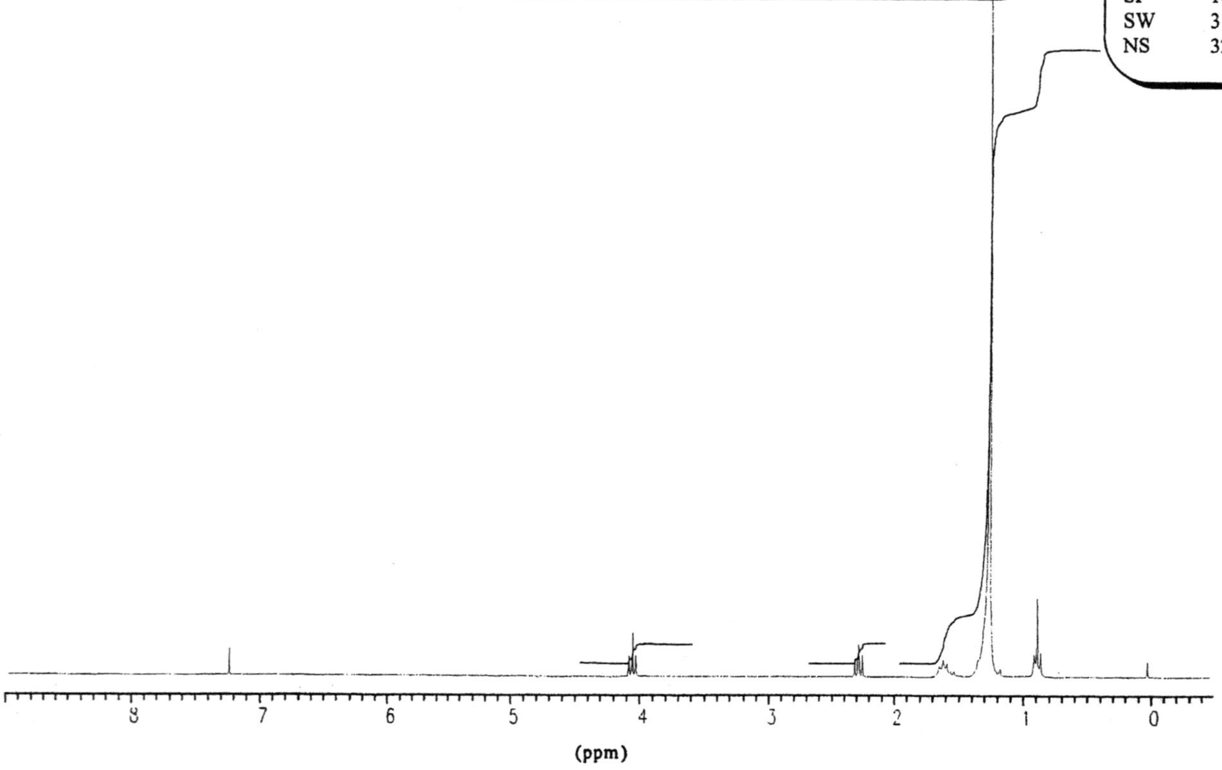

Acrylic acid, 2-tert-butyl-6-(3-tert-butyl-2-hydroxy-5-methylbenzyl)-4-methylphenyl ester

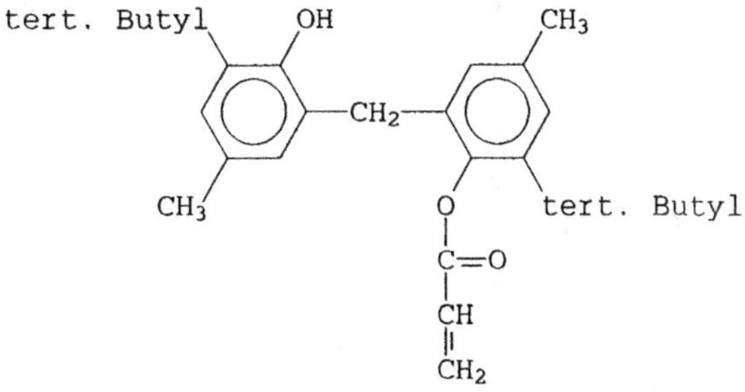

CAS No.	- 61167-58-6
PM Ref. No.	- 31520
Restrictions	- SML = 6.0 mg/kg
Formula	- $C_{26}H_{34}O_3$
Molecular weight	- 394

<u>Alternative names</u> - 2-(1,1-Dimethylethyl)-6-[[3-(1,1-di-methylethyl)-2-hydroxy-5-methylphenyl]methyl]-6-methyl-phenyl-acrylate

<u>Physical Characteristics</u>
- White granules, mp 131.9°C.
- Soluble in acetone, chloroform, dichloromethane.

<u>Handling</u> - Store at room temperature.

<u>Safety</u> -

<u>Availability</u> - Standard sample was supplied.

<u>Current uses</u> - Antioxidant.

<u>Applications</u> - Non polyolefins.

<u>GC Retention time</u> - 47.1 min.

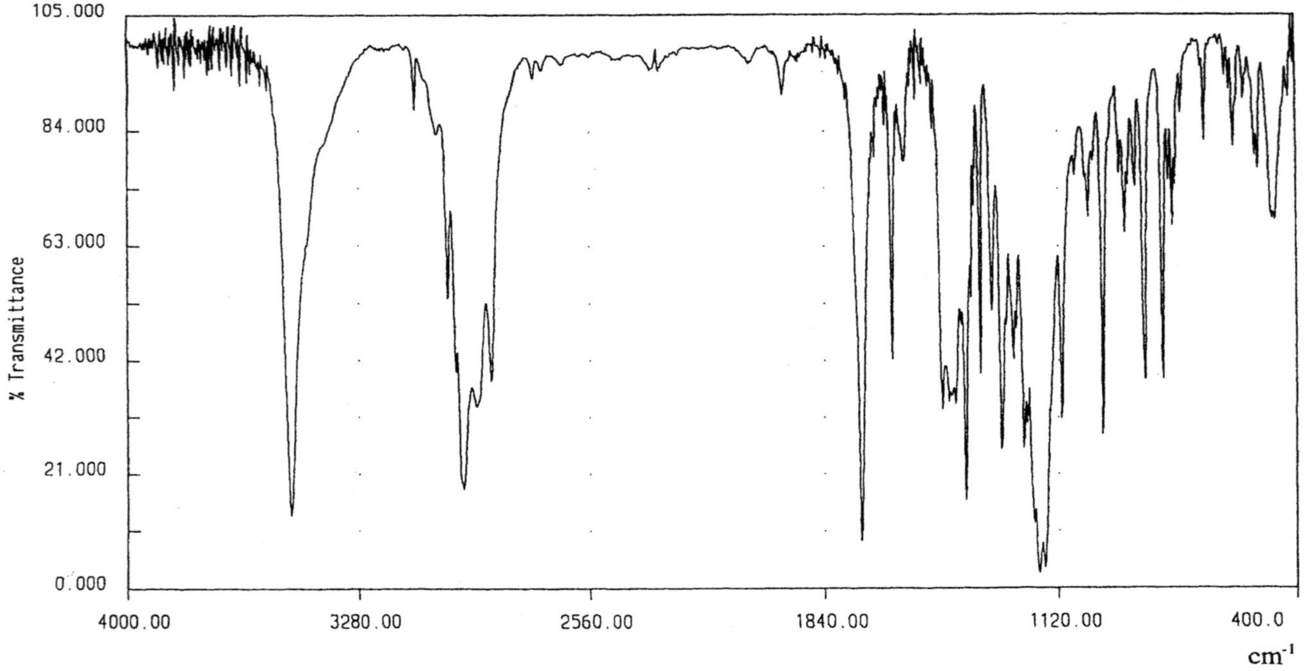

Name : ACRYLIC ACID, 2-TERT-BUTYL-6-(3-TERT-BUTYL-2-HYDROXY-5-METHYLBENZYL)-4-METHYLPHENYL ESTER

PM Nr : 31520 CAS Nr : 61167-58-6

KBr pellet

ACRYLIC ACID,2-TERT-BUTYL-6-(3-TERT-BUTYL-2-HYDROXY-5-METHYLBENZYL)-4-METHYLPHENYL ESTER

Peak wavenumber	% T	Relative intensity
420.5	90.3	10.0
464.9	67.7	33.2
524.7	79.7	20.9
775.5	66.7	34.2
787.1	75.0	25.7
804.4	38.5	63.2
658.4	38.4	63.3
885.4	73.9	26.8
916.3	65.4	35.6
931.7	76.4	24.3
985.7	28.3	73.7
1012.8	78.6	22.0
1028.2	68.3	32.6
1068.7	75.9	24.8
1111.1	31.3	70.6
1163.2	3.7	99.0
1180.6	2.7	100.0
1217.2	30.2	71.7
1226.9	25.7	76.4
1257.7	42.3	59.3
1296.3	25.5	76.6
1325.3	51.1	50.3
1360.0	39.6	62.1
1379.3	70.3	30.5
1388.9	53.5	47.8
1406.3	16.1	86.2
1439.1	34.0	67.8
1477.7	33.0	68.9
1508.5	84.4	16.0
1599.2	78.3	22.3
1633.9	42.1	59.5
1655.1	85.1	15.3
1686.0	79.1	21.5
1730.4	8.6	93.9
2868.5	38.1	63.6
2916.7	33.4	68.4
2957.2	18.2	84.1
3005.5	53.5	47.8
3493.5	13.3	89.1

Acrylic acid, 2-tert-butyl-6-(3-tert-butyl-2-hydroxy-5-methylbenzyl)-4-methylphenyl ester

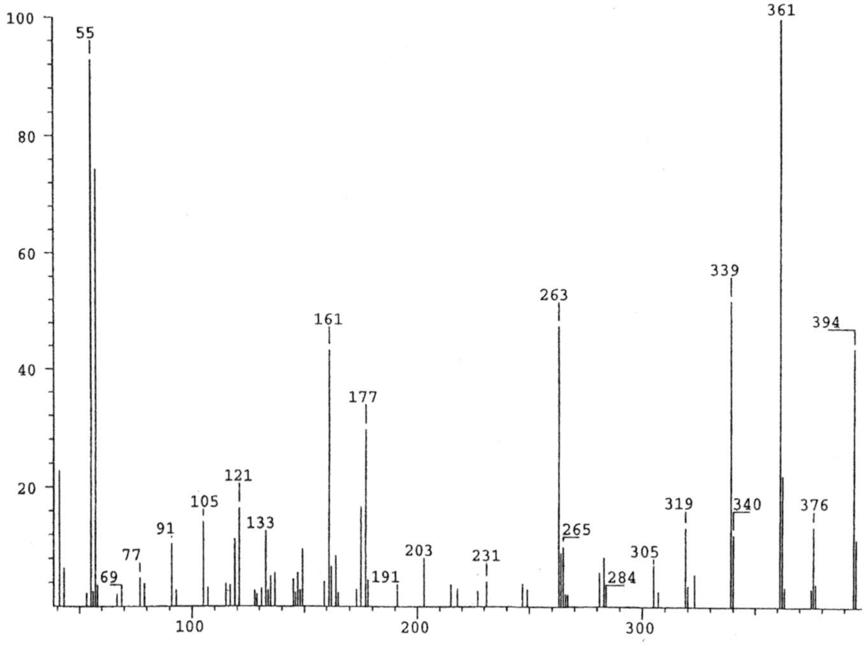

M/Z	Ion Intensity (%)	M/Z	Ion Intensity (%)
41	22.7	265	10.0
55	92.8	319	13.2
57	74.2	339	51.9
91	10.4	340	12.1
105	14.1	361	100.0
121	16.5	362	22.1
161	43.3	376	13.5
177	29.6	394	43.7
263	47.4	395	11.4

Spectrometer : Finnigan Mat SSQ 700
Inletsystem : Capillary GC/MS
Source Temperature : 180°C
Electron Energy : 70 eV
Scan Range : 40-400

PM n°	CAS n°	Name
31520	61167-58-6	ACRYLIC ACID, 2-TERT-BUTYL-6-(3-TERT-BUTYL-2-HYDROXY-5-METHYLBENZYL)-4-METHYLPHENYL ESTER

Solvent	$CDCL_3$
SF	250.13
SI	16384
SW	3125
NS	16

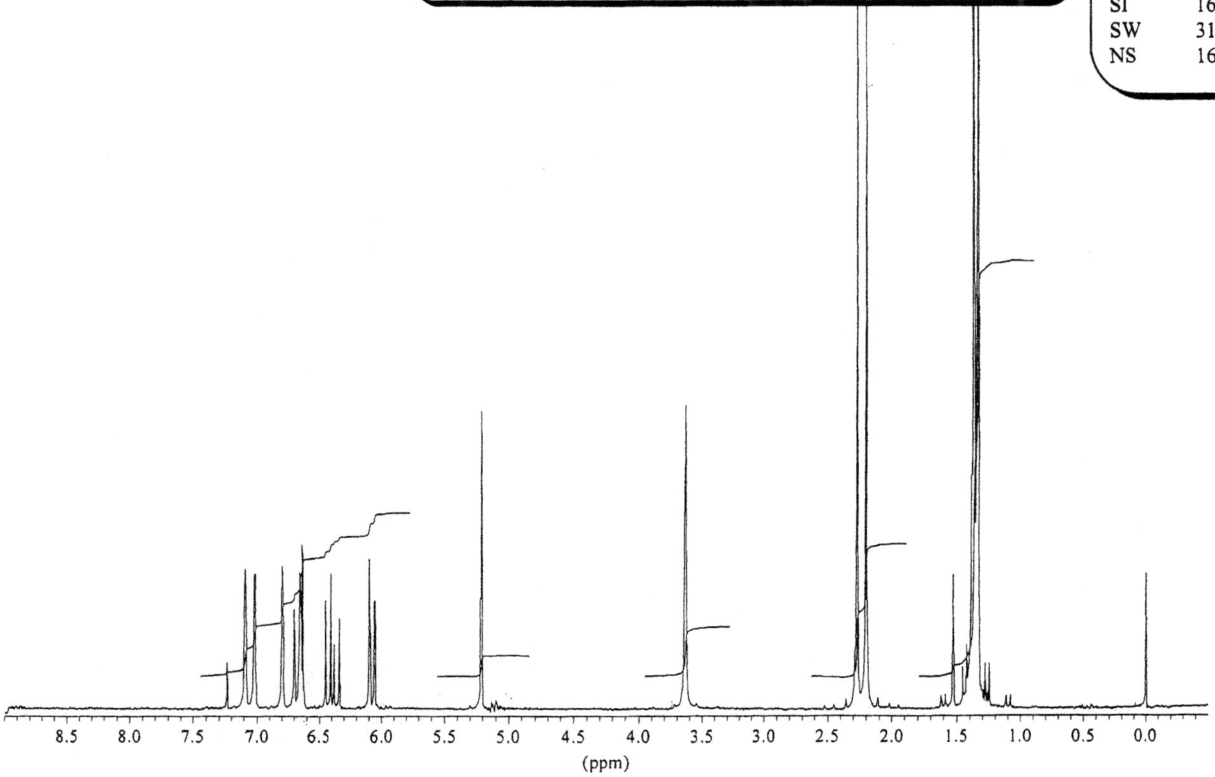

Adipic acid, bis(2-ethylhexyl) ester

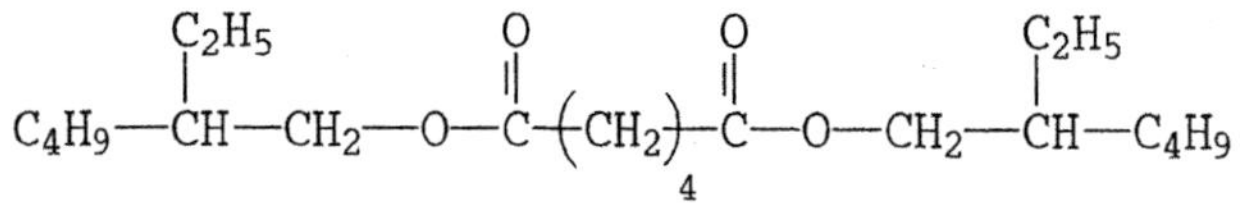

CAS No.	- 00103-23-1
PM Ref. No.	- 31920
Restrictions	- SML = 18 mg/kg
Formula	- $C_{22}H_{42}O_4$
Molecular weight	- 370

Alternative names	- Hexanedioic acid, bis(2-ethylhexyl) ester, Di(2-ethylhexyl) adipate, Dioctyl adipate
Physical Characteristics	- Colourless liquid, mp -67.8°C, bp 360°C. - Soluble in alcohol, ether, acetone, acetic acid.
Handling	- Store at room temperature.
Safety	-
Availability	- Standard sample was supplied.
Current uses	- Plasticizer.
Applications	- PVC, PVDC, coatings, paper.
GC Retention time	- 45.0 min.

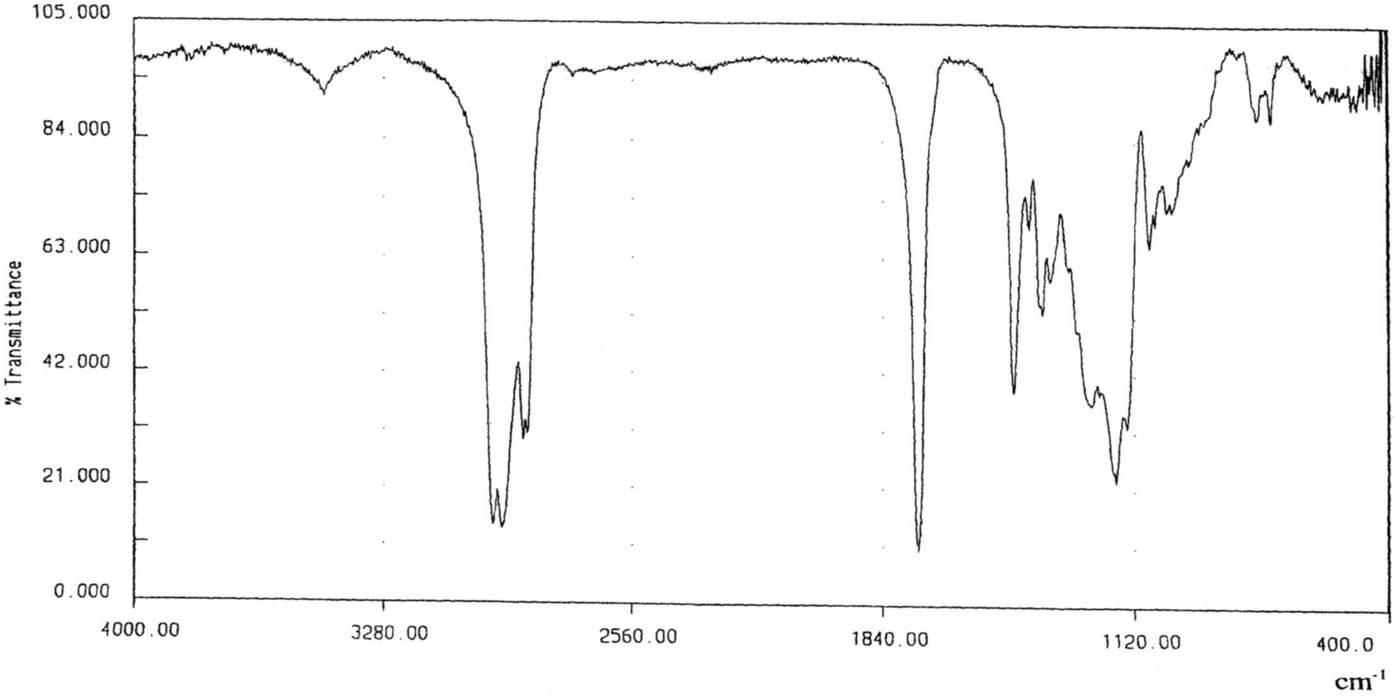

Name : ADIPIC ACID, BIS(2-ETHYLHEXYL) ESTER

PM Nr : 31920 CAS Nr : 00103-23-1

KBr pellet

CAS No. 00103-23-1 PM Ref. No. 31920

ADIPIC ACID, BIS(2-ETHYLHEXYL) ESTER

Peak wavenumber	% T	Relative intensity
455.3	92.0	8.8
484.2	93.6	7.2
501.6	94.6	6.0
731.1	88.0	13.3
771.6	86.6	14.8
904.7	89.3	11.9
966.5	84.1	17.6
1014.7	77.7	24.7
1062.9	77.6	24.9
1078.3	74.0	28.8
1142.0	37.0	69.9
1172.9	25.4	82.8
1242.3	40.5	66.0
1360.0	64.0	39.9
1383.1	57.2	47.5
1419.8	73.3	29.6
1464.2	40.9	65.6
1738.1	9.85	100.0
2862.7	31.63	75.8
2874.3	30.47	77.1
2934.1	13.80	95.6
2959.2	14.48	94.9
3454.9	90.07	11.0

Adipic acid, bis(2-ethylhexyl) ester

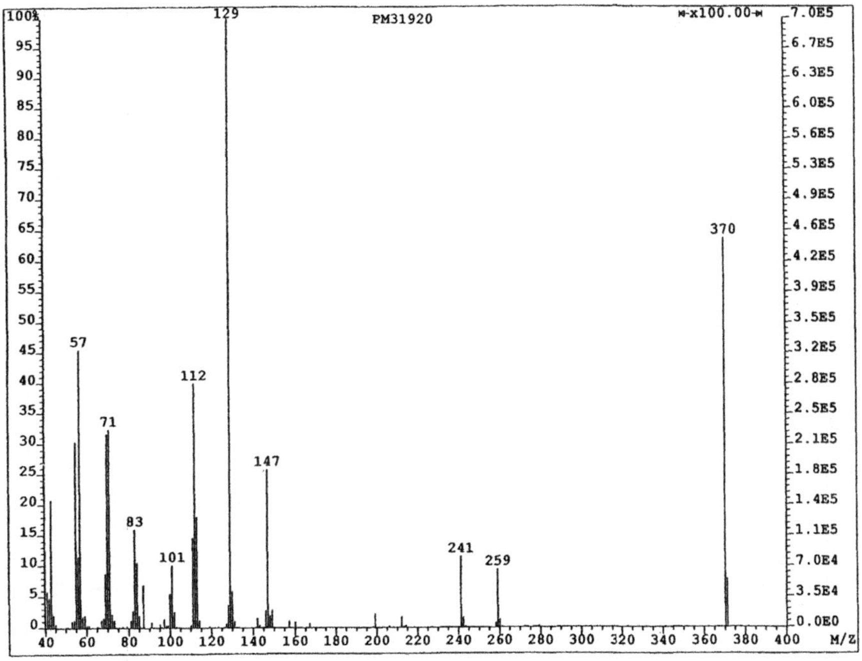

M/Z	Ion Intensity (%)	M/Z	Ion Intensity (%)
43	20.8	100	5.5
55	30.4	101	10.1
56	11.6	111	14.7
57	45.6	112	40.1
69	8.7	113	18.1
70	31.7	129	100.0
71	32.4	130	5.9
83	16.0	147	25.9
84	10.5	241	11.5
87	7.0	259	9.4

Spectrometer	: VG Autospec
Inletsystem	: Flow injection particle beam MS
Source Temperature	: 250°C
Electron Energy	: 70 eV
Scan Range	: 40-1000

PM n° 31920

CAS n° 00103-23-1

Name ADIPIC ACID, BIS(2-ETHYLHEXYL) ESTER

Solvent	$CDCL_3$
SF	250.13
SI	16384
SW	3125
NS	32

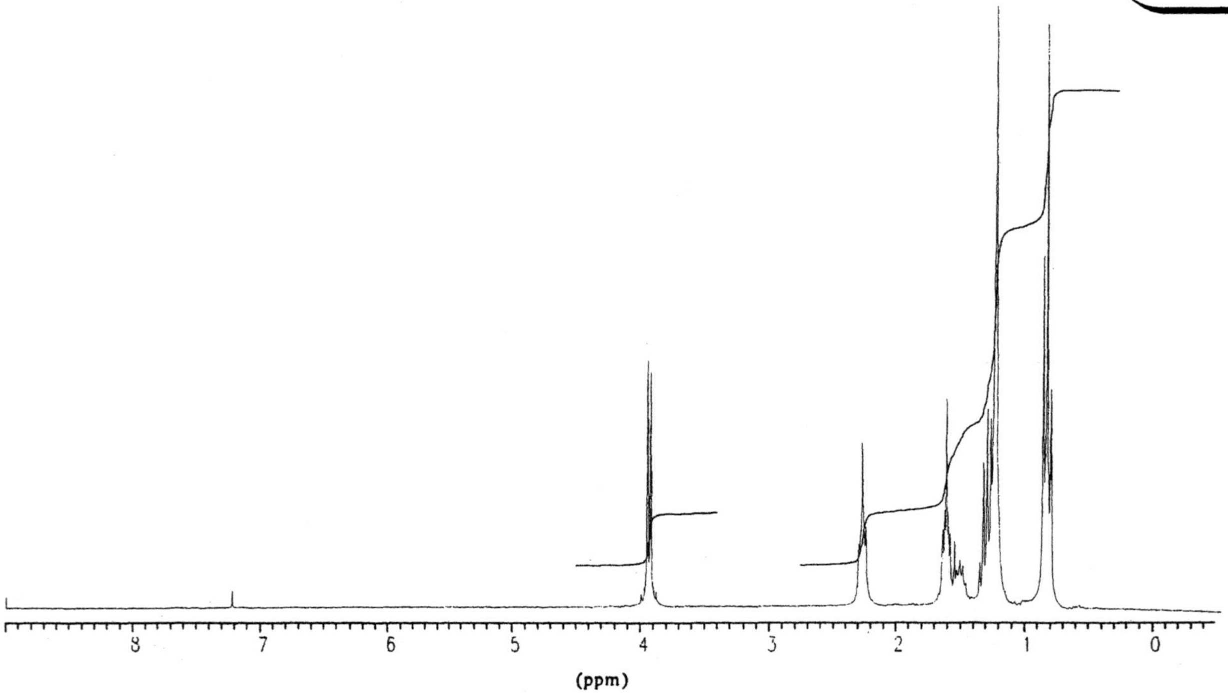

Adipic acid, dibutyl ester

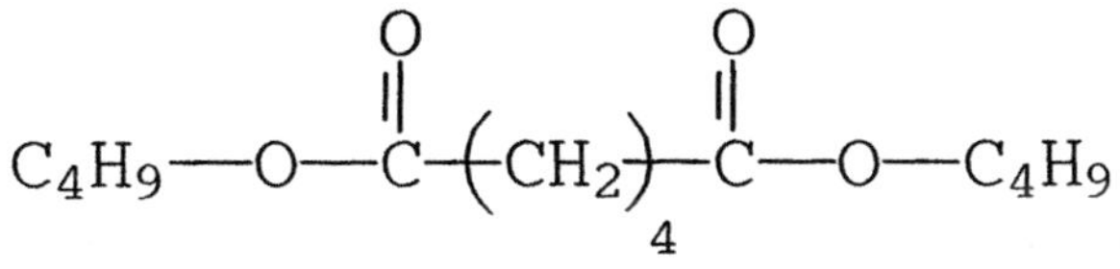

CAS No.	- 00105-99-7
PM Ref. No.	- 32240
Restrictions	- SML = 3.0 mg/kg
Formula	- $C_{14}H_{26}O_4$
Molecular weight	- 258
Alternative names	- Hexanedioic acid, dibutyl ester Dibutyl adipate
Physical Characteristics	- Colourless liquid, mp -32.4°C, bp 305°C. - Soluble in alcohol, ether, acetone, acetic acid.
Handling	- Store at room temperature.
Safety	-
Availability	- Standard sample was supplied.
Current uses	- Plasticizer.
Applications	- General use.
GC Retention time	- 33.2 min.

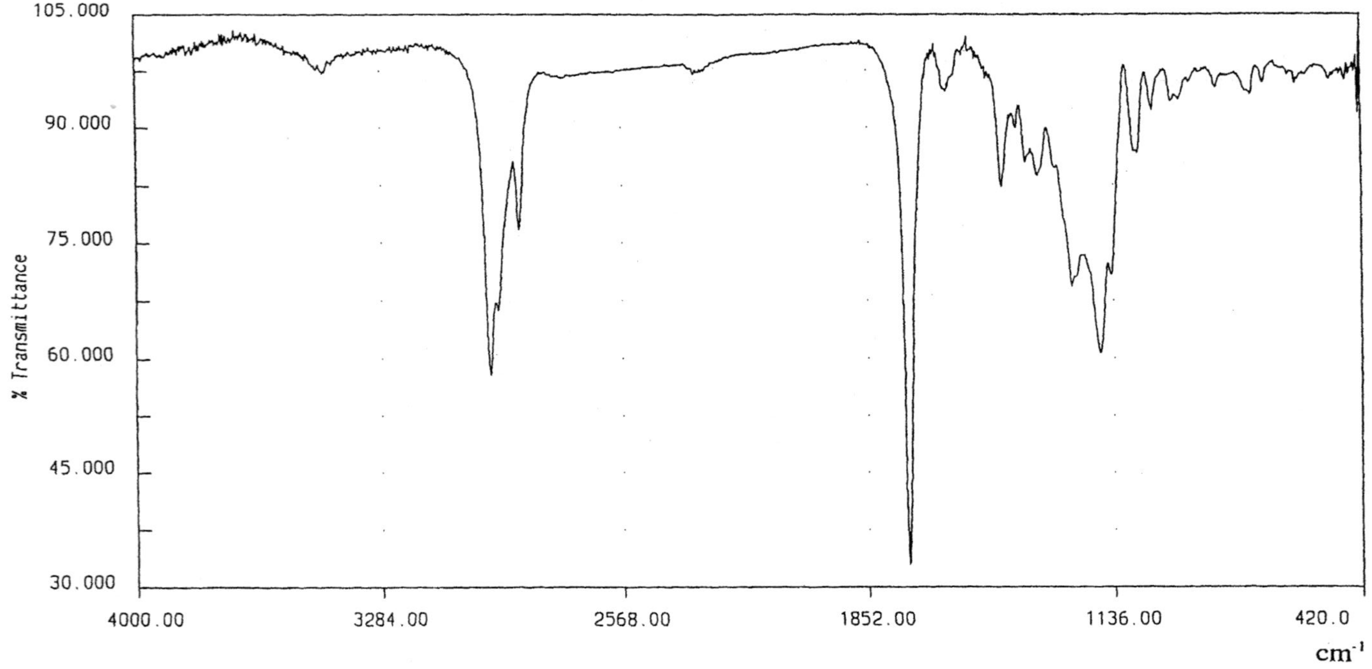

Name : ADIPIC ACID, DI-N-BUTYL ESTER

PM Nr : 32240 CAS Nr : 00105-99-7

KBr pellet

CAS No. 00105-99-7 PM Ref. No. 32240

ADIPIC ACID, DI-N-BUTYL ESTER

Peak wavenumber	% T	Relative intensity
414.8	77.2	25.8
434.0	78.2	24.6
455.3	94.2	6.6
470.7	93.1	7.8
511.2	84.0	18.1
586.4	82.4	19.9
667.5	88.8	12.7
738.8	83.0	19.2
947.2	78.7	24.0
1024.3	73.9	29.5
1076.4	55.6	50.2
1147.8	32.3	76.5
1180.6	19.0	91.5
1242.3	28.2	81.1
1360.0	48.6	58.0
1392.8	50.9	55.4
1421.7	61.4	43.6
1462.2	42.0	65.5
1504.7	88.9	12.5
1730.4	11.5	100.0
2766.3	91.2	10.0
2874.3	34.9	73.5
2945.7	20.8	89.5
3115.4	68.8	35.2
3138.6	71.2	32.5
3206.1	67.4	36.8
3233.1	65.2	39.4
3269.7	63.2	41.6
3294.8	61.0	44.0
3304.5	57.8	47.7
3323.8	58.8	46.5
3341.1	59.1	46.2
3379.7	53.0	53.1
3418.3	49.5	57.1
3443.4	46.9	60.0
3472.3	46.9	60.0
3634.3	77.2	25.8
3659.4	80.1	22.5
3686.4	81.9	20.4
3705.7	87.1	14.6

Adipic acid, dibutyl ester

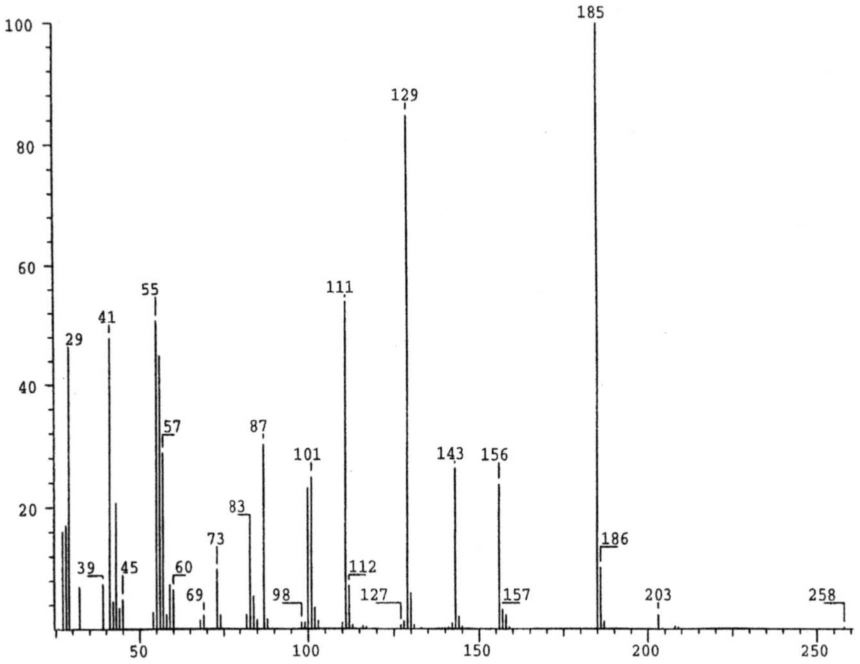

M/Z	Ion Intensity (%)	M/Z	Ion Intensity (%)
29	45.4	100	23.0
41	47.8	101	24.9
43	20.7	111	53.9
55	50.7	129	84.7
56	44.9	143	26.3
57	28.8	156	23.7
83	17.8	185	100.0
87	30.1	186	10.0

Spectrometer	: Finnigan Mat SSQ 700
Inletsystem	: Capillary GC/MS
Source Temperature	: 180°C
Electron Energy	: 70 eV
Scan Range	: 25-500

PM n° 32240

CAS n° 00105-99-7

Name ADIPIC ACID, DI-n-BUTYL ESTER

Solvent	$CDCL_3$
SF	250.13
SI	16384
SW	3125
NS	32

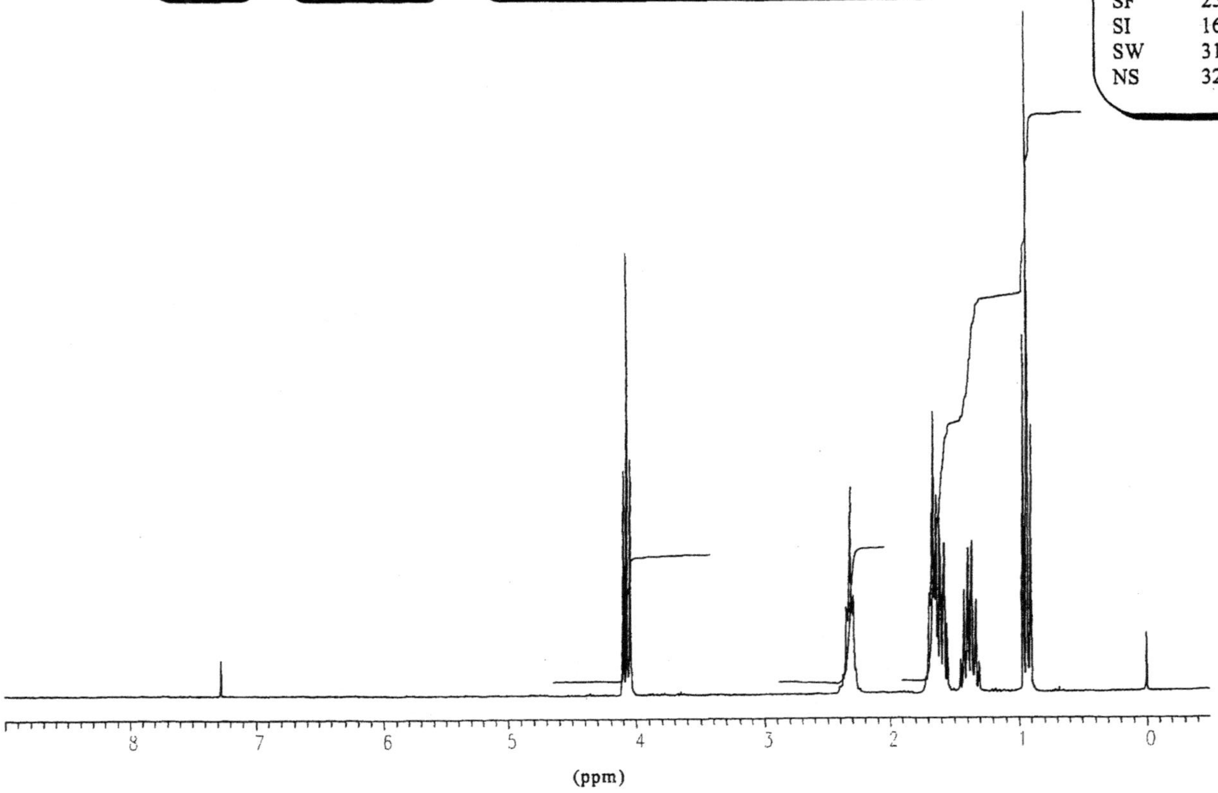

Alkyl (C_{10}-C_{13})benzenesulphonic acid, sodium salt

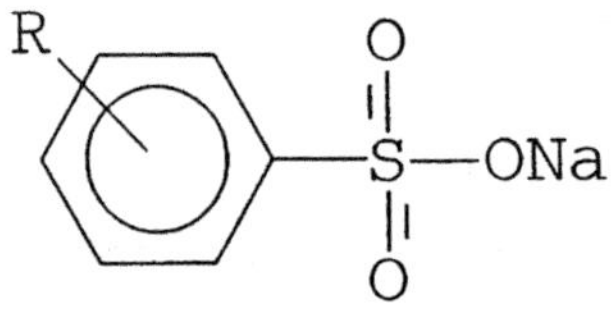

CAS No.	-
PM Ref. No.	- 33800
Restrictions	- None
Formula	-
Molecular weight	-

Alternative names -

Physical Characteristics
- Light beige past.
- Soluble in water, alcohol, ether.

Handling - Store at room temperature.
Safety - Irritant.
Availability - Preparation in aqueous solution.

Current uses - Antistatic agent, emulsifier.
Applications - PP, PVC.

GC Retention time - Alkyl (C_{10}-C_{13})benzenesulphonic acid cluster 28.3 - 39.9 min.

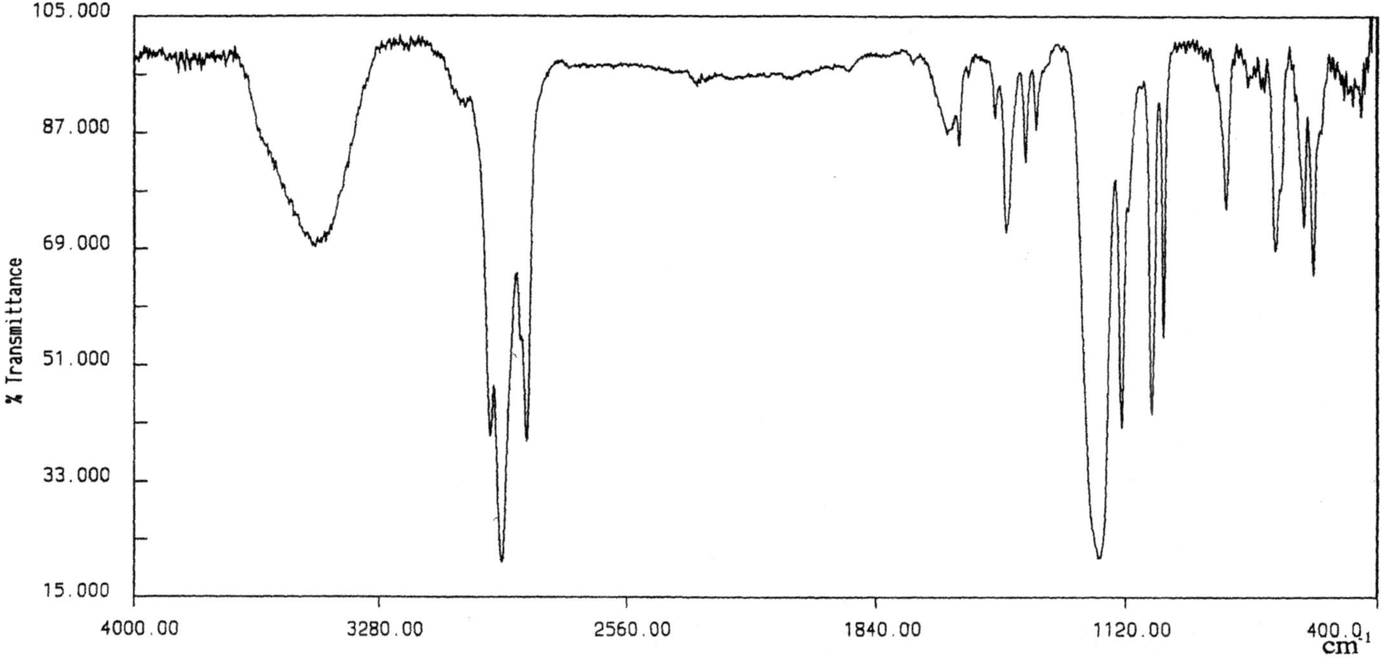

Name : ALKYL (C10-C13) BENZENESULPHONIC ACID

PM Nr : 33800 CAS Nr :

KBr pellet

CAS No. PM Ref. No. 33800

ALKYL(C_{10}-C_{13}) BENZENESULPHONIC ACID

Peak wavenumber	% T	Relative intensity
445.6	89.5	13.2
468.8	91.1	11.2
584.5	65.1	43.9
611.5	72.5	34.6
692.5	68.8	39.3
833.4	75.3	31.1
1012.8	55.5	56.0
1045.6	43.6	71.0
1132.4	41.4	73.7
1197.9	21.1	99.3
1466.1	71.6	35.7
1496.9	89.2	13.6
1603.0	85.1	18.8
2855.0	39.5	76.2
2924.4	20.5	100.0
2957.2	40.1	75.4
3437.6	70.1	37.6

Alkyl (C_{10}-C_{13})benzenesulphonic acid, sodium salt

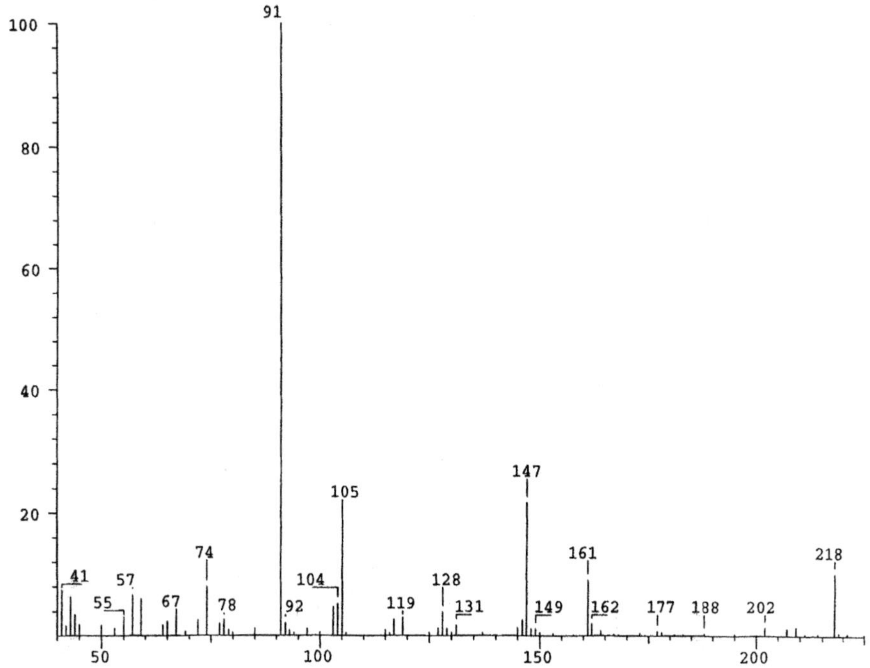

M/Z	Ion Intensity (%)	M/Z	Ion Intensity (%)
41	7.5	104	5.3
43	6.4	105	21.1
57	6.7	147	21.7
59	6.1	161	9.2
74	8.1	218	10.2
91	100.0	219	0.4

Spectrometer	: Finnigan Mat SSQ 700
Inletsystem	: Capillary GC/MS
Source Temperature	: 180°C
Electron Energy	: 70 eV
Scan Range	: 40-500

Alkyl (C_{10}-C_{13})benzenesulphonic acid, sodium salt

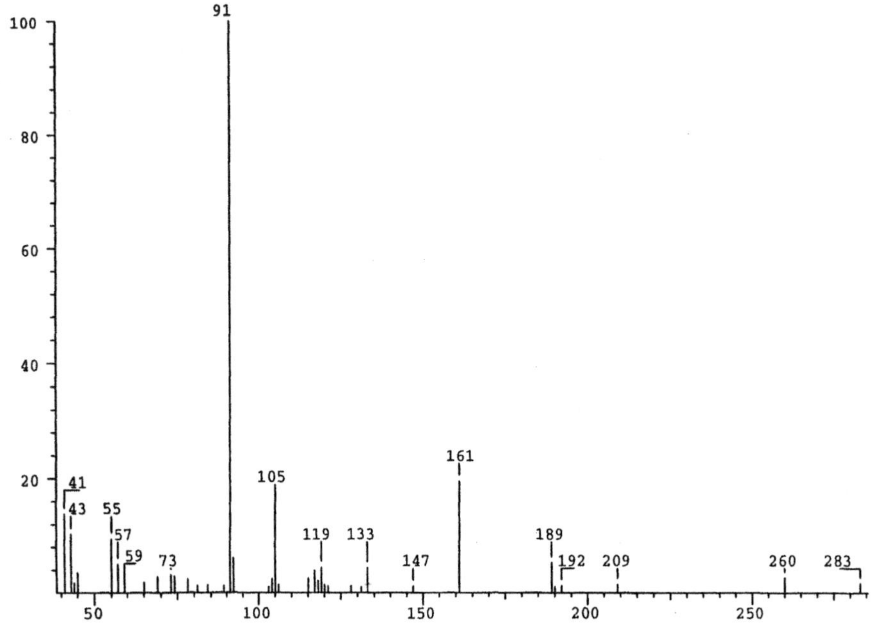

M/Z	Ion Intensity (%)	M/Z	Ion Intensity (%)
41	13.9	105	17.9
43	10.3	119	4.5
55	9.5	133	4.4
57	5.0	161	19.6
91	100.0	189	5.3
92	6.1		

Spectrometer	: Finnigan Mat SSQ 700
Inletsystem	: Capillary GC/MS
Source Temperature	: 180°C
Electron Energy	: 70 eV
Scan Range	: 40-500

Alkyl (C_{10}-C_{13})benzenesulphonic acid, sodium salt

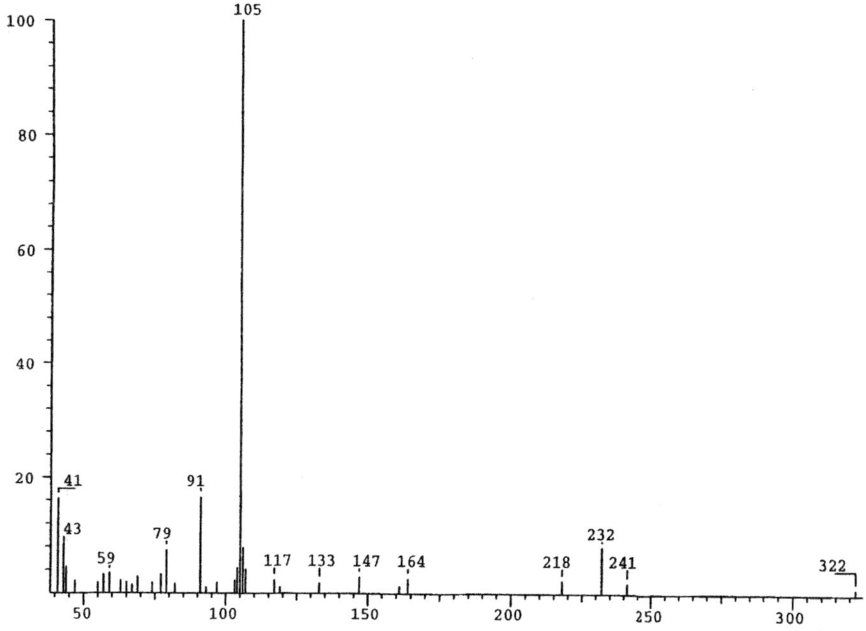

M/Z	Ion Intensity (%)	M/Z	Ion Intensity (%)
41	16.28	105	100.0
43	8.6	106	7.9
79	7.4	232	8.1
91	16.5		

Spectrometer	: Finnigan Mat SSQ 700
Inletsystem	: Capillary GC/MS
Source Temperature	: 180°C
Electron Energy	: 70 eV
Scan Range	: 40-500

PM n° 33800

CAS n°

Name
ALKYL (C_{10}-C_{13}) BENZENSULPHONIC ACID

Solvent	CD_3OD
SF	250.13
SI	16384
SW	3125
NS	36

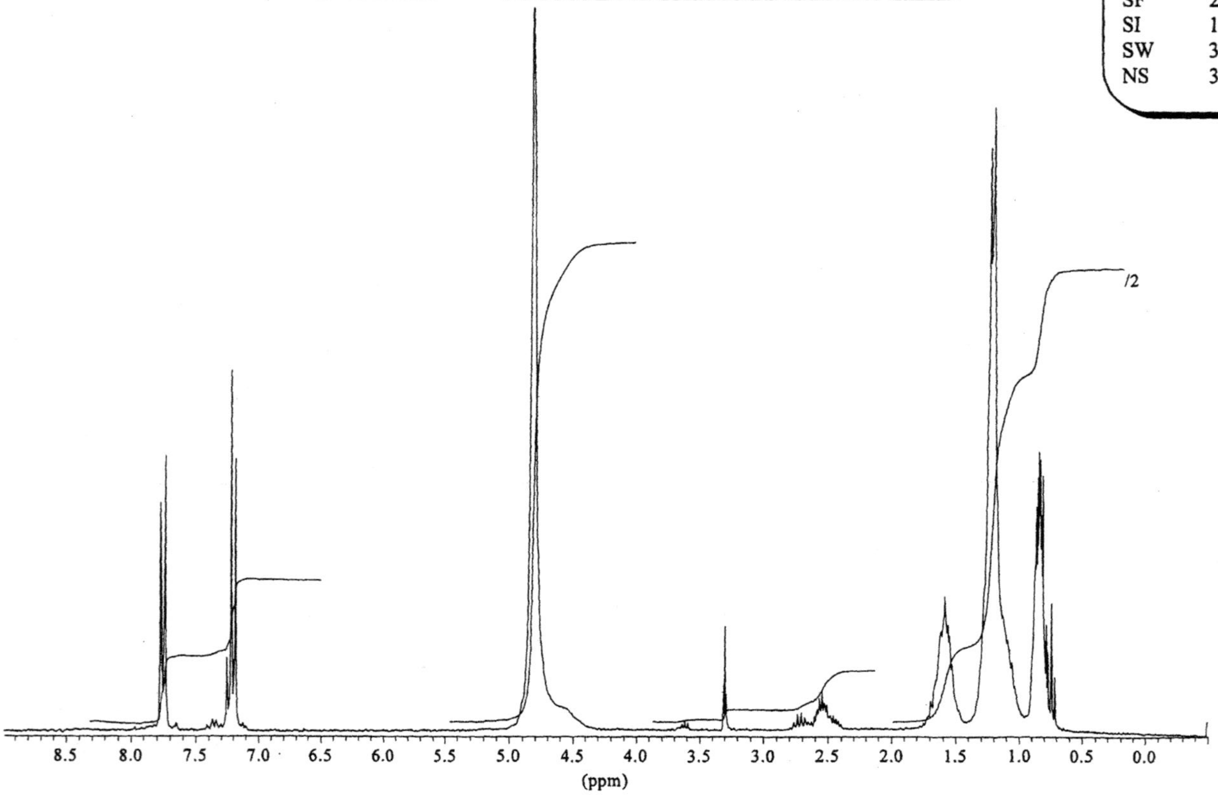

Alkyl (C_{10}-C_{20})sulphonic acid, esters with phenols

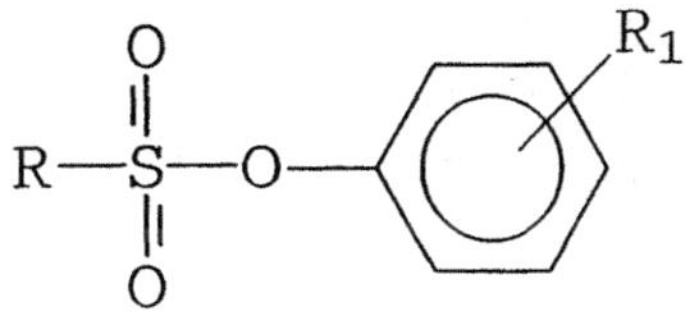

CAS No.	-
PM Ref. No.	- 34240
Restrictions	- SML = 6.0 mg/kg
Formula	-
Molecular weight	-

Alternative names	-
Physical Characteristics	- Colourless liquid. - Soluble in water, alcohol, ether.
Handling	- Store at room temperature.
Safety	-
Availability	- Standard sample was supplied.
Current uses	- Plasticizer.
Applications	- PVC.
GC Retention time	- Cluster 40.9 - 52.0 min.

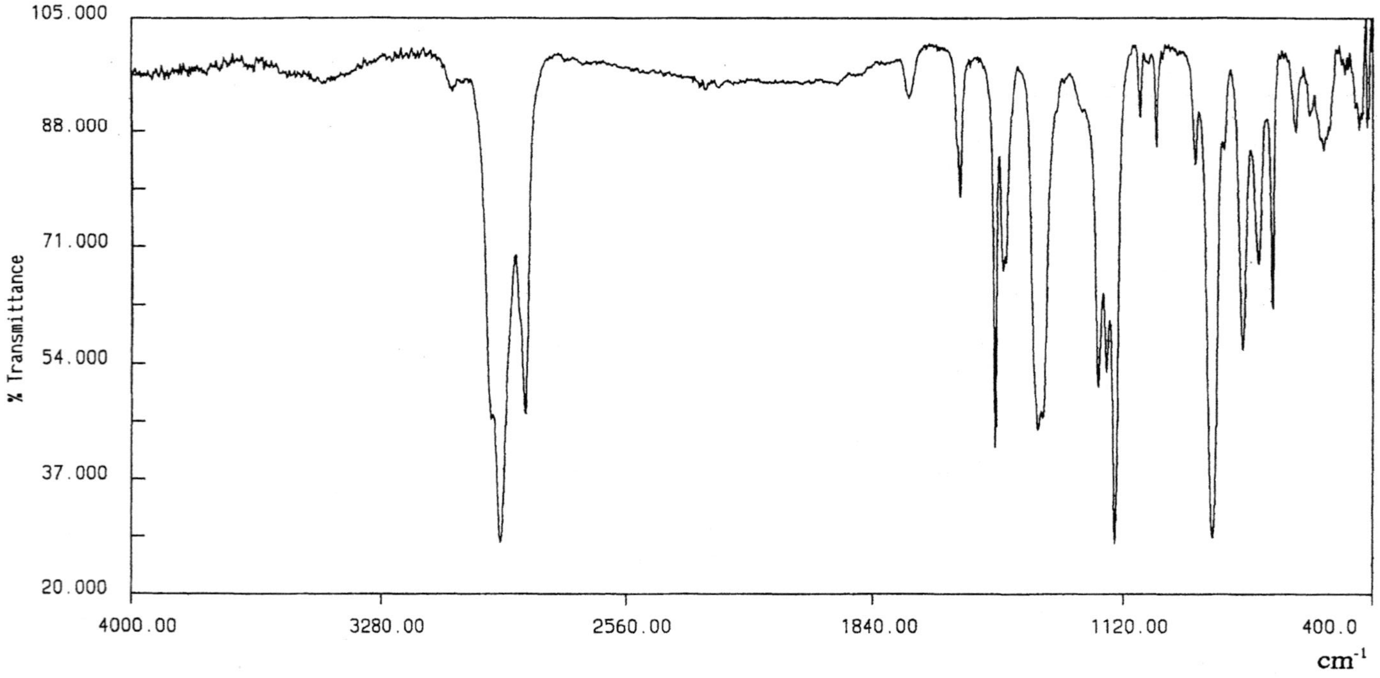

Name : ALKYL (C10-C20) SULPHONIC ACID, ESTERS WITH PHENOLS

PM Nr : 34240 CAS Nr :

KBr pellet

CAS No. PM Ref. No. 34240

ALKYL (C 10-C 20) SULPHONIC ACID, ESTERS WITH PHENOLS

Peak wavenumber	% T	Relative intensity
414.8	88.7	15.6
439.8	88.2	16.3
542.1	85.2	20.4
621.2	88.1	16.4
688.7	62.0	52.3
729.2	68.5	43.4
775.5	56.0	60.6
860.4	28.2	98.8
910.5	83.1	23.3
1024.3	85.8	19.6
1145.9	27.4	100.0
1169.0	52.7	65.1
1192.2	50.6	68.0
1352.3	45.9	74.4
1489.2	41.6	80.4
1589.5	78.3	29.9
2855.0	46.6	73.5
2928.3	27.60	99.7

Alkyl (C_{10}-C_{20})sulphonic acid, esters with phenols

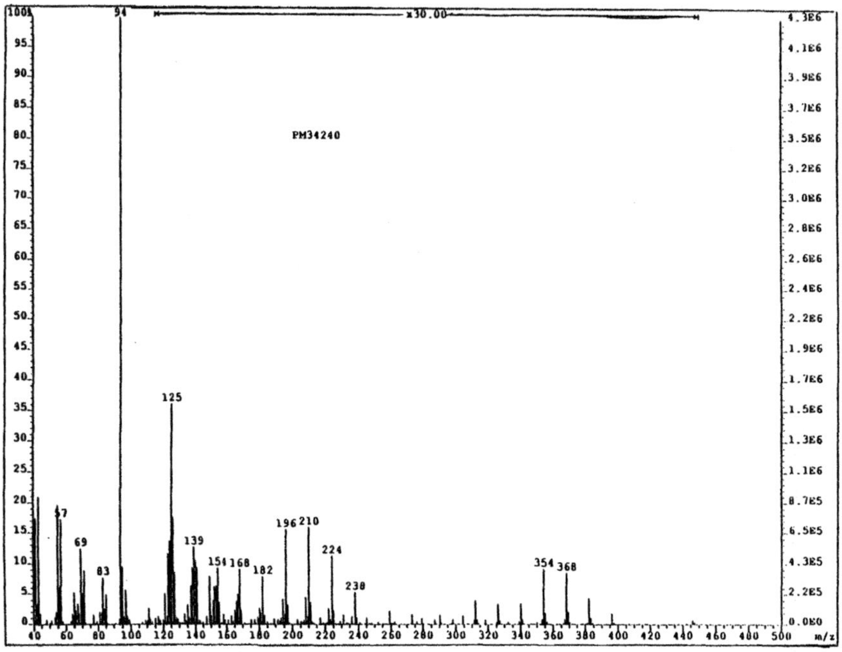

M/Z	Ion Intensity (%)	M/Z	Ion Intensity (%)
41	17.5	70	5.4
43	21.0	71	9.0
55	19.7	83	7.8
56	6.5	85	5.0
57	17.3	94	100.0
65	5.3	95	9.7
69	12.5	97	5.9

Spectrometer : VG Autospec
Inletsystem : Flow injection particle beam MS
Source Temperature : 250°C
Electron Energy : 70 eV
Scan Range : 40-1000

PM n°
34240

CAS n°

Name
ALKYL (C_{10}-C_{20}) SULPHONIC ESTER WITH PHENOLS

Solvent	$CDCL_3$
SF	250.13
SI	16384
SW	3125
NS	32

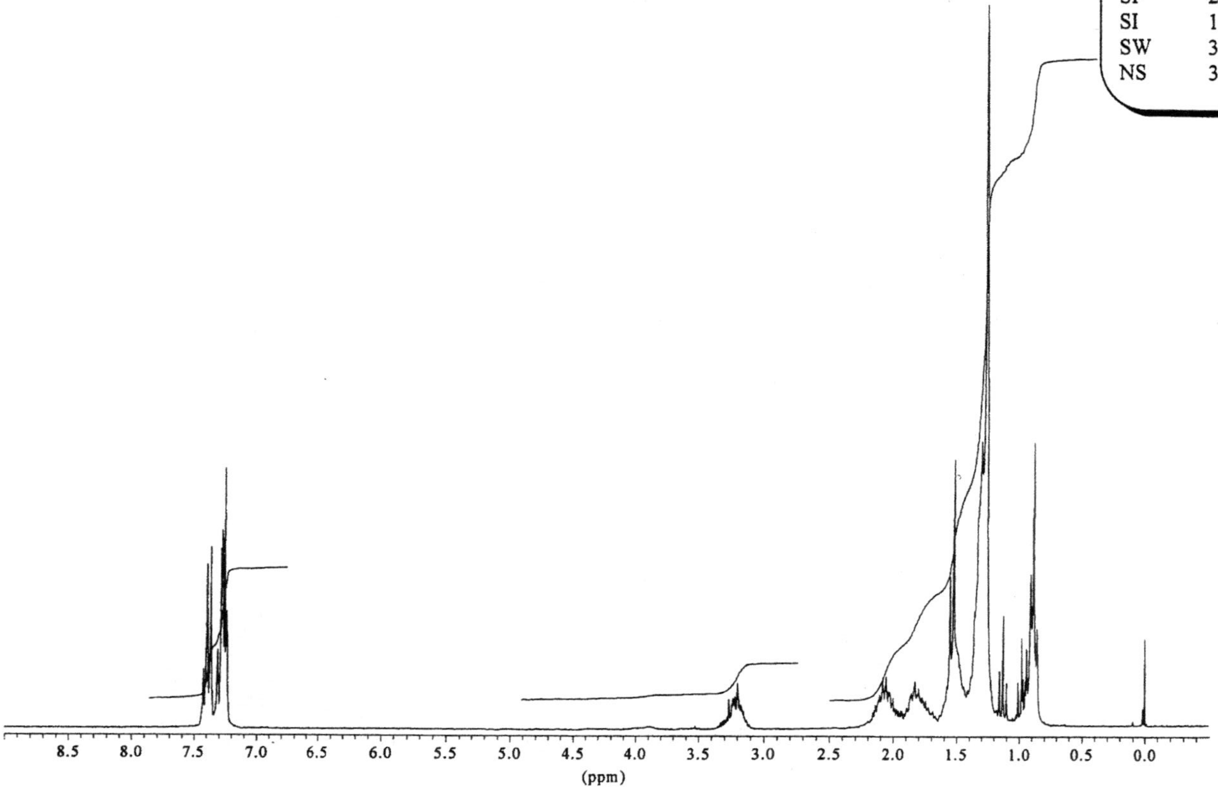

Azelaic acid, bis(2-ethylhexyl) ester

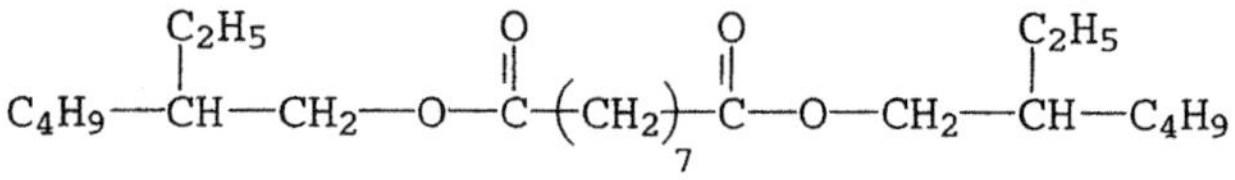

CAS No.	- 00103-24-2
PM Ref. No.	- 36320
Restrictions	- SML = 3.0 mg/kg
Formula	- $C_{25}H_{48}O_4$
Molecular weight	- 412

Alternative names - Nonanedioic acid, di(2-ethylhexyl) ester, Di(2-ethylhexyl) azelate

Physical Characteristics - Colourless liquid, mp -78°C, bp 208 - 210°C (4 mm Hg).
- Soluble in alcohol, acetone, benzene.

Handling - Store at room temperature.

Safety -

Availability - Standard sample was supplied.

Current uses - Plasticizer.

Applications - PVC, PVDC, inks.

GC Retention time - 49.6 min.

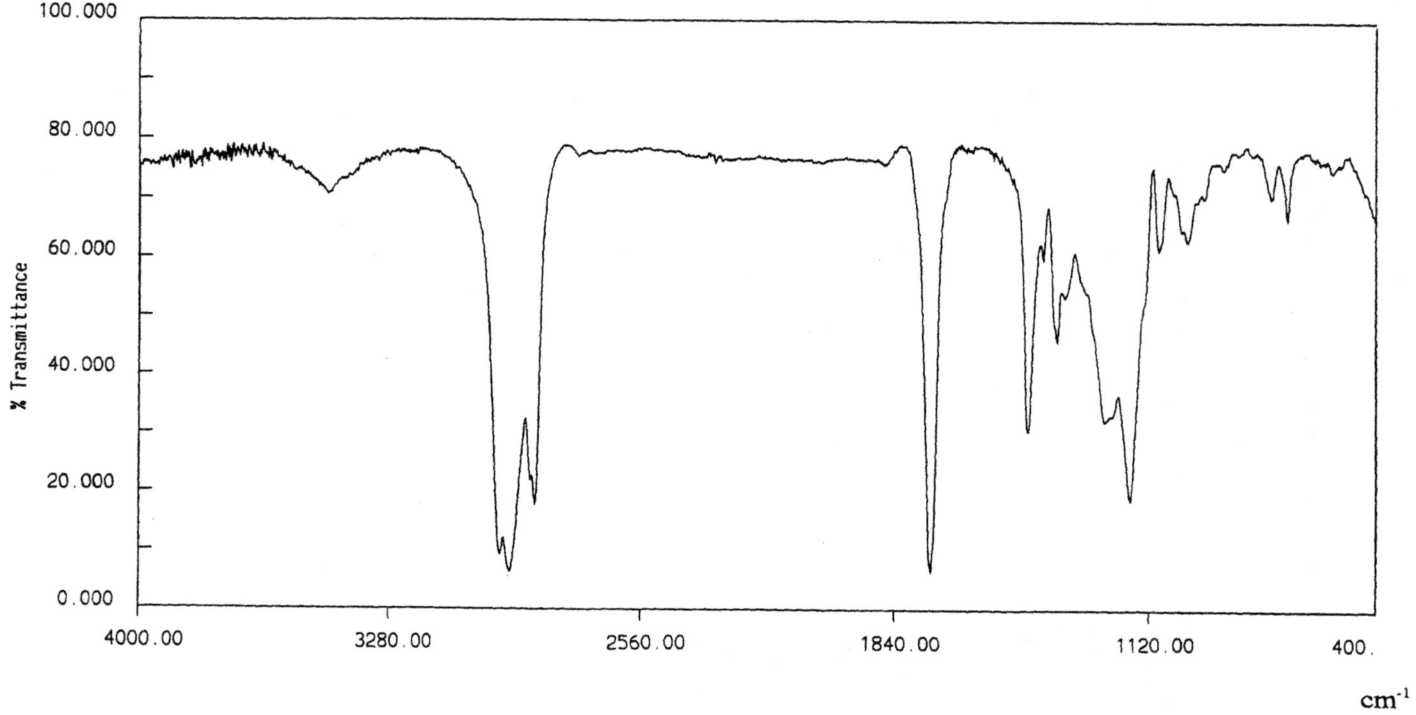

Name : AZELAIC ACID, BIS(2-ETHYLHEXYL) ESTER

PM Nr : 36320 CAS Nr : 00103-24-2

KBr pellet

CAS No. 00103-24-2 PM Ref. No. 36320

AZELAIC ACID, BIS(2-ETHYLHEXYL) ESTER

Peak wavenumber	% T	Relative intensity
410.9	73.4	28.4
437.9	66.9	35.3
729.2	66.6	35.7
773.6	70.2	31.8
962.6	70.1	31.9
1010.8	62.8	39.7
1093.8	61.4	41.2
1174.8	18.7	86.8
1228.8	32.9	71.6
1360.0	53.3	49.9
1381.2	45.8	57.8
1421.7	59.7	43.1
1464.2	30.4	74.3
1738.1	6.4	99.9
1863.5	75.7	26.0
2860.8	17.9	87.6
2932.2	6.3	100.0
3738.5	75.8	25.8

Azelaic acid, bis(2-ethylhexyl) ester

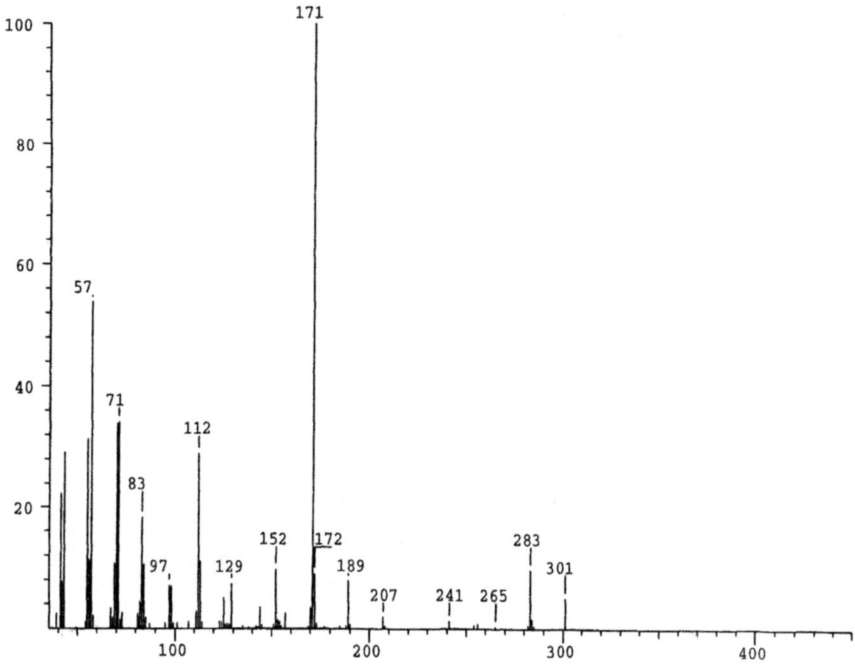

M/Z	Ion Intensity (%)	M/Z	Ion Intensity (%)
41	22.3	84	10.5
43	29.1	112	29.9
55	31.4	113	11.1
56	11.3	152	9.7
57	53.8	171	100.0
70	33.9	172	9.1
71	34.1	283	9.7
83	18.3	301	5.0

Spectrometer	: Finnigan Mat SSQ 700
Inletsystem	: Capillary GC/MS
Source Temperature	: 180°C
Electron Energy	: 70 eV
Scan Range	: 25-500

PM n°	CAS n°	Name
36320	00103-24-2	AZELAIC ACID, BIS(2-ETHYLHEXYL) ESTER

Solvent	$CDCL_3$
SF	250.13
SI	16384
SW	3125
NS	32

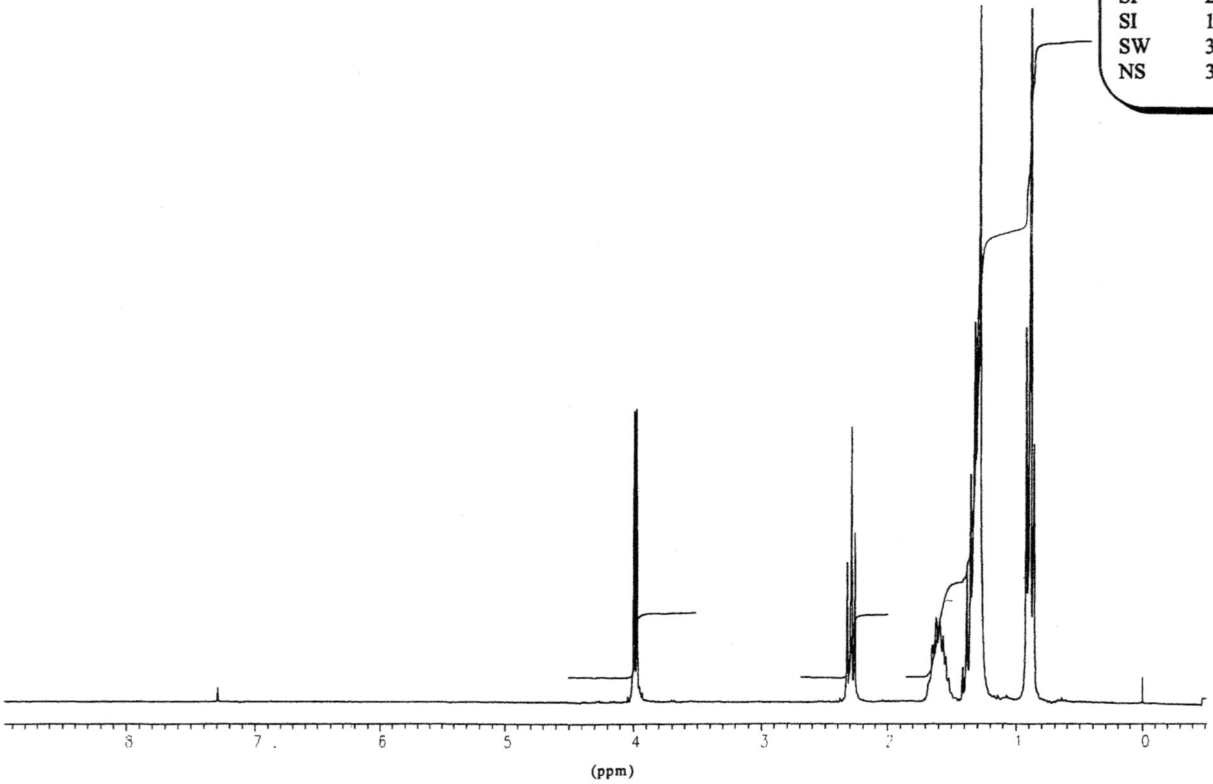

Azelaic acid, bis(6-methylheptyl) ester

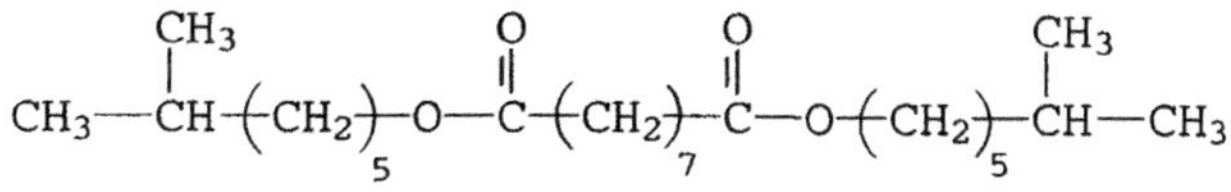

CAS No.	- 00106-03-6
PM Ref. No.	- 36400
Restrictions	- SML = 3.0 mg/kg
Formula	- $C_{25}H_{48}O_4$
Molecular weight	- 412

Alternative names	- Di(6-methylheptyl) azelate, Dioctyl azelate, Nonanedioic acid, di(6-methylheptyl) ester
Physical Characteristics	- Colourless liquid. - Soluble in alcohol, acetone, benzene.
Handling	- Store at room temperature.
Safety	-
Availability	- Standard sample was supplied.
Current uses	- Plasticizer.
Applications	- PVC, PVDC.
GC Retention time	- Cluster 48.8 - 51.5 min.
Note	- Contains other octyl isomers.

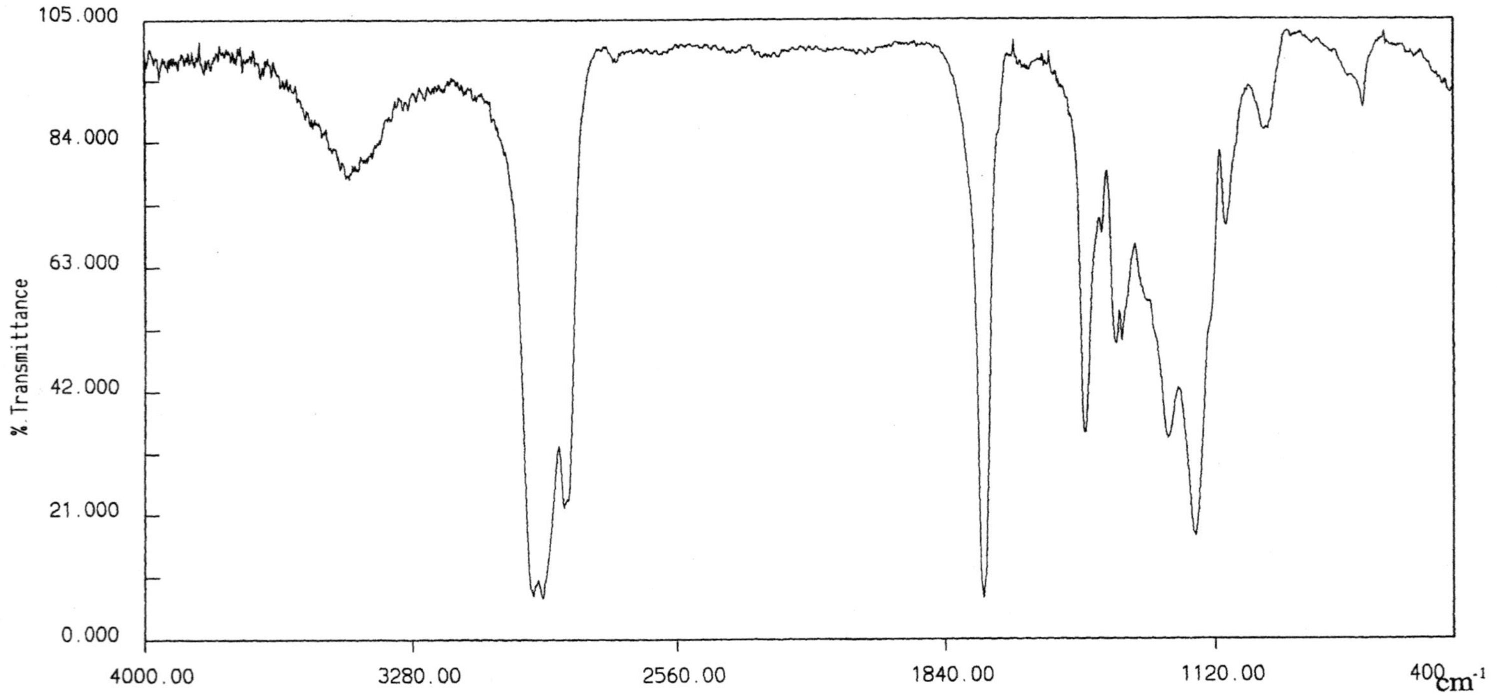

Name : AZELAIC ACID, BIS(6-METHYLHEPTYL) ESTER

PM Nr : 36400 CAS Nr : 00106-03-6

KBr pellet

CAS No. 00106-03-6 PM Ref. No. 36400

AZELAIC ACID, BIS(6-METHYLHEPTYL) ESTER

Peak wavenumber	% T	Relative intensity
414.8	76.2	25.6
426.3	92.1	8.5
451.4	90.1	10.6
727.3	89.8	11.0
993.5	86.1	15.0
1093.8	68.8	33.5
1172.9	17.5	88.7
1246.2	34.0	70.9
1365.8	50.5	53.2
1381.2	49.9	53.9
1421.7	68.6	33.8
1464.2	34.9	69.9
1738.1	6.9	100.0
2872.4	22.2	83.6
2930.2	7.0	99.9
3337.3	87.5	13.4
3391.3	81.3	20.1

Azelaic acid, bis(6-methylheptyl) ester

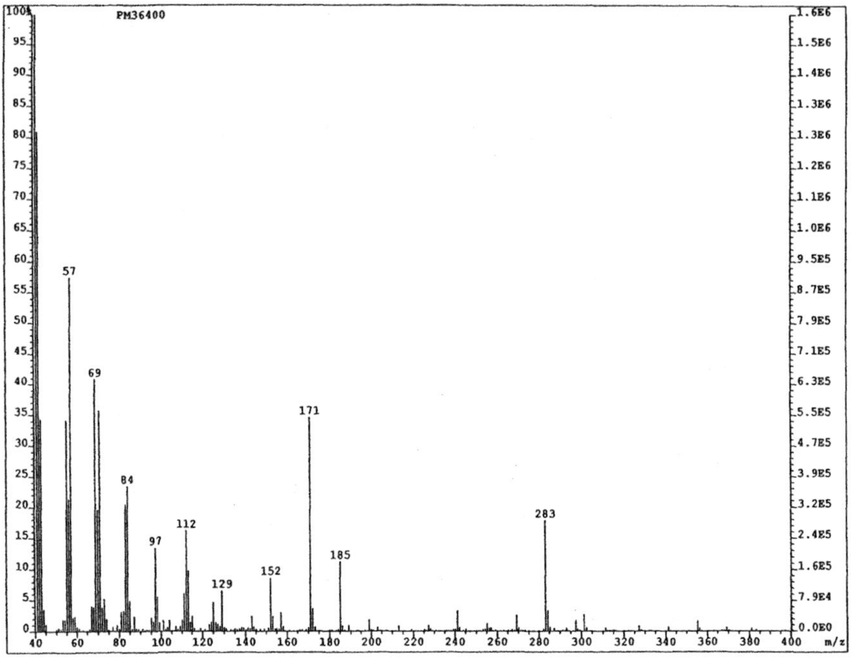

M/Z	Ion Intensity (%)	M/Z	Ion Intensity (%)
41	100.0	84	23.3
42	81.0	97	13.5
43	34.4	111	6.1
55	33.7	112	16.6
56	21.5	113	10.4
57	57.7	129	6.7
69	41.1	152	8.6
70	19.6	171	34.4
71	35.6	185	11.7
83	20.2	283	17.8

Spectrometer	: VG Autospec
Inletsystem	: Flow injection particle beam MS
Source Temperature	: 250°C
Electron Energy	: 70 eV
Scan Range	: 40-1000

PM n° 36400

CAS n° 00106-03-6

Name
AZELAIC ACID, BIS(6-METHYLHEPTYL) ESTER

Solvent	$CDCL_3$
SF	250.13
SI	16384
SW	3125
NS	32

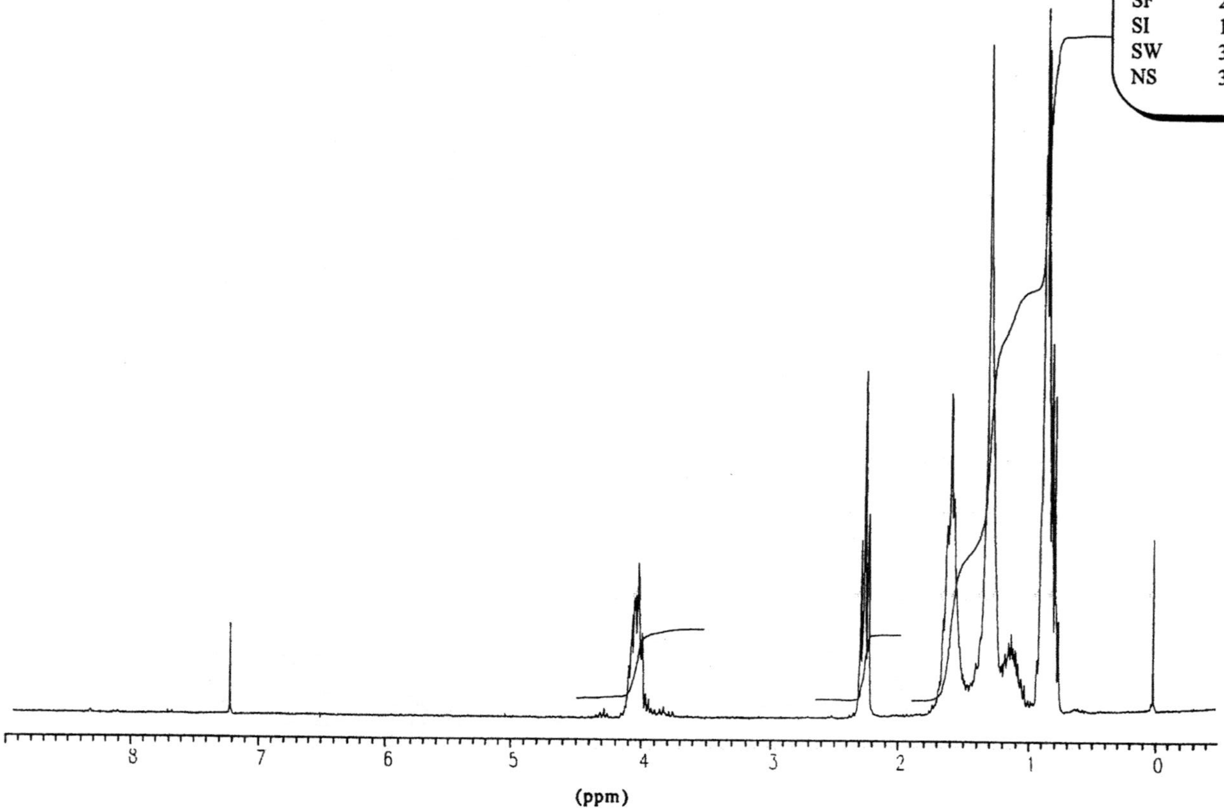

2,5-Bis(5-tert-butyl-2-benzoxazolyl)thiophene

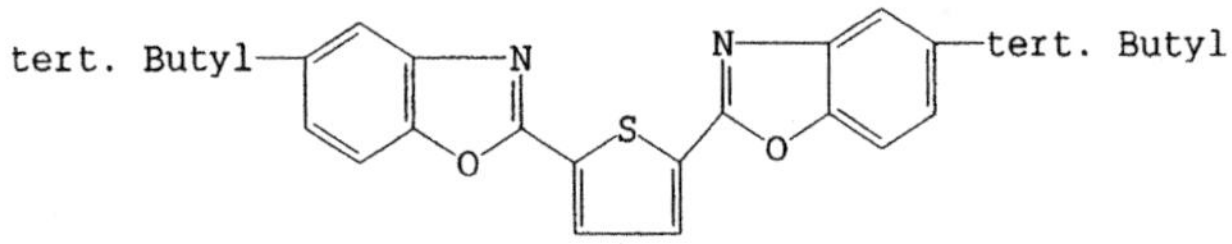

CAS No.	- 07128-64-5
PM Ref. No.	- 38560
Restrictions	- SML = 0.6 mg/kg
Formula	- $C_{26}H_{26}N_2O_2S$
Molecular weight	- 430

Alternative names	- 2,2'-(2,5-Thiophenediyl)bis(5-tert-butyl-benzoxazole)
Physical Characteristics	- Yellow powder, mp 196 - 203°C. - Soluble in acetone, toluene.
Handling	- Store at room temperature.
Safety	-
Availability	- Standard sample was supplied.
Current uses	- UV-stabilizer, optical brightner.
Applications	- General use.
GC Retention time	- 64.8 min.

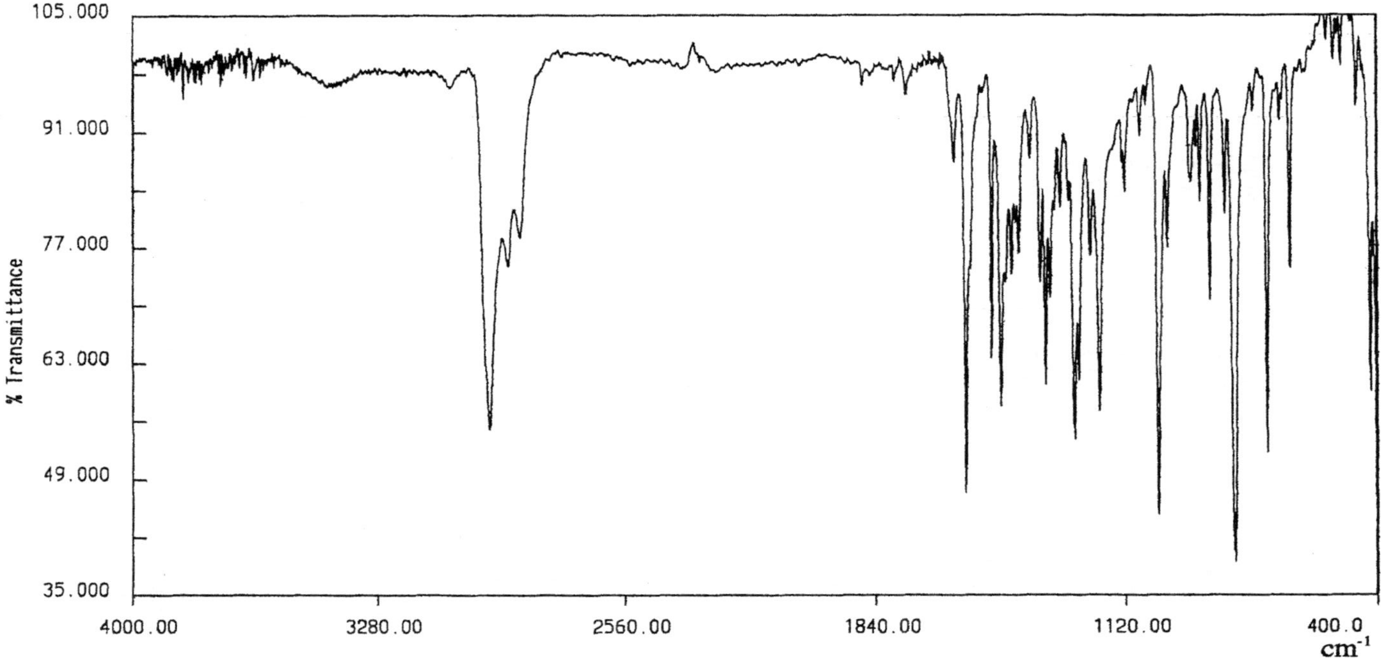

Name : 2,5-BIS(5-TERT-BUTYL-2-BENZOXAZOLYL)THIOPHENE

PM Nr : 38560 CAS Nr : 07128-64-5

KBr pellet

CAS No. 07128-64-5 PM Ref. No. 38560

2,5-BIS(5-tert-BUTYL-2-BENZOXAZOLYL)THIOPHENE

Peak wavenumber	% T	Relative intensity
418,6	72,5	45,1
459,1	89,9	16,6
501,6	92,1	13,0
522,8	93,4	10,8
648,2	73,9	42,8
677,1	91,7	13,6
715,7	52,0	78,7
756,2	92,9	11,6
808,3	38,9	100,0
835,3	80,8	31,5
877,7	70,5	48,4
906,7	82,4	28,9
918,2	89,1	17,9
931,7	84,8	24,9
999,3	77,0	37,7
1026,3	44,9	90,4
1122,7	83,5	27,1
1196,0	57,0	70,5
1221,1	75,4	40,3
1253,9	60,4	64,9
1265,5	53,4	76,4
1307,9	80,9	31,3
1334,9	70,1	49,0
1348,4	59,7	66,1
1363,8	71,7	46,4
1392,8	86,4	22,3
1425,6	75,0	41,0
1446,8	72,4	45,3
1464,2	71,4	46,9
1477,7	56,7	71,0
1504,7	62,4	61,7
1579,9	46,2	88,2
1612,7	15,0	139,4
2866,6	74,4	42,0
2901,3	72,2	45,6
2955,3	54,3	74,9

2,5-Bis(5-tert-butyl-2-benzoxazolyl)thiophene

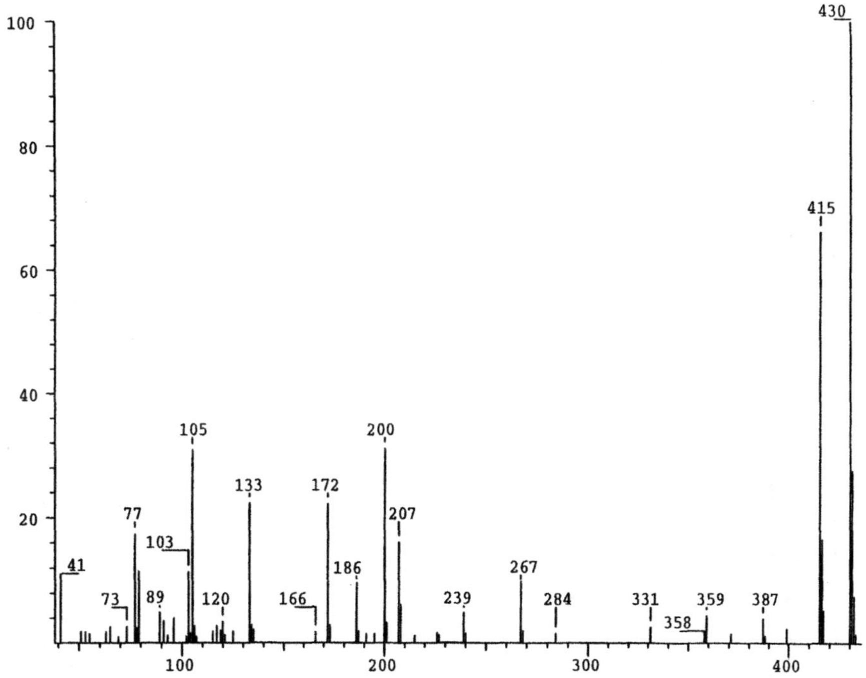

M/Z	Ion Intensity (%)	M/Z	Ion Intensity (%)
41	10.1	267	9.8
77	17.4	415	66.3
79	11.5	416	16.5
103	11.3	417	5.2
105	30.9	430	100.0
133	22.3	431	27.6
172	22.2	432	7.5
186	9.6	433	1.5
200	31.1		

Spectrometer	: Finnigan Mat SSQ 700
Inletsystem	: Capillary GC/MS
Source Temperature	: 180°C
Electron Energy	: 70 eV
Scan Range	: 40-500

PM n°	CAS n°	Name
38560	07128-64-5	2,5-BIS(5-tert-BUTYL-2-BENZOXAZOLYL) THIOPHENE

Solvent	$CDCl_3$
SF	250.13
SI	16384
SW	3125
NS	20

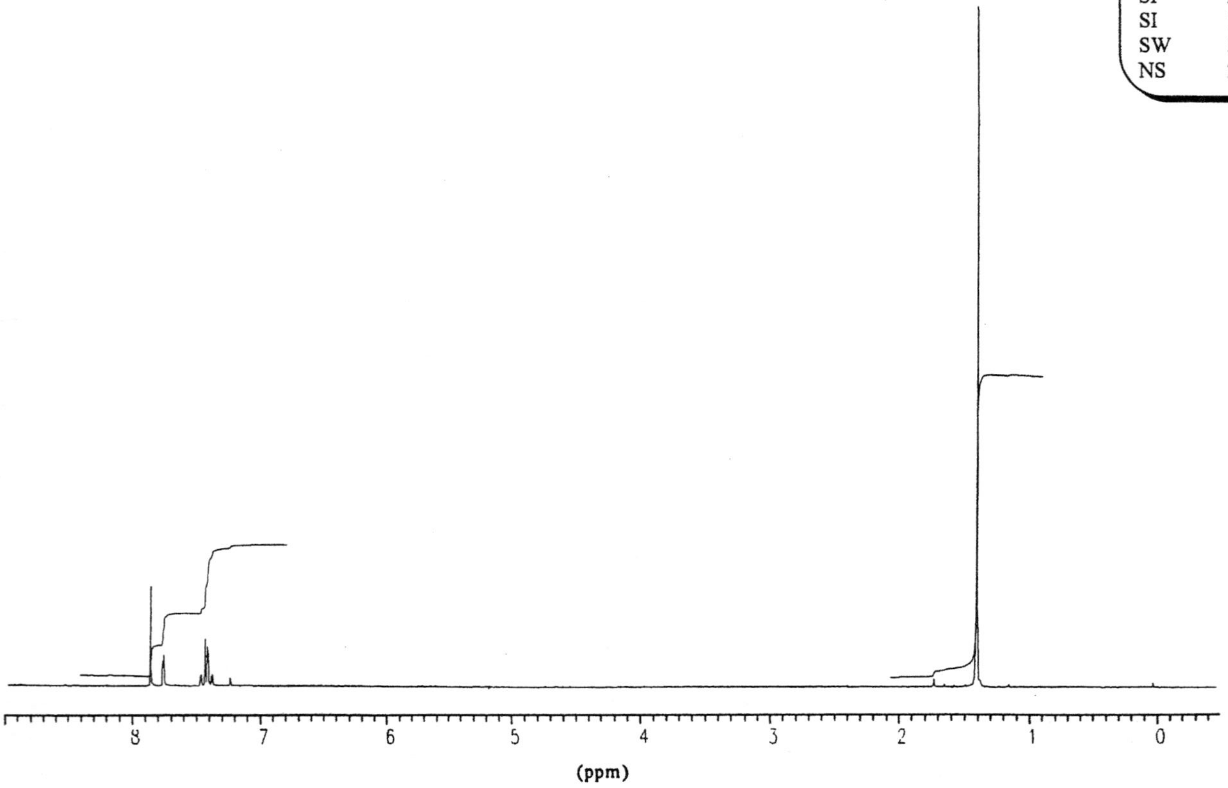

N,N'-Bis(3-(3,5-di-tert-butyl-4-hydroxyphenyl)propionyl) hydrazide

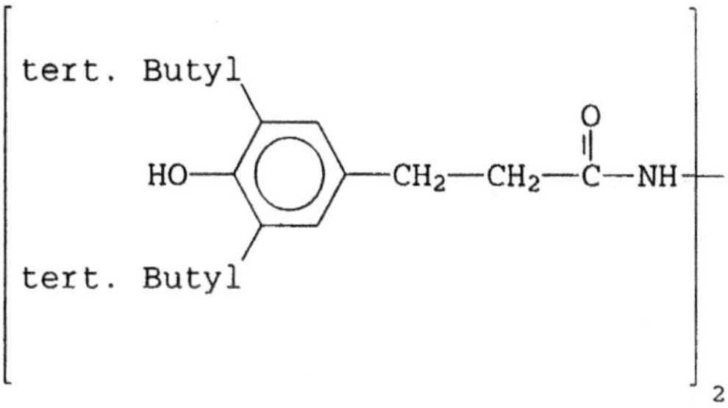

CAS No.	- 32687-78-8
PM Ref. No.	- 38800
Restrictions	- SML = 15 mg/kg
Formula	- $C_{34}H_{52}N_2O_4$
Molecular weight	- 552

Alternative names - 1,2-Bis(3,5-di-tert-butyl-4-hydroxy-hydrocinnamoyl)-hydrazine

Physical Characteristics - White powder, mp 225 - 227°C.
- Soluble in most organic solvents.

Handling - Store at room temperature.
Safety - Harmful.
Availability - Standard sample was supplied.

Current uses - Antioxidant.
Applications - Polyolefins, styrene co-polymers, polyacetals.

GC Retention time - 65.4 min.

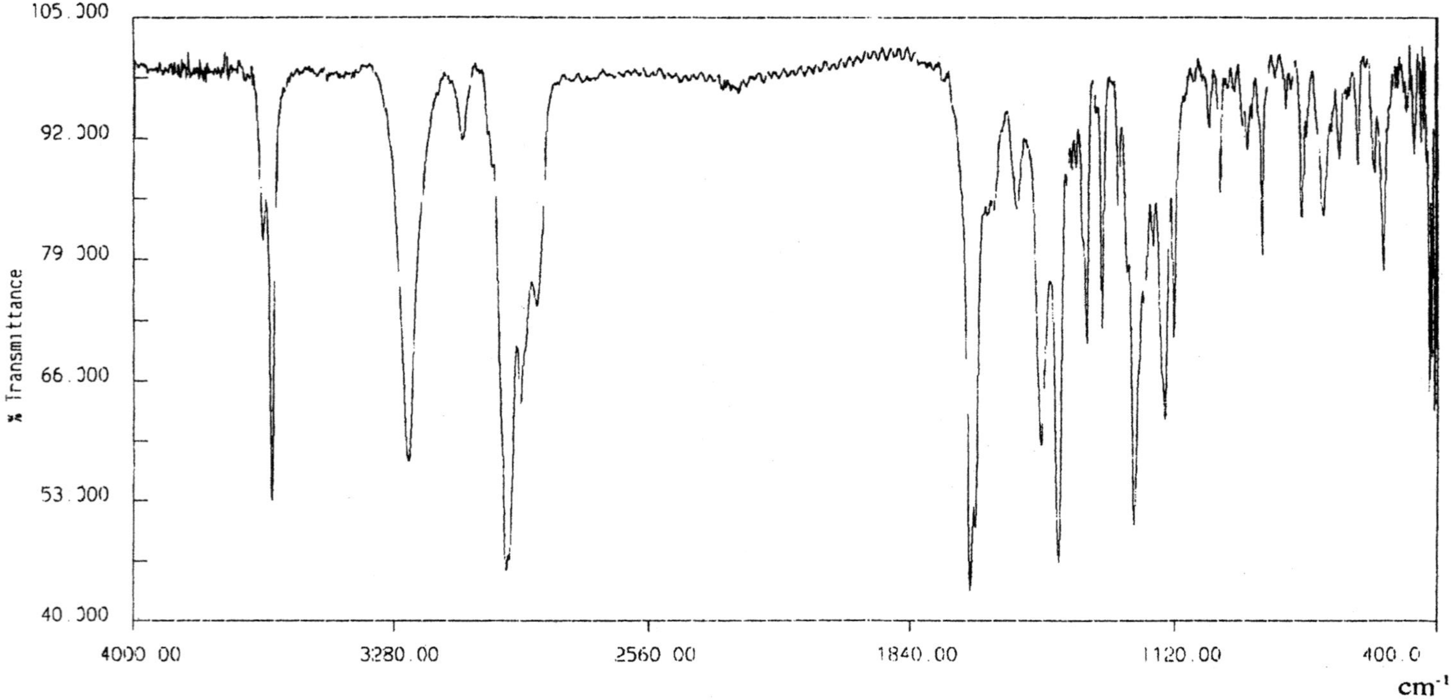

Name : N, N'- BIS(3- (3,5-DI-Tert-BUTYL-4-HYDROXYPHENYL) PROPIONYL) HYDRAZIDE

PM Nr : 38800 CAS Nr : 32687-78-8

KBr pellet

CAS No. 32687-78-8 PM Ref. No. 38800

N,N'-BIS(3-(3,5-DI-tert-BUTYL-4-HYDROXYPHENYL)PROPIONYL) HYDRAZIDE

Peak wavenumber	% T	Relative intensity
420,5	65,0	61,6
569,1	88,2	20,8
615,4	89,2	19,0
665,5	89,8	17,9
708,0	83,7	28,7
769,7	83,5	29,0
877,7	79,4	36,2
918,2	90,9	16,0
993,5	86,2	24,3
1122,7	70,7	51,5
1147,8	61,9	67,0
1180,6	80,5	34,3
1205,7	74,2	45,4
1232,7	50,4	87,2
1275,1	84,4	27,4
1319,5	71,7	49,8
1360,0	70,0	52,8
1388,9	88,8	19,7
1414,0	86,8	23,2
1435,2	46,3	94,5
1485,4	59,0	72,1
1551,0	84,4	27,4
1612,7	84,3	27,6
1664,8	50,0	88,0
1676,3	43,1	100,0
2920,6	63,4	64,4
2961,1	45,4	96,0
3238,9	57,2	75,3
3618,9	53,0	82,7
3644,0	81,0	33,4

N,N'-Bis(3-(3,5-di-tert-butyl-4-hydroxyphenyl)propionyl) hydrazide

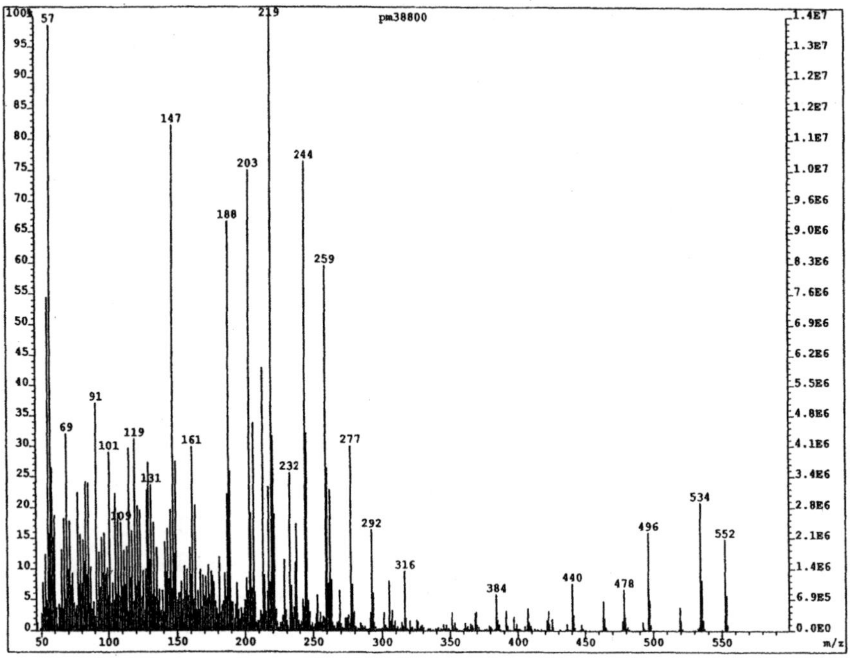

M/Z	Ion Intensity (%)	M/Z	Ion Intensity (%)
55	54.6	213	42.9
57	98.8	219	100.0
69	32.5	220	31.9
91	37.4	224	76.7
101	28.8	259	60.1
119	31.3	277	30.7
147	82.8	496	15.9
161	30.1	534	21.1
188	66.9	552	15.3
203	75.5	553	5.8

Spectrometer : VG Autospec
Inletsystem : Flow injection particle beam MS
Source Temperature : 250°C
Electron Energy : 70 eV
Scan Range : 40-1000

PM n°	CAS n°	Name
38800	32687-78-8	N, N'-BIS(3-(3,5-DI-TERT-BUTYL-4-HYDROXYPHENYL) PROPIONYL) HYDRAZIDE

Solvent	$CDCL_3$
SF	250.13
SI	16384
SW	3125
NS	24

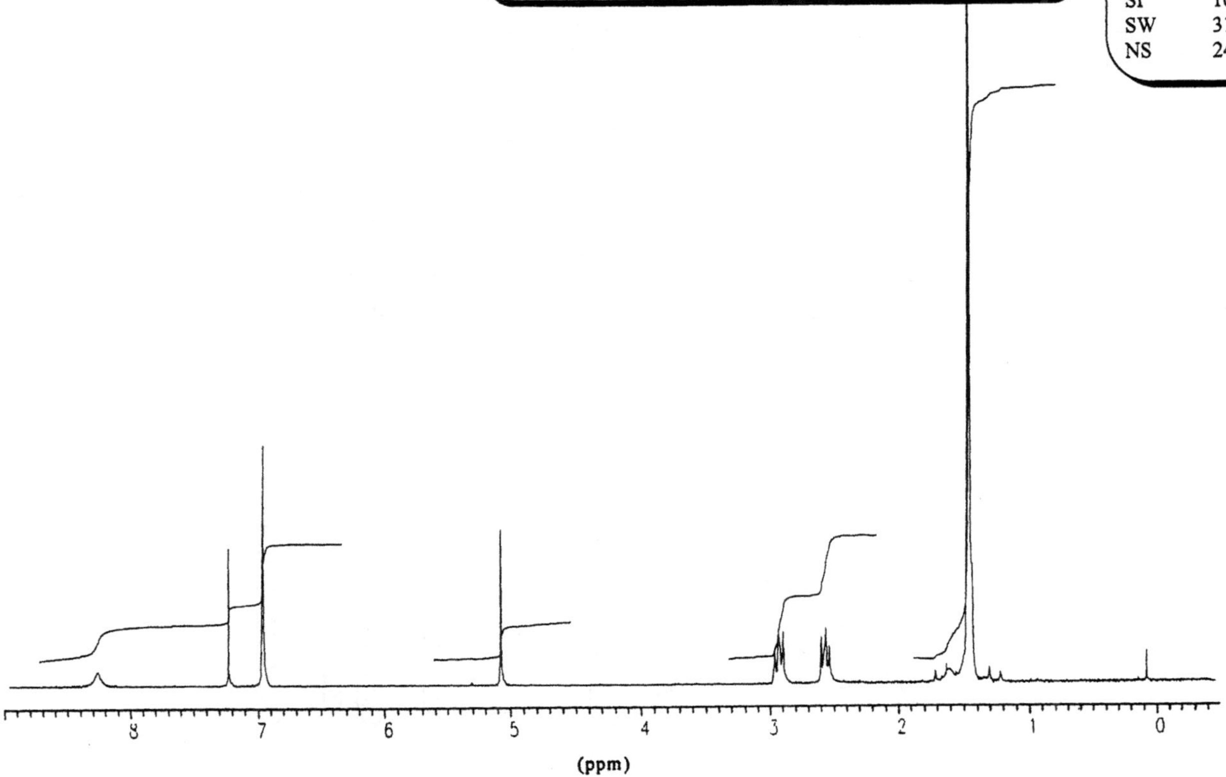

Bis(2,4-di-tert-butylphenyl)pentaerythritol diphosphite

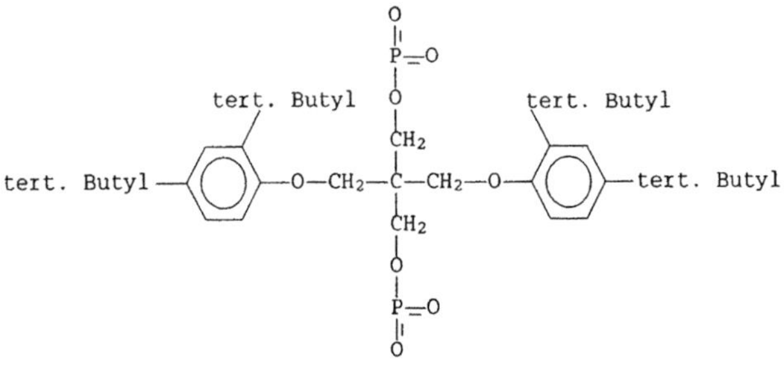

CAS No.	- 26741-53-7
PM Ref. No.	- 38820
Restrictions	- SML = 0.6 mg/kg
Formula	- $C_{33}H_{50}O_8P_2$
Molecular weight	- 636

Alternative names	-
Physical Characteristics	- White powder, mp 160 - 175°C. - Soluble in toluene, acetone, dichloromethane.
Handling	- Store at room temperature.
Safety	- Irritant.
Availability	- Standard sample was supplied.
Current uses	- Stabilizer.
Applications	- PET, PC, PVC, polyolefins.
GC Retention time	- 65.1 min.

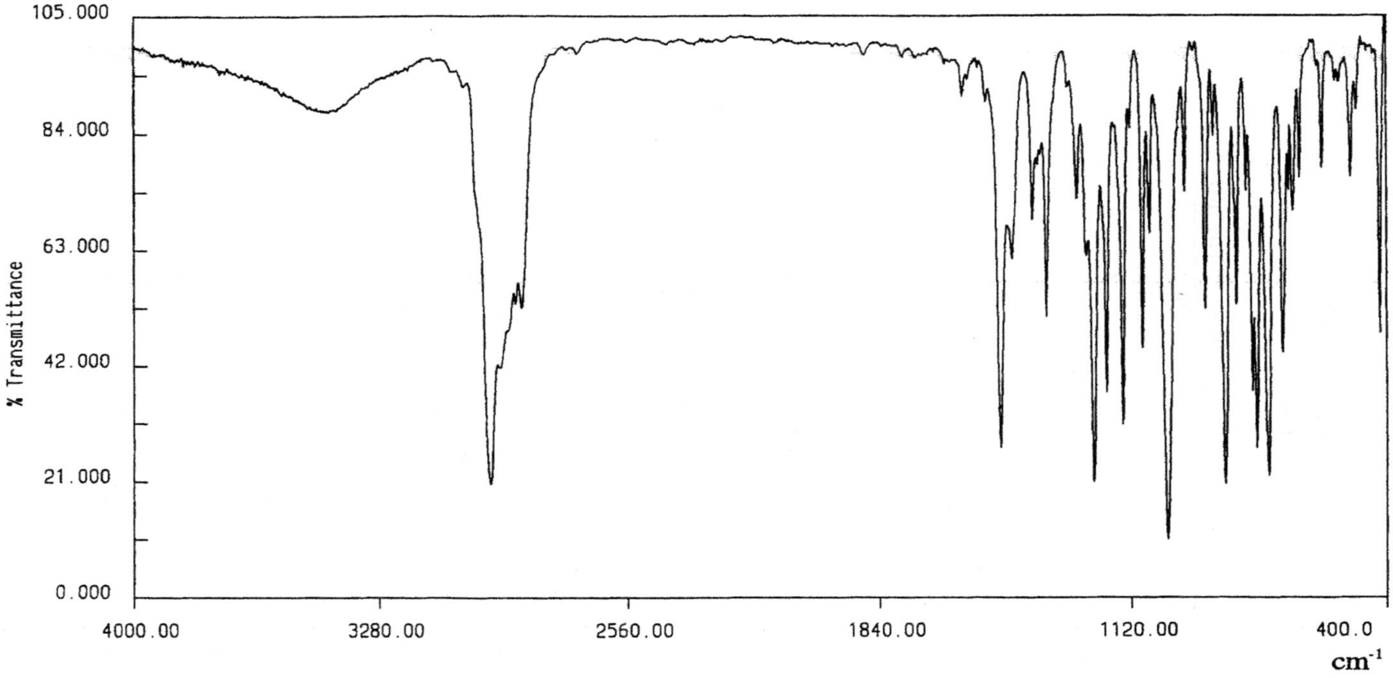

Name : BIS(2,4-DI-TERT-BUTYLPHENYL)PENTAERYTHRITOL DIPHOSPHITE

PM Nr : 38820 CAS Nr : 26741-53-7

KBr pellet

CAS No. 26741-53-7 PM Ref. No. 38820

BIS(2,4-DI-TERT-BUTYLPHENYL)PENTAERYTHRITOL DIPHOSPHITE

Peak wavenumber	% T	Relative intensity
418.6	47.9	58.2
482.3	88.3	13.1
499.6	76.1	26.7
580.6	77.8	24.8
644.3	75.9	26.9
663.6	70.0	33.5
677.1	73.7	29.4
694.5	44.3	62.2
735.0	21.9	87.3
767.8	26.9	81.6
779.3	37.4	70.0
796.7	73.5	29.6
823.7	53.0	52.5
856.5	20.5	88.8
889.3	83.5	18.4
910.5	52.2	53.4
968.4	73.4	29.7
1016.6	10.48	100.0
1068.7	66.00	38.0
1088.0	45.14	61.3
1145.9	31.21	76.8
1192.2	36.99	70.4
1226.9	20.98	88.3
1273.2	72.21	31.0
1361.9	50.75	55.0
1387.0	78.32	24.2
1402.4	68.42	35.3
1460.3	61.33	43.2
1495.0	27.09	81.4
1605.0	90.55	10.6
2868.5	52.48	53.1
2930.2	41.48	65.4
2957.2	20.55	88.7
3404.8	88.05	13.4

Bis(2,4-di-tert-butylphenyl)pentaerythritol diphosphite

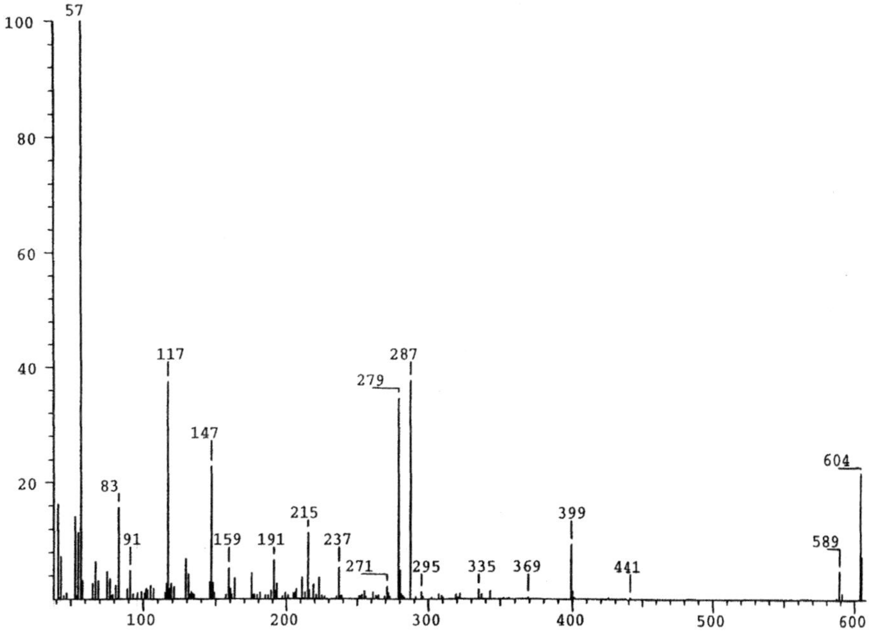

M/Z	Ion Intensity (%)	M/Z	Ion Intensity (%)
41	16.2	191	6.7
53	14.1	215	11.4
55	11.4	237	5.4
57	100.0	279	34.6
83	15.6	287	37.7
91	4.8	399	9.4
117	37.5	589	4.9
147	22.8	604	21.8
159	5.3	605	7.4

Spectrometer	: Finnigan Mat SSQ 700
Inletsystem	: Capillary GC/MS
Source Temperature	: 180°C
Electron Energy	: 70 eV
Scan Range	: 40-700

PM n°
38820

CAS n°
26741-53-7

Name:
BIS (2,4-DI-TERT-BUTYLPHENYL)
PENTAERYTHRITOL DIPHOSPHITE

Solvent	$CDCL_3$
SF	250.13
SI	16384
SW	3125
NS	32

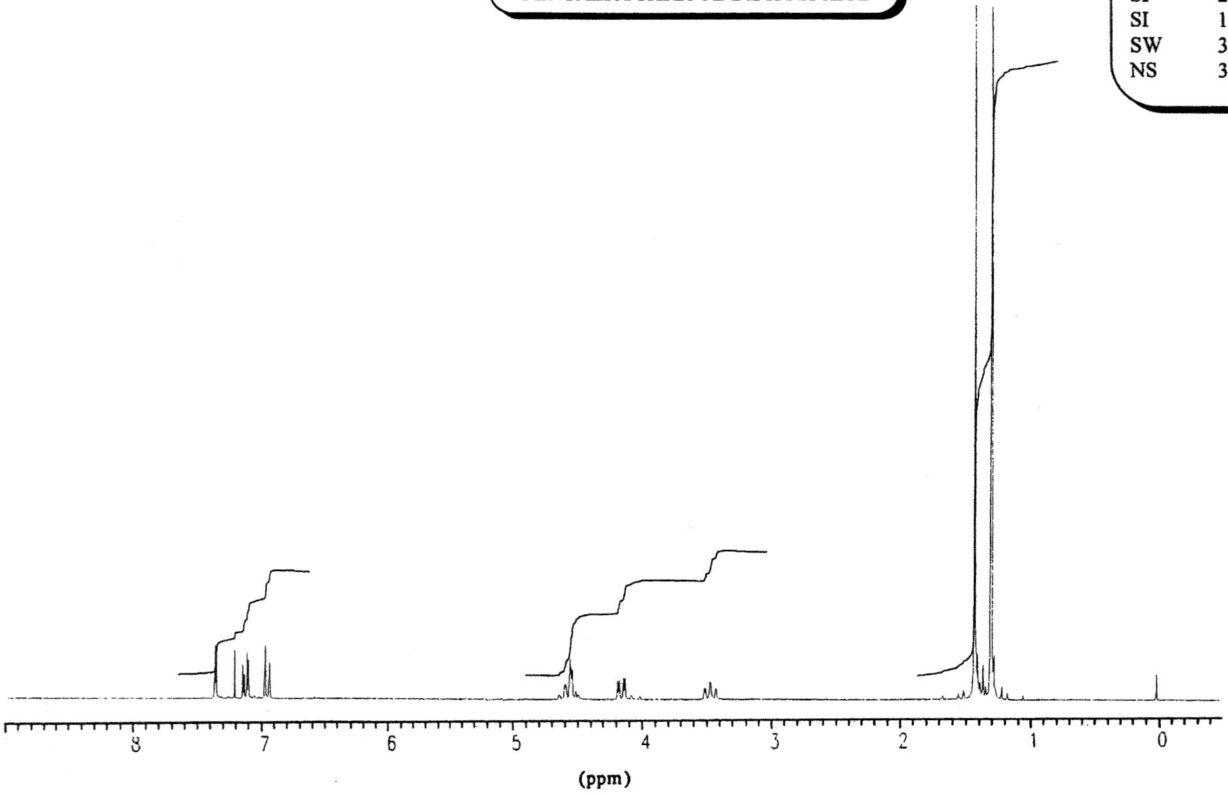

Bis(4-ethylbenzylidene) sorbitol

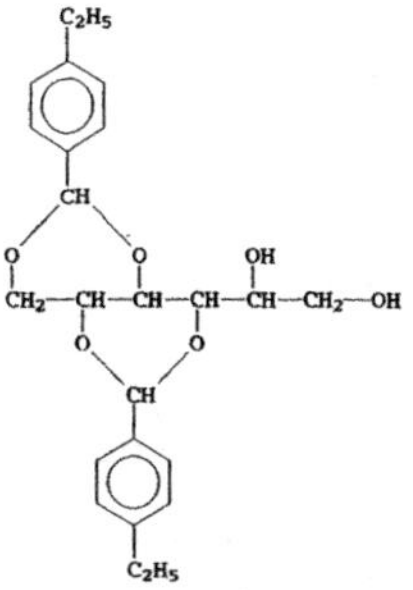

CAS No.	- 79072-96-1
PM Ref. No.	- 38950
Restrictions	- None
Formula	- $C_{24}H_{30}O_6$
Molecular weight	- 414

Alternative names	-
Physical Characteristics	- White powder, mp 212 - 226°C.
Handling	- Store at room temperature.
Safety	-
Availability	- Standard sample was supplied.
Current uses	- Nucleating agent.
Applications	- PP.
GC Retention time	-

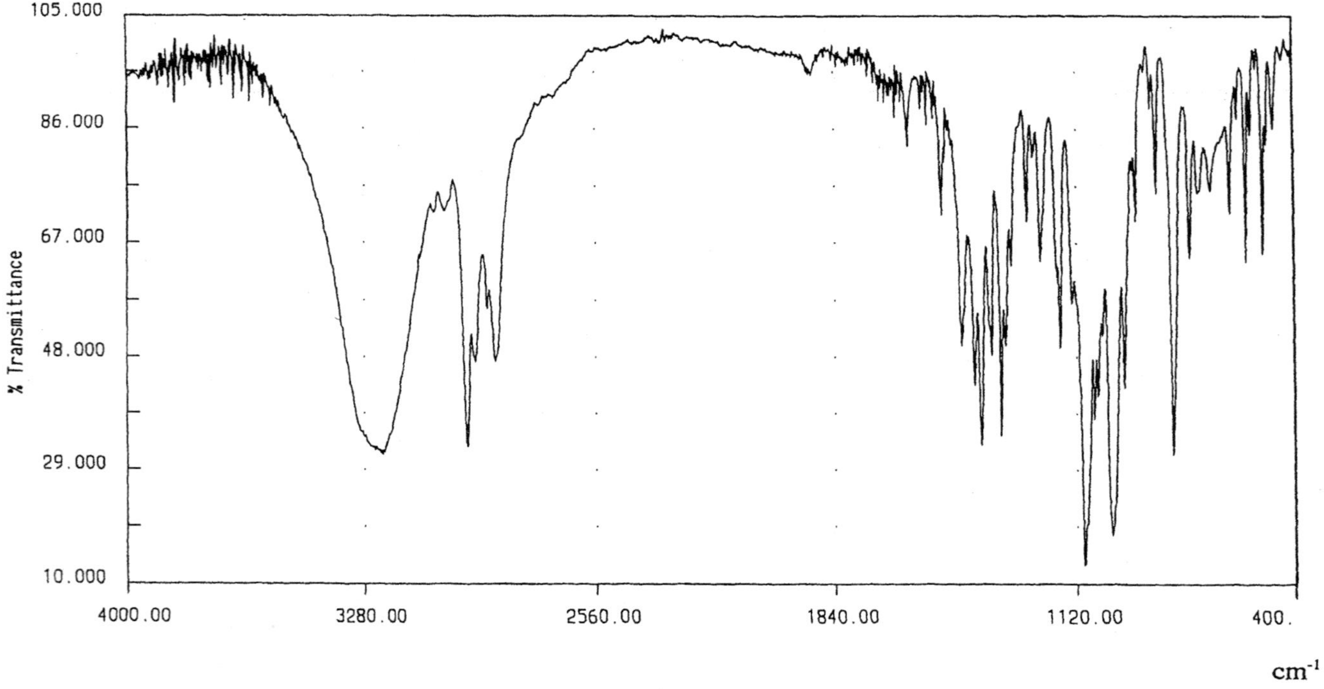

Name : BIS(4-ETHYLBENZYLIDENE)SORBITOL

PM Nr : 38950 CAS Nr : 79072-96-1

KBr pellet

CAS No. 79072-96-1 PM Ref. No. 38950

BIS(4-ETHYLBENZYLIDENE)SORBITOL

Peak wavenumber	% T	Relative intensity
414.8	91.0	10.3
451.4	95.0	6.1
466.8	93.4	8.0
430.5	86.1	17.0
551.7	83.4	20.2
563.3	65.3	42.3
513.4	63.9	44.0
661.7	72.0	34.1
721.5	75.7	29.6
758.1	75.4	30.0
783.2	64.6	43.2
835.3	31.5	83.5
883.5	75.2	30.2
900.9	89.5	12.8
945.2	70.7	35.7
980.0	42.8	69.8
1016.8	18.0	100.0
1045.6	51.5	59.1
1049.1	41.3	71.6
1068.7	37.5	76.2
1099.6	13.0	106.1
1134.3	56.9	52.6
1169.0	45.5	66.5
1225.0	63.9	44.0
1246.2	81.2	22.9
1263.5	70.4	36.1
1311.8	63.1	45.0
1329.1	49.8	61.2
1342.6	34.6	79.8
1371.6	48.2	63.2
1400.5	33.0	81.7
1421.7	43.1	69.4
1458.4	49.8	61.2
1418.2	71.6	34.6
1618.5	82.9	20.9
1701.4	90.6	11.5
2870.4	46.9	64.8
2932.2	47.0	64.6
2959.2	32.7	82.1
3030.5	72.1	34.0
3063.3	71.9	34.3
3225.4	31.4	83.7
3590.0	90.9	11.1

Bis(4-ethylbenzylidene) sorbitol

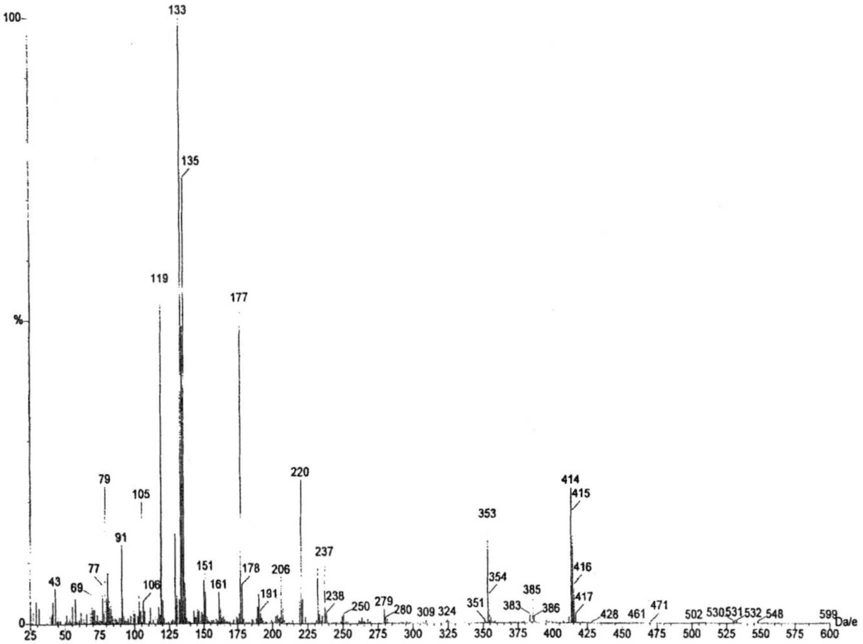

M/Z	Ion Intensity (%)	M/Z	Ion Intensity (%)
79	22.5	150	7.1
81	8.4	151	8.1
91	12.9	177	52.5
105	20.0	206	7.5
119	55.7	220	23.5
129	14.7	232	10.4
133	100.0	237	10.4
134	49.1	353	16.7
135	73.6	414	22.2
136	6.6	415	18.4

Spectrometer	: VG Autospec
Inletsystem	: Probe inlet
Source Temperature	: 350-400°C
Electron Energy	: 70 eV
Scan Range	: 25-1000

PM n°	CAS n°	Name
38950	79072-96-1	BIS(4-ETHYLBENZYLIDENE)SORBITOL

Solvent	DMSO-d_6
SF	250.13
SI	16384
SW	3125
NS	32

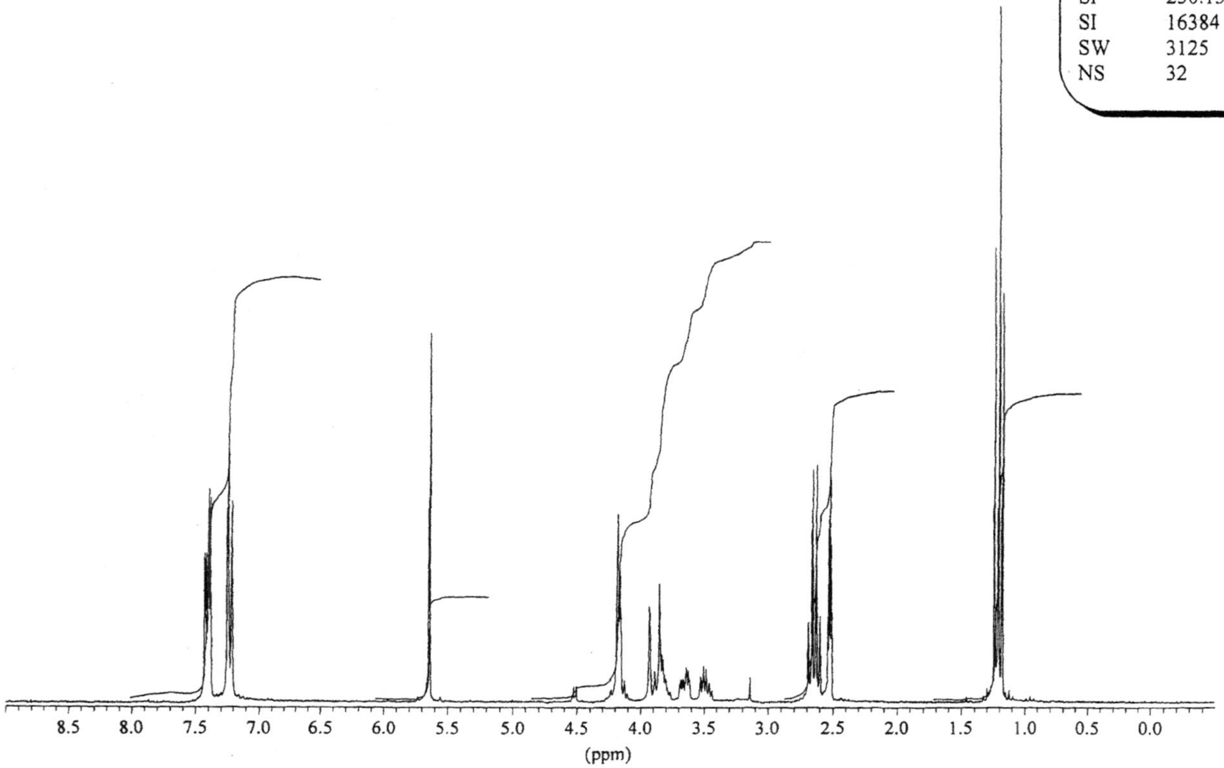

N,N-Bis(2-hydroxyethyl)alkyl(C_8-C_{18})amine

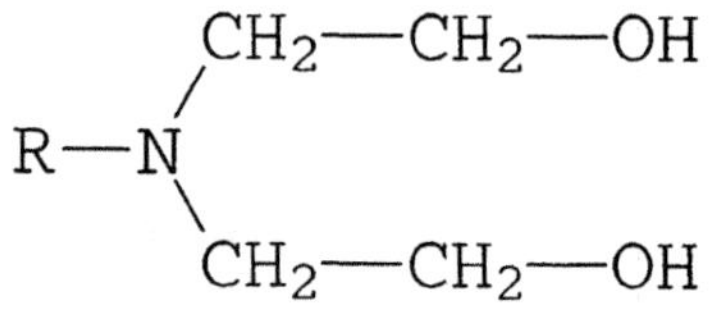

CAS No.	- 71786-60-2
PM Ref. No.	- 39090
Restrictions	- SML = 1.2 mg/kg as free amine
Formula	-
Molecular weight	-

Alternative names -

Physical Characteristics
- Slightly yellow liquid, mp -2°C.
- Soluble in acetone, hexane.

Handling - Store at room temperature.
Safety - Strongly alkaline.
Availability - Standard sample was supplied.

Current uses - Antistatic agent.
Applications - PS, PVC, PE, PP.

GC Retention time -

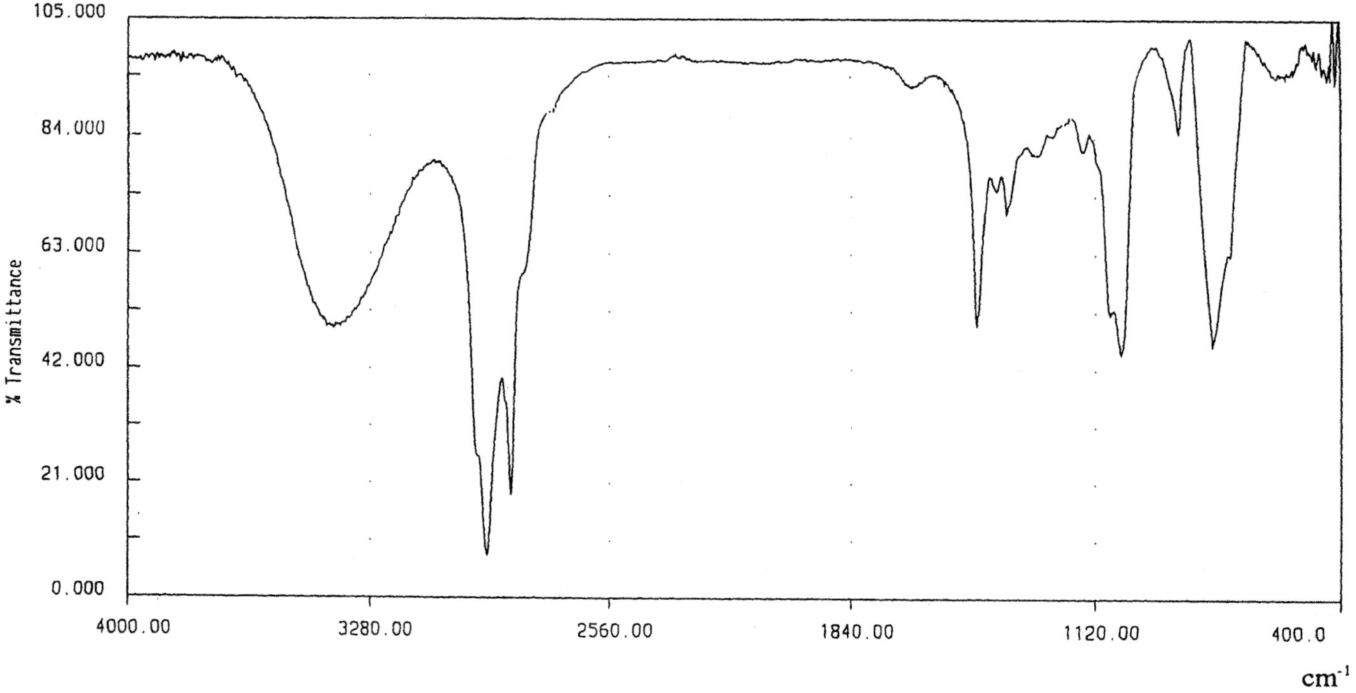

Name : N,N-BIS(2-HYDROXYETHYL)ALKYL(C_8-C_{18})AMINE

PM Nr : 39090 CAS Nr : 71786-60-2

KBr pellet

CAS No. 71786-60-2 PM Ref. No. 39090

N,N-BIS(2-HYDROXYETHYL)ALKYL (C8-C18AMINE)

Peak wavenumber	% T	Relative intensity
414.8	93.2	7.4
437.9	93.9	6.6
451.4	94.9	5.5
534.4	94.9	5.5
559.8	94.6	5.9
559.4	94.0	6.5
723.4	62.7	40.4
779.3	46.4	58.1
1045.6	44.7	59.9
1155.5	81.5	20.0
1240.4	84.1	17.2
1288.6	80.7	20.9
1377.3	70.2	32.3
1468.0	49.9	54.3
2735.4	88.2	12.8
2855.0	18.7	88.1
2924.4	7.7	100.0
3345.0	51.0	53.1

N,N-Bis(2-hydroxyethyl)alkyl(C_8-C_{18})amine

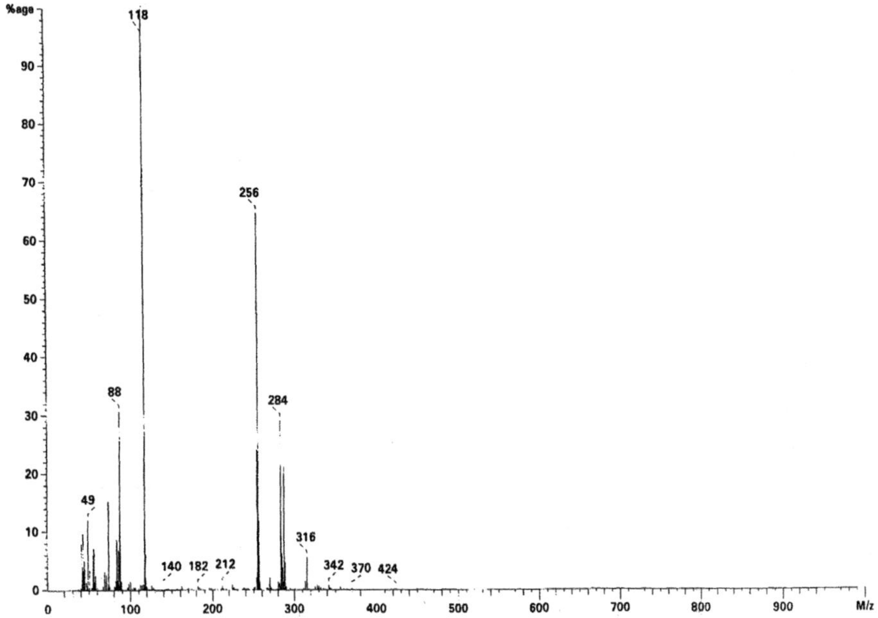

M/Z	Ion Intensity (%)	M/Z	Ion Intensity (%)
41	8.0	255	24.7
43	100.0	256	65.3
49	12.0	284	29.3
74	15.3	285	6.0
88	31.3	288	23.3
118	100.0		

Spectrometer	: VG Autospec
Inletsystem	: Probe inlet
Source Temperature	: 350-400°C
Electron Energy	: 70 eV
Scan Range	: 40-1000

PM n°
39090

CAS n°
71786-60-2

Name
N,N-BIS(2-HYDROXYETHYL)ALKYL(C_8-C_{18})AMINE

Solvent	$CDCL_3$
SF	250.13
SI	16384
SW	6494
NS	24

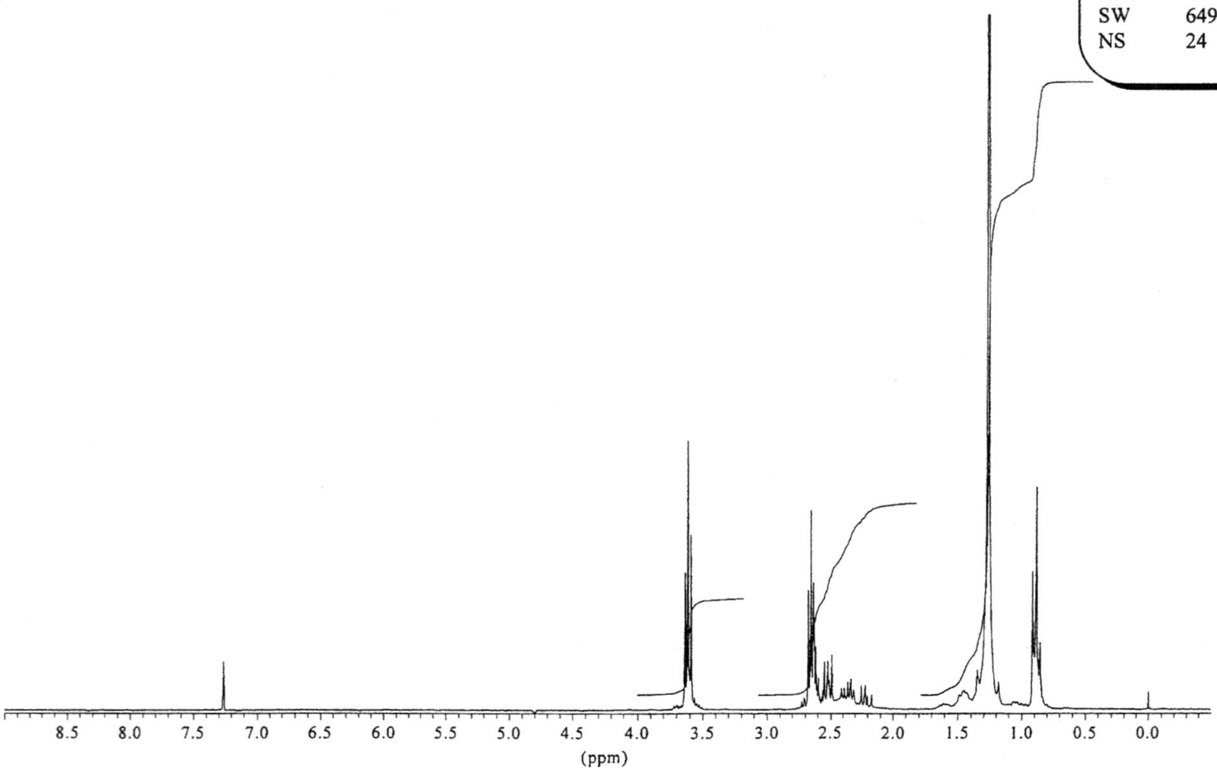

N,N-Bis(2-hydroxyethyl)lauramide

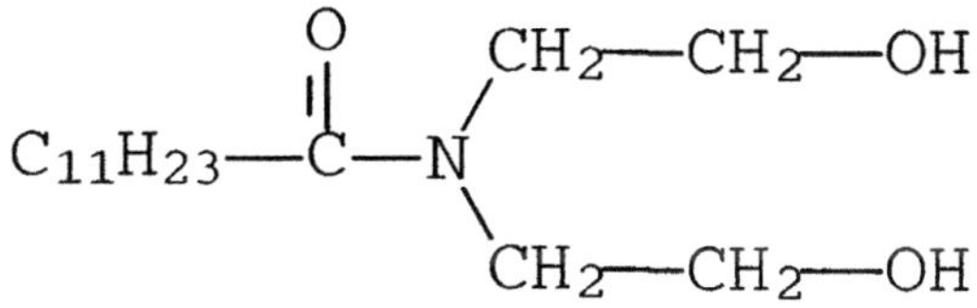

CAS No.	- 00120-40-1
PM Ref. No.	- 39280
Restrictions	- None
Formula	- $C_{16}H_{33}NO_3$
Molecular weight	- 287

Alternative names	- Lauryl diethanolamide
Physical Characteristics	- Colourless viscous liquid, mp 39°C. - Soluble in alcohol, methanol, ether.
Handling	- Store at room temperature.
Safety	-
Availability	- Standard sample was supplied.
Current uses	- Antistatic agent.
Applications	- General use.
GC Retention time	- Degradation product Lauric acid 28.8 min.

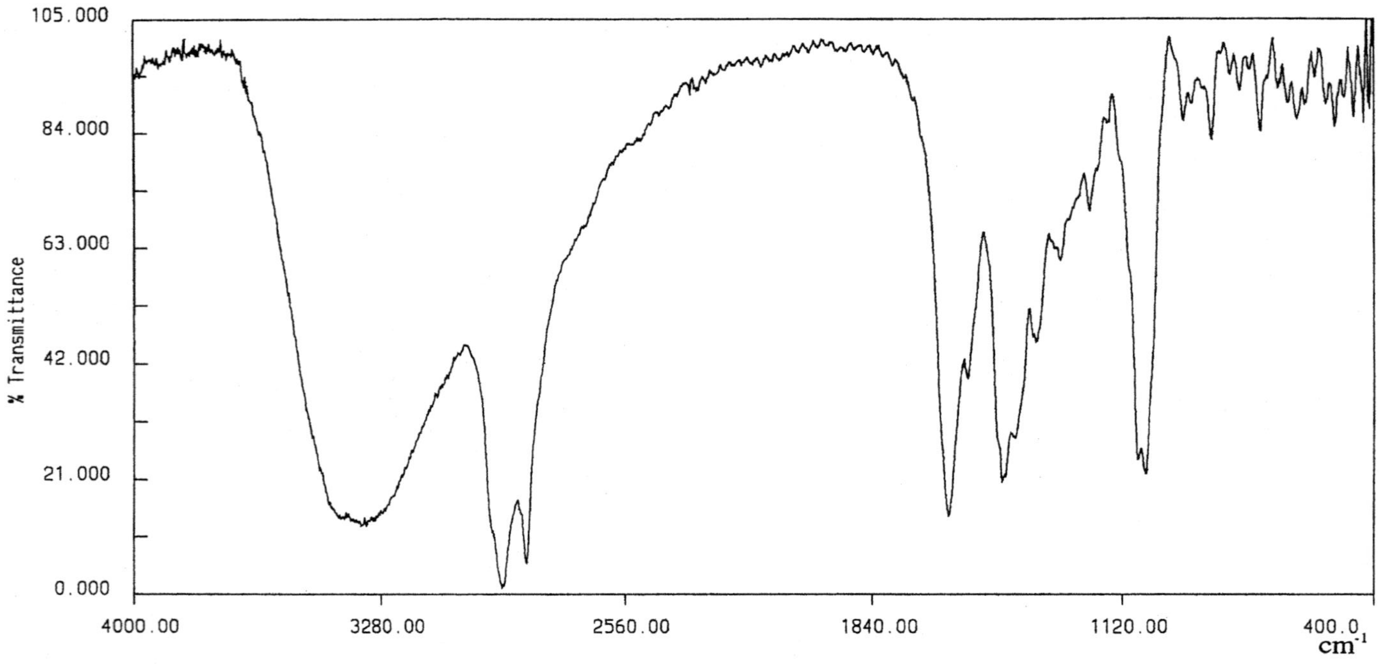

Name : N,N-BIS(2-HYDROXYETHYL) LAURAMIDE

PM Nr : 39280 CAS Nr : 00120-40-1

KBr pellet

CAS No. 00120-40-1 PM Ref. No. 39280

N,N-BIS(2-HYDROXYETHYL)LAURAMIDE

Peak wavenumber	% T	Relative intensity
410.9	88.5	11.6
428.3	86.2	14.0
457.2	87.1	13.0
486.1	90.7	9.4
513.1	85.4	14.8
538.2	89.4	10.7
598.0	89.4	10.7
619.2	86.7	13.4
644.3	89.7	10.4
723.4	84.5	15.6
862.3	83.0	17.1
920.2	89.5	10.6
943.3	86.5	13.6
1053.3	22.0	78.8
1163.2	86.1	14.1
1213.4	69.8	30.5
1296.3	60.6	39.8
1369.6	45.85	54.8
1427.5	28.42	72.4
1468.0	20.54	80.3
1566.4	39.24	61.4
1622.3	14.25	86.7
2513.6	82.54	17.7
2687.2	66.60	33.8
2853.1	5.52	95.5
2924.4	1.11	100.0
3090.3	39.16	61.5
3115.4	35.93	64.8
3275.5	14.74	86.2

N,N-Bis(2-hydroxyethyl)lauramide

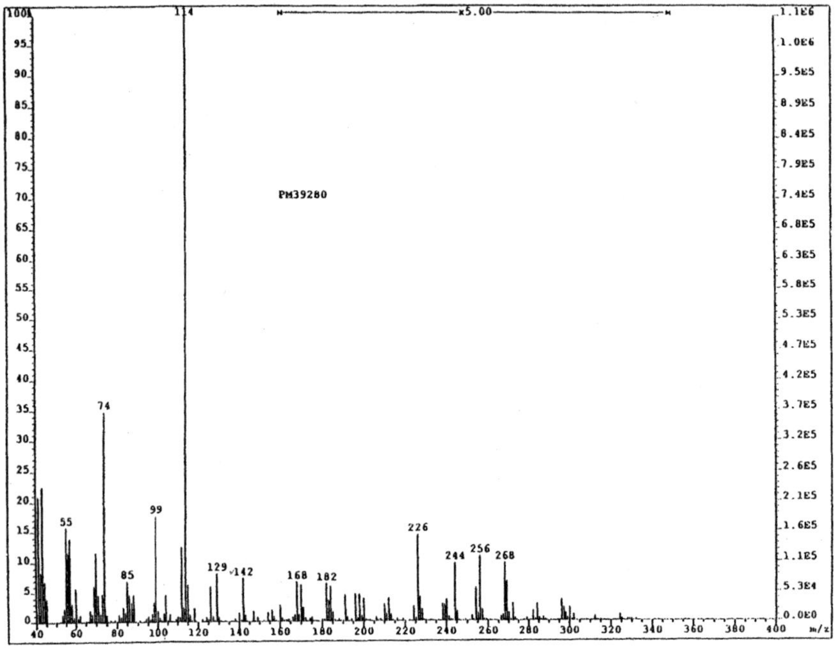

M/Z	Ion Intensity (%)	M/Z	Ion Intensity (%)
41	21.0	99	17.8
43	22.7	112	12.6
55	15.8	114	100.0
57	14.0	115	6.5
69	6.3	126	6.2
70	11.7	129	8.2
85	7.1	142	7.5

Spectrometer : VG Autospec
Inletsystem : Flow injection particle beam MS
Source Temperature : 250°C
Electron Energy : 70 eV
Scan Range : 40-1000

Lauric acid

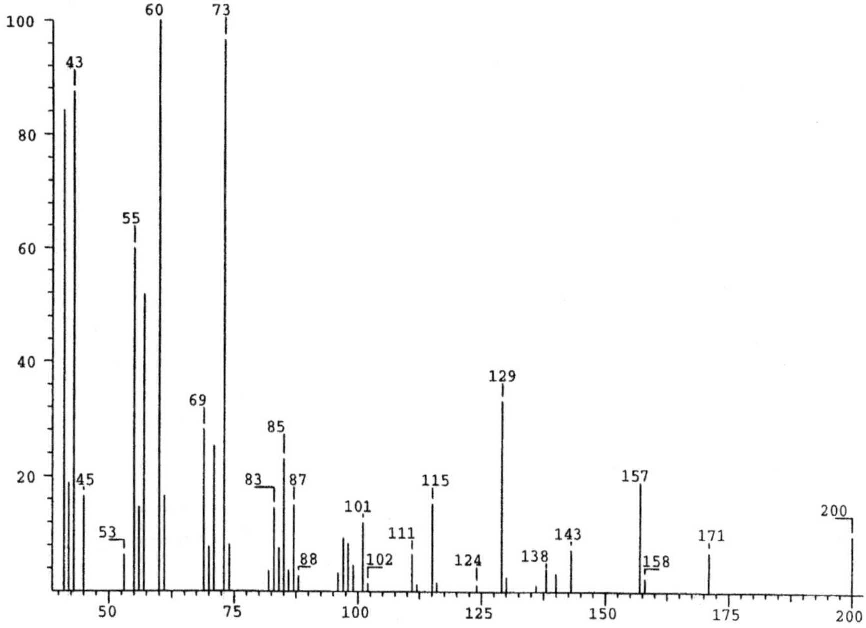

M/Z	Ion Intensity (%)	M/Z	Ion Intensity (%)
41	84.1	73	96.6
42	18.9	83	14.5
43	87.4	85	23.0
55	59.9	87	15.0
56	14.6	97	9.3
57	51.8	101	11.9
60	100.0	115	15.24
61	16.6	129	33.1
69	28.0	157	18.0
71	25.2	200	10.0

Spectrometer	: Finnigan Mat SSQ 700
Inletsystem	: Capillary GC/MS
Source Temperature	: 180°C
Electron Energy	: 70 eV
Scan Range	: 40-500

PM n°	CAS n°	Name:
39280	00120-40-1	N,N-BIS (2-HYDROXYETHYL) LAURAMIDE

Solvent	$CDCL_3$
SF	250.13
SI	16384
SW	3125
NS	32

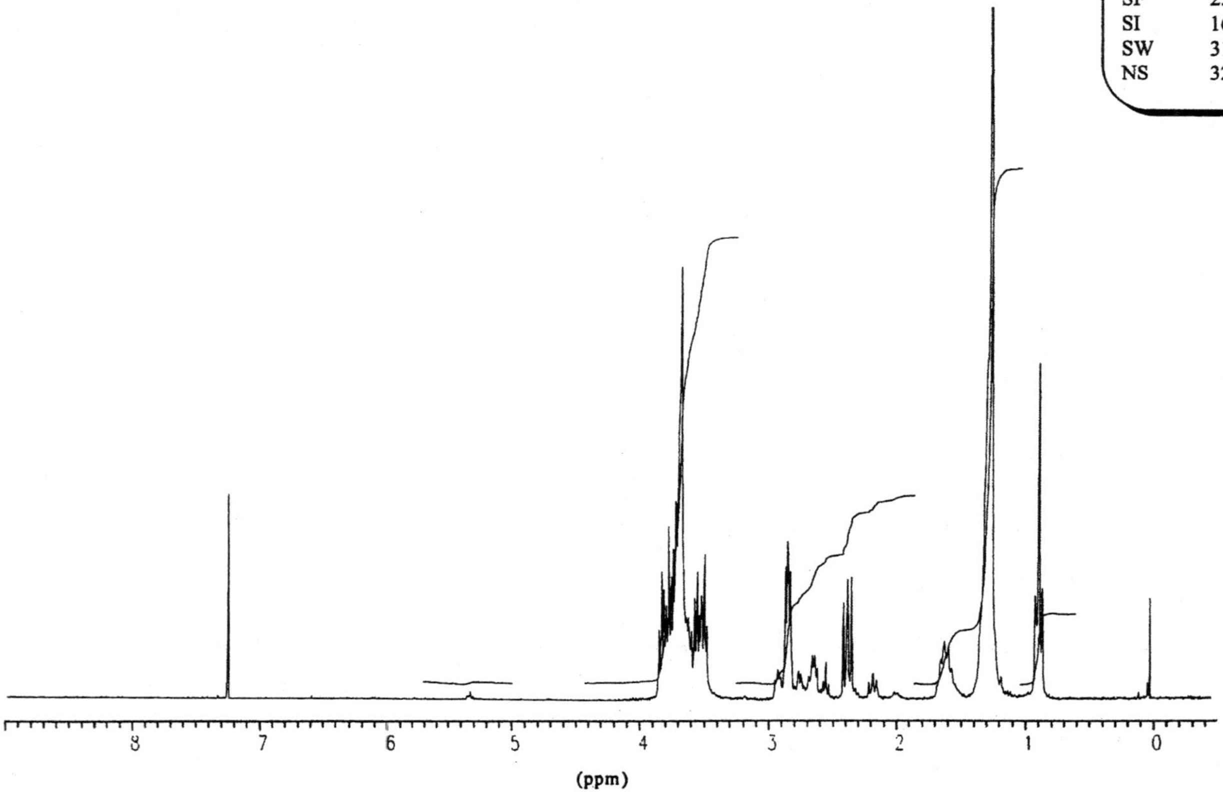

2,4-Bis(octylmercapto)-6-(4-hydroxy-3,5-di-tert-butylanilino)-1,3,5-triazine

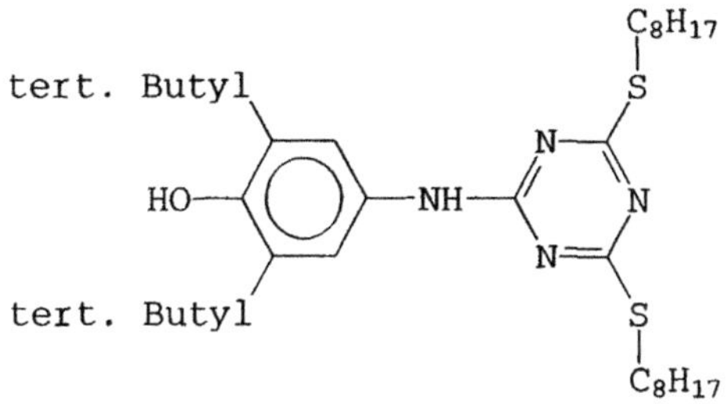

CAS No.	- 00991-84-4
PM Ref. No.	- 40000
Restrictions	- SML = 30 mg/kg
Formula	- $C_{33}H_{56}N_4OS_2$
Molecular weight	- 588

Alternative names	- Phenol, 4-((4,6-bis(octylthio)-1,3,5-triazin-2-yl)amino)-2,6-bis(1,1-di-methylethyl)-
Physical Characteristics	- White to light yellow crystalline powder, mp 91 - 96°C. - Soluble in benzene, acetone, hexane.
Handling	- Store at room temperature.
Safety	-
Availability	- Standard sample was supplied.
Current uses	- Antioxidant.
Applications	- Styrene co-polymers, PP, rubber.
GC Retention time	- 79.4 min.

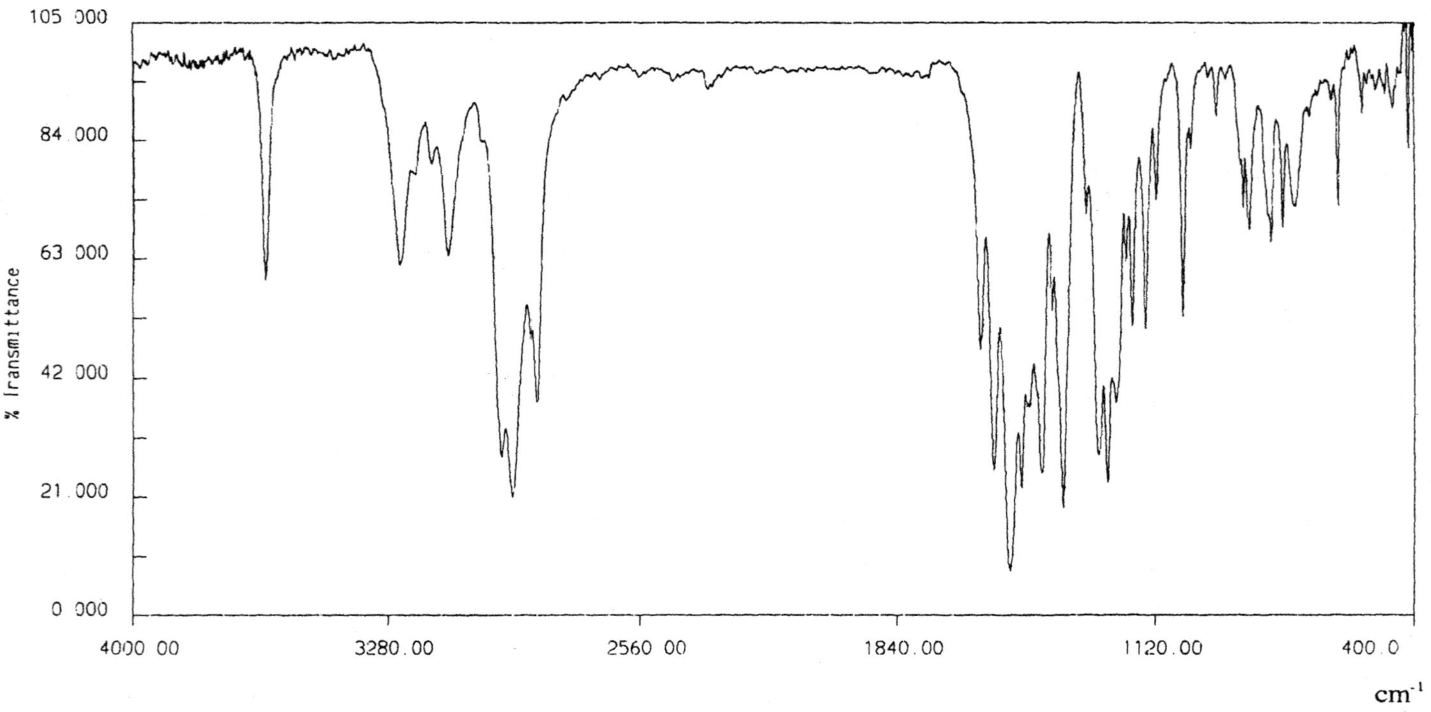

Name : 2,4-BIS(OCTYLMERCAPTO)-6-(4-HYDROXY-3,5-DI-Tert-BUTYLANILINO)-1,3,5-TRIAZINE

PM Nr : 40000 CAS Nr : 00991-84-4

KBr pellet

CAS No. 00991-84-4 PM Ref. No. 40000

2,4-BIS(OCTYLMERCAPTO)-6-(4-HYDROXY-3,5-DI-tert- -BUTYLANILINO)-1,3,5-TRIAZINE

Peak wavenumber	% T	Relative intensity
414,8	82,8	18,6
457,2	89,9	10,9
478,4	92,4	8,2
505,4	93,2	7,4
530,5	94,3	6,2
542,1	89,0	11,9
609,6	72,4	29,9
688,7	88,5	12,5
733,0	72,1	30,2
765,8	68,7	33,9
798,6	66,0	36,9
860,4	68,1	34,6
875,8	72,0	30,4
1043,6	52,7	51,3
1116,9	73,5	28,7
1145,9	50,5	53,7
1184,4	50,8	53,3
1201,8	62,5	40,7
1230,7	37,5	67,8
1253,9	23,4	83,0
1279,0	28,4	77,6
1311,8	70,9	31,5
1377,3	19,2	87,6
1406,3	53,7	50,2
1439,1	25,1	81,2
1477,7	36,8	68,5
1493,1	22,5	84,0
1524,0	7,7	100,0
1570,3	25,4	80,9
1606,9	46,9	57,6
2764,3	91,1	9,6
2853,1	37,5	67,8
2924,4	21,0	85,6
2955,3	27,9	78,2
3107,7	63,5	39,6
3155,9	79,7	22,0
3200,3	77,8	24,1
3246,6	61,6	41,6
3628,5	59,0	44,4

2,4-Bis(octylmercapto)-6-(4-hydroxy-3,5-di-tert-butylanilino)-1,3,5-triazine

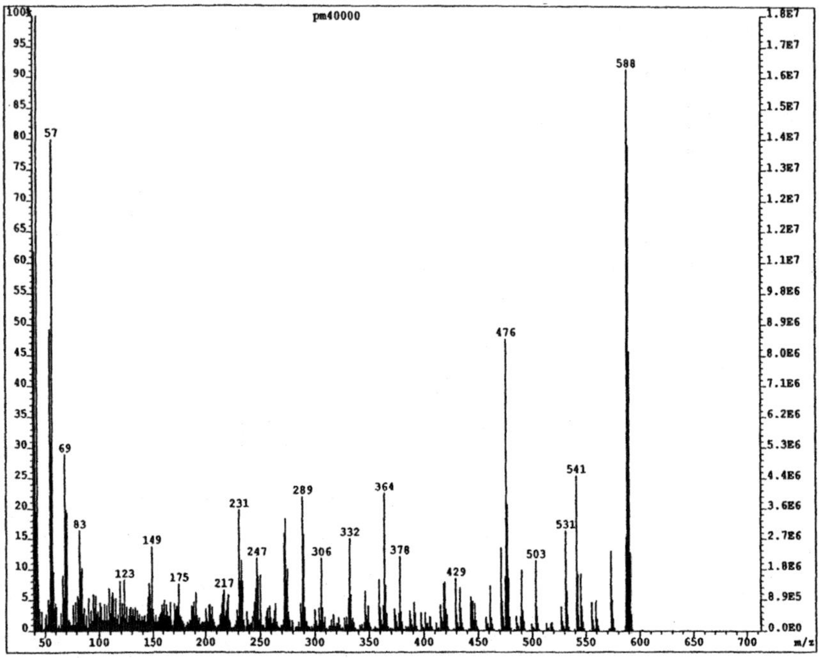

M/Z	Ion Intensity (%)	M/Z	Ion Intensity (%)
41	62.6	273	79.8
43	100.0	289	22.1
55	49.1	364	22.7
56	29.4	476	48.5
57	79.8	477	20.9
69	29.4	531	17.8
70	20.2	541	26.4
71	19.6	588	92.0
83	17.2	589	79.8
231	20.2	590	46.0

Spectrometer	: VG Autospec
Inletsystem	: Flow injection particle beam MS
Source Temperature	: 250°C
Electron Energy	: 70 eV
Scan Range	: 40-1000

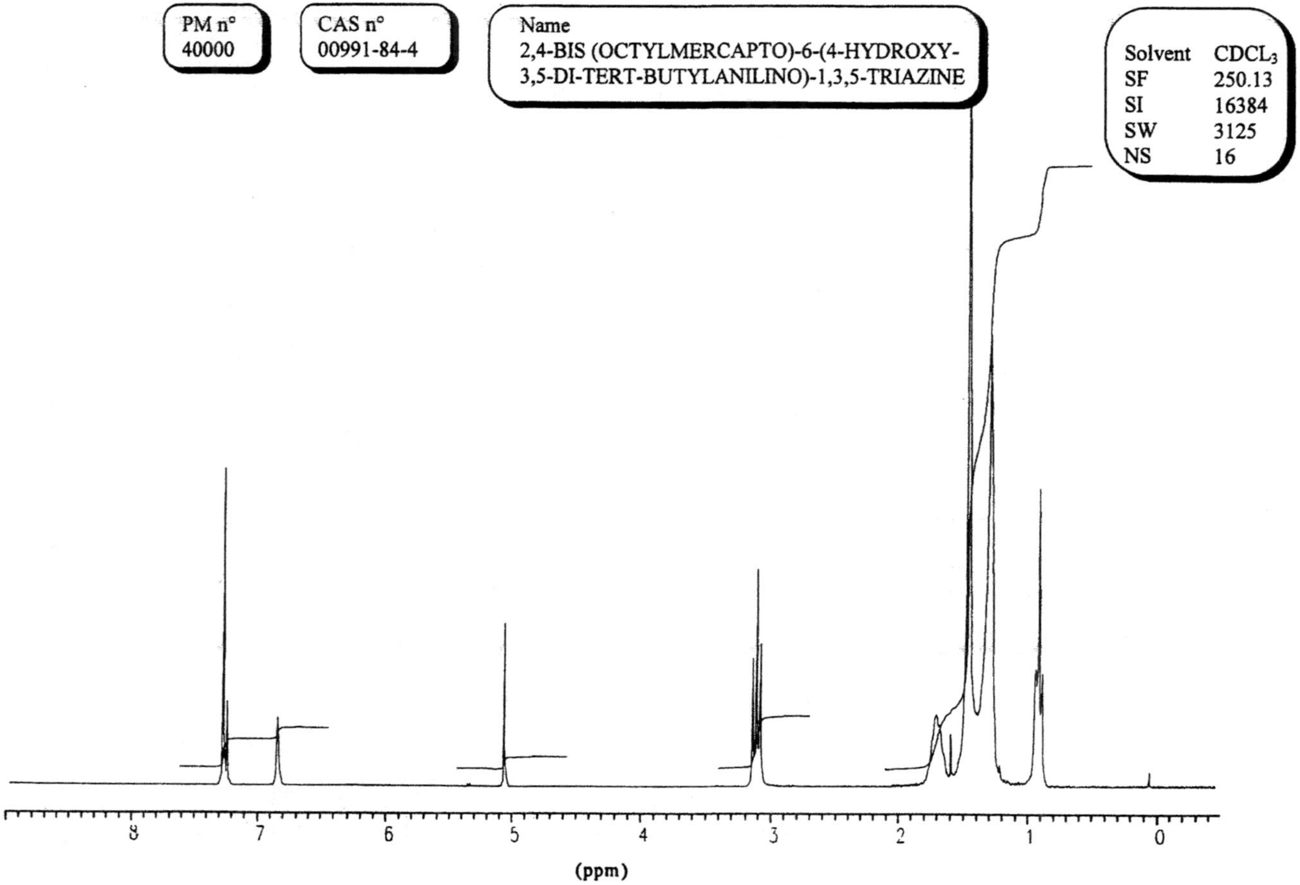
PM n°
40000
CAS n°
00991-84-4
Name
2,4-BIS (OCTYLMERCAPTO)-6-(4-HYDROXY-3,5-DI-TERT-BUTYLANILINO)-1,3,5-TRIAZINE
Solvent CDCL3
SF 250.13
SI 16384
SW 3125
NS 16
8
7
6
5
4
3
2
1
0
(ppm)

2,4-Bis(octylthiomethyl)-6-methylphenol

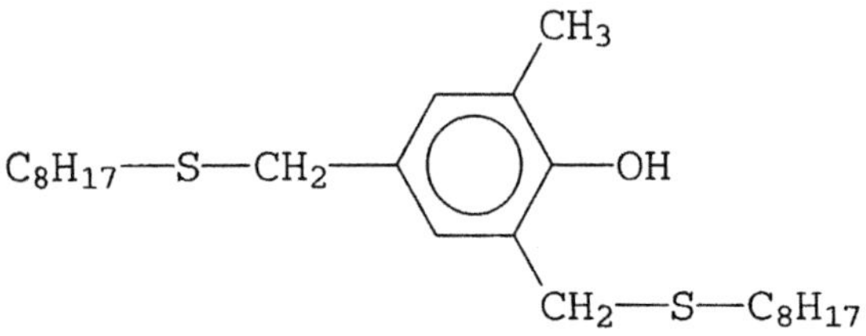

CAS No.	- 110553-27-0
PM Ref. No.	- 40020
Restrictions	- SML = 6.0 mg/kg
Formula	- $C_{25}H_{44}OS_2$
Molecular weight	- 424

Alternative names	- 2,4-Bis((octylthio)methyl)-o-cresol
Physical Characteristics	- Pale yellow liquid, mp approx. 14°C, bp > 178°C (0.08 mm Hg). - Soluble in most organic solvents.
Handling	- Store at room temperature.
Safety	-
Availability	- Standard sample was supplied.
Current uses	- Antioxidant.
Applications	- Styrene co-polymers, paraffin waxes, coatings.
GC Retention time	-

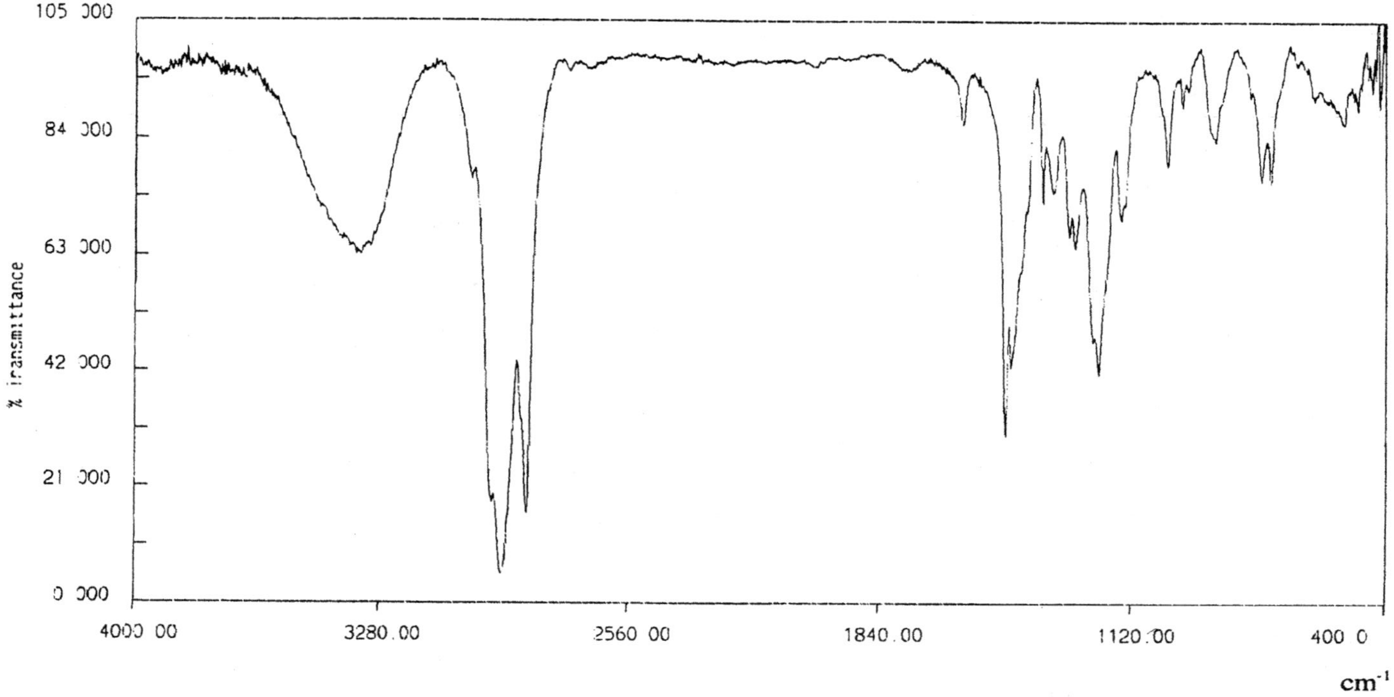

Name : 2,4-BIS(OCTYLTHIOMETHYL)-6-METHYLPHENOL

PM Nr : 40020 CAS Nr : 110553-27-0

KBr pellet

CAS No. 1105532-27-0 PM Ref. No. 40020

2,4-BIS(OCTYLTHIOMETHYL)-6-METHYLPHENOL

Peak wavenumber	% T	Relative intensity
414,8	89,7	10,9
436,0	92,4	8,0
447,5	95,0	5,3
476,5	89,2	11,4
515,1	86,7	14,0
723,4	76,0	25,3
748,5	76,5	24,8
879,7	83,6	17,3
974,2	89,7	10,9
1020,5	79,1	22,0
1153,6	69,5	32,2
1217,2	41,4	61,8
1286,7	64,4	37,6
1304,0	66,6	35,2
1350,3	74,3	27,1
1377,3	72,6	28,9
1468,0	42,6	60,6
1483,4	30,4	73,4
2855,0	16,1	88,5
2926,4	5,2	100,0
3308,3	64,6	37,3

2,4-Bis(octylthiomethyl)-6-methylphenol

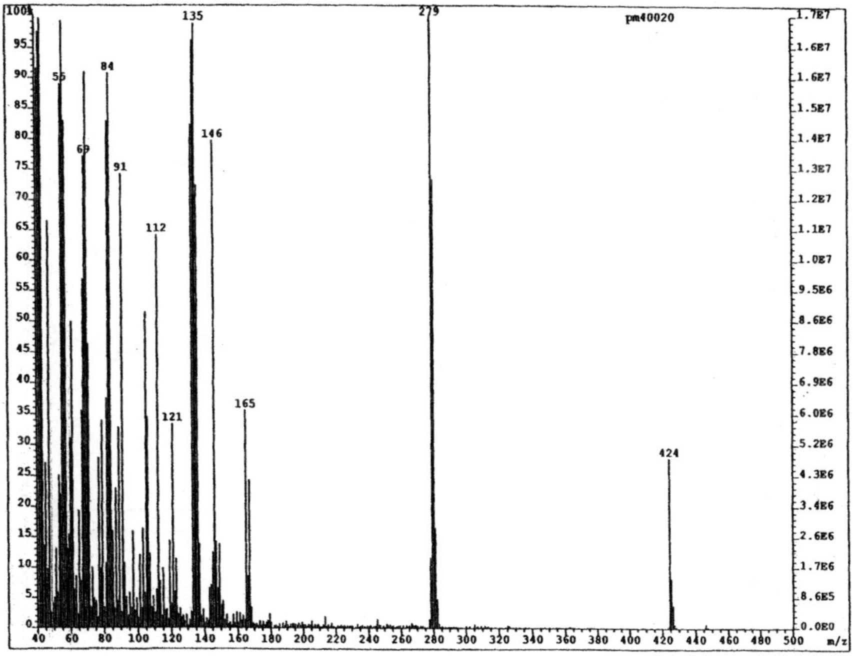

M/Z	Ion Intensity (%)	M/Z	Ion Intensity (%)
41	92.0	112	64.4
43	99.8	133	82.8
47	66.9	134	96.3
55	89.0	135	98.8
56	99.4	136	72.4
57	82.8	146	79.8
69	77.3	279	100.0
70	91.4	280	73.6
83	83.4	424	27.7
91	74.8	425	8.5

Spectrometer	: VG Autospec
Inletsystem	: Flow injection particle beam MS
Source Temperature	: 250°C
Electron Energy	: 70 eV
Scan Range	: 40-1000

PM n°	CAS n°	Name
40020	110553-27-0	2,4-BIS (OCTYLTHIOMETHYL)-6-METHYLPHENOL

Solvent	$CDCL_3$
SF	250.13
SI	16384
SW	3125
NS	42

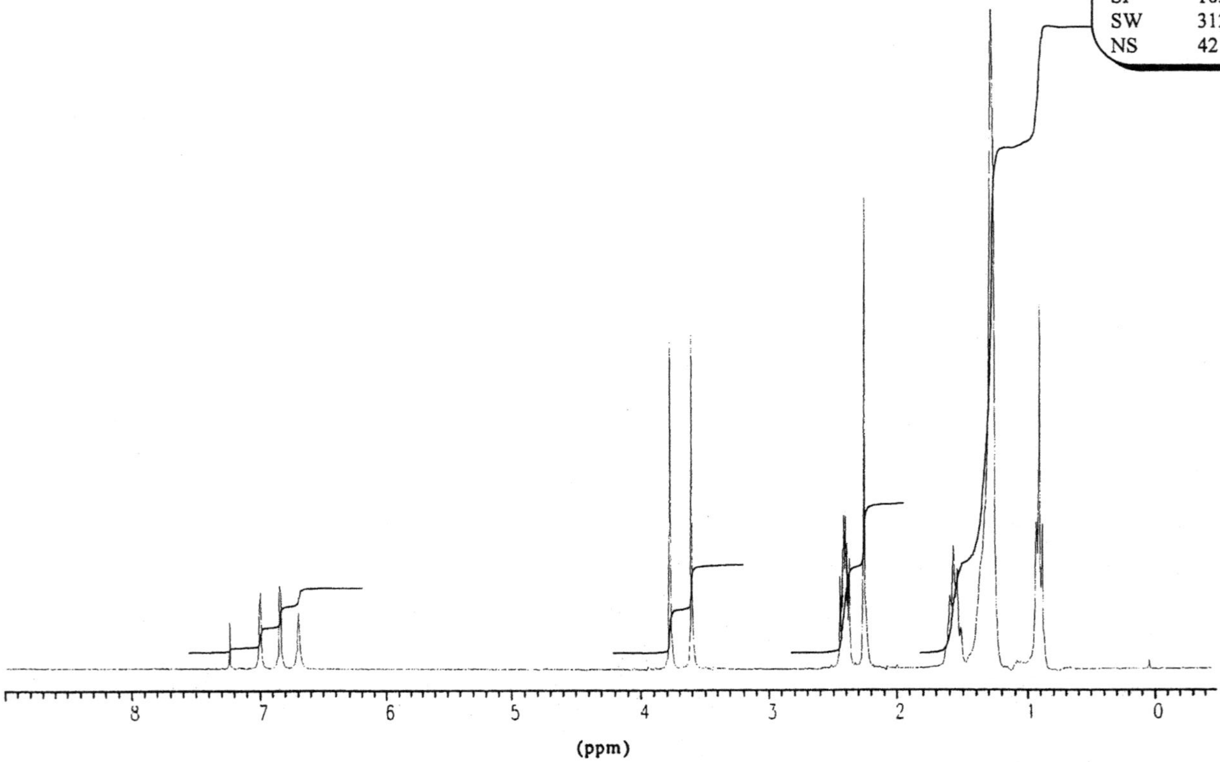

Bisphenol A

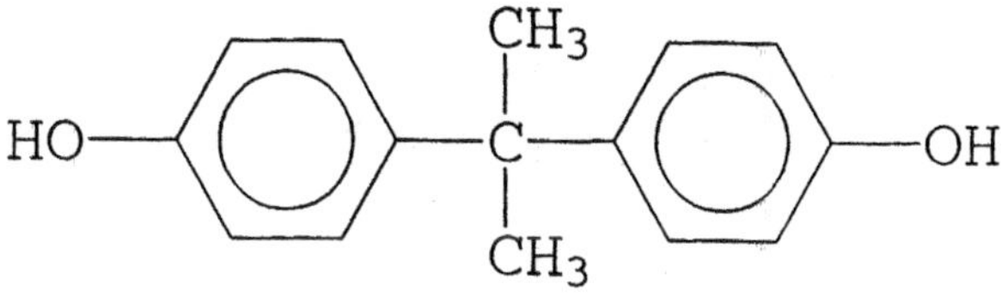

CAS No.	- 00080-05-7
PM Ref. No.	- 40060
Restrictions	- SML = 3.0 mg/kg
Formula	- $C_{15}H_{16}O_2$
Molecular weight	- 228

Alternative names	- 2,2 Bis(4-hydroxyphenyl) propane
Physical Characteristics	- Coarse powder, mp 150 - 155°C, bp 250 - 252°C (13 mm Hg). - Soluble in most organic solvents.
Handling	- Store at room temperature.
Safety	- Irritant.
Availability	- Standard sample was supplied.
Current uses	- PVC, PC, epoxy phenolic coatings.
Applications	- Monomer, stabilizer.
GC Retention time	- 40.5 min.

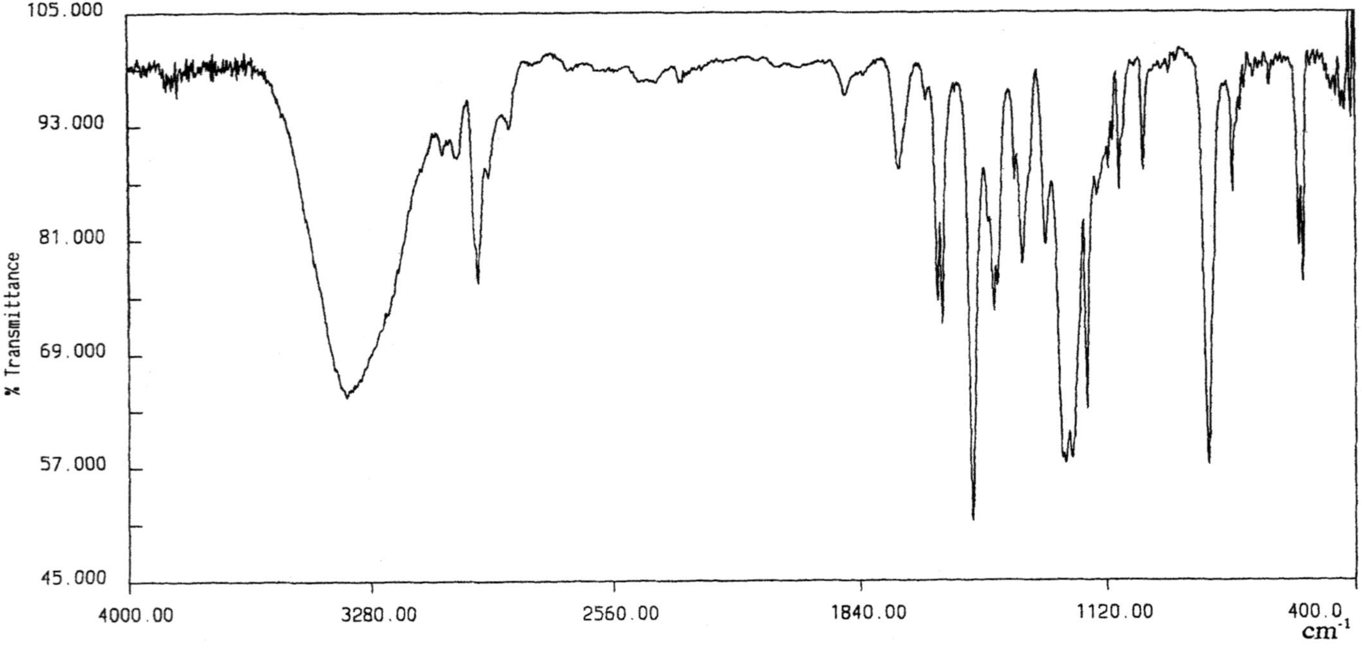

Name : BISPHENOL A

PM Nr : 40060 CAS Nr : 00080-05-7

KBr pellet

BISPHENOL A

Peak wavenumber	% T	Relative intensity
430.2	94.4	11.4
439.8	94.9	10.5
551.7	76.5	48.0
565.2	80.3	40.2
827.6	57.3	87.5
1113.1	88.3	24.1
1147.8	85.5	29.6
1176.7	63.2	75.4
1219.2	57.9	86.3
1294.4	80.5	40.0
1361.9	78.4	44.2
1435.2	76.2	48.8
1446.8	73.5	54.3
1510.5	51.2	100.0
1599.2	72.1	57.2
1612.7	74.5	52.2
2932.2	87.5	25.7
2965.0	76.55	48.0
3235.0	73.00	55.3
3318.0	66.18	69.2

Bisphenol A

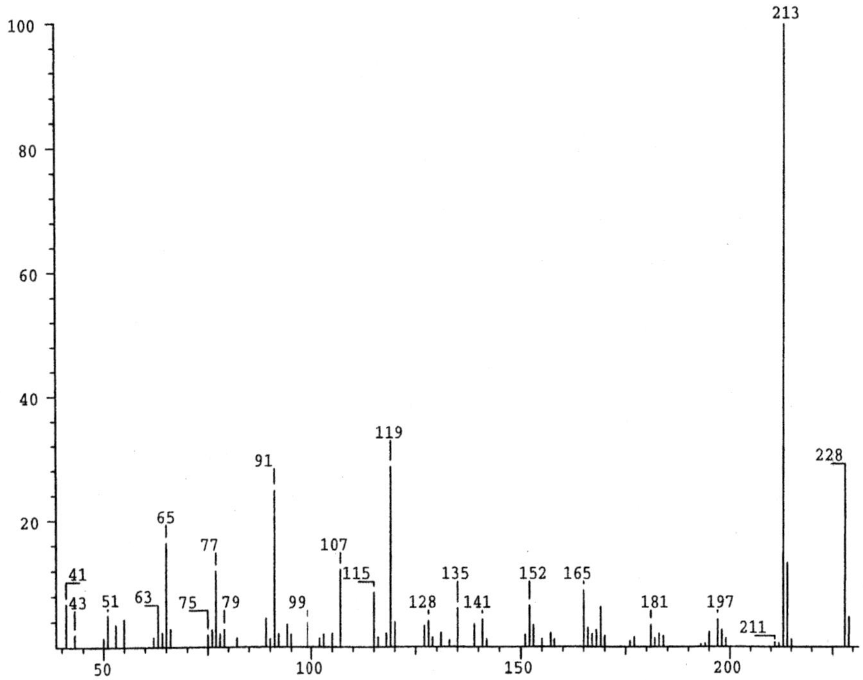

M/Z	Ion Intensity (%)	M/Z	Ion Intensity (%)
41	6.7	115	8.6
43	2.0	119	28.6
51	5.0	135	6.1
53	3.6	165	8.9
55	4.3	213	100.0
65	16.4	214	13.2
77	11.9	228	28.2
91	24.8	229	4.8
107	12.2		

Spectrometer	: Finnigan Mat SSQ 700
Inletsystem	: Capillary GC/MS
Source Temperature	: 180°C
Electron Energy	: 70 eV
Scan Range	: 40-500

PM n°	CAS n°	Name
40060	00080-05-7	BISPHENOL A

Solvent	$CDCl_3$
SF	250.13
SI	8192
SW	2380
NS	32

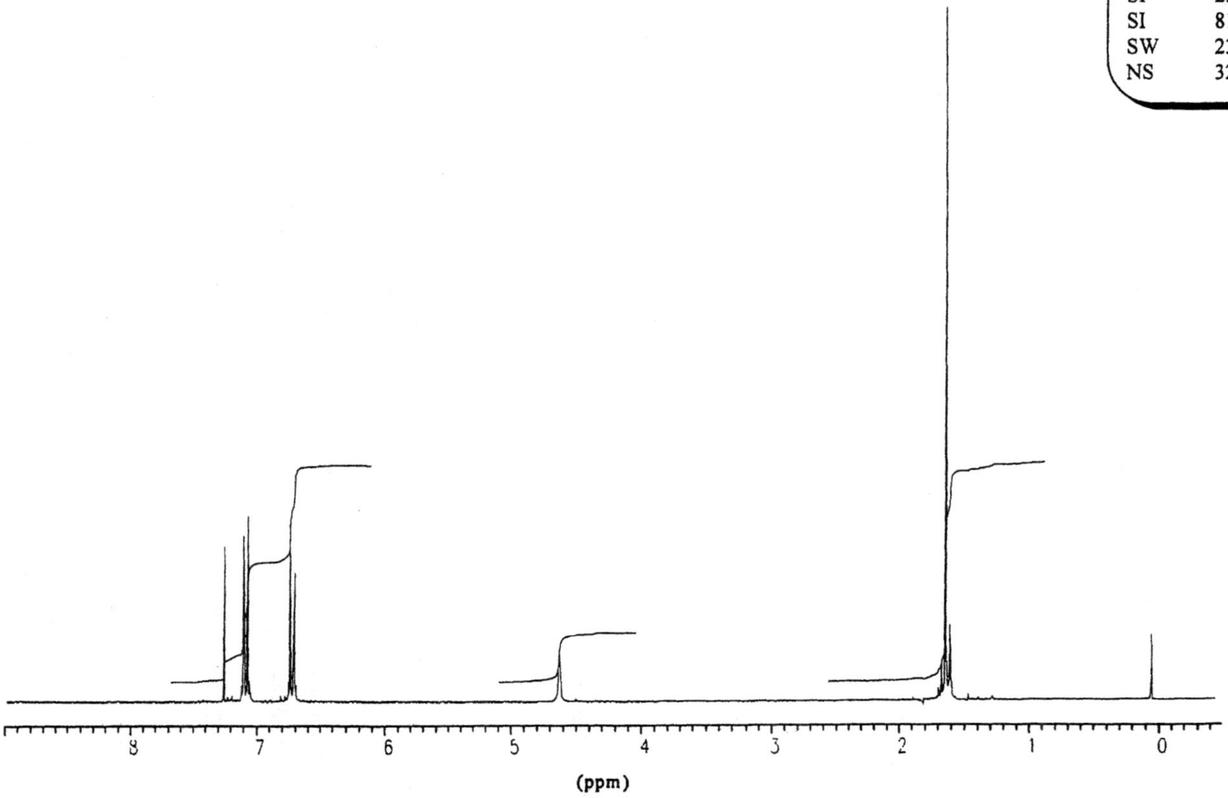

tert-Butyl-4-hydroxyanisole

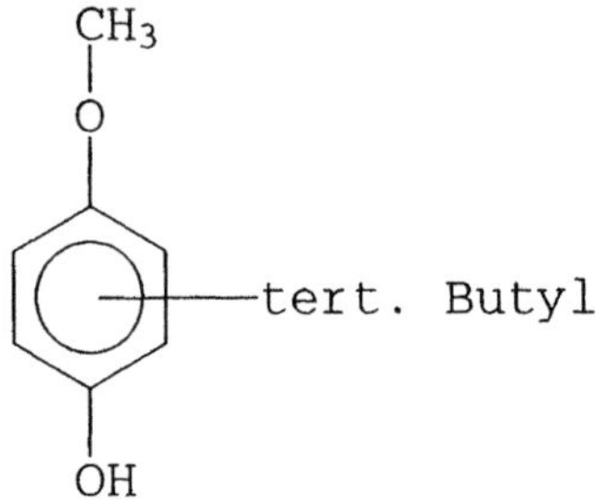

CAS No.	- 25013-16-5
PM Ref. No.	- 40720
Restrictions	- SML = 3.0 mg/kg
Formula	- $C_{11}H_{16}O_2$
Molecular weight	- 180

Alternative names - Butylated hydroxyanisole (BHA)

Physical Characteristics - Slightly yellow flakes, mp 48 - 55°C, bp 264 - 270°C (733 mm Hg).
- Soluble in most organic solvents.

Handling - Store at room temperature.
Safety - Harmful.
Availability - Standard sample supplied.

Current uses - Antioxidant.
Applications - General use.

GC Retention time - 26.9 min.

Note - Mixture of 2-(main) and 3-(minor) isomers.

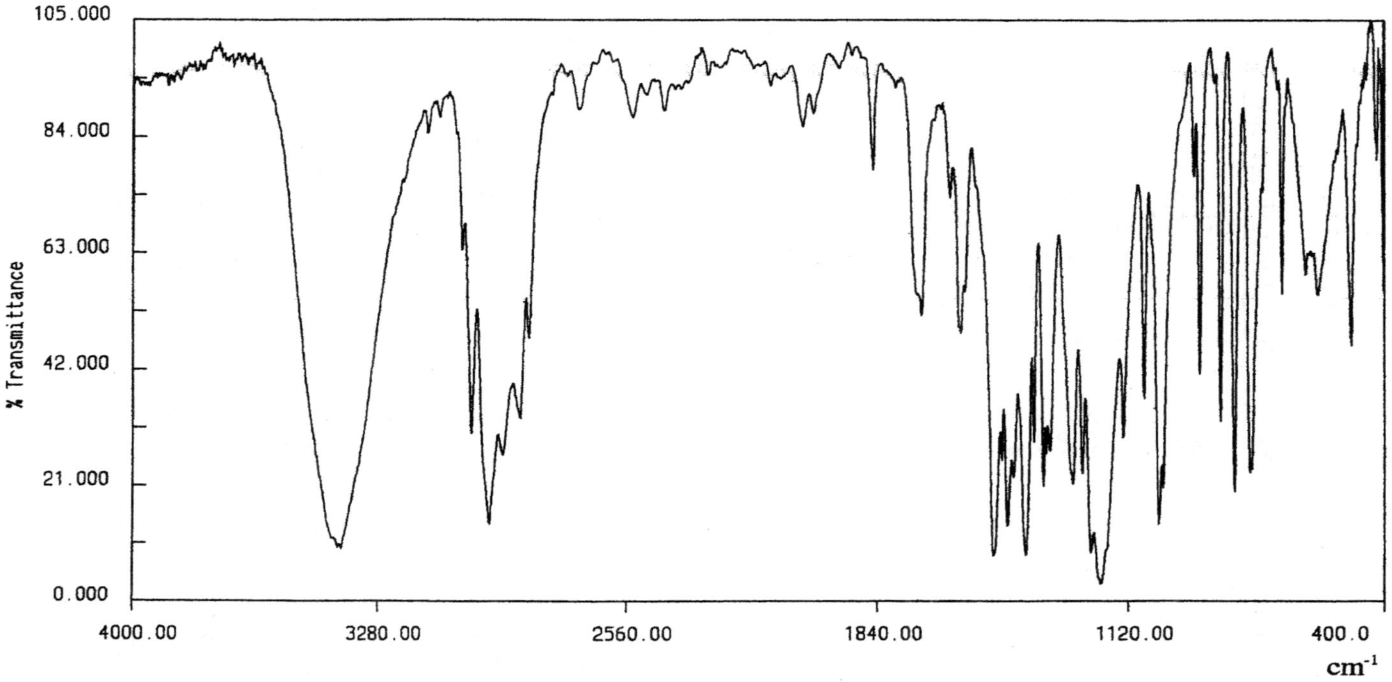

Name : TERT-BUTYL-4-HYDROXYANISOLE (=BHA)

PM Nr : 40720 CAS Nr : 25013-16-5

KBr pellet

CAS No. 25013-16-5 PM Ref. No. 40720

tert-BUTYL-4-HYDROXYANISOLE (BHA)

Peak wavenumber	% T	Relative intensity
420,5	70,5	30,7
491,9	50,5	51,6
582,6	75,1	25,9
615,4	86,3	14,3
682,9	58,1	43,7
773,6	23,5	79,7
816,0	20,1	83,3
856,5	32,8	70,0
914,4	41,3	61,2
929,8	77,3	23,7
1020,5	23,9	79,3
1034,0	16,7	86,8
1074,5	46,5	55,7
1136,2	38,5	64,1
1199,9	4,0	100,0
1226,9	11,2	92,5
1252,0	29,5	73,5
1280,9	26,8	76,3
1346,5	34,0	68,8
1358,1	33,2	69,6
1365,8	25,8	77,3
1392,8	35,5	67,2
1415,9	10,4	93,4
1452,6	27,2	75,9
1469,9	16,6	86,9
1487,3	31,0	71,9
1508,5	10,0	93,8
1632,0	80,5	20,3
1713,0	54,6	47,3
1851,9	77,8	23,1
2050,6	84,1	16,6
2453,8	89,2	11,3
2542,5	88,1	12,4
2698,7	87,0	13,5
2775,9	89,9	10,5
2841,5	47,6	54,6
2866,6	33,4	69,4
2914,8	27,0	76,1
2953,4	14,4	89,2
3003,5	31,5	71,4
3030,5	66,4	35,0
3130,9	88,8	11,7
3391,3	9,9	93,9

tert-Butyl-4-hydroxyanisole

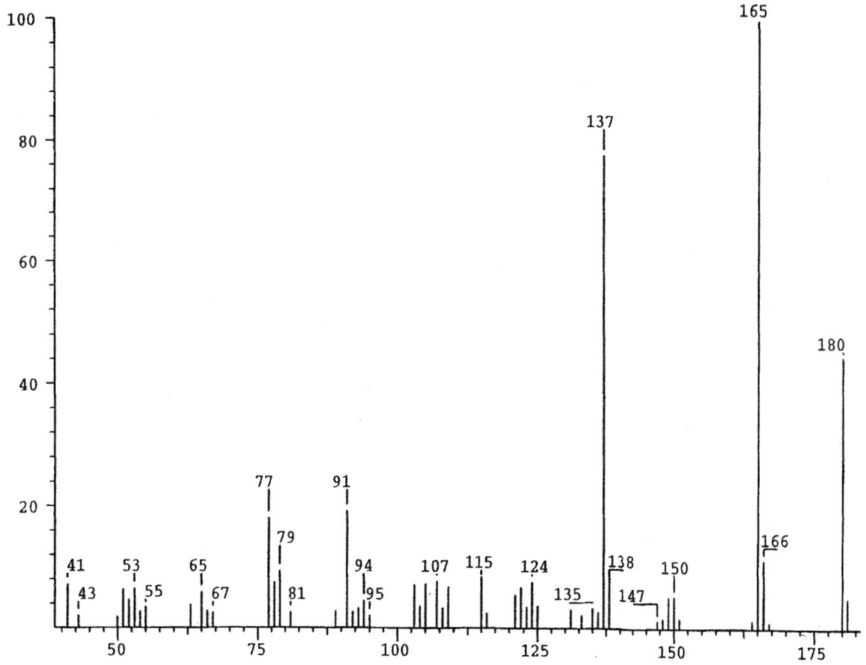

M/Z	Ion Intensity (%)	M/Z	Ion Intensity (%)
53	6.6	124	7.6
65	5.9	137	77.8
77	18.0	138	8.8
79	9.4	165	100.0
91	19.2	166	11.3
105	7.3	180	44.7
107	7.7	181	5.0
115	8.5		

Spectrometer	: Finnigan Mat SSQ 700
Inletsystem	: Capillary GC/MS
Source Temperature	: 180°C
Electron Energy	: 70 eV
Scan Range	: 40-500

PM n°
40720

CAS n°
25013-16-5

Name
tert-BUTYL-4-HYDROXYANISOLE (=BHA)

Solvent	$CDCl_3$
SF	250.13
SI	16384
SW	3125
NS	32

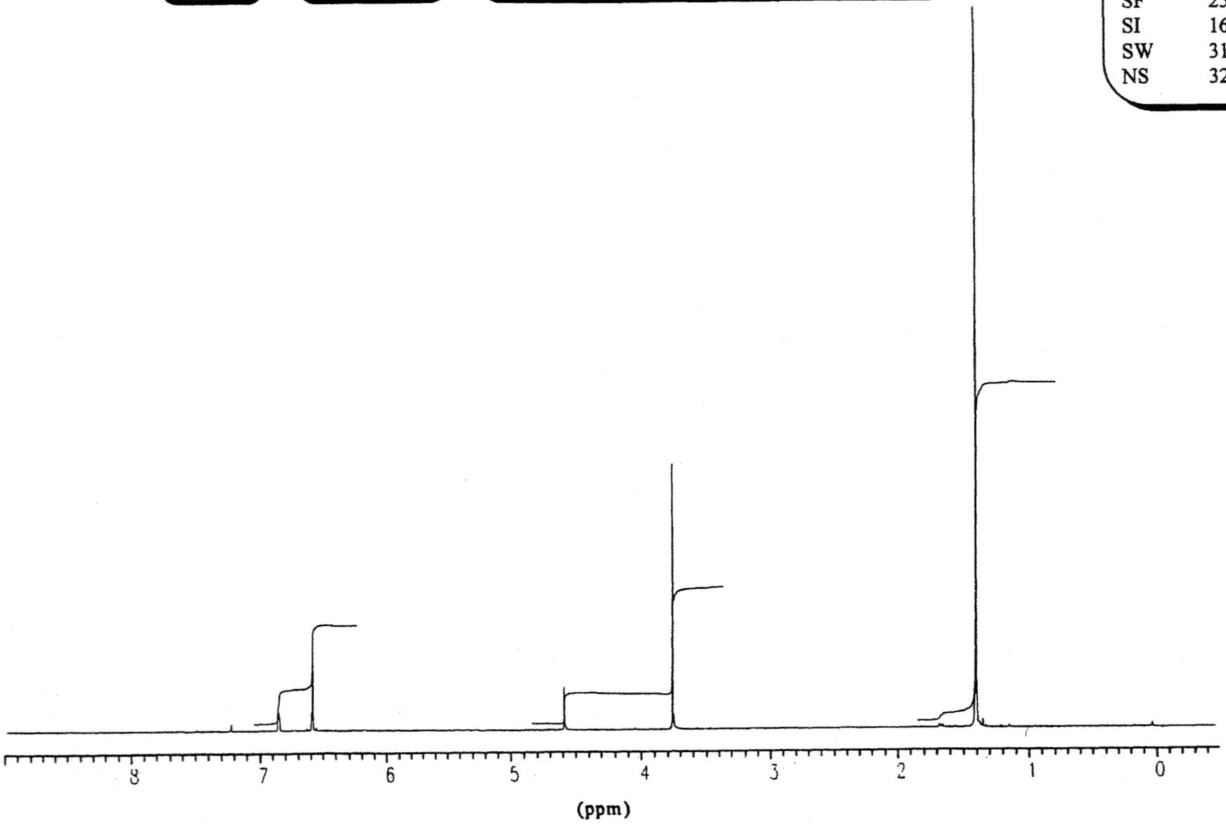

Citric acid, triethyl ester

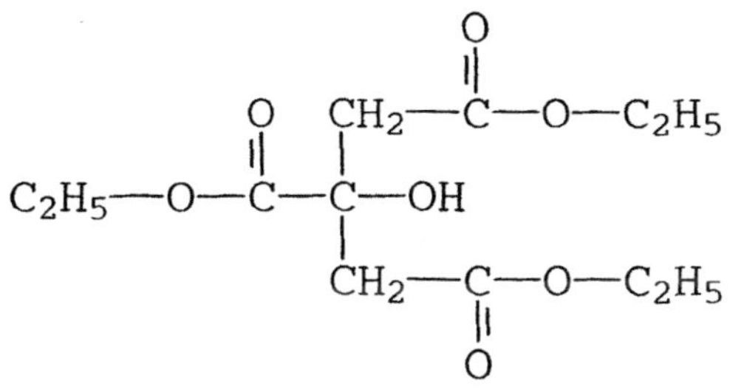

CAS No.	- 00077-93-0
PM Ref. No.	- 44640
Restrictions	- None
Formula	- $C_{12}H_{20}O_7$
Molecular weight	- 276

Alternative names	- Triethyl citrate
Physical Characteristics	- Slightly yellow liquid, mp 78.5 - 79°C, bp 287°C decomposes. - Soluble in alcohol, ether, chloroform.
Handling	- Store at room temperature.
Safety	-
Availability	- Standard sample was supplied.
Current uses	- Plasticizer.
Applications	- PVC, PVDC, CA, paper, inks, epoxy polymers.
GC Retention time	- 30.6 min.

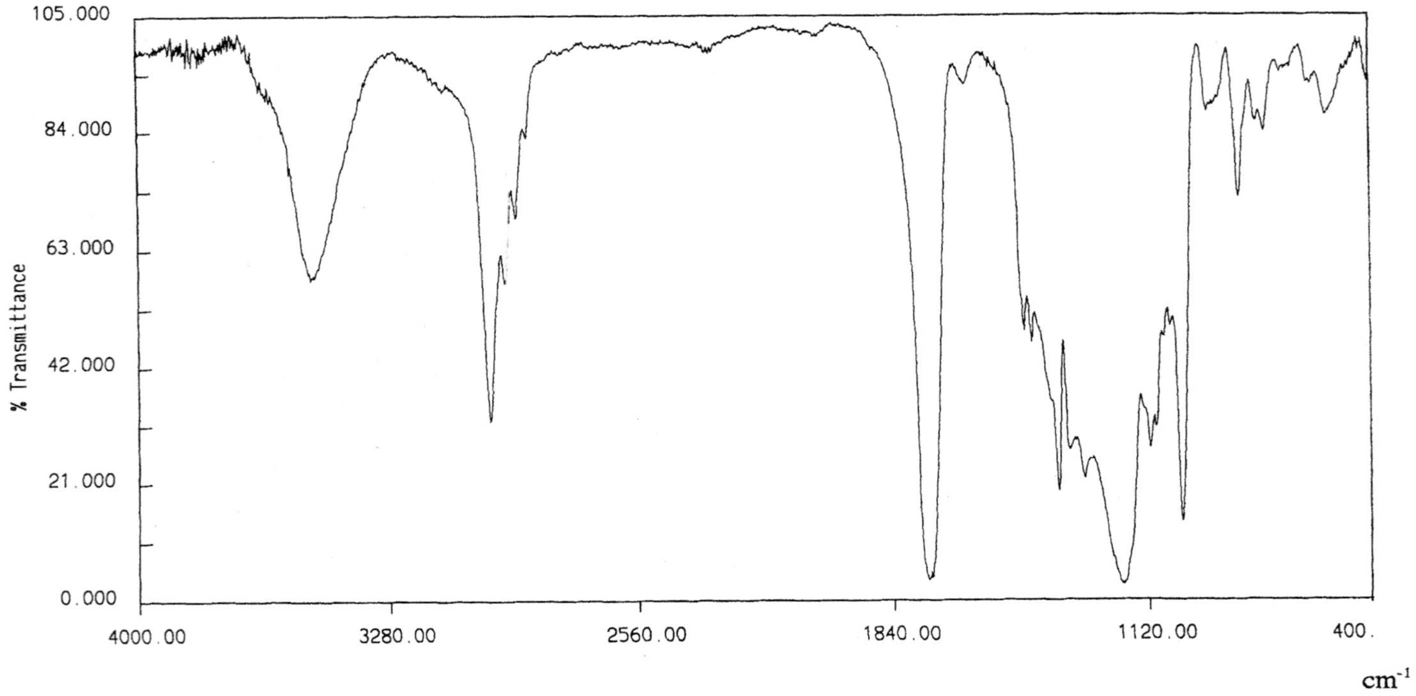

Name : CITRIC ACID, TRIETHYL ESTER

PM Nr : 44640 CAS Nr : 0007-93-0

KBr pellet

CAS No. 00077-93-0 PM Ref. No. 44640

CITRIC ACID, TRIETHYL ESTER

Peak wavenumber	% T	Relative intensity
611.5	86.8	13.6
787.1	83.8	16.6
862.3	72.1	28.7
933.7	88.6	11.7
1026.3	14.1	88.3
1076.4	47.4	54.1
1097.6	31.2	70.8
1115.0	27.4	74.7
1192.2	2.7	100.0
1300.2	21.9	80.3
1371.6	19.6	82.6
1446.8	46.3	55.2
1468.0	48.3	53.2
1741.9	3.6	99.2
2909.0	68.6	32.3
2941.8	57.1	44.1
2986.2	32.3	69.6
3497.4	58.0	43.2

Citric acid, triethyl ester

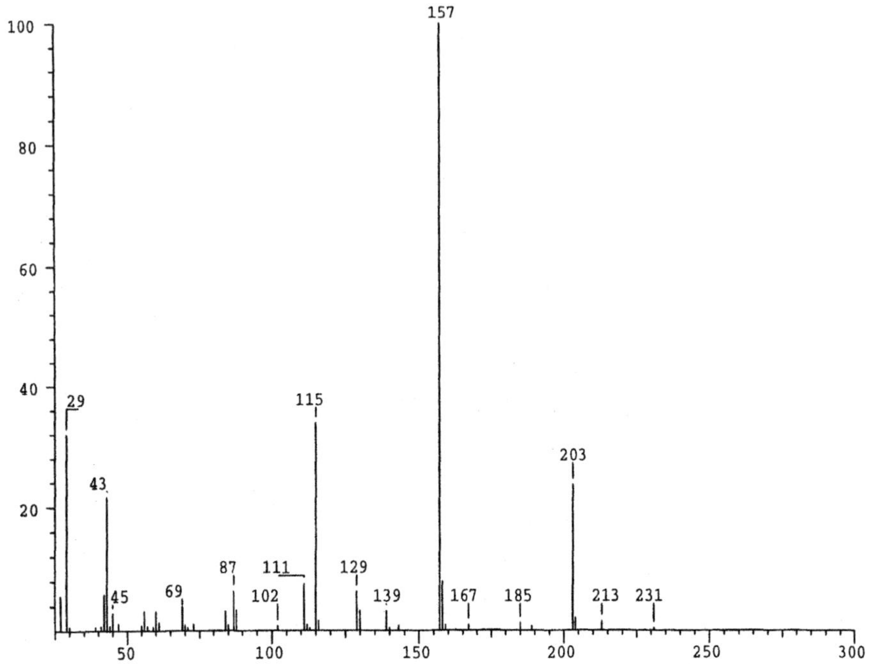

M/Z	Ion Intensity (%)	M/Z	Ion Intensity (%)
27	5.6	115	33.9
29	32.0	129	6.3
42	5.9	130	3.3
43	21.6	139	3.2
56	3.1	157	100.0
60	3.2	158	7.9
69	4.0	203	23.6
84	3.1	204	2.1
87	6.4	213	1.6
111	5.6	231	0.4

Spectrometer : Finnigan Mat SSQ 700
Inletsystem : Capillary GC/MS
Source Temperature : 180°C
Electron Energy : 70 eV
Scan Range : 25-500

PM n°	CAS n°	Name
44640	00077-93-0	CITRIC ACID, TRIETHYL ESTER

Solvent	$CDCL_3$
SF	250.13
SI	16384
SW	3125
NS	32

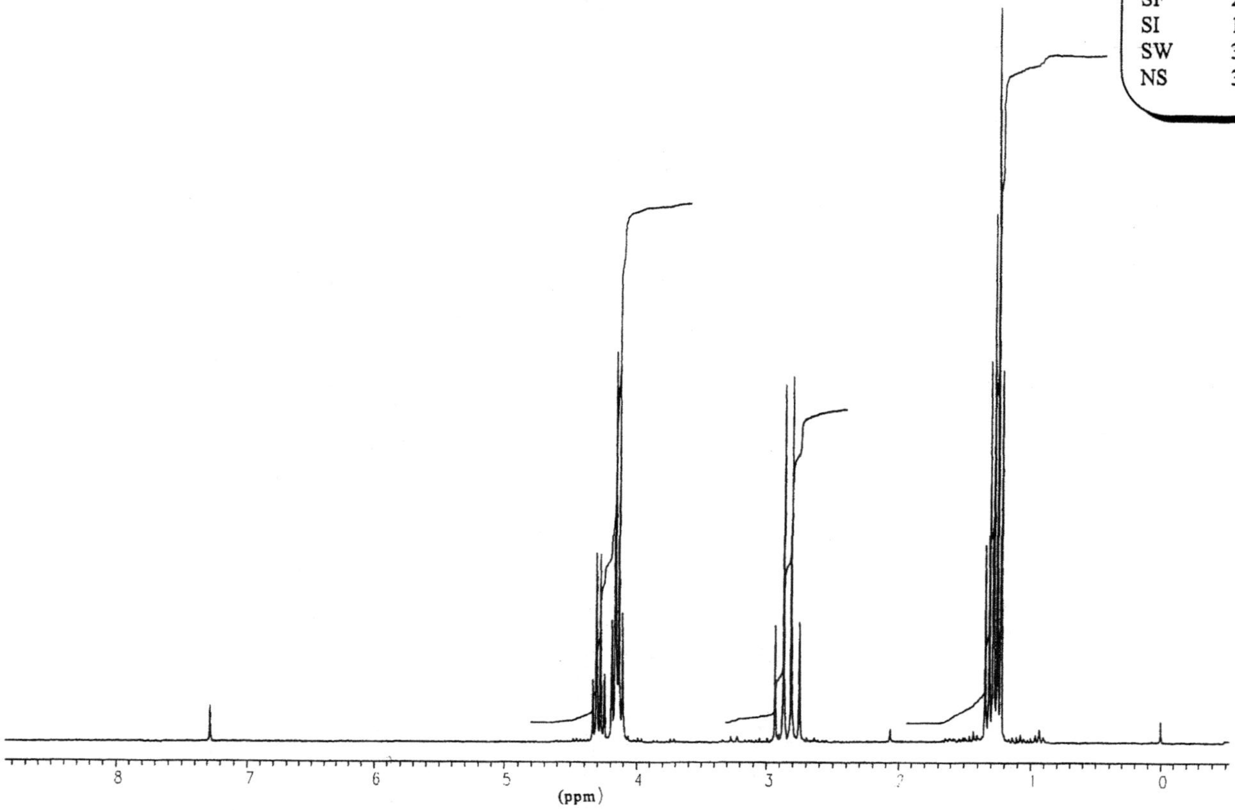

Dibenzothiazyl disulphide

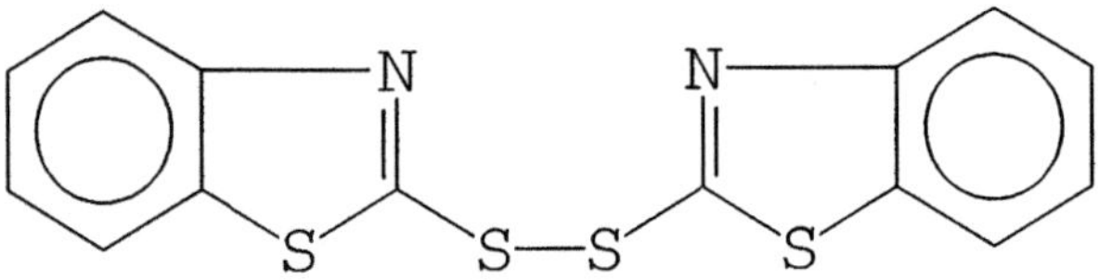

CAS No.	- 00120-78-5
PM Ref. No.	- 46400
Restrictions	- None
Formula	- $C_{14}H_8N_2S_4$
Molecular weight	- 332

Alternative names	- 2,2'-Dithiobis[benzothiazole]
Physical Characteristics	- Pale yellow powder, mp 177 - 180°C. - Soluble in dichloromethane, benzene, chloroform, acetone.
Handling	- Store at room temperature.
Safety	-
Availability	- Standard sample was supplied.
Current uses	- Accelerator, stabilizer.
Applications	- Rubber, PVC.
GC Retention time	- Degradation product 2-Mercaptobenzothiazole 53.9 min.

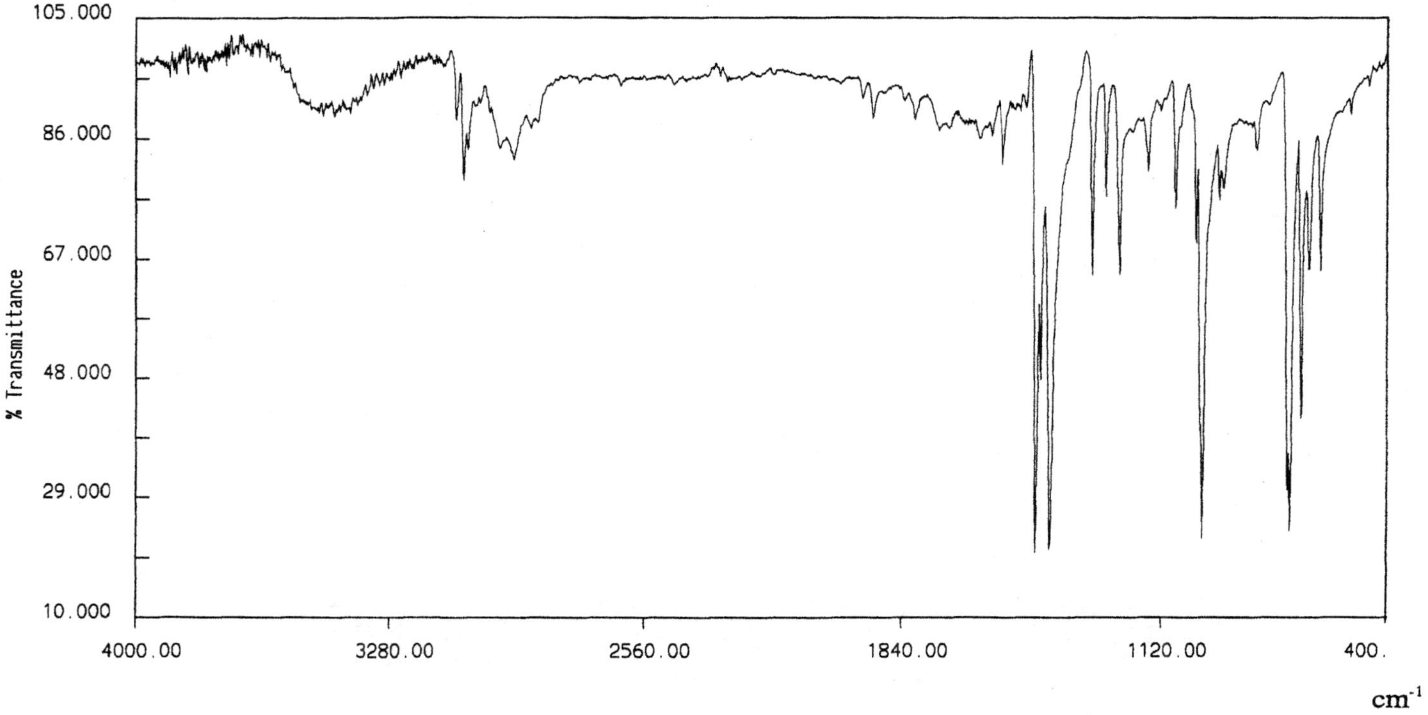

Name : DIBENZOTHIAZYL DISULPHIDE

PM Nr : 46400 CAS Nr : 00120-78-5

KBr pellet

CAS No. 00120-78-5 PM Ref. No. 46400

DIBENZOTHIAZYL DISULPHIDE

Peak wavenumber	% T	Relative intensity
407.0	76.0	30.1
434.0	72.9	34.0
534.4	94.2	7.2
584.5	89.7	12.9
671.3	64.9	44.0
402.2	65.2	43.7
725.3	41.5	73.3
758.1	23.6	95.8
850.7	84.0	20.1
945.2	78.1	27.4
956.8	76.2	29.9
1005.0	22.4	97.3
1020.5	69.5	38.2
1080.3	74.9	31.5
1120.8	90.2	12.3
1157.4	80.9	24.0
1238.5	64.4	44.7
1275.1	76.8	29.1
1311.8	64.3	44.8
1429.4	20.8	99.3
1454.5	47.5	65.8
1468.0	20.2	100.0
1560.6	81.9	22.7
1589.5	86.4	17.1
1801.7	88.9	13.9
1919.4	89.1	13.6
2853.1	88.4	14.6
2922.5	82.6	21.9
3051.8	84.2	19.8
3063.3	79.3	25.9
3372.0	91.0	11.3

Dibenzothiazyl disulphide

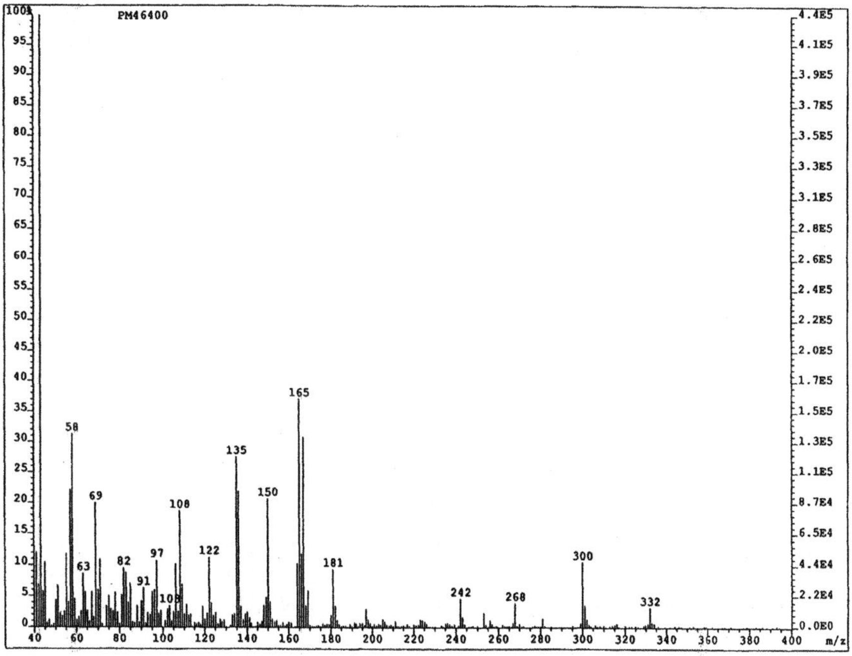

M/Z	Ion Intensity (%)	M/Z	Ion Intensity (%)
41	11.7	135	46.0
43	100.0	136	21.5
57	22.1	150	20.9
58	30.7	165	36.8
69	20.2	166	11.7
71	11.0	167	30.7
82	9.2	181	9.2
97	11.0	300	10.4
108	18.4	332	3.0
122	11.0		

Spectrometer	: VG Autospec
Inletsystem	: Flow injection particle beam MS
Source Temperature	: 250°C
Electron Energy	: 70 eV
Scan Range	: 40-1000

2-Mercaptobenzothiazole

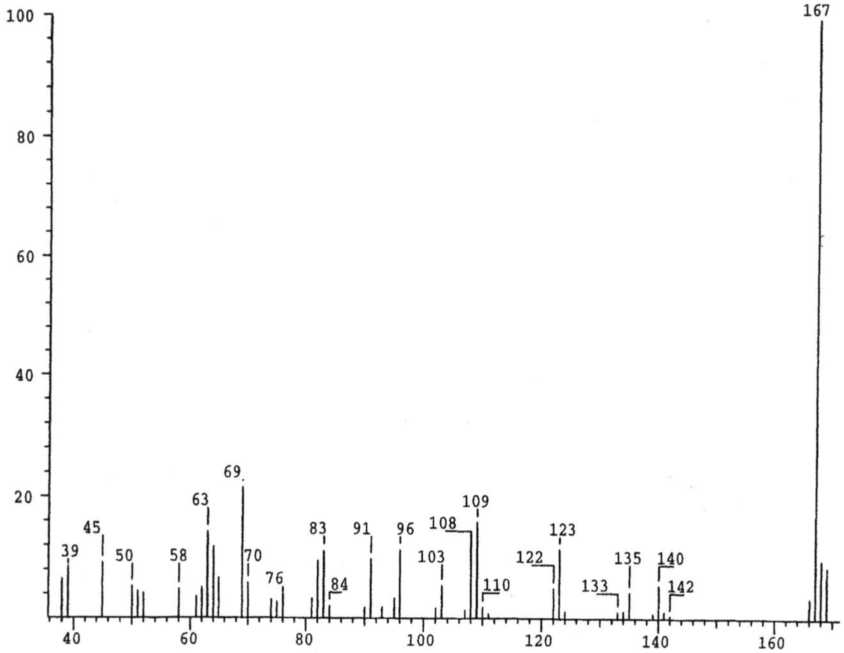

M/Z	Ion Intensity (%)	M/Z	Ion Intensity (%)
45	9.1	109	15.8
63	14.2	123	11.4
64	11.7	135	4.4
69	21.5	140	5.6
83	11.1	167	100.0
91	9.8	168	9.8
96	11.3	169	8.5
108	13.4		

Spectrometer	: Finnigan Mat SSQ 700
Inletsystem	: Capillary GC/MS
Source Temperature	: 180°C
Electron Energy	: 70 eV
Scan Range	: 40-500

PM n°	CAS n°	Name
46400	00120-78-5	DIBENZOTHIAZYL DISULFIDE

Solvent	C_6D_6
SF	250.13
SI	16384
SW	3125
NS	60

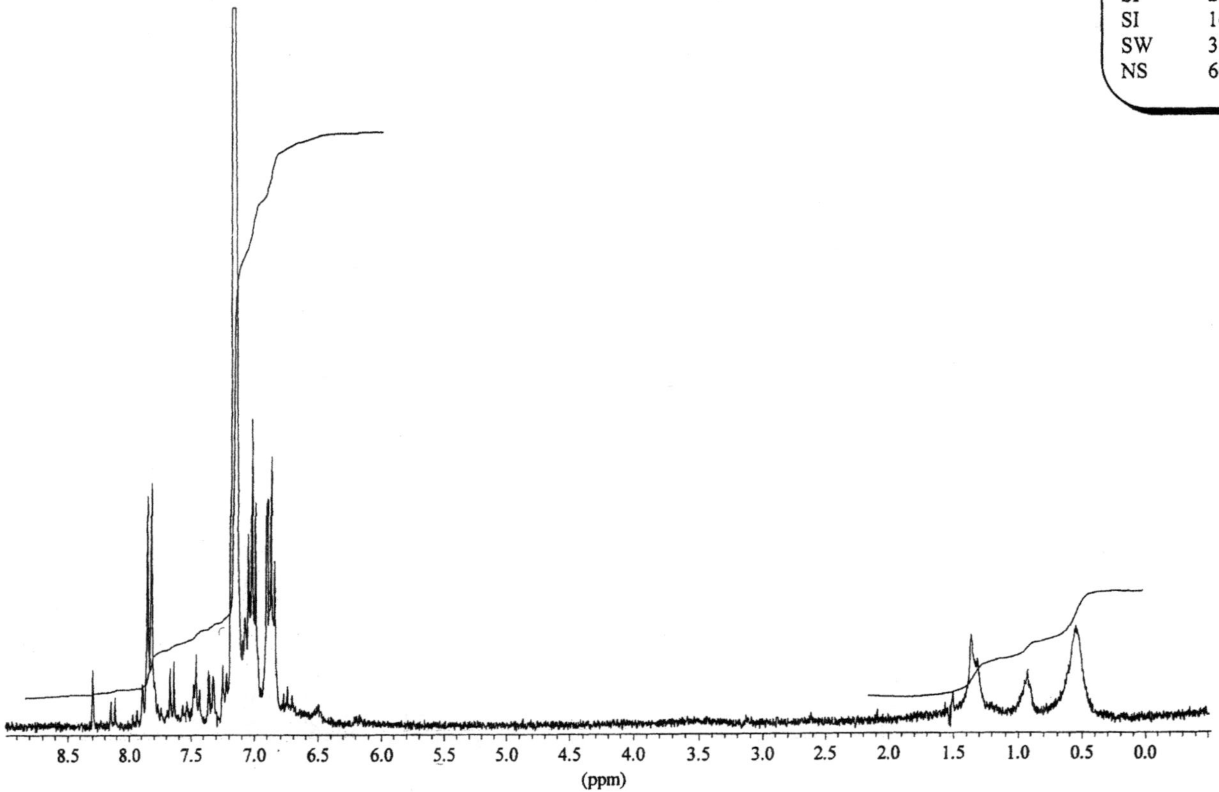

2,6-Di-tert-butyl-p-cresol

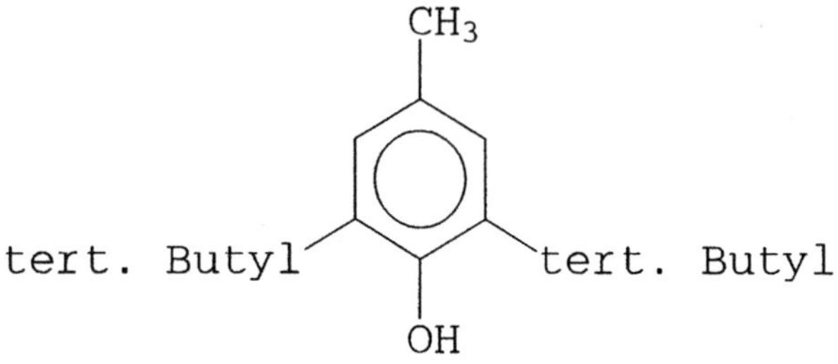

CAS No.	- 00128-37-0
PM Ref. No.	- 46640
Restrictions	- SML = 3.0 mg/kg
Formula	- $C_{15}H_{24}O$
Molecular weight	- 220

Alternative names - Butylated hydroxytoluene (BHT)

Physical Characteristics - White crystalline solid, mp 69-70°C, bp 265°C.
- Soluble in most organic solvents.

Handling - Store at room temperature.
Safety - Harmful.
Availability - Standard sample supplied.

Current uses - Antioxidant.
Applications - General use.

GC Retention time - 27.8 min.

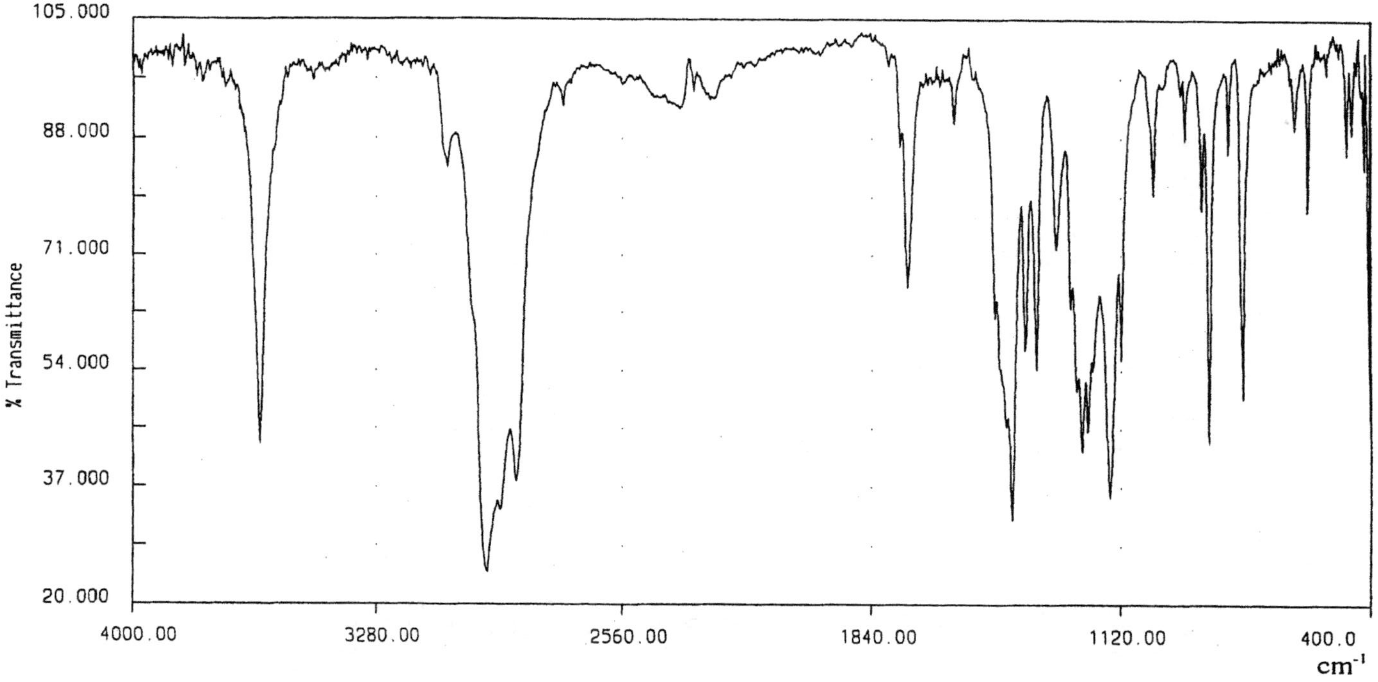

Name : 2,6-DI-TERT-BUTYL-p-CRESOL (BHT)

PM Nr : 46640 CAS Nr : 00128-37-0

KBr pellet

CAS No. 00128-37-0 PM Ref. No. 46640

2,6-DI-TERT-BUTYL-p-CRESOL(BHT)

Peak wavenumber	% T	Relative intensity
418.6	77.3	29.9
468.8	82.2	23.4
580.6	78.1	28.8
619.2	91.1	11.7
769.7	49.3	66.6
812.1	85.1	19.7
866.1	44.2	73.4
887.4	78.9	27.8
1026.3	79.9	26.5
1120.7	55.8	58.1
1151.6	35.8	84.4
1199.9	54.8	59.5
1213.4	45.9	71.2
1230.7	42.9	75.0
1306.0	73.9	34.4
1361.9	55.8	58.2
1396.6	58.6	54.5
1433.3	33.06	88.0
2733.5	90.31	12.7
2872.4	36.65	83.3
2916.7	32.64	88.6
2955.3	23.96	100.0
3067.2	81.31	24.6
3626.6	43.99	73.7

2,6-Di-tert-butyl-p-cresol

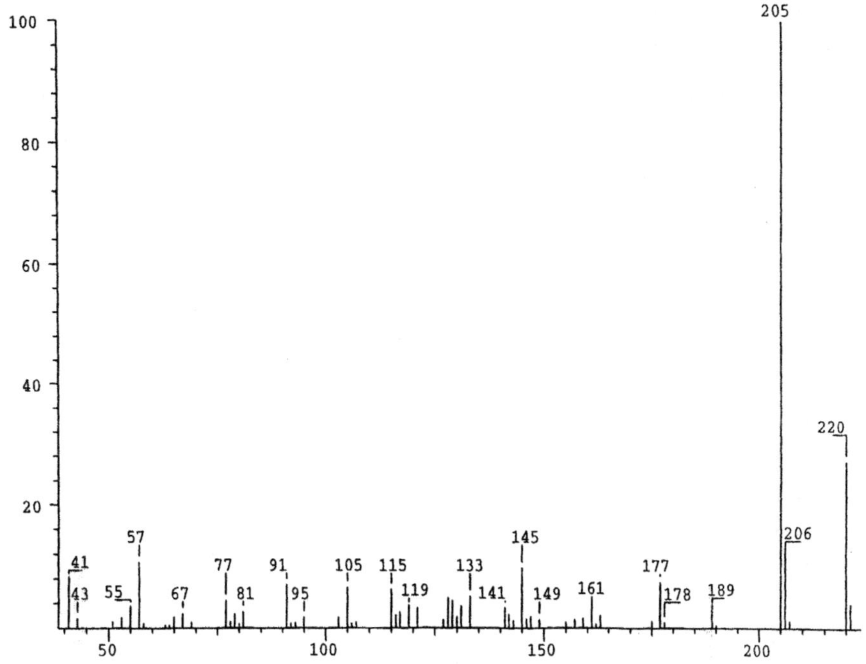

M/Z	Ion Intensity (%)	M/Z	Ion Intensity (%)
41	8.4	161	4.1
57	10.7	177	7.5
77	4.5	189	4.0
91	7.0	205	100.0
105	6.6	206	13.2
115	6.3	220	27.3
133	5.2	221	4.0
145	9.8		

Spectrometer : Finnigan Mat SSQ 700
Inletsystem : Capillary GC/MS
Source Temperature : 180°C
Electron Energy : 70 eV
Scan Range : 40-500

PM n°
46640

CAS n°
00128-37-0

Name
2,6-DI-TERT-BUTYL-p-CRESOL (BHT)

Solvent	$CDCL_3$
SF	250.13
SI	16384
SW	2380
NS	32

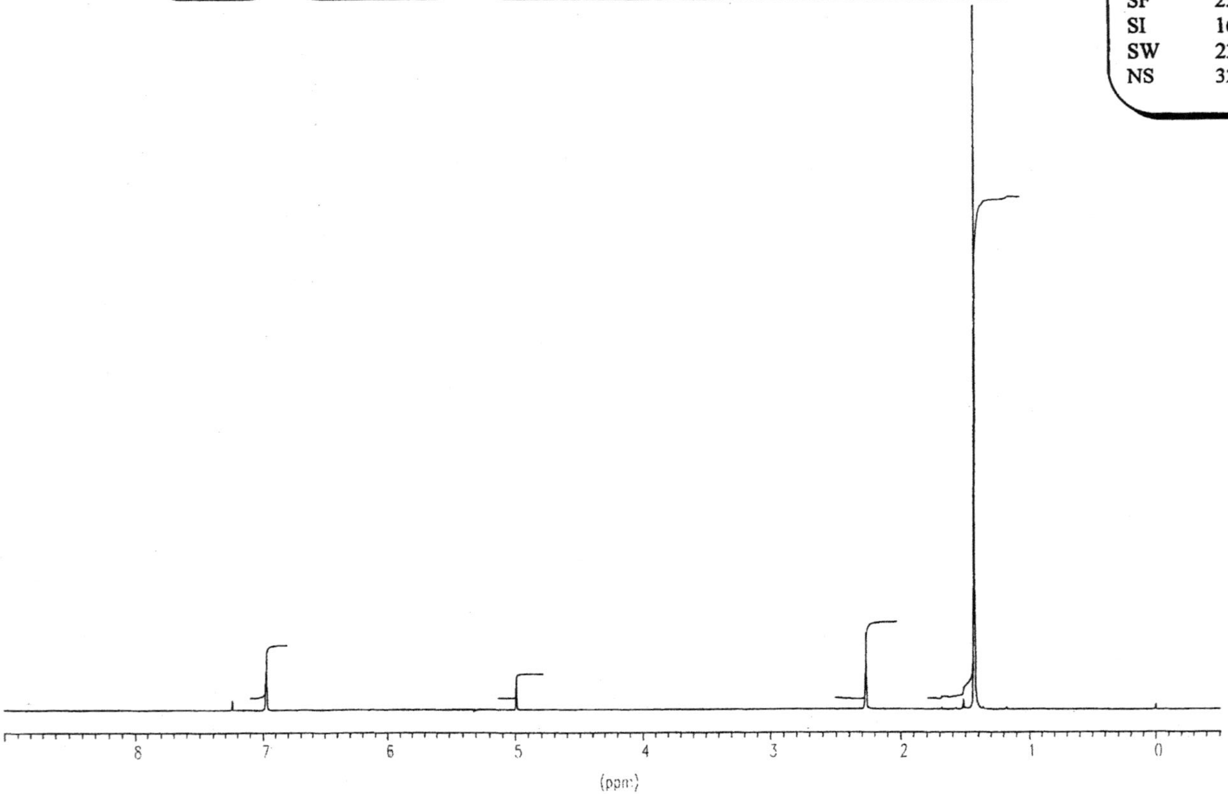

3,5-Di-tert-butyl-4-hydroxybenzylphosphonic acid, monoethyl ester, calcium salt

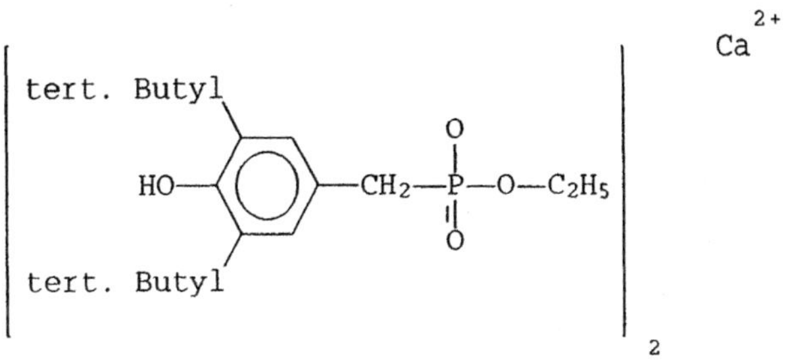

CAS No. - 65140-91-2
PM Ref. No. - 46880
Restrictions - SML = 6.0 mg/kg
Formula - $C_{34}H_{56}O_8P_2Ca$
Molecular weight - 694

Alternative names -

Physical Characteristics - White crystalline powder, mp > 260°C.
- Soluble in most organic solvents.

Handling - Store at room temperature.
Safety -
Availability - Standard sample was supplied.

Current uses - Antioxidant.
Applications - Polyolefins, rubber, paper.

GC Retention time -

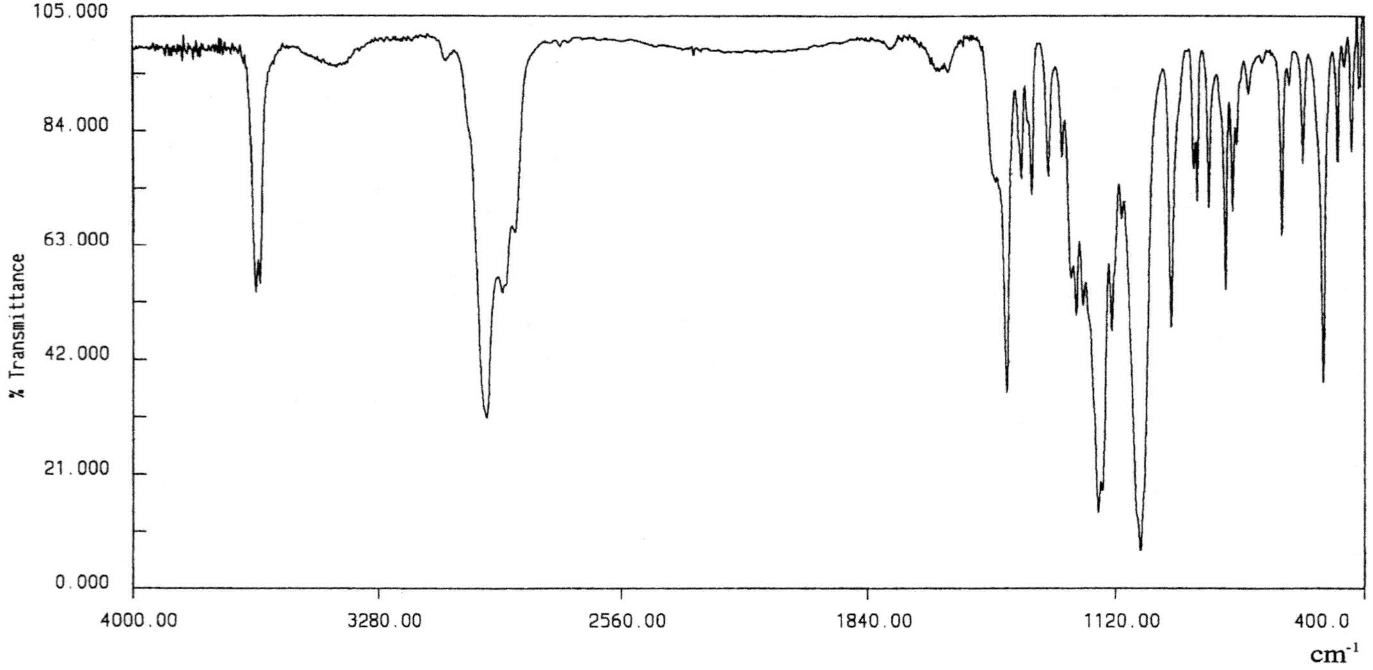

Name : 3,5-DI-TERT-BUTYL-4-HYDROXYBENZYLPHOSPHONIC ACID, MONOETHYL ESTER, Ca SALT

PM Nr : 46880 CAS Nr : 65140-91-2

KBr pellet

CAS No. 65140-91-2 PM Ref. No. 46880

3,5-DI-TERT-BUTYL-4-HYDROXYBENZYLPHOSPHONIC ACID, MONOETHYL, Ca SALT

Peak wavenumber	% T	Relative intensity
414.8	91.6	9.0
436.0	80.1	21.4
476.5	78.3	23.3
518.9	37.8	66.8
574.9	78.0	23.6
617.3	92.2	8.4
638.5	64.9	37.7
733.0	90.8	9.9
767.8	81.6	19.8
779.3	69.2	33.1
800.6	54.9	48.5
848.8	70.0	32.2
881.6	71.1	31.0
891.2	77.1	24.6
958.7	48.0	55.9
1045.6	6.9	100.0
1101.5	67.8	34.6
1130.4	47.20	56.7
1159.4	17.83	88.3
1170.9	13.73	92.7
1213.4	51.92	51.7
1232.7	50.00	53.7
1273.2	79.23	22.3
1313.7	75.52	26.3
1361.9	72.15	29.9
1390.9	75.11	26.7
1435.2	35.93	68.8
2874.3	65.30	37.3
2910.9	54.12	49.3
2959.2	31.17	73.9
3644.0	54.20	49.2

3,5-Di-tert-butyl-4-hydroxybenzylphosphonic acid, monoethyl ester, calcium salt

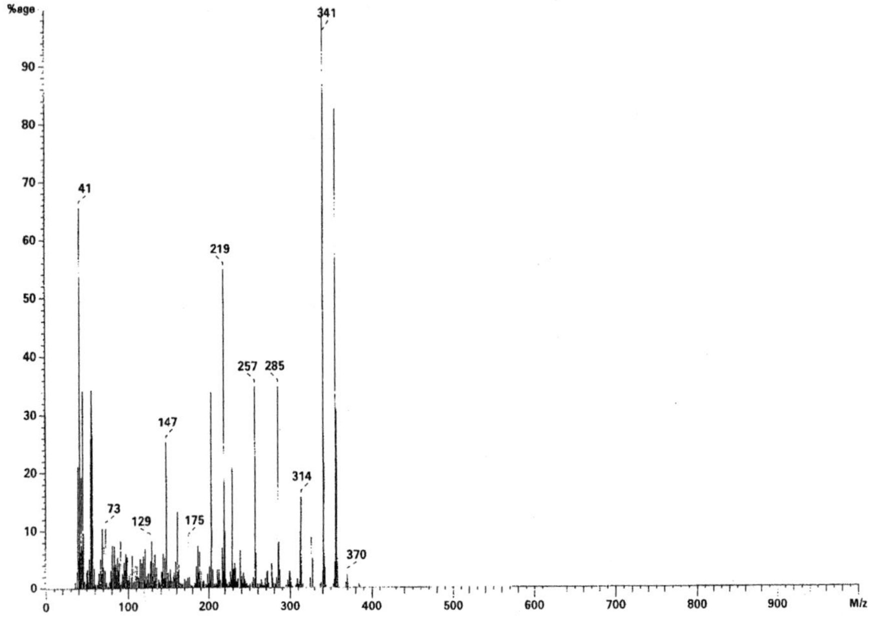

M/Z	Ion Intensity (%)	M/Z	Ion Intensity (%)
40	21.3	203	33.3
41	65.3	219	54.7
43	18.7	229	20.7
45	33.3	257	34.7
55	25.3	285	34.7
56	34.0	314	15.3
57	24.7	341	100.0
73	10.0	342	21.3
147	25.3	356	82.0
161	12.7	357	30.7

Spectrometer	: VG Autospec
Inletsystem	: Probe inlet
Source Temperature	: 350-400°C
Electron Energy	: 70 eV
Scan Range	: 40-1000

PM n° 46880

CAS n° 65140-91-2

Name:
3,5-DI-TERT-BUTYL-4-HYDROXYBENZYLPHOSPHONIC ACID, ETHYLESTER, Ca SALT

Solvent	CD_3OD
SF	250.13
SI	16384
SW	3125
NS	96

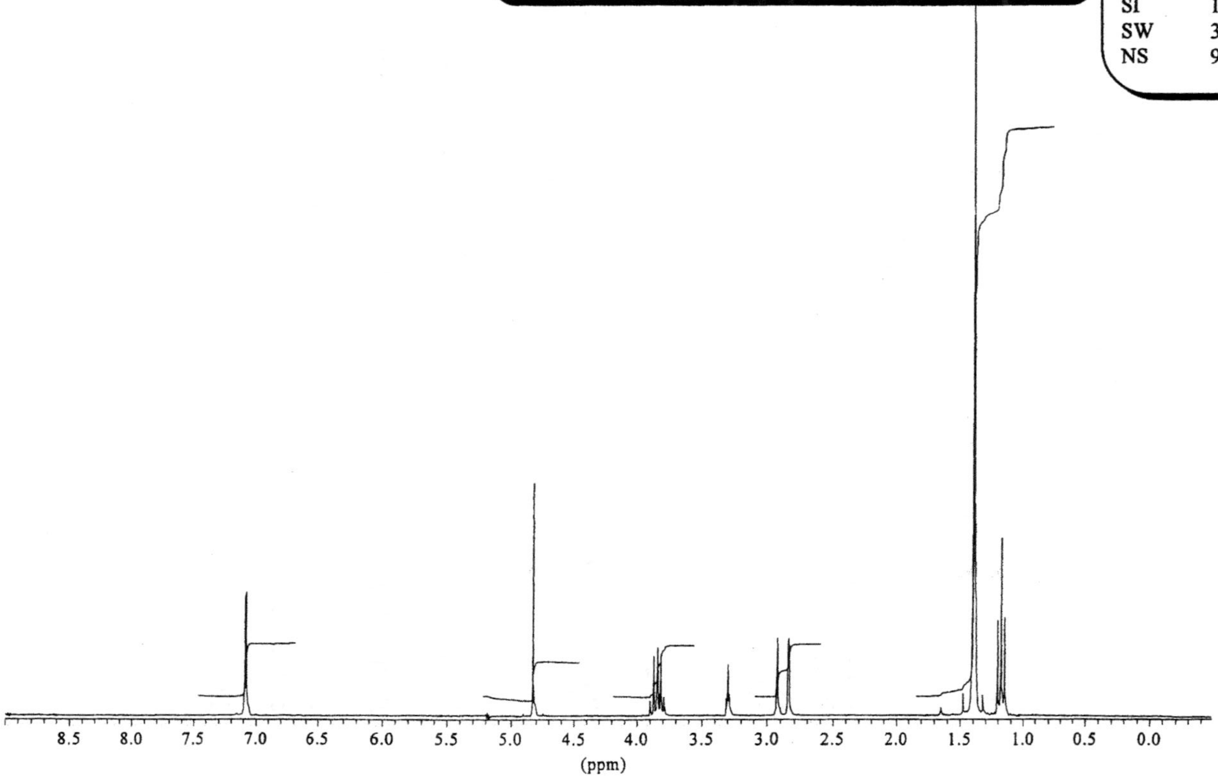

2,4-Dihydroxybenzophenone

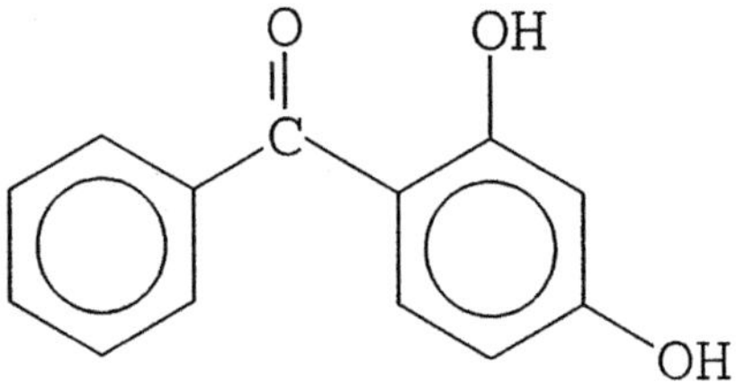

CAS No.	- 00131-56-6
PM Ref. No.	- 48640
Restrictions	- SML = 6.0 mg/kg
Formula	- $C_{13}H_{10}O_3$
Molecular weight	- 214

Alternative names	- Benzoresorcinol
Physical Characteristics	- White powder, mp 144.5 - 147°C. - Soluble in alcohol, benzene, ether, glacial acetic acid.
Handling	- Store at room temperature.
Safety	-
Availability	- Standard sample was supplied.
Current uses	- UV-stabilizer.
Applications	- Polyolefins.
GC Retention time	- 39.5 min.

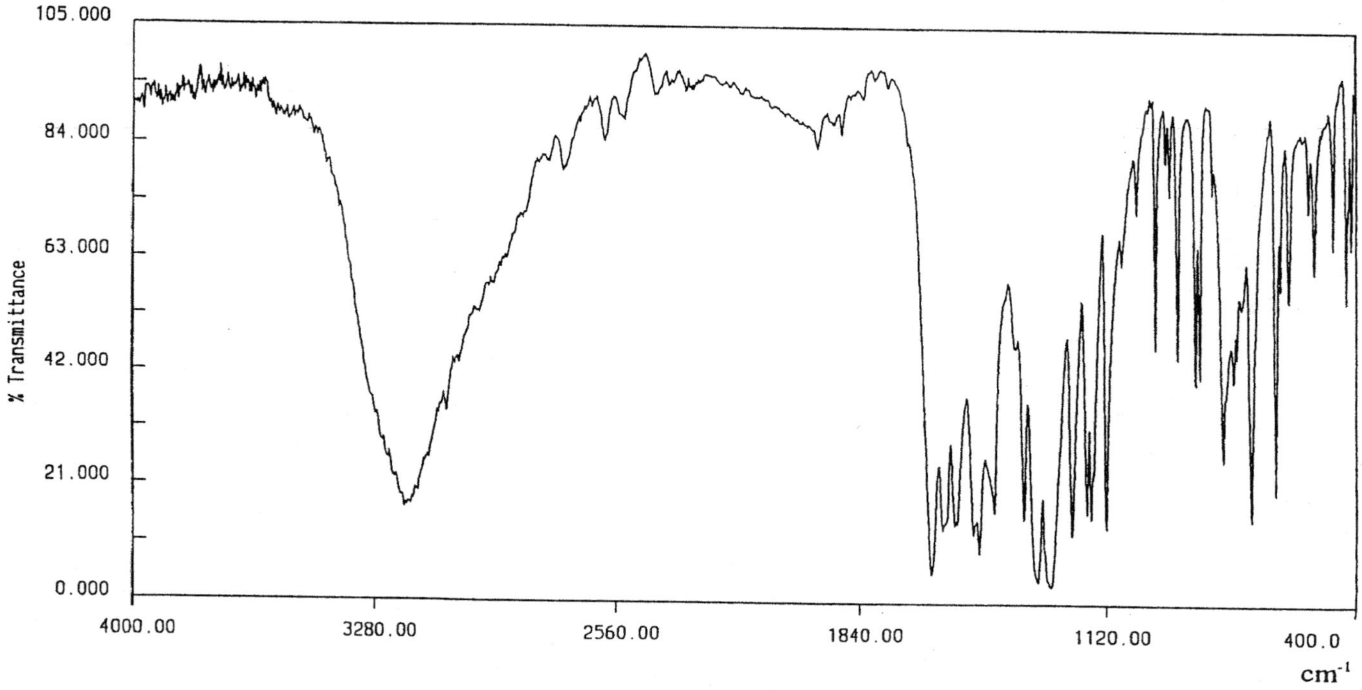

Name : 2,4-DIHYDROXYBENZOPHENONE

PM Nr : 48640 CAS Nr : 00131-56-6

KBr pellet

CAS No. 00131-56-6 PM Ref. No. 48640

2,4-DIHYDROXYBENZOPHENONE

Peak wavenumber	% T	Relative intensity
412.8	65.8	35.3
428.3	55.8	45.6
464.9	65.8	35.3
517.0	61.1	40.2
532.4	72.5	28.3
590.3	55.9	45.5
628.9	20.3	82.2
700.3	15.5	87.2
731.1	54.6	46.9
742.7	45.5	56.3
752.3	41.0	60.9
783.2	26.2	76.1
814.1	75.7	25.1
850.7	41.6	60.3
862.3	40.4	61.5
914.4	45.2	56.6
937.5	75.3	25.5
949.1	81.3	19.3
980.0	47.0	54.7
1034.0	72.0	28.9
1078.3	62.3	38.9
1122.7	13.9	88.8
1167.1	15.7	87.0
1178.7	16.6	86.0
1221.1	12.7	90.1
1280.9	3.1	100.0
1321.4	4.1	98.9
1360.0	15.6	87.1
1446.8	16.7	86.0
1491.2	9.3	93.6
1562.5	14.3	88.4
1599.2	13.5	89.2
1630.1	5.4	97.7
1894.3	85.9	14.6
1965.7	83.4	17.1
2536.7	88.4	11.9
2594.6	84.5	16.0
2720.0	79.2	21.5
2764.3	80.7	19.9
2837.6	70.5	30.4
2883.9	63.0	38.2
2920.6	58.3	43.1
2965.0	53.0	48.5
3024.8	43.7	58.1
3063.3	34.5	67.6
3119.3	26.0	76.4
3152.1	19.9	82.7
3970.0	89.6	10.8

2,4-Dihydroxybenzophenone

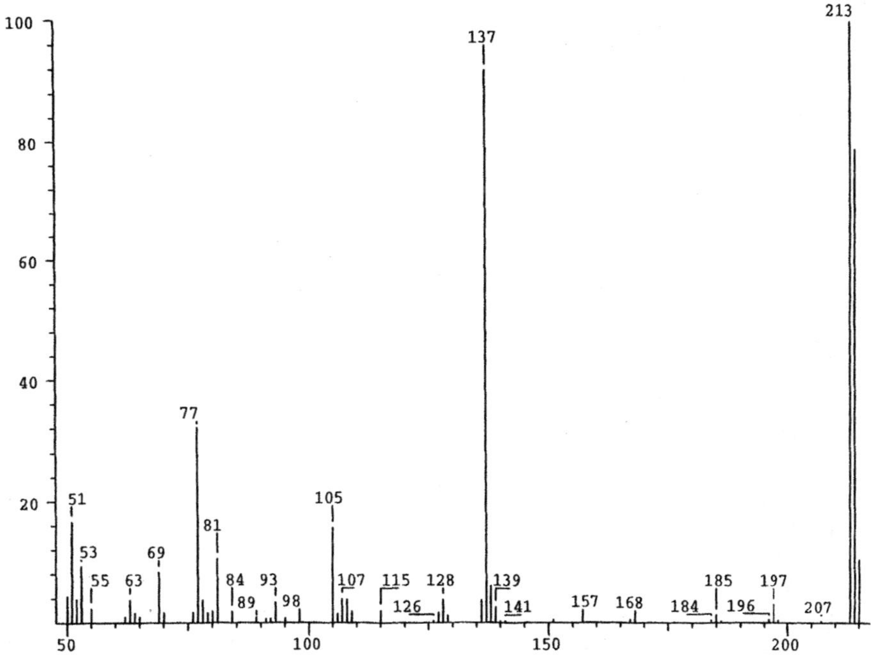

M/Z	Ion Intensity (%)	M/Z	Ion Intensity (%)
51	16.7	137	91.9
53	9.4	138	6.2
69	8.4	213	100.0
77	32.2	214	78.9
81	10.5	215	10.5
105	15.7		

Spectrometer	: Finnigan Mat SSQ 700
Inletsystem	: Capillary GC/MS
Source Temperature	: 180°C
Electron Energy	: 70 eV
Scan Range	: 40-500

PM n°	CAS n°	Name
48640	00131-56-6	2,4-DIHYDROXYBENZOPHENONE

Solvent	$CDCL_3$
SF	250.13
SI	16384
SW	4000
NS	32

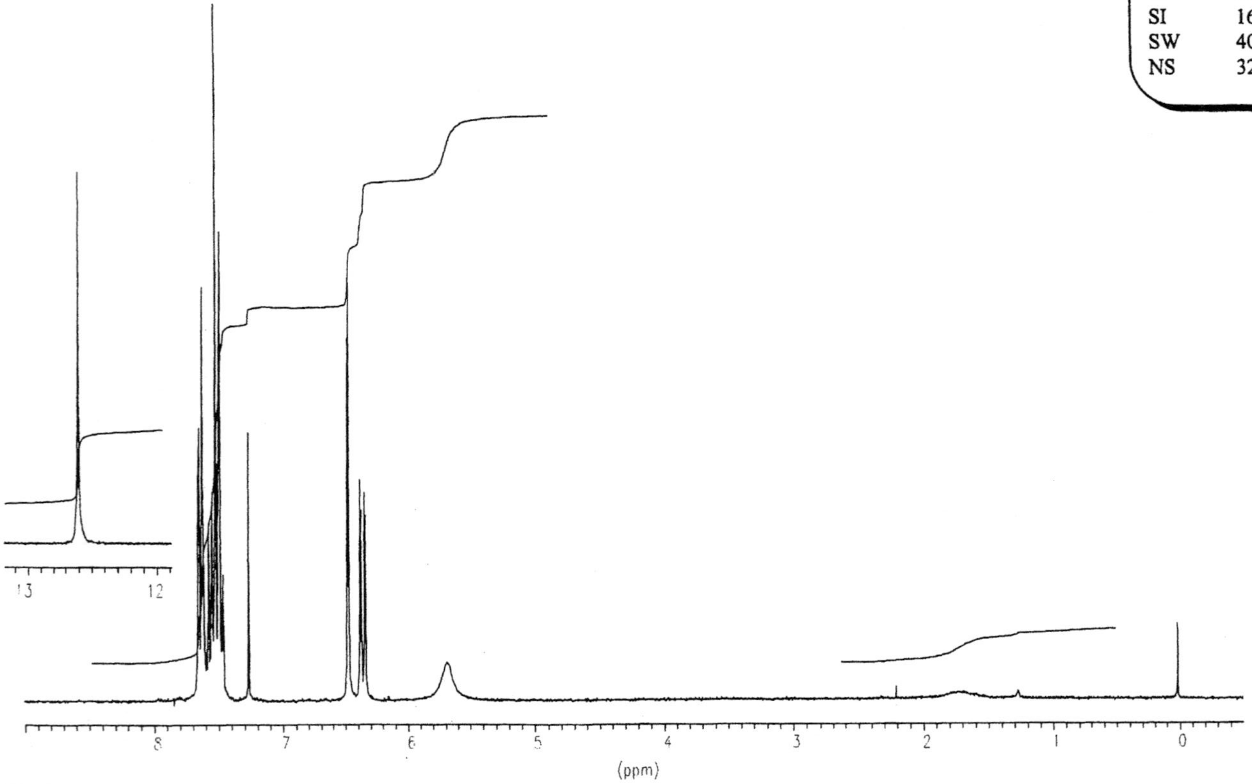

4,4'-Dihydroxybenzophenone

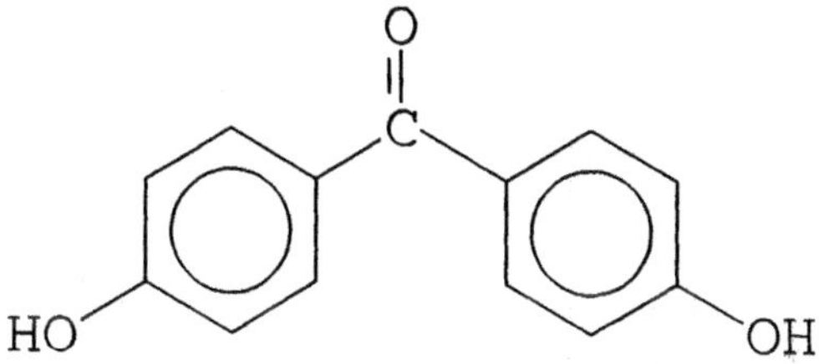

CAS No.	- 00611-99-4
PM Ref. No.	- 48720
Restrictions	- SML = 6.0 mg/kg
Formula	- $C_{13}H_{10}O_3$
Molecular weight	- 214

Alternative names	- Methanone, bis(4-hydroxyphenyl)
Physical Characteristics	- Yellow powder, mp 213 - 215°C, bp 350°C. - Soluble in most organic solvents.
Handling	- Store at room temperature.
Safety	- Irritant.
Availability	- Standard sample was supplied.
Current uses	- UV-stabilizer.
Applications	- General use.
GC Retention time	- 43.7 min.

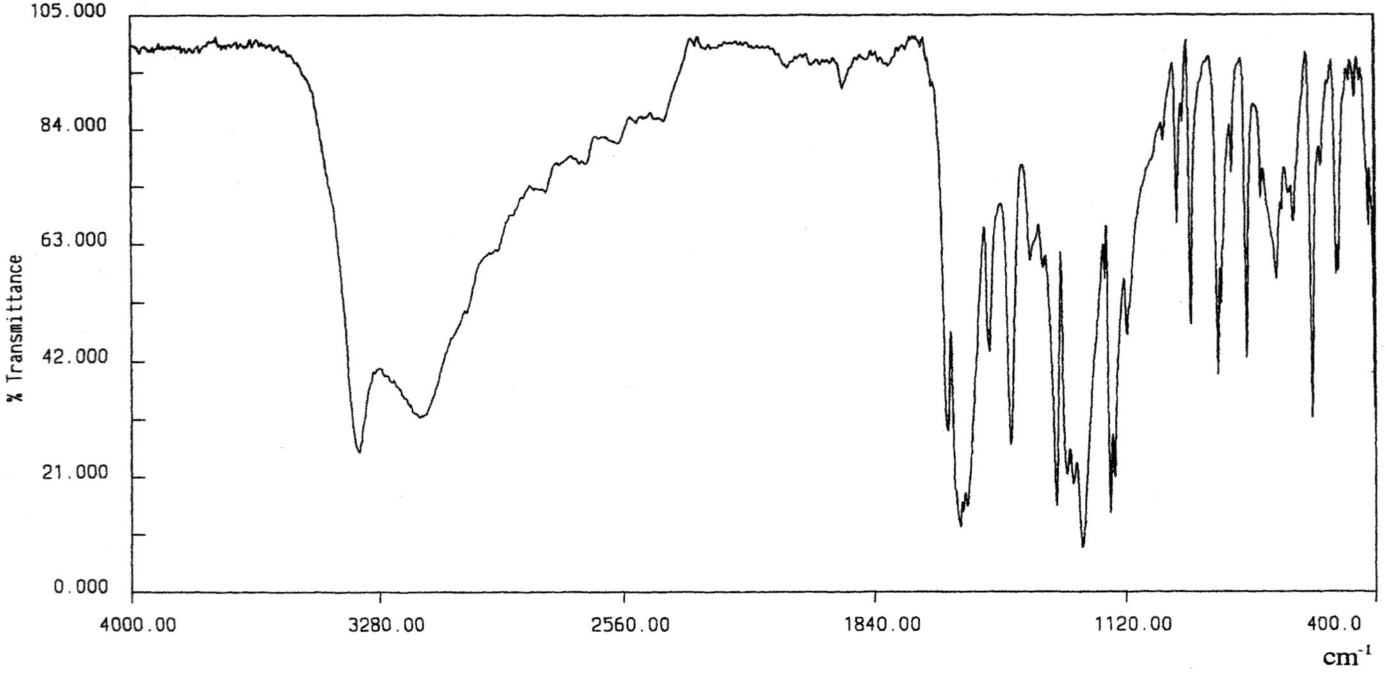

Name : 4,4-DIHYDROXYBENZOPHENONE

PM Nr : 48720 CAS Nr : 00611-99-4

KBr pellet

CAS No. 00611-99-4 PM Ref. No. 48720

4,4-DIHYDROXYBENZOPHENONE

Peak wavenumber	% T	Relative intensity
418.6	71.4	31.2
513.1	58.7	44.9
582.6	31.9	74.0
634.7	67.5	35.3
671.3	70.1	32.5
686.7	57.7	46.1
731.1	72.9	29.4
771.6	43.9	61.1
814.1	78.5	23.3
854.6	40.9	64.3
929.8	48.4	56.1
970.3	67.1	35.8
1010.8	82.7	18.8
1115.0	47.0	57.6
1151.6	21.0	86.0
1165.2	14.4	93.1
1178.7	57.0	46.7
1244.2	8.1	100.0
1269.3	19.7	87.3
1286.7	21.6	85.3
1317.5	15.8	91.6
1388.9	60.3	43.2
1446.8	26.9	79.6
1508.5	43.5	61.4
1574.1	15.6	91.9
1595.3	11.9	95.9
1628.1	29.2	77.0
1929.1	89.4	11.5
2442.2	82.4	19.2
2573.4	78.7	23.2
2667.9	75.1	27.1
2789.4	70.2	32.4
2885.9	66.1	36.8
2930.2	59.9	43.6
3024.8	49.2	55.3
3163.6	31.0	75.0
3339.2	24.9	81.7

4,4'-Dihydroxybenzophenone

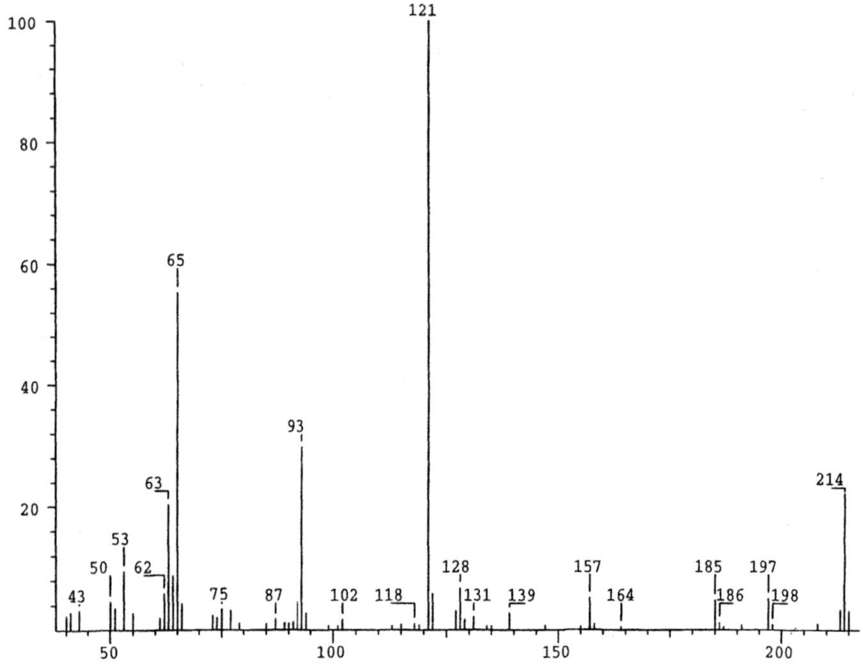

M/Z	Ion Intensity (%)	M/Z	Ion Intensity (%)
50	4.6	121	100.0
53	9.5	128	6.7
63	20.3	157	5.2
64	8.8	185	5.2
65	55.3	214	22.3
93	29.8	215	3.1

Spectrometer	: Finnigan Mat SSQ 700
Inletsystem	: Capillary GC/MS
Source Temperature	: 180°C
Electron Energy	: 70 eV
Scan Range	: 40-500

PM n°	CAS n°	Name
48720	00611-99-4	4,4'-DIHYDROXYBENZOPHENONE

Solvent	acetone-d_6
SF	250.13
SI	16384
SW	3125
NS	32

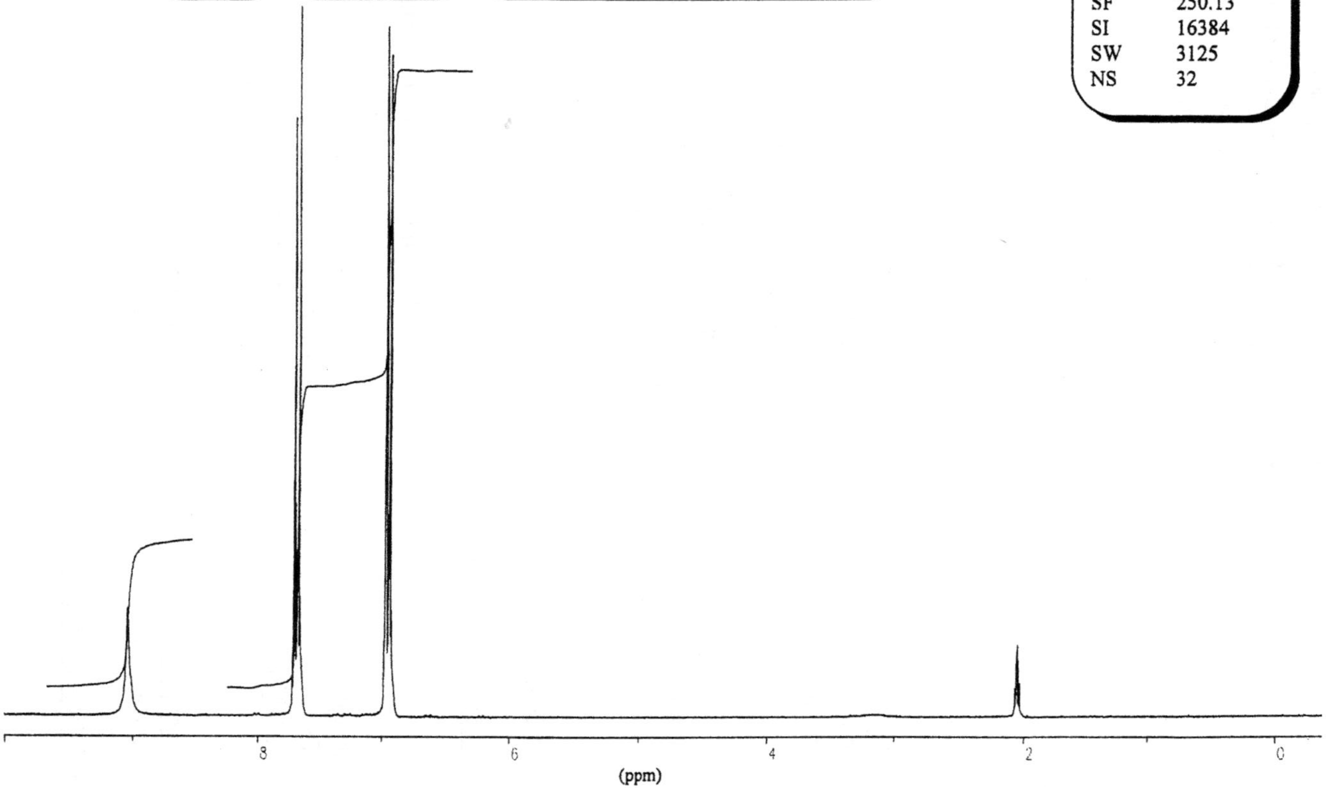

Di-n-octyltin bis(2-ethylhexyl mercaptoacetate) and

Mono-n-octyltin tris(2-ethylhexyl mercaptoacetate)

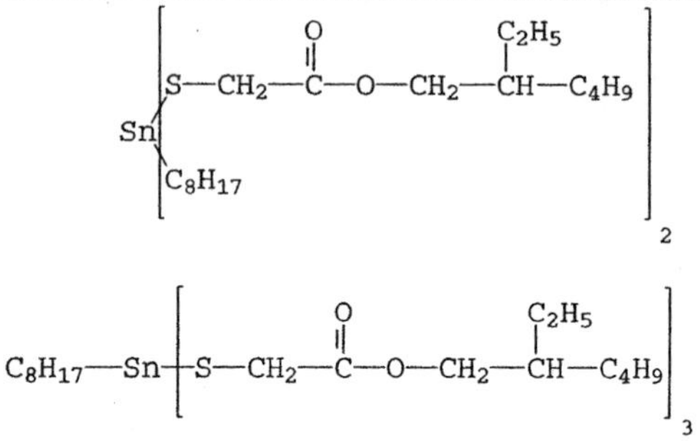

CAS No.	- 15571-58-1 and 27107-89-7
PM Ref. No.	- 50320 and 67680
Restrictions	- SML = 0.02 mg/kg for all di-n-octyltin derivatives. - SML = 1.2 mg/kg for all mono-n-octyltin derivatives.
Formula	- $C_{36}H_{72}O_4S_2Sn$ and $C_{38}H_{74}O_6S_3Sn$
Molecular weight	- 750.7 and 840.7
Alternative names	-
Physical Characteristics	- Viscous colourless liquid, mp < 0°C, bp >250°C. - Soluble in toluene, hexane, tetrahydrofuran.
Handling	- Store at room temperature.
Safety	- Irritant.
Availability	- Liquid preparation containing PM Ref. No. 50320 and 67680.
Current uses	- Stabilizer.
Applications	- PVC.
GC Retention time	- Octyl thioglycolate 25.4 min.

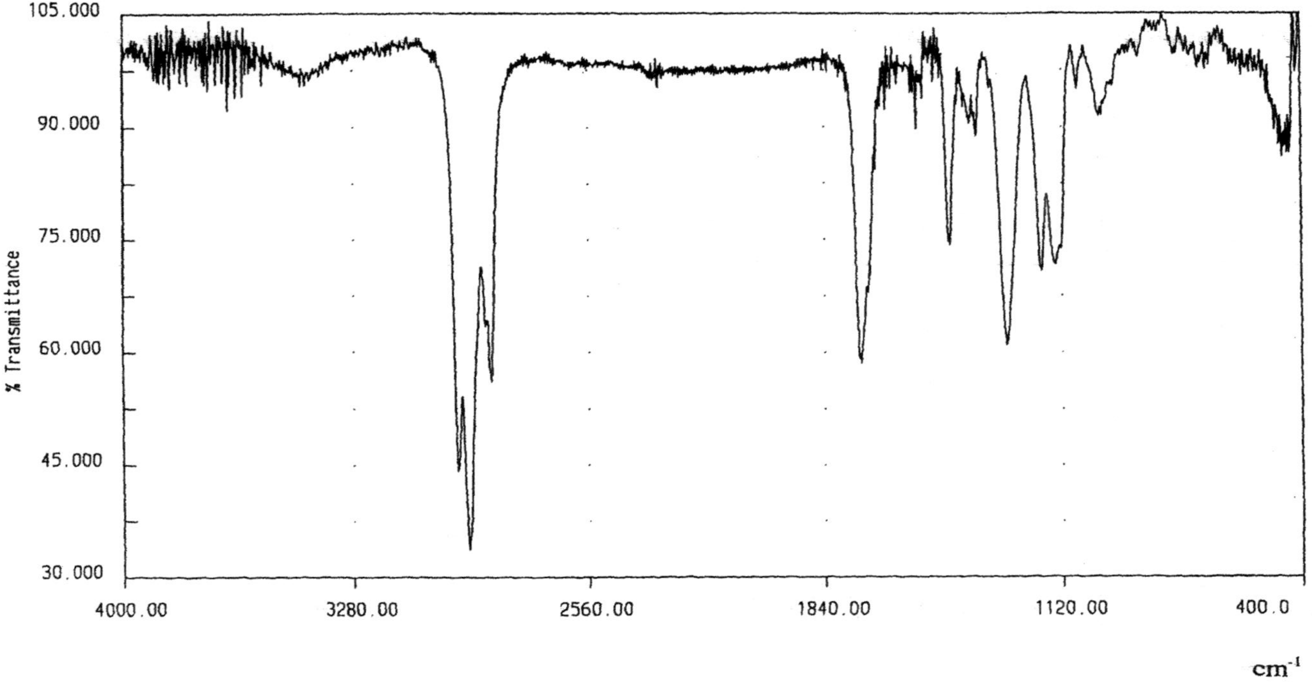

Name : DI-N-OCTYLTIN BIS(2-ETHYLHEXYLMERCAPTOACETATE) + MONO-N-OCTTYLTIN TRIS(2-ETHYLHEXYLMERCAPTOACETATE)

PM Nr : 50320 + 67680 CAS Nr : 15571-58-1 / 27107-89-7

DI-N-OCTYLTIN BIS(2-ETHYLHEXYLMERCAPTOACETATE)
+ MONO-N-OCTYLTIN TRIS(2-ETHYLHEXYLMERCAPTOACETATE)

Peak wavenumber	% T	Relative intensity
434.0	86.5	21.3
464.9	87.6	19.6
1130.4	73.6	41.6
1182.5	70.7	46.2
1290.5	60.8	61.8
1400.5	90.3	15.3
1460.3	74.1	40.9
1730.4	58.3	65.8
2855.0	56.0	69.4
2926.4	36.6	100.0
2961.1	44.1	88.2

Di-n-octyltin bis(2-ethylhexyl mercaptoacetate) and Mono-n-octyltin tris(2-ethylhexyl mercaptoacetate)

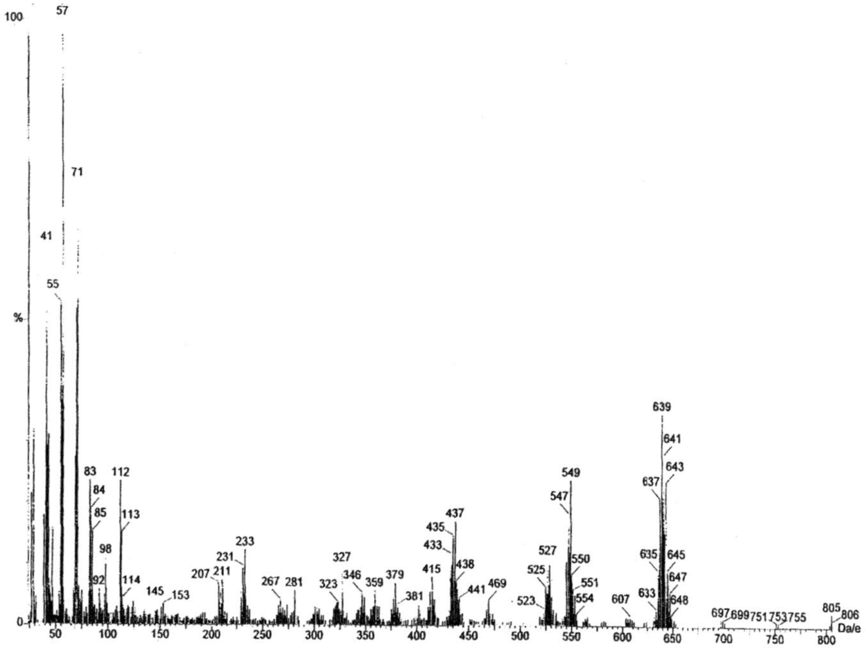

M/Z	Ion Intensity (%)	M/Z	Ion Intensity (%)
27	21.4	71	73.2
29	31.9	83	23.5
41	62.5	84	18.9
42	29.3	112	23.4
43	31.1	549	23.8
55	53.0	637	21.0
56	43.7	639	34.9
57	100.0	640	21.2
69	27.4	641	28.1
70	52.0	643	23.5

Spectrometer	: VG Autospec
Inletsystem	: Probe inlet
Source Temperature	: 350-400°C
Electron Energy	: 70 eV
Scan Range	: 25-1000

Octyl thioglycolate

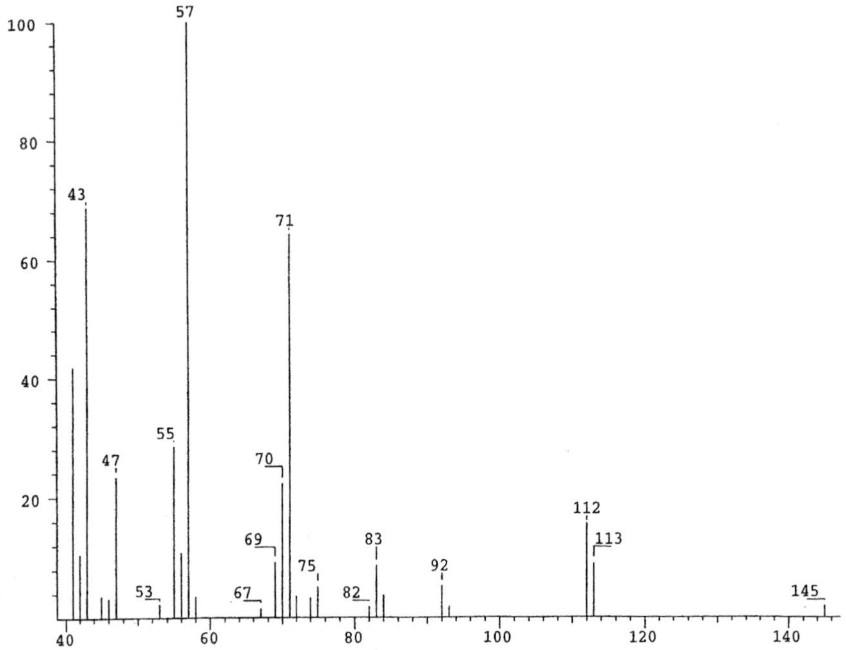

M/Z	Ion Intensity (%)	M/Z	Ion Intensity (%)
41	41.8	70	22.2
42	10.6	71	64.2
43	68.8	83	8.7
47	23.3	84	3.7
55	28.4	92	5.3
56	10.8	112	15.7
57	100.0	113	9.0
69	9.2		

Spectrometer	: Finnigan Mat SSQ 700
Inletsystem	: Capillary GC/MS
Source Temperature	: 180°C
Electron Energy	: 70 eV
Scan Range	: 40-700

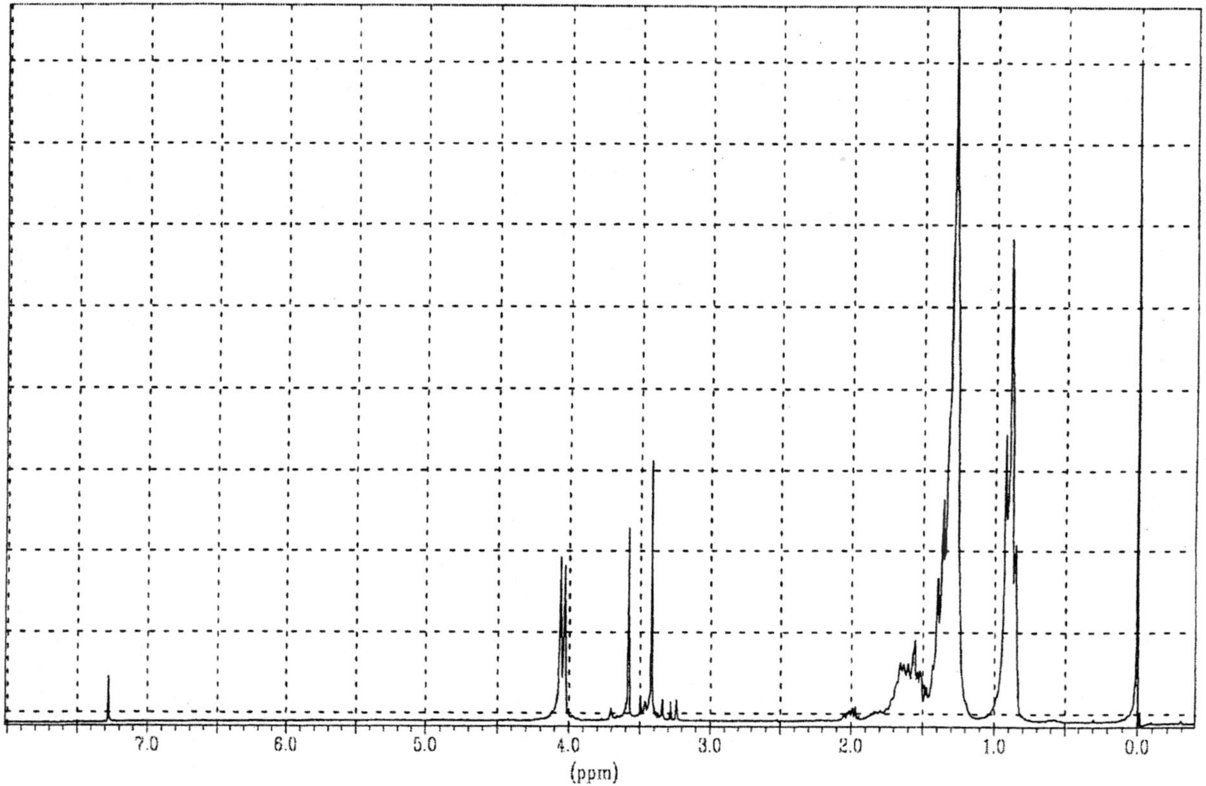
7.0
6.0
5.0
4.0
3.0
2.0
1.0
0.0
(ppm)

Di-n-octyltin bis(isooctyl mercaptoacetate) and

Mono-n-octyltin tris(isooctyl mercaptoacetate)

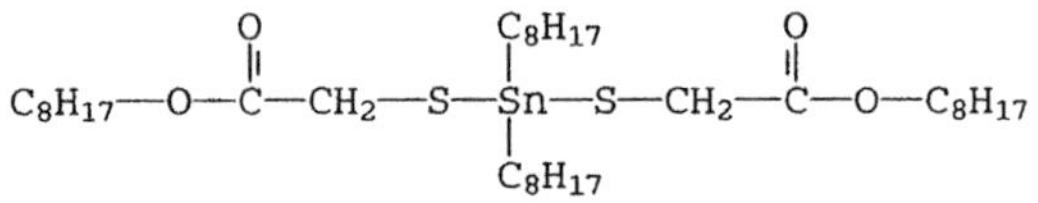

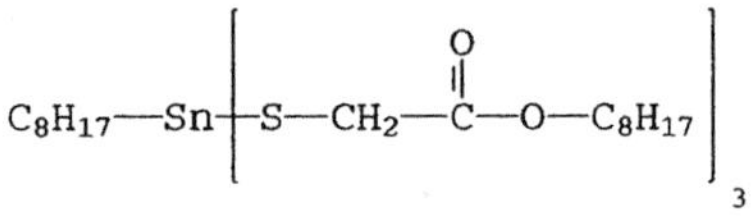

CAS No.	- 26401-97-8 and 26401-86-5
PM Ref. No.	- 50480 and 67760
Restrictions	- SML = 0.02 mg/kg for all di-n-octyltin derivatives. - SML = 1.2 mg/kg for all mono-n-octyltin derivatives.
Formula	- $C_{36}H_{72}O_4S_2Sn$ and $C_{38}H_{74}O_6S_3Sn$
Molecular weight	- 750.7 and 840.7
Alternative names	-
Physical Characteristics	- Viscous colourless liquid, mp <0°C, bp > 250°C. - Soluble in toluene, hexane, tetrahydrofuran.
Handling	- Store at room temperature.
Safety	- Irritant.
Availability	- Liquid preparation containing Pm Ref. No. 50480 and 67760.
Current uses	- Stabilizer.
Applications	- PVC.
GC Retention time	- Isomers of octyl thioglycolate cluster 24.6 - 26.4 min.

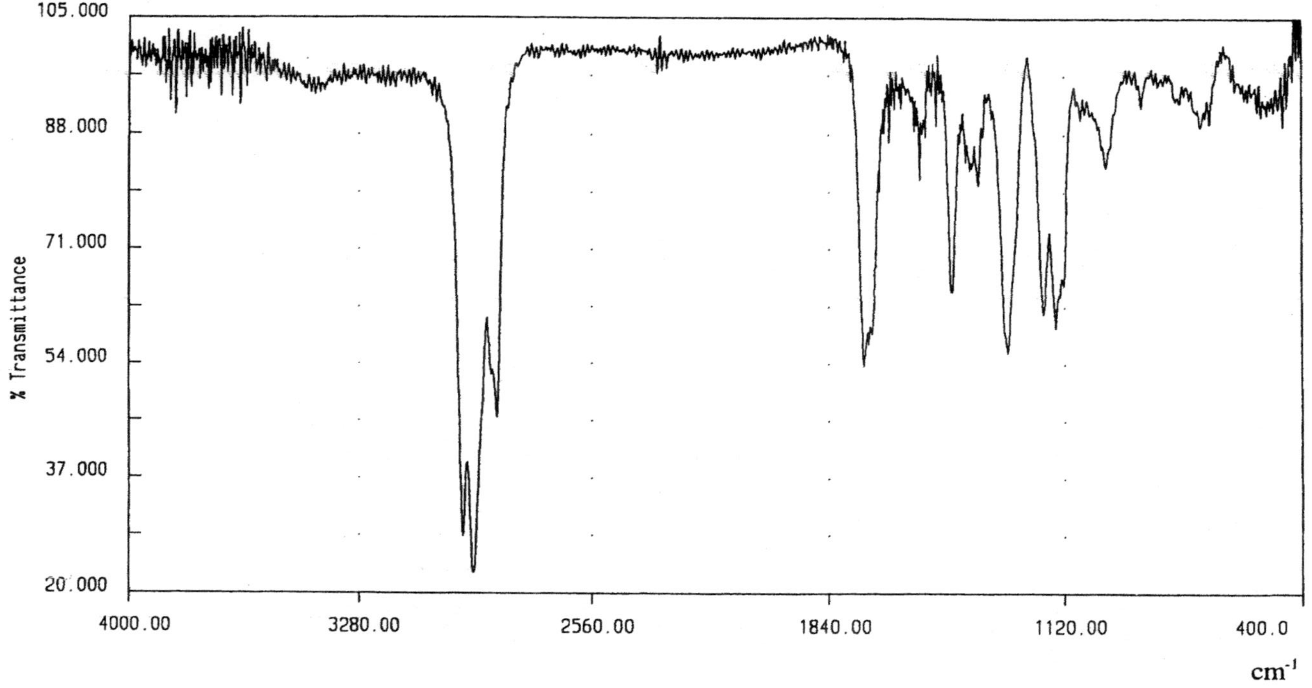

Name : DI- N-OCTYLTIN BIS(ISOOCTYLMERCAPTOACETATE) + MONO-N-OCTYLTIN TRIS(ISOOCTYLMERCAPTOACETATE)

PM Nr : 50480 + 67760 CAS Nr : 26401-97-8 / 26401-86-5

KBr pellet

CAS No. 26401-97-8 and 26401-86-5 PM Ref. No. 50480 and 67760

DI-N-OCTYLTIN BIS(ISOOCTYLMERCAPTOACETATE)
+ MONO-N-OCTYLTIN TRIS(ISOOCTYLMERCAPTOACETATE)

Peak wavenumber	% T	Relative intensity
443.7	92.0	86.0
455.3	89.1	14.1
470.7	91.7	10.8
484.2	91.3	11.3
679.0	89.5	13.6
991.5	83.1	21.9
1070.6	90.1	12.8
1122.7	65.6	44.6
1047.8	59.3	52.8
1184.4	61.4	50.1
1292.5	55.6	57.6
4367.7	87.5	16.2
1381.2	80.4	25.4
1466.1	64.7	45.8
1541.3	87.5	16.2
1578.0	88.4	15.0
1709.1	58.4	54.0
1437.2	53.6	60.2
2855.0	45.8	70.3
2928.3	22.9	100.0
2959.2	28.3	93.0
3856.2	90.7	12.1
3454.9	90.7	12.1

Di-n-octyltin bis(isooctyl mercaptoacetate) and Mono-n-octyltin tris(isooctyl mercaptoacetate)

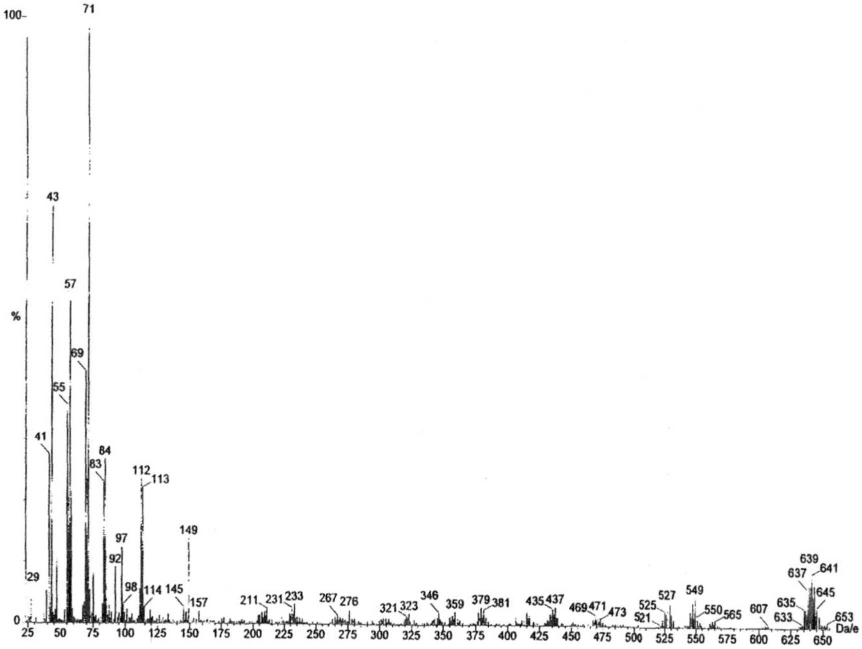

M/Z	Ion Intensity (%)	M/Z	Ion Intensity (%)
41	27.5	71	100.0
42	9.3	83	23.0
43	69.0	84	26.8
47	12.0	85	11.8
55	35.8	92	9.1
56	20.4	97	12.5
57	54.5	112	23.6
58	22.7	113	22.1
69	41.4	149	13.9
70	23.6	639	10.3

Spectrometer	: VG Autospec
Inletsystem	: Probe inlet
Source Temperature	: 350-400°C
Electron Energy	: 70 eV
Scan Range	: 25-1000

Isomers of octyl thioglycolate

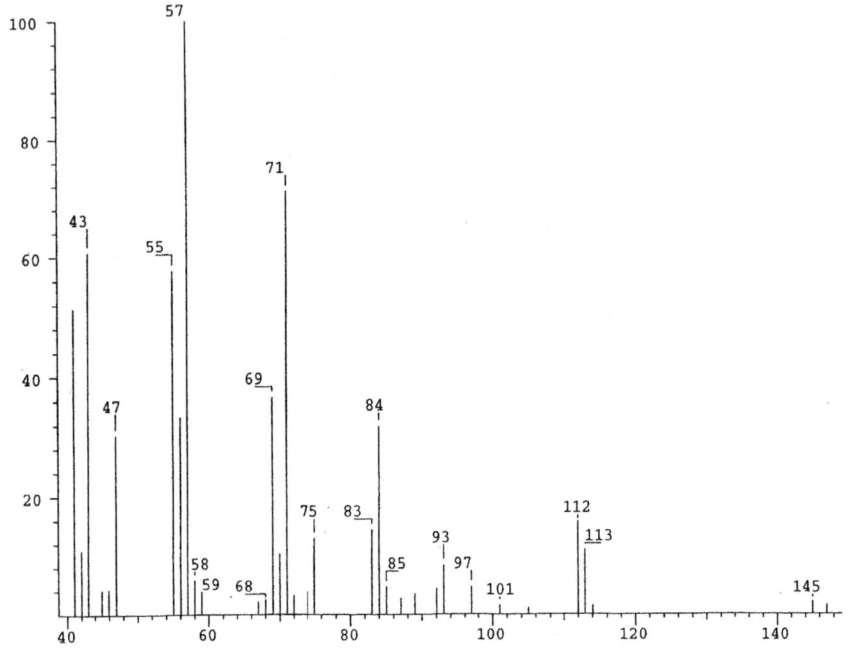

M/Z	Ion Intensity (%)	M/Z	Ion Intensity (%)
41	51.3	70	10.2
42	10.9	71	71.1
43	60.8	75	12.9
47	30.3	83	14.3
55	57.7	84	31.7
56	33.3	112	15.7
57	100.0	113	10.9
69	36.6		

Spectrometer	: Finnigan Mat SSQ 700
Inletsystem	: Capillary GC/MS
Source Temperature	: 180°C
Electron Energy	: 70 eV
Scan Range	: 40-700

PM n°	CAS n°	Name
50480	26401-97-8	DI-n-OCTYLTIN BIS(ISOOCTYLMERCAPTOACETATE)
67760	26401-86-5	MONO-n-OCTYLTIN TRIS(ISOOCTYLMERCAPTOACETATE)

Solvent	$CDCL_3$
SF	250.13
SI	16384
SW	3125
NS	24

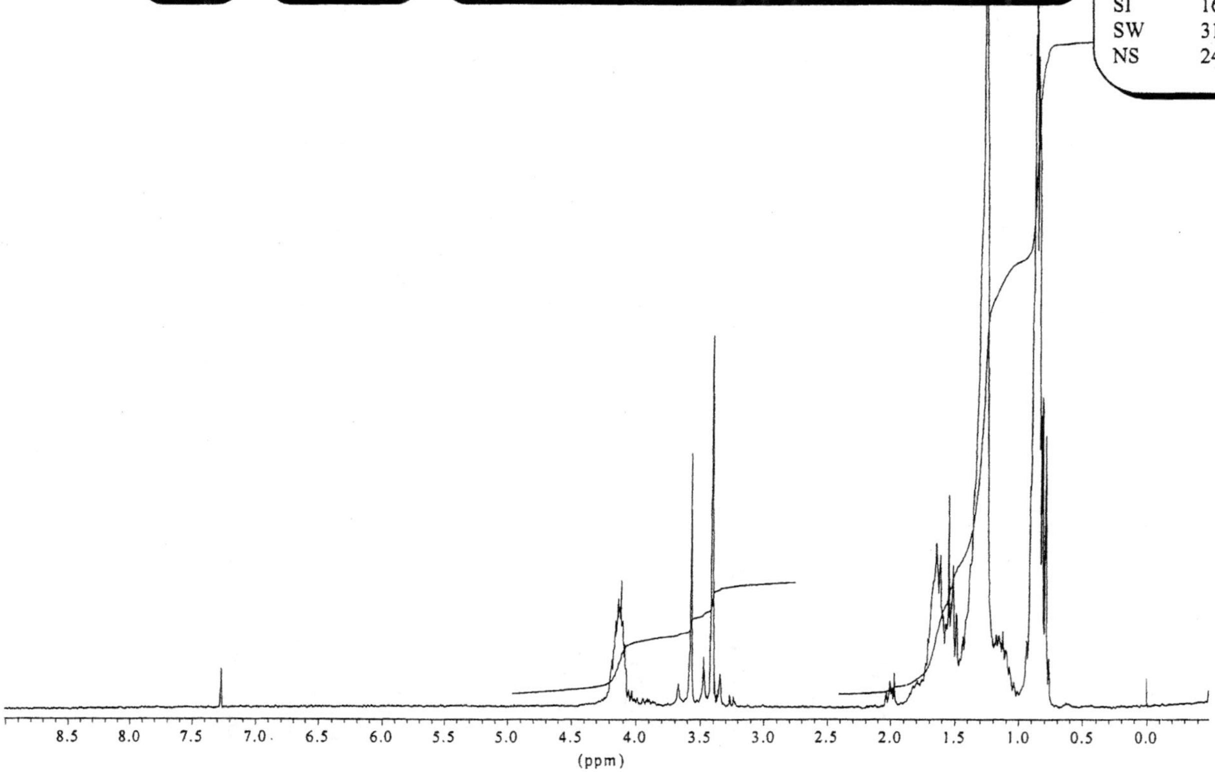

N,N'-Diphenylthiourea

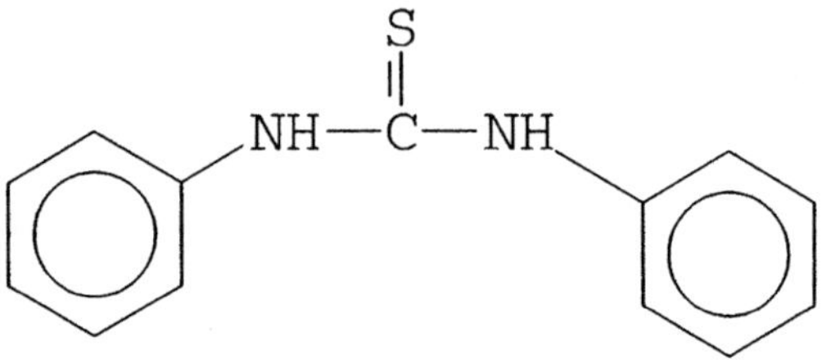

CAS No.	- 00102-08-9
PM Ref. No.	- 51680
Restrictions	- SML = 3.0 mg/kg
Formula	- $C_{13}H_{12}N_2S$
Molecular weight	- 228

Alternative names	- Thiocarbanilide
Physical Characteristics	- White crystalline powder, mp 152 - 155°C. - Soluble in alcohol, ether, chloroform.
Handling	- Store at room temperature.
Safety	- Toxic.
Availability	- Standard sample was supplied.
Current uses	- Vulcanizing, stabilizer.
Applications	- PVC, PVDC, rubber.
GC Retention time	- Degradation products 11.8, 19.1, 33.3 min.

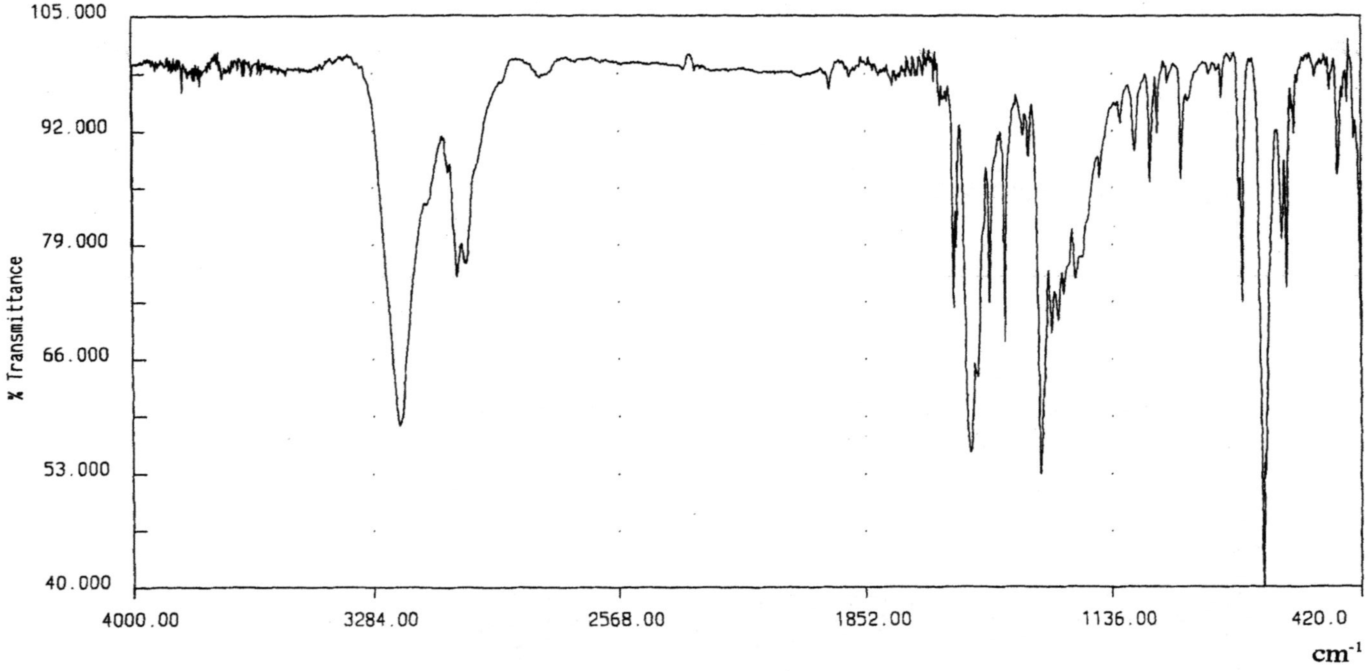

Name : N,N-DIPHENYLTHIOUREA

PM Nr : 51680 CAS Nr : 00120-08-9

KBr pellet

CAS No. 00102-08-9 PM Ref. No. 51680

N,N'DIPHENYLTHIOUREA

Peak wavenumber	% T	Relative intensity
418,6	71,9	35,5
486,1	75,1	31,4
611,5	79,7	25,6
630,8	53,7	58,5
644,3	61,5	48,6
698,3	20,8	100,0
758,1	53,9	58,2
933,7	76,3	29,9
1003,1	78,8	26,8
1024,3	74,0	32,8
1070,6	73,1	34,0
1136,2	83,5	20,8
1172,9	76,0	30,3
1223,0	65,5	43,6
1244,2	60,4	50,0
1294,4	50,4	62,6
1315,6	53,4	58,9
1344,6	38,8	77,3
1495,0	62,1	47,9
1529,7	48,6	64,9
1551,0	43,6	71,2
1591,5	70,5	37,3
1599,2	62,7	47,1
2905,2	86,5	17,1
3013,2	59,1	51,7
3036,3	57,1	54,2
3125,1	67,8	40,7
3206,1	40,8	74,8

N,N'-Diphenylthiourea

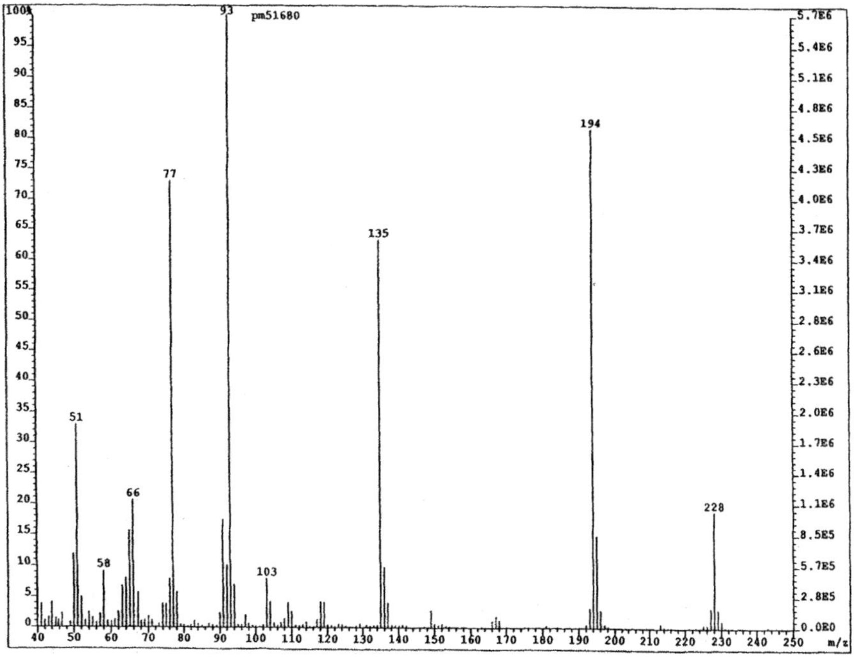

M/Z	Ion Intensity (%)	M/Z	Ion Intensity (%)
50	12.3	78	6.1
51	33.1	91	17.8
58	9.2	92	10.4
63	7.4	93	100.0
64	8.0	94	7.4
65	16	135	63.8
66	20.9	136	10.4
67	6.1	194	82.2
76	8.6	195	15.3
77	73.6	228	19.0

Spectrometer	: VG Autospec
Inletsystem	: Flow injection particle beam MS
Source Temperature	: 250°C
Electron Energy	: 70 eV
Scan Range	: 40-1000

Degradation product

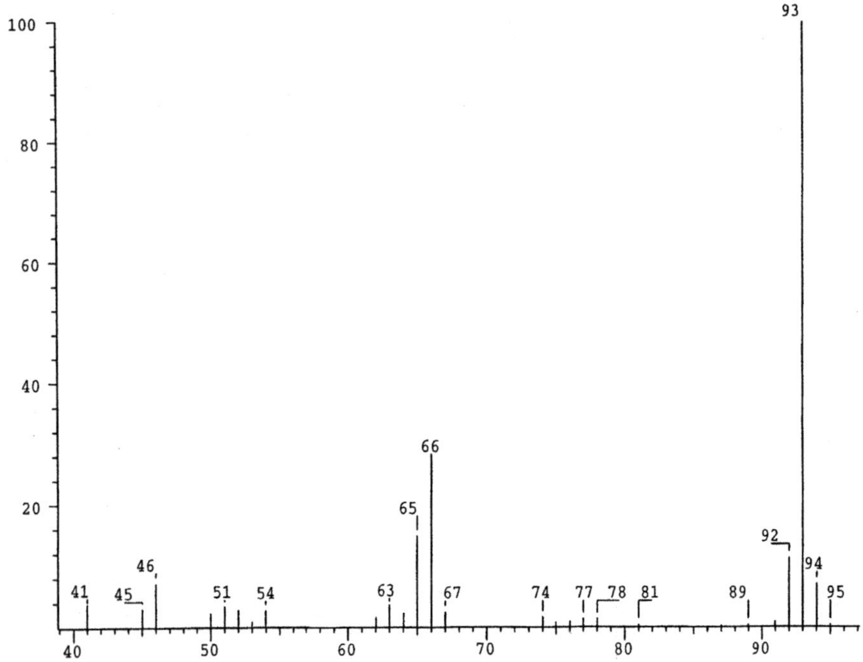

M/Z	Ion Intensity (%)	M/Z	Ion Intensity (%)
41	3.8	66	27.2
45	3.1	92	11.2
46	7.2	93	100.0
51	3.4	94	7.0
63	3.6	95	0.4
65	14.8		

Spectrometer	: Finnigan Mat SSQ 700
Inletsystem	: Capillary GC/MS
Source Temperature	: 180°C
Electron Energy	: 70 eV
Scan Range	: 40-500

Degradation product

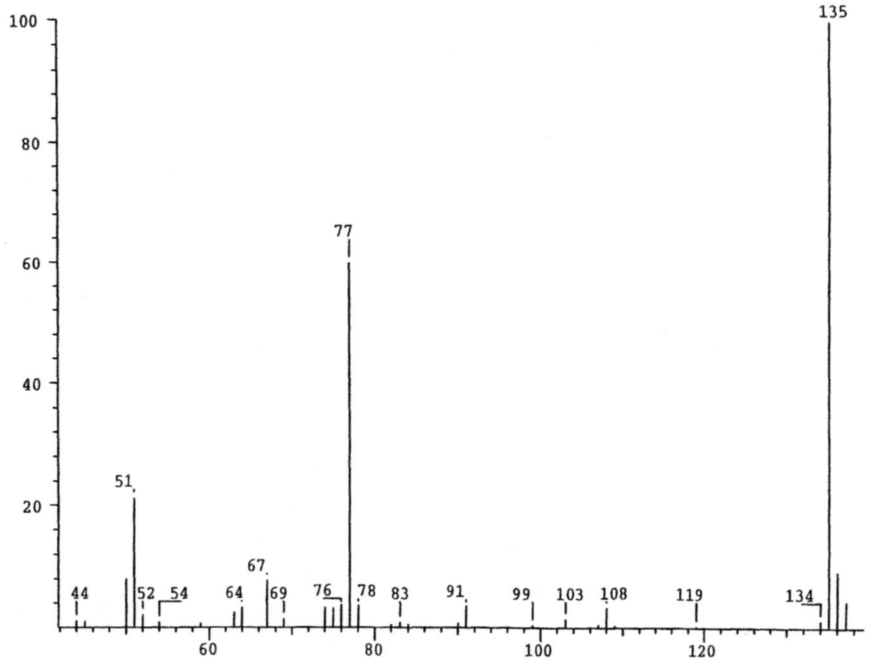

M/Z	Ion Intensity (%)	M/Z	Ion Intensity (%)
50	8.0	78	3.6
51	21.1	91	3.7
67	7.7	103	1.5
74	3.2	135	100.0
75	3.2	136	9.3
76	3.7	137	4.4
77	59.9		

Spectrometer	: Finnigan Mat SSQ 700
Inletsystem	: Capillary GC/MS
Source Temperature	: 180°C
Electron Energy	: 70 eV
Scan Range	: 40-500

Degradation product

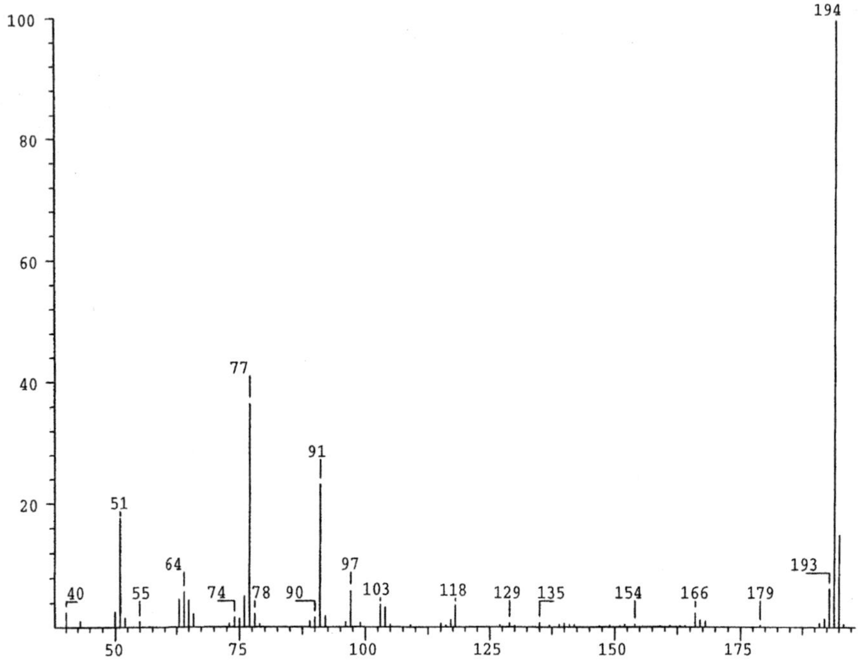

M/Z	Ion Intensity (%)	M/Z	Ion Intensity (%)
51	17.7	97	6.0
63	4.6	118	3.7
64	5.7	193	6.3
76	5.1	194	100.0
77	36.5	195	15.0
91	23.1	196	0.6

Spectrometer : Finnigan Mat SSQ 700
Inletsystem : Capillary GC/MS
Source Temperature : 180°C
Electron Energy : 70 eV
Scan Range : 40-500

PM n°	CAS n°	Name
51680	00102-08-9	N,N'DIPHENYLTHIOUREA

Solvent	$CDCl_3$
SF	250.13
SI	16384
SW	3125
NS	16

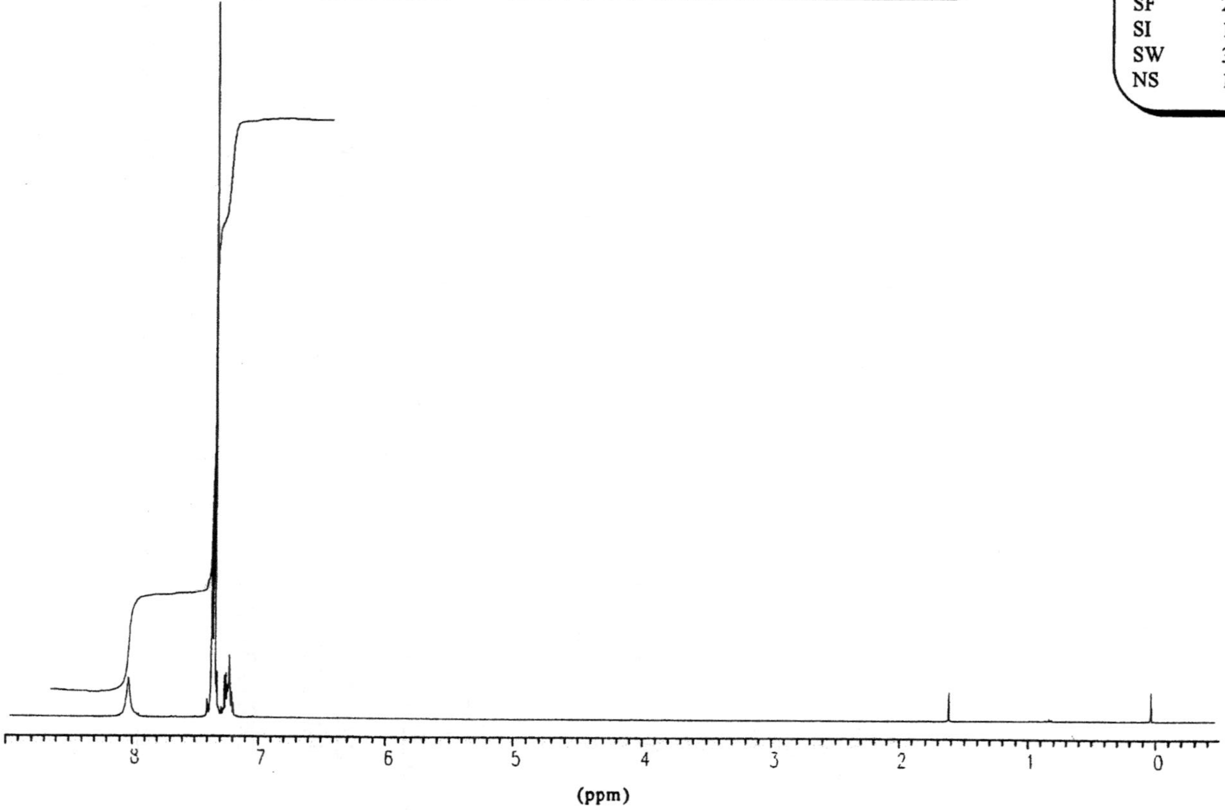

Dodecylbenzenesulphonic acid

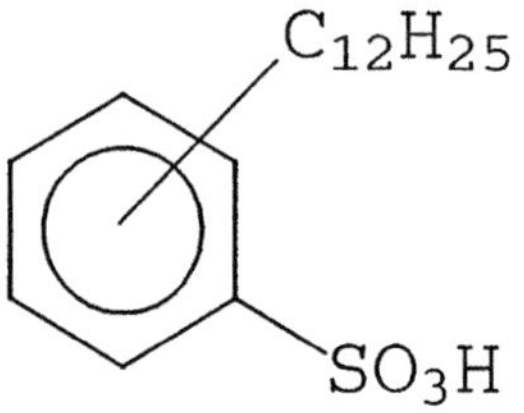

CAS No.	- 27176-87-0
PM Ref. No.	- 52000
Restrictions	- SML = 30 mg/kg
Formula	- $C_{18}H_{30}O_3S$
Molecular weight	- 326

Alternative names	- Laurylbenzenesulfonic acid
Physical Characteristics	- Viscous brownish liquid, mp 10°C, bp 315°C. - Soluble in water, alcohol, ether.
Handling	- Store at room temperature.
Safety	- Hygroscopic.
Availability	- Standard sample was supplied.
Current uses	- Antistatic agent, emulsifier.
Applications	- General use.
GC Retention time	-
Note	- Current use as Dodecylbenzene sulphonic acid, Na-salt.

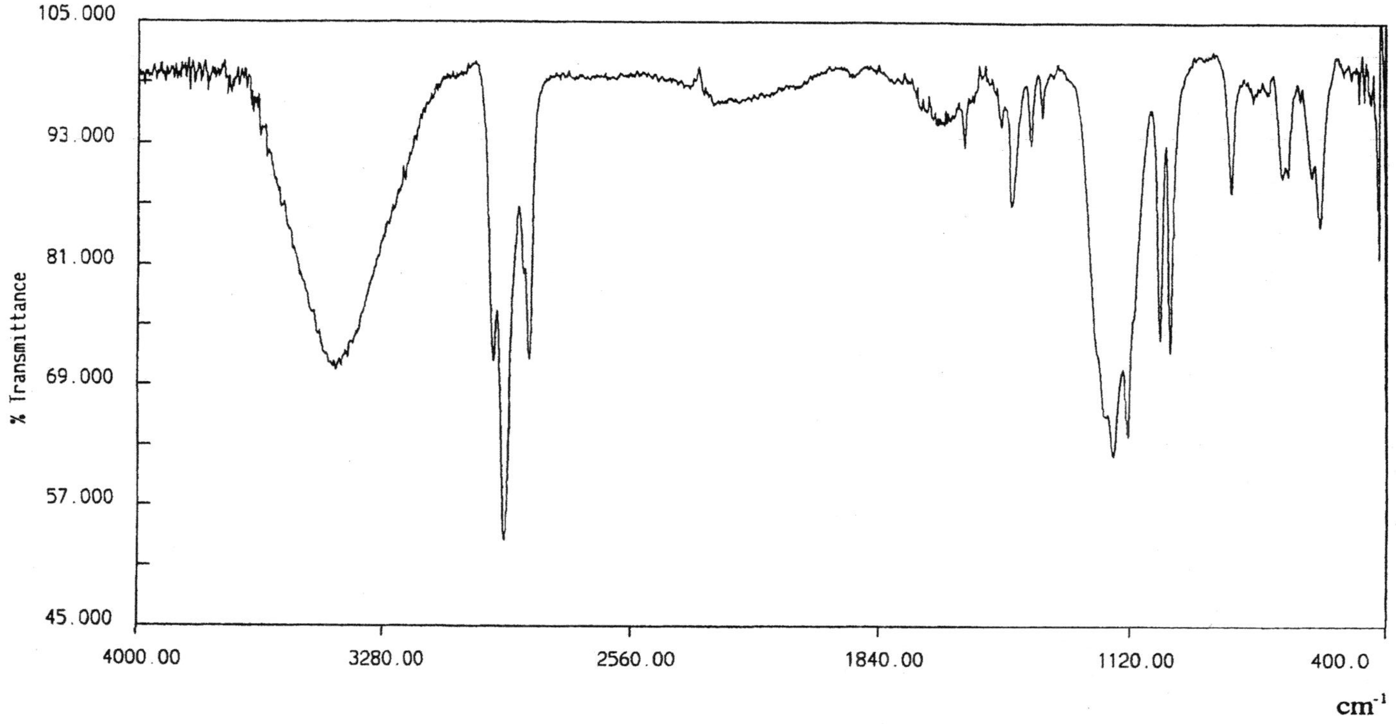

Name : DODECYLBENZENESULPHONIC ACID

PM Nr : 52000 CAS Nr : 27176-87-0

KBr pellet

CAS No. 27176-87-0 PM Ref. No. 52000

DODECYLBENZENE SULPHONIC ACID

Peak wavenumber	% T	Relative intensity
420,5	81,8	39,1
584,5	84,9	32,5
1008,9	72,3	59,6
1037,8	73,5	57,0
1130,4	63,8	77,9
1172,9	61,9	82,0
2926,4	53,5	100,0
2957,2	71,2	61,9
3213,8	89,3	23,0
3250,5	86,2	29,7
3372,0	73,0	58,1

PM n°	CAS n°	Name
52000	27176-87-0	DODECYLBENZENE SULPHONIC ACID

Solvent	$CDCl_3$
SF	250.13
SI	16384
SW	4000
NS	32

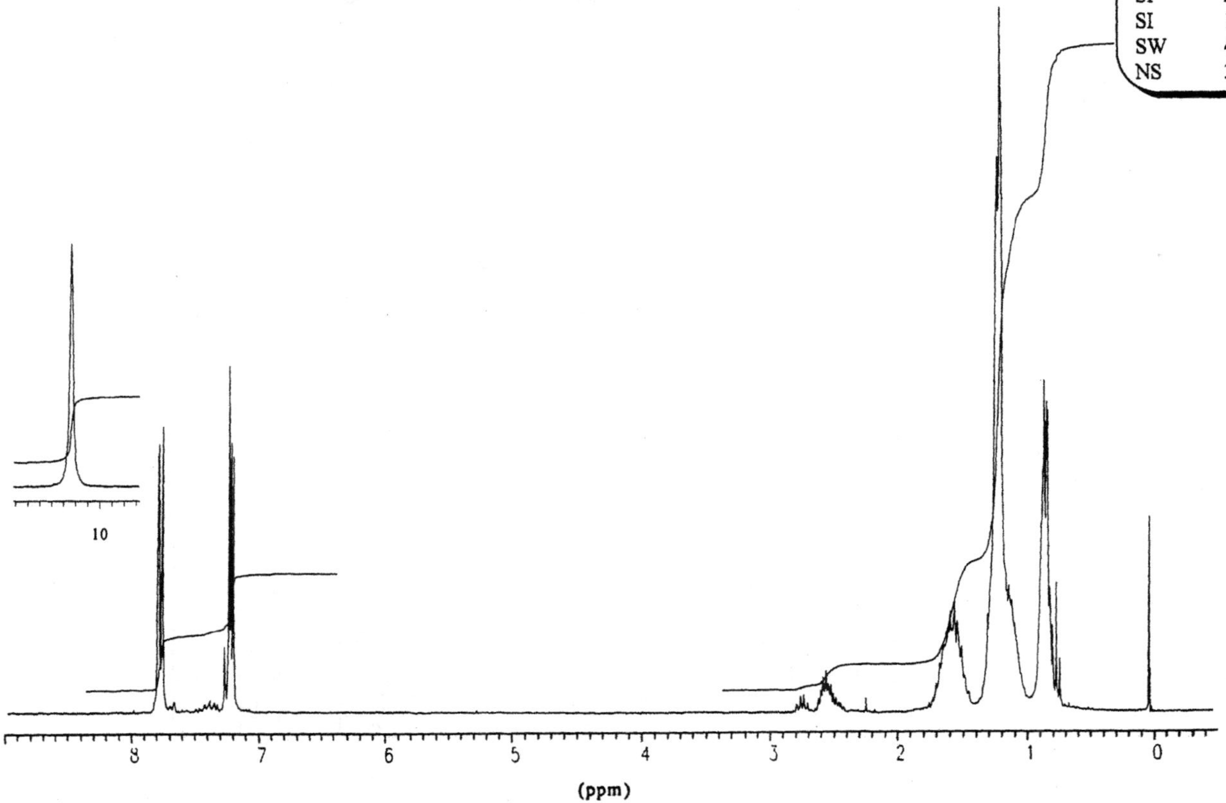

Erucamide

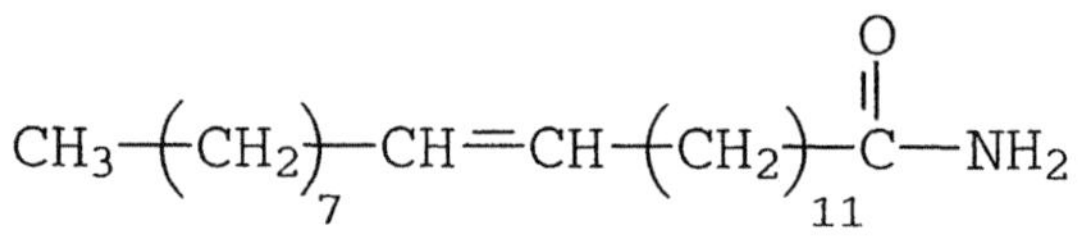

CAS No.	- 00112-84-5
PM Ref. No.	- 52720
Restrictions	- None
Formula	- $C_{22}H_{43}NO$
Molecular weight	- 337

Alternative names - Cis-13-docosenoamide

Physical Characteristics
- White flakes or beads, mp 75 - 80°C, bp decomposes.
- Soluble in isopropanol, slightly in alcohol, acetone.

Handling - Store at room temperature.

Safety -

Availability - Standard sample was supplied.

Current uses - Slip and anti-blocking agent.

Applications - PVC, polyolefins.

GC Retention time - 50.4 min.

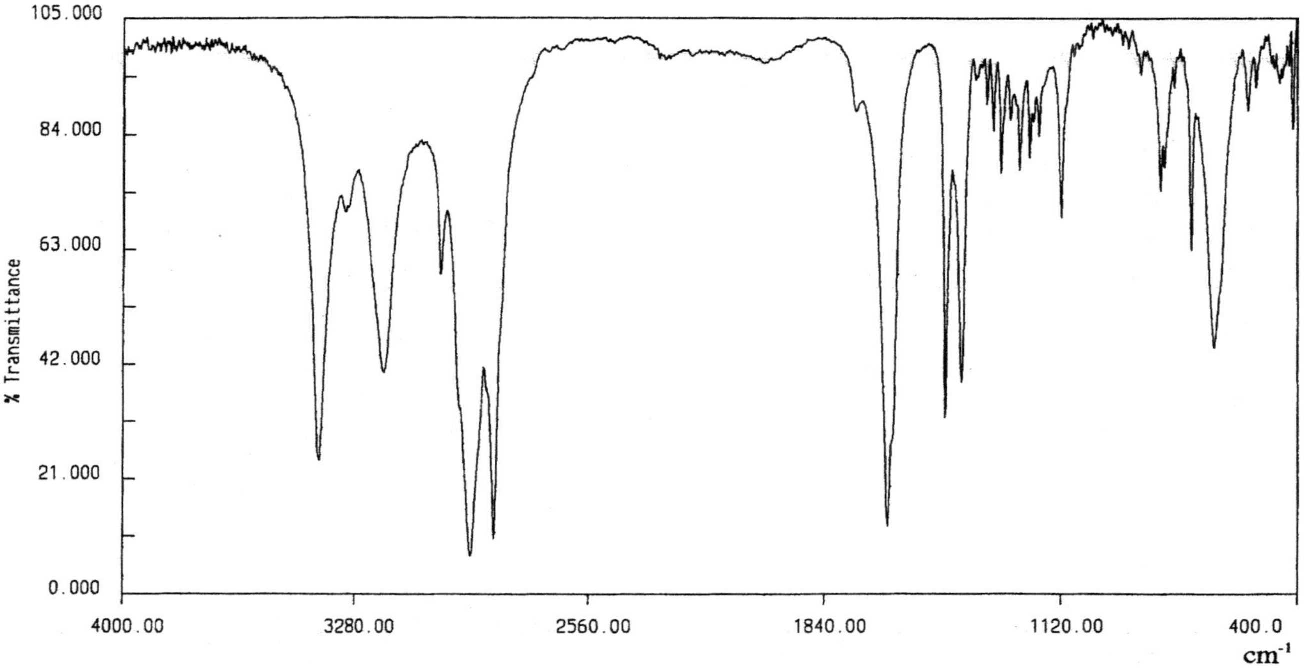

Name : ERUCAMIDE

PM Nr : 52720

CAS Nr : 00112-84-5

KBr pellet

CAS No. 00112-84-5 PM Ref. No. 52720

ERUCAMIDE

Peak wavenumber	% T	Relative intensity
412.8	85.0	16.1
451.4	93.3	7.2
522.8	92.6	7.9
545.9	88.3	12.5
648.2	45.1	59.0
721.5	62.8	39.9
804.4	78.1	23.5
816.0	73.7	28.2
1118.9	68.8	33.5
1205.7	86.3	14.7
1215.3	79.7	21.8
1246.2	77.5	24.2
1273.2	86.6	14.4
1302.1	76.9	24.8
1325.3	84.6	16.6
1344.6	89.4	11.3
1419.8	38.5	66.1
1469.9	32.11	72.9
1645.5	12.38	94.1
2849.2	10.09	96.6
2918.7	6.89	100.0
3011.3	58.54	44.5
3186.8	40.44	64.0
3312.2	69.92	32.3
3393.2	24.47	81.1

Erucamide

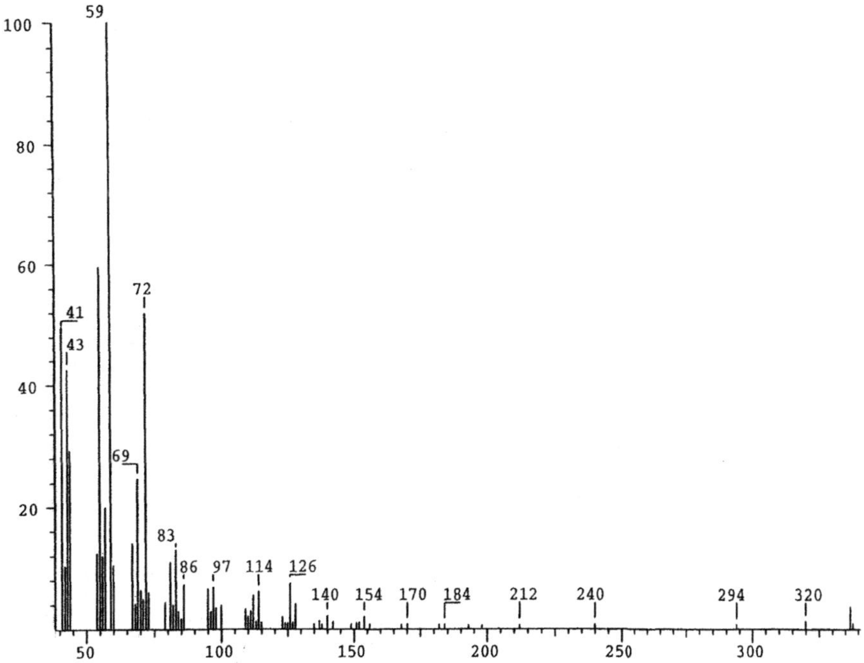

M/Z	Ion Intensity (%)	M/Z	Ion Intensity (%)
41	49.8	69	24.7
42	10.3	72	51.9
43	42.5	81	11.0
44	29.2	83	13.0
55	59.6	86	7.3
56	12.0	114	6.2
57	20.0	126	7.5
59	100.0	320	1.3
60	10.4	337	3.7
67	14.1	338	0.9

Spectrometer	: Finnigan Mat SSQ 700
Inletsystem	: Capillary GC/MS
Source Temperature	: 180°C
Electron Energy	: 70 eV
Scan Range	: 40-500

PM n°	CAS n°	Name
52720	00112-84-5	ERUCAMIDE

Solvent	$CDCL_3$
SF	250.13
SI	16384
SW	3125
NS	32

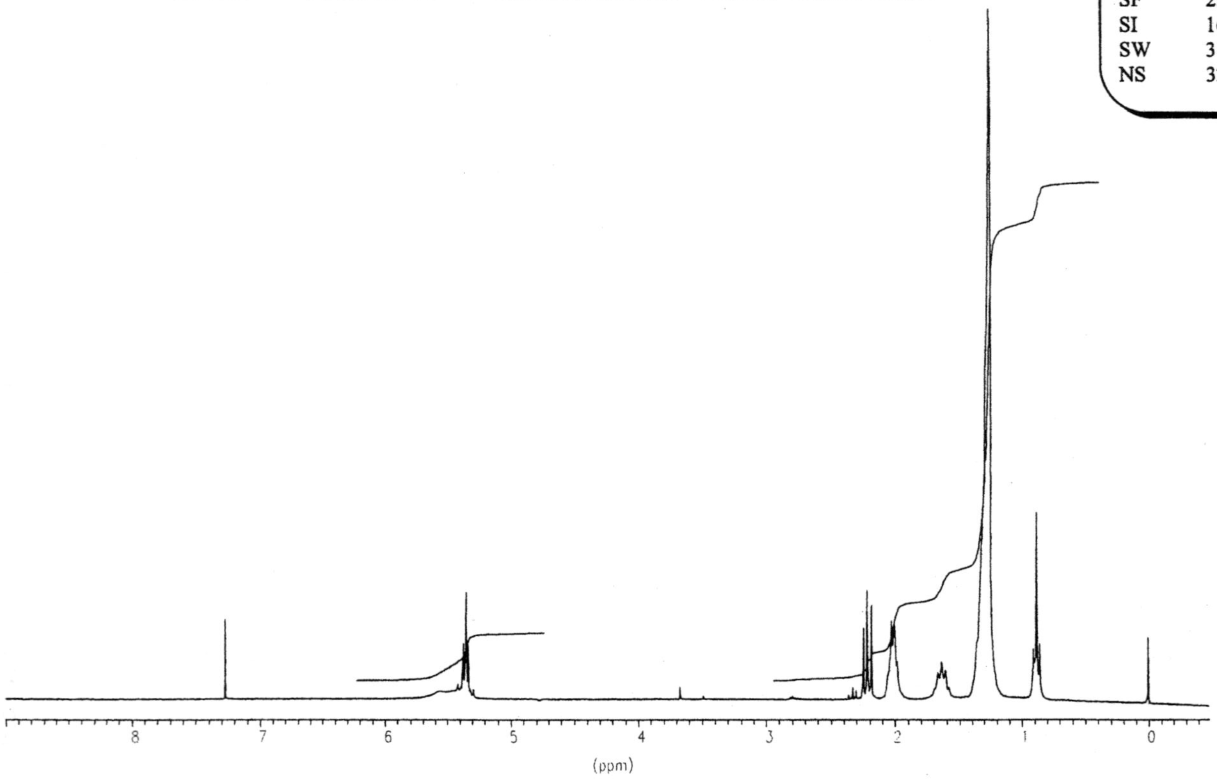

2-Ethoxy-2'-ethyloxanilide

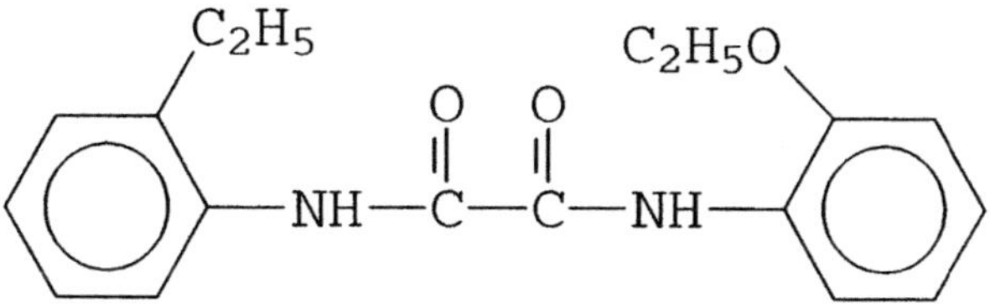

CAS No.	- 23949-66-8
PM Ref. No.	- 53200
Restrictions	- SML = 30 mg/kg
Formula	- $C_{18}H_{20}N_2O_3$
Molecular weight	- 312

Alternative names	- 2-ethoxy-2'-ethyl-oxalic acid-bis anilide
Physical Characteristics	- White to slightly yellow powder, mp 124 -127°C, bp > 400°C. - Soluble in acetone, chloroform, dichloromethane.
Handling	- Store at room temperature.
Safety	-
Availability	- Standard sample was supplied.
Current uses	- UV-stabilizer.
Applications	- PVC, PMMA, PE.
GC Retention time	- 48.1 min.

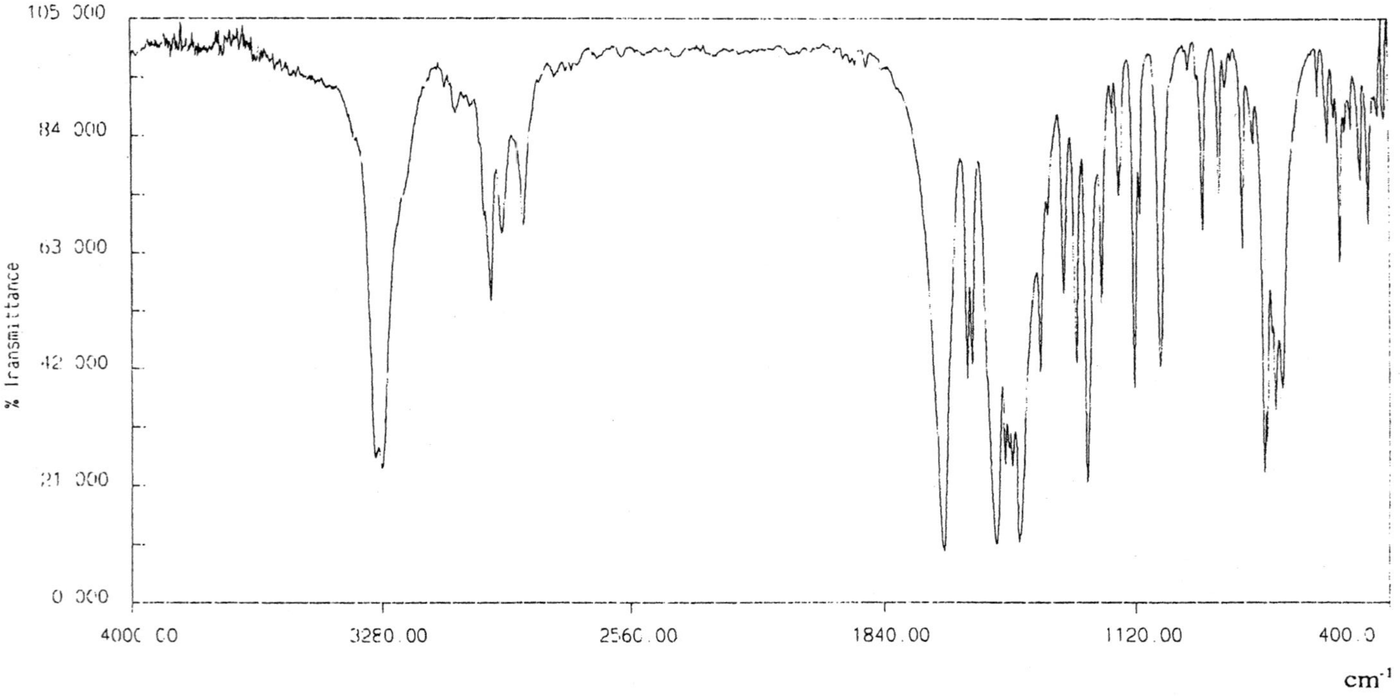

Name : 2-ETHOXY-2'-ETHYLOXANILIDE

PM Nr : 53200 CAS Nr : 23949-66-8

KBr pellet

CAS No. 23949-66-8 PM Ref. No. 53200

2-ETHOXY-2'-ETHYLOXANILIDE

Peak wavenumber	% T	Relative intensity
414,8	87,1	14,2
430,2	87,6	13,7
459,1	68,0	35,2
480,3	76,1	26,3
507,3	85,2	16,3
524,7	84,8	16,7
540,1	61,3	42,6
572,9	82,7	19,0
702,2	38,6	67,6
721,5	34,6	72,0
754,3	23,1	84,7
785,1	82,7	19,0
816,0	64,0	39,6
881,6	73,8	28,8
929,8	66,1	37,3
1049,4	42,5	63,3
1105,4	70,1	32,9
1122,7	38,4	67,8
1167,1	73,4	29,3
1184,4	88,1	13,1
1217,2	54,0	50,6
1259,7	21,5	86,4
1288,6	43,2	62,5
1325,3	55,5	49,0
1369,6	69,8	33,3
1392,8	41,5	64,4
1454,5	10,8	98,2
1471,9	24,4	83,2
1493,1	24,6	83,0
1516,2	10,4	98,6
1587,6	42,6	63,2
1601,1	40,2	65,8
1668,6	9,2	100,0
2870,4	67,9	35,3
2932,2	66,4	37,0
2965,0	54,2	50,4
3065,3	88,1	13,1
3281,3	24,2	83,5

2-Ethoxy-2'-ethyloxanilide

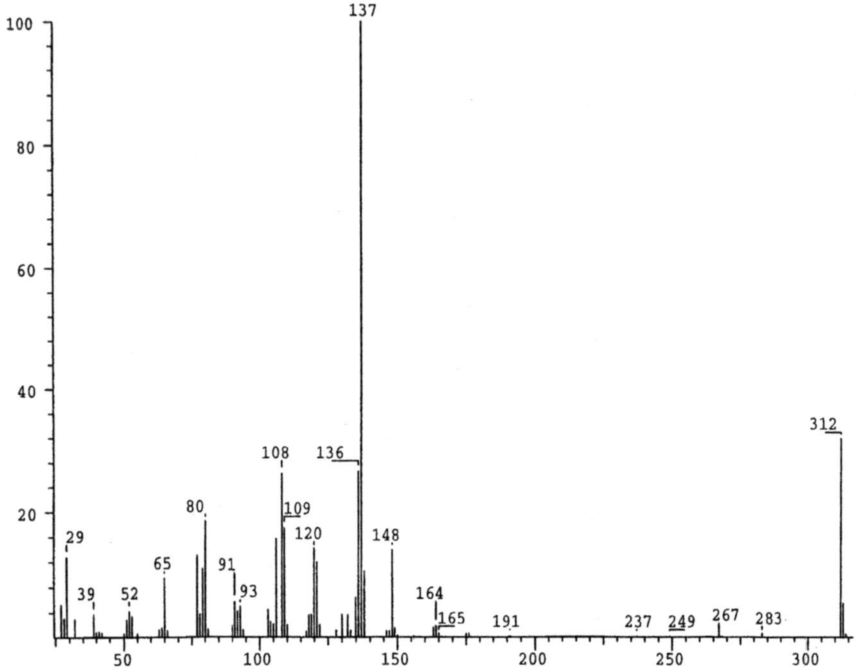

M/Z	Ion Intensity (%)	M/Z	Ion Intensity (%)
29	12.8	120	14.2
65	9.5	121	12.0
77	13.0	136	26.6
79	10.9	137	100.0
80	18.6	138	10.5
106	15.7	148	14.0
108	26.3	312	32.1
109	17.4	313	5.5

Spectrometer	: Finnigan Mat SSQ 700
Inletsystem	: Capillary GC/MS
Source Temperature	: 180°C
Electron Energy	: 70 eV
Scan Range	: 25-450

PM n° 53200

CAS n° 23949-66-8

Name 2-ETHOXY-2'-ETHYLOXANILIDE

Solvent	$CDCL_3$
SF	250.13
SI	16384
SW	3125
NS	55

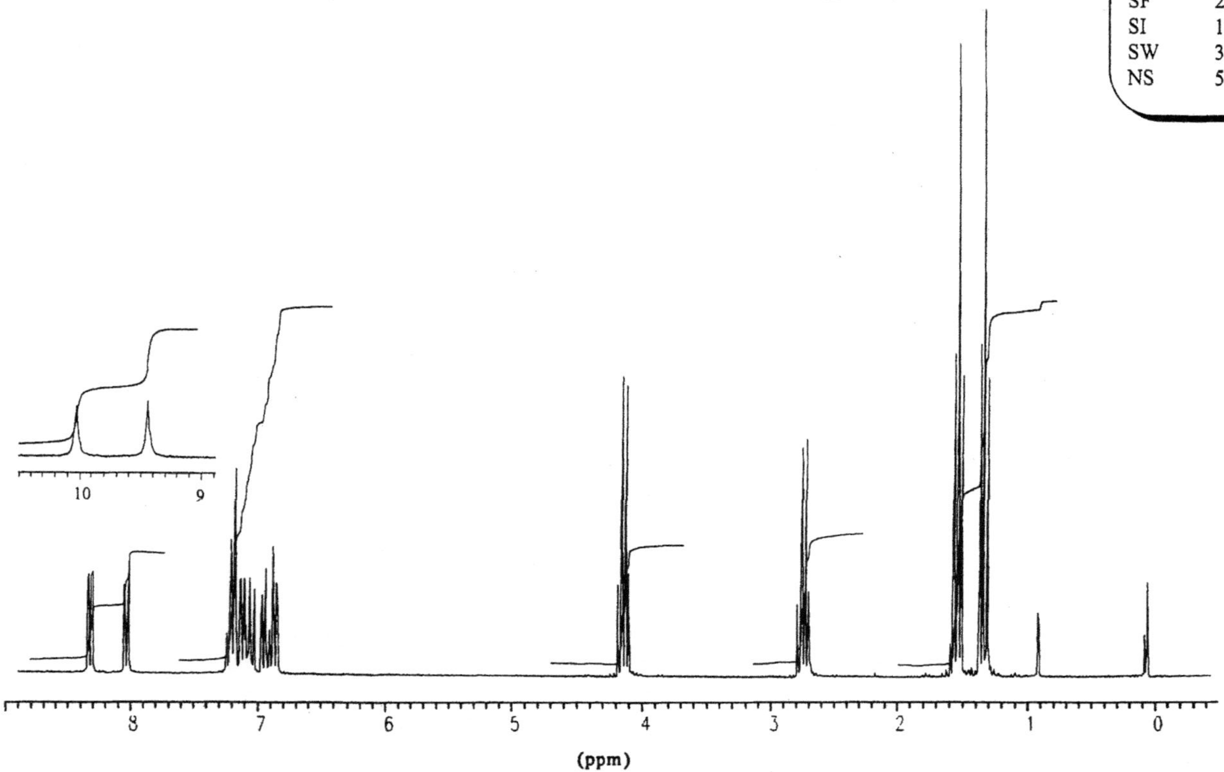

Ethylbenzene

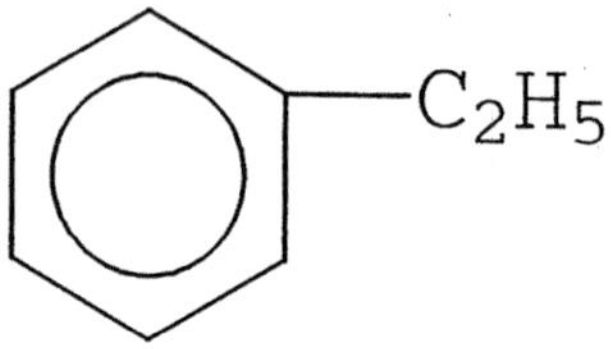

CAS No.	- 00100-41-4
PM Ref. No.	- 53255
Restrictions	- SML = 0.6 mg/kg
Formula	- C_8H_{10}
Molecular weight	- 106

Alternative names - Ethylbenzol

Physical Characteristics - Colourless liquid, mp -95°C, bp 136°C.
- Soluble in most organic solvents.

Handling - Store at room temperature.
Safety - Harmful.
Availability - Standard sample was supplied.

Current uses - Solvent.
Applications - PS.

GC Retention time - 8.3 min.

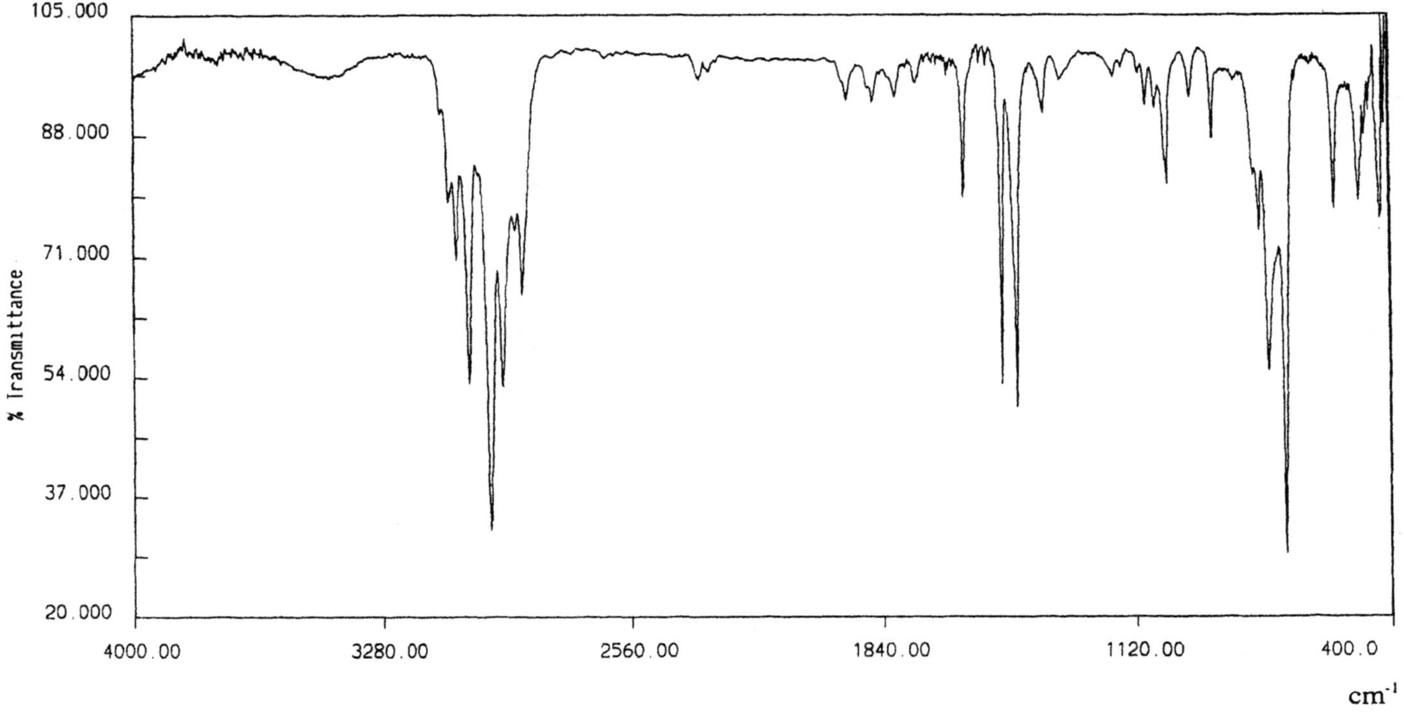

Name : ETHYLBENZENE

PM Nr : 53255 CAS Nr : 00100-41-4

KBr pellet

CAS No. 00100-41-4 PM Ref. No. 53255

ETHYLBENZENE

Peak wavenumber	% T	Relative intensity
412,8	89,6	14,6
428,3	76,4	33,2
455,3	91,4	12,1
468,8	88,2	16,6
486,1	78,8	29,8
555,6	77,8	31,2
696,4	28,8	100,0
746,5	55,3	62,8
771,6	74,7	35,5
1062,9	91,9	11,4
1375,4	91,2	12,4
1454,5	49,8	70,5
1496,9	53,1	65,9
1605,0	79,3	29,1
2874,3	65,6	48,3
2934,1	52,8	66,3
2966,9	32,3	95,1
3028,6	53,1	65,9
3065,3	70,7	41,2

Ethylbenzene

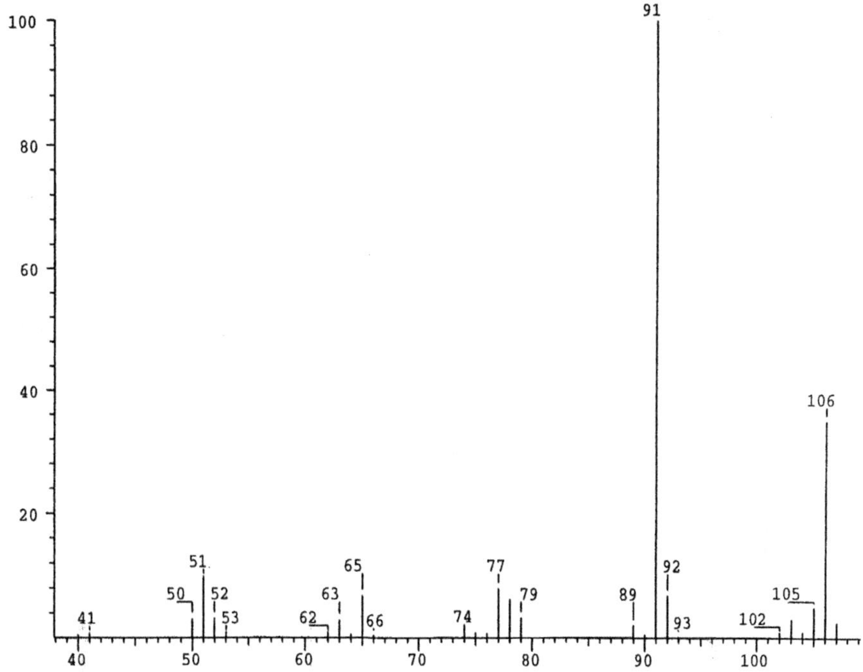

M/Z	Ion Intensity (%)	M/Z	Ion Intensity (%)
50	3.1	89	2.0
51	9.9	91	100.0
52	3.1	92	6.8
63	2.9	103	3.0
65	6.8	104	0.9
74	1.1	105	5.0
77	7.9	106	35.1
78	6.2	107	2.4
79	3.3		

Spectrometer : Finnigan Mat SSQ 700
Inletsystem : Capillary GC/MS
Source Temperature : 180°C
Electron Energy : 70 eV
Scan Range : 40-500

PM n° 53255

CAS n° 00100-41-4

Name ETHYLBENZENE

Solvent	$CDCl_3$
SF	250.13
SI	16384
SW	3125
NS	24

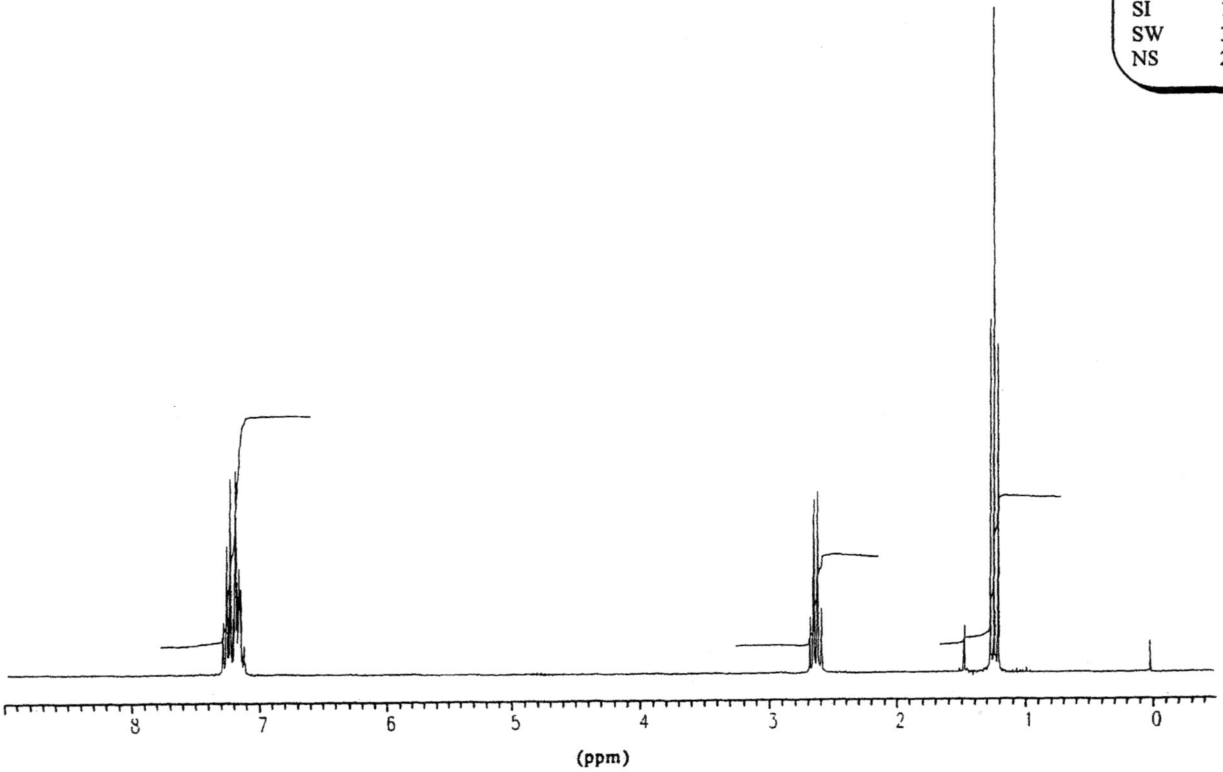

N,N'-Ethylenebisstearamide

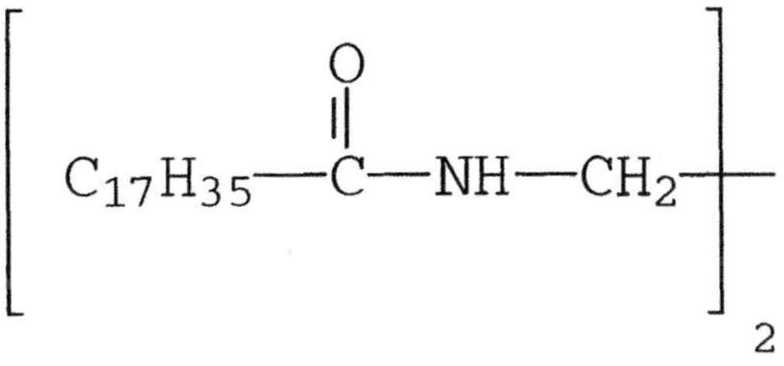

CAS No.	- 00110-30-5
PM Ref. No.	- 53520
Restrictions	- None
Formula	- $C_{38}H_{76}N_2O_2$
Molecular weight	- 592

Alternative names	- Bisstearoylethylenediamine
Physical Characteristics	- Slightly yellow powder, mp 144 - 146°C. - Not soluble.
Handling	- Store at room temperature.
Safety	-
Availability	- Standard sample was supplied.
Current uses	- Antistatic agent, lubricant, release agent.
Applications	- General use.
GC Retention time	-

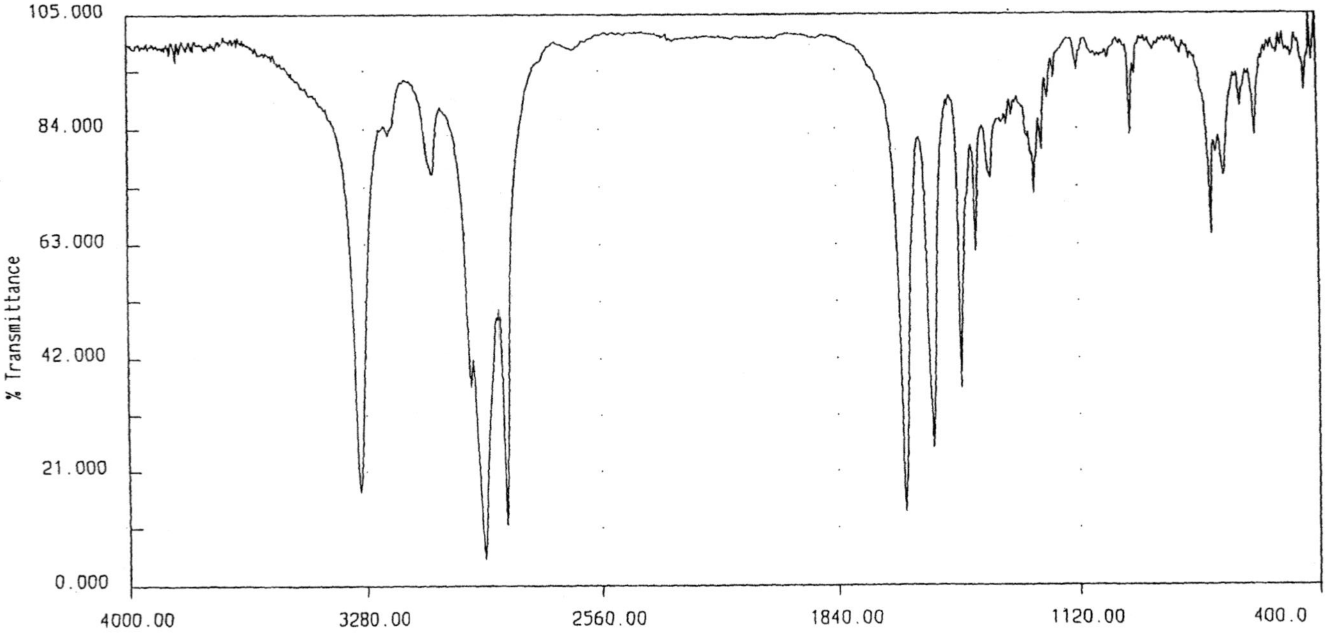

Name : N,N'-ETHYLENEBISSTEARAMIDE

PM Nr : 53520 CAS Nr : 0010-30-52

KBr pellet

CAS No. 00110-30-5 PM Ref. No. 53520

N,N'-ETHYLENEBISSTEARAMIDE

Peak wavenumber	% T	Relative intensity
436.0	90.7	9.8
584.5	82.5	18.4
526.9	87.9	12.7
679.0	75.3	26.0
702.2	79.4	21.7
717.6	64.8	37.0
1225.0	80.1	20.9
1250.0	72.1	29.3
1315.6	86.2	14.5
1331.0	84.5	16.3
1381.2	74.9	26.4
1425.6	61.5	40.5
1471.9	36.4	66.9
1456.7	25.1	78.8
1639.7	13.4	91.1
2849.2	10.9	93.7
2916.7	4.9	100.0
2957.2	36.7	66.6
3074.9	75.6	25.7
3209.9	82.7	18.2
3300.6	17.0	87.3

N,N'-Ethylenebisstearamide

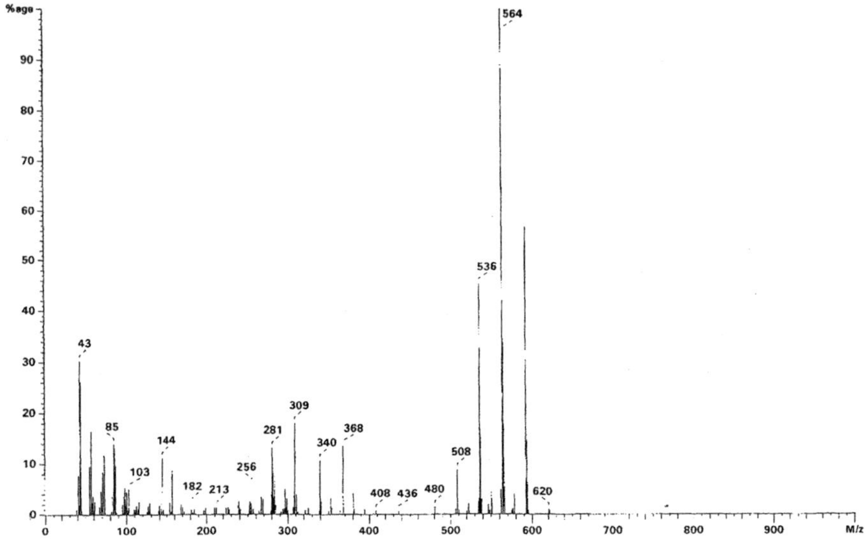

M/Z	Ion Intensity (%)	M/Z	Ion Intensity (%)
43	30.0	309	18.0
44	32.7	310	12.7
55	9.3	340	10.0
57	16.7	368	13.3
73	12.0	536	45.3
85	13.3	537	16.0
86	12.7	564	100.0
144	11.3	565	34.7
281	14.0	592	56.7
282	10.0	593	22.0

Spectrometer	: VG Autospec
Inletsystem	: Flow injection particle beam MS
Source Temperature	: 250°C
Electron Energy	: 70 eV
Scan Range	: 40-1000

PM n° 53520

CAS n° 00110-30-5

Name N,N'-ETHYLENEBISSTEARAMIDE

Solvent	$CDCL_3$
SF	250.13
SI	16384
SW	6494
NS	188

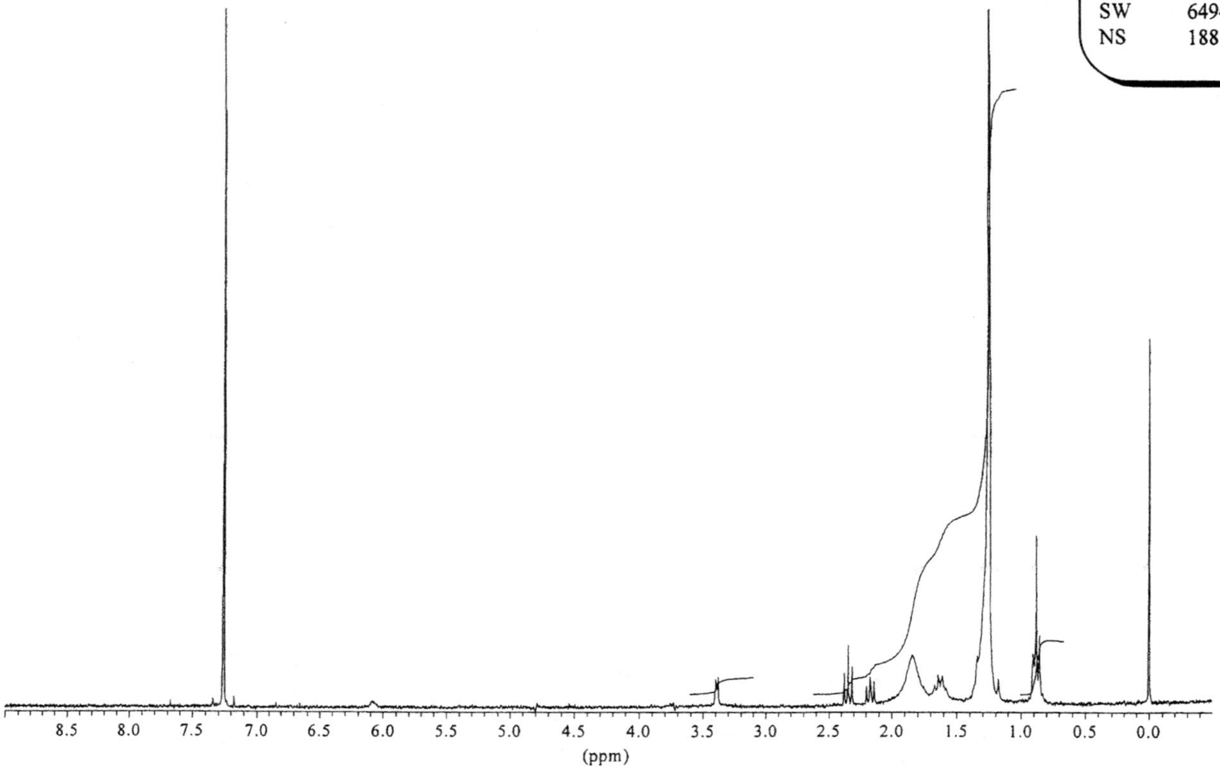

Ethyleneglycol bis(3,3-bis(3-tert-butyl-4-hydroxyphenyl)butyrate)

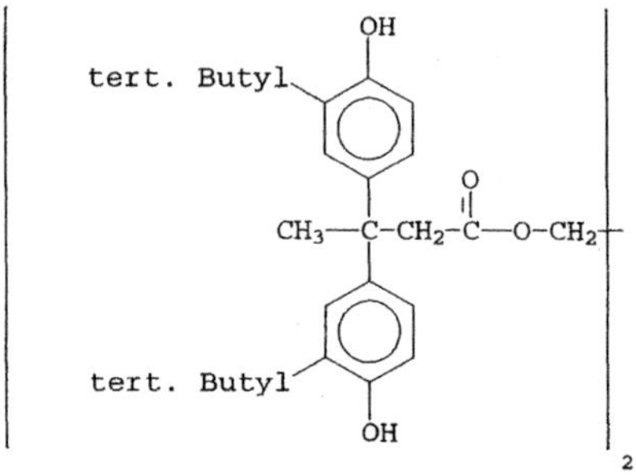

CAS No.	- 32509-66-3
PM Ref. No.	- 53670
Restrictions	- SML = 6.0 mg/kg
Formula	- $C_{50}H_{66}O_8$
Molecular weight	- 794

Alternative names -

Physical Characteristics
- Slightly yellow crystalline powder, mp 169°C.
- Soluble in acetone, methanol, chloroform.

Handling - Store at room temperature.

Safety -

Availability - Standard sample was supplied.

Current uses - Antioxidant.

Applications - PP, PE

GC Retention time -

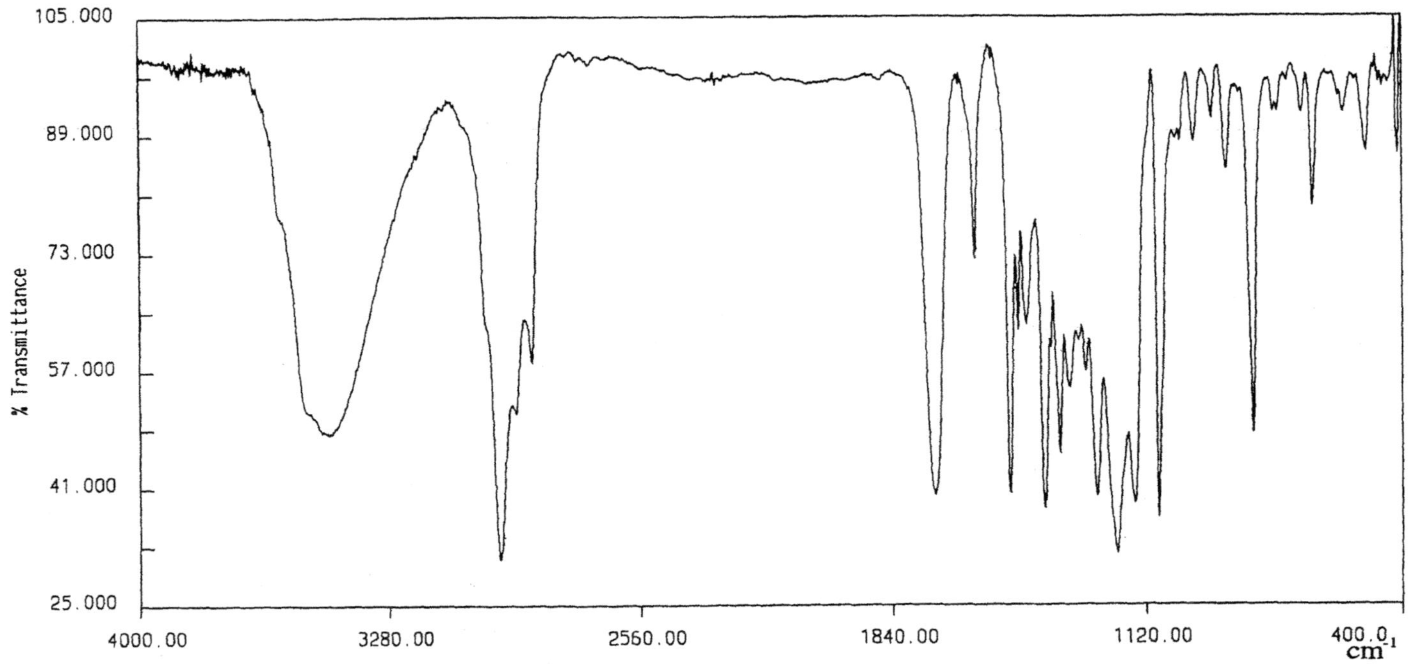

Name : ETHYLENE GLYCOL BIS(3,3-BIS(3-TERT-BUTYL-4-HYDROXYPHENYL) BUTYRATE)

PM Nr : 53670 CAS Nr : 32509-66-3

KBr pellet

CAS No. 32509-66-3 PM Ref. No. 53670

ETHYLENE GLYCOL BIS(3,3-BIS(3-TERT-BUTYL-4-HYDROXYPHENYL) BUTYRATE)

Peak wavenumber	% T	Relative intensity
410.9	86.3	19.9
499.6	86.8	19.2
563.3	91.9	11.8
650.1	79.3	30.2
817.9	48.5	74.9
891.2	84.4	22.7
983.8	88.0	17.5
1022.4	88.0	17.4
1086.1	36.9	91.9
1153.6	38.9	89.0
1203.7	32.0	99.0
1259.7	39.8	87.7
1292.5	56.9	62.8
1336.8	54.7	66.0
1365.8	45.6	79.2
1408.2	38.2	90.0
1462.2	63.4	53.4
1485.4	62.51	54.6
1508.5	40.29	86.9
1720.7	39.99	87.4
2868.5	58.21	60.8
2912.9	51.12	71.2
2959.2	31.32	100.0
3198.4	85.84	20.6
3451.1	48.25	75.3

Ethyleneglycol bis(3,3-bis(3-tert-butyl-4-hydroxyphenyl)butyrate)

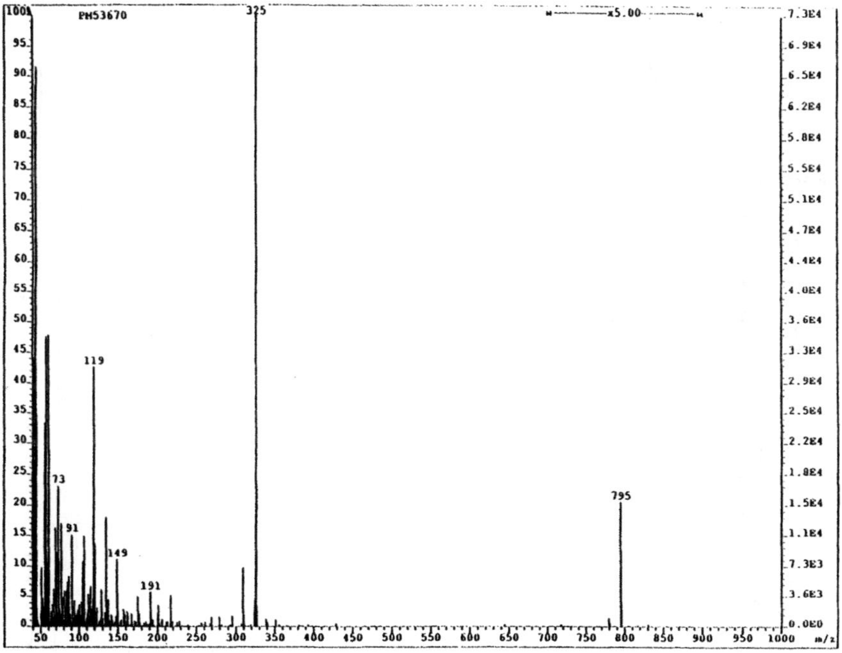

M/Z	Ion Intensity (%)	M/Z	Ion Intensity (%)
41	44.3	91	15.3
43	88.5	107	15.0
55	33.6	119	42.6
57	47.7	121	13.9
69	16.4	135	18.1
70	12.4	325	100.0
73	23.2	326	24.0
77	17.2		

Spectrometer	: VG Autospec
Inletsystem	: Flow injection particle beam MS
Source Temperature	: 250°C
Electron Energy	: 70 eV
Scan Range	: 40-1000

PM n° 53670

CAS n° 32509-66-3

Name: ETHYLENEGLYCOL BIS (3,3-BIS (3-TERT-BUTYL-4-HYDROXYPHENYL) BUTYRATE)

Solvent	$CDCL_3$
SF	250.13
SI	16384
SW	3125
NS	32

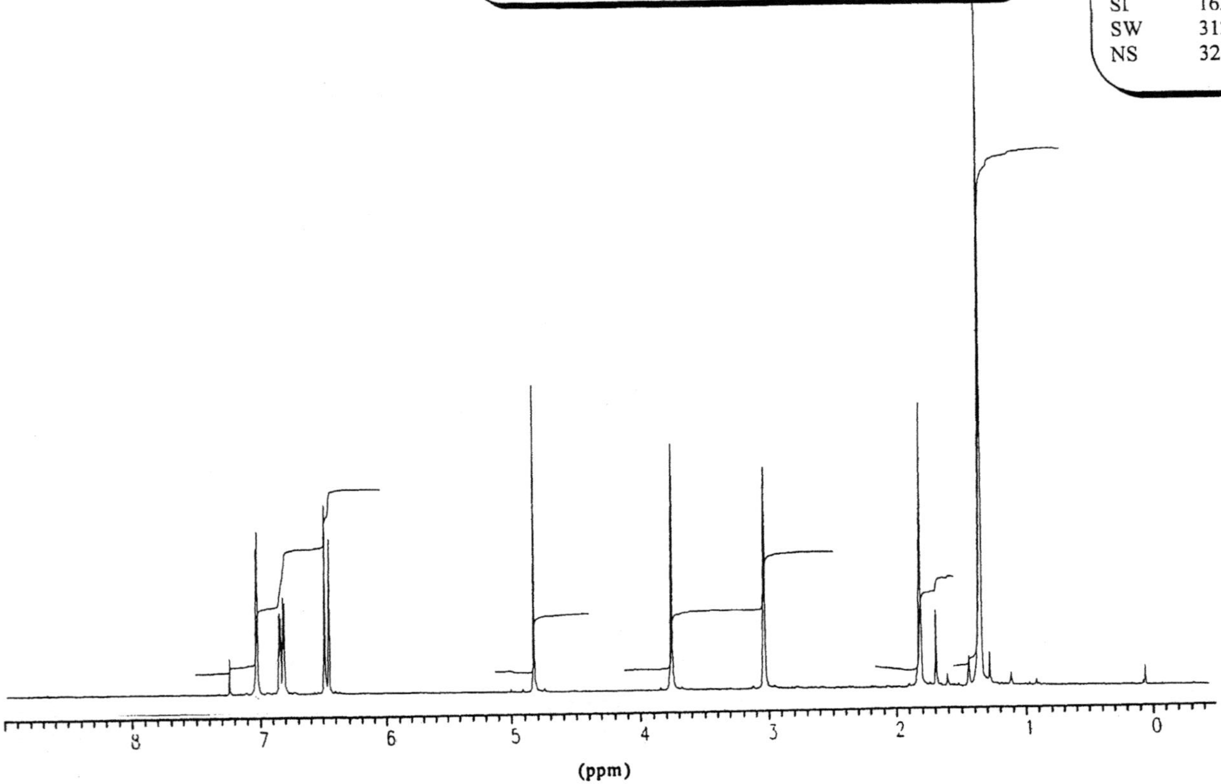

2-Ethylhexanoic acid

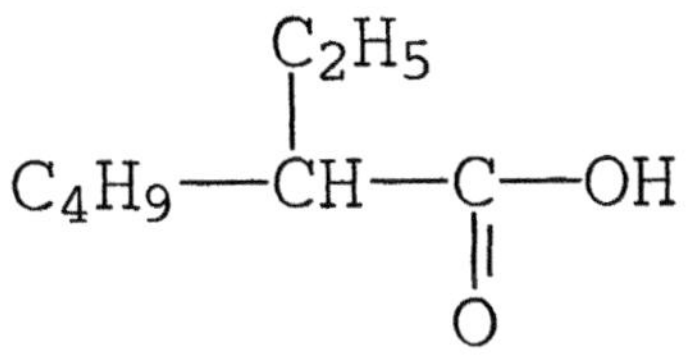

CAS No.	- 00149-57-5
PM Ref. No.	- 54120
Restrictions	- None
Formula	- $C_8H_{16}O_2$
Molecular weight	- 144

Alternative names	- 2-Ethylcaproic acid
Physical Characteristics	- Coulourless liquid, mp -59°C, bp 228°C. - Soluble in water, alcohol, ether.
Handling	- Store at room temperature.
Safety	-
Availability	- Standard sample was supplied.
Current uses	- Stabilizer.
Applications	- PVC.
GC Retention time	- 17.4 min.
Note	- Current use as calcium- or zinc salt.

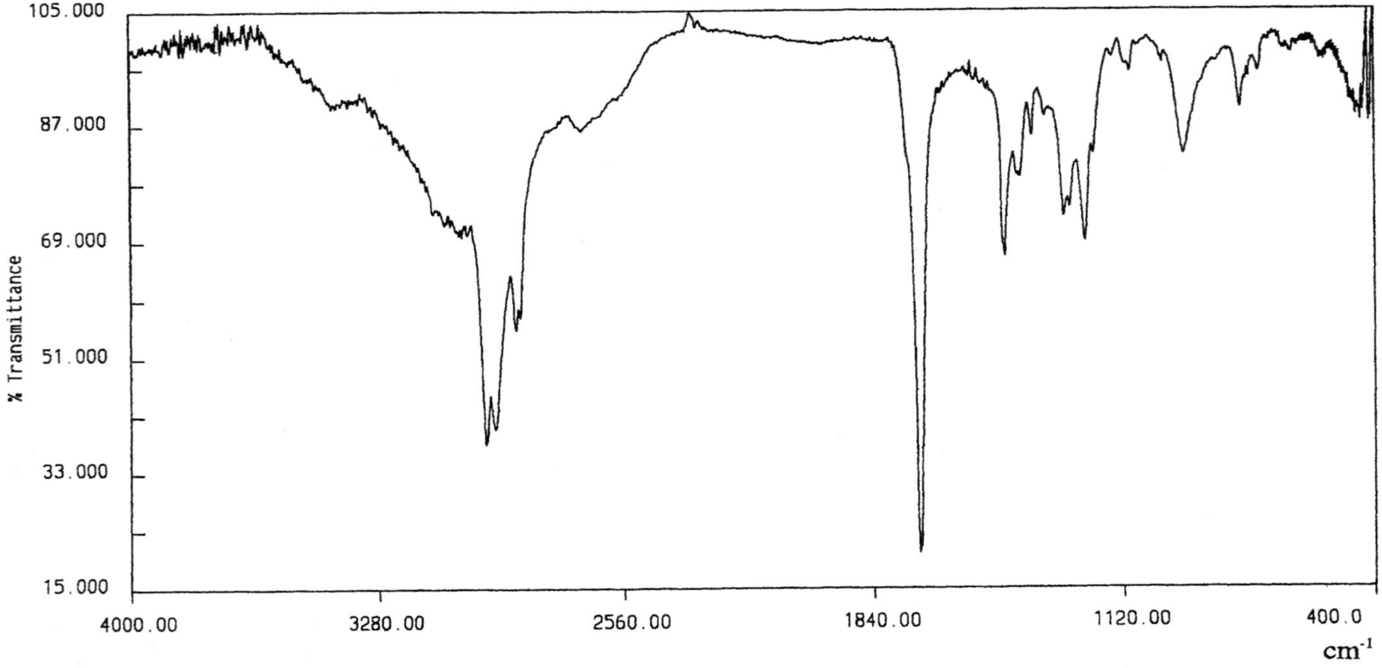

Name : 2-ETHYLHEXANOIC ACID

PM Nr : 54120 CAS Nr : 00149-57-5

KBr pellet

CAS No. 00149-57-5 PM Ref. No. 54120

2-ETHYLHEXANOIC ACID

Peak wavenumber	% T	Relative intensity
414.8	87.5	15.7
439.8	87.5	15.7
945.2	82.3	22.3
1205.7	82.3	22.3
1228.8	68.9	39.1
1273.2	74.3	32.4
1383.1	85.2	18.6
1415.9	78.7	26.8
1462.2	66.6	42.0
1709.1	20.5	100.0
2638.9	88.1	15.0
2862.7	57.3	53.7
2876.3	55.4	56.1
2936.0	39.9	75.6
2965.0	37.5	78.6

2-Ethylhexanoic acid

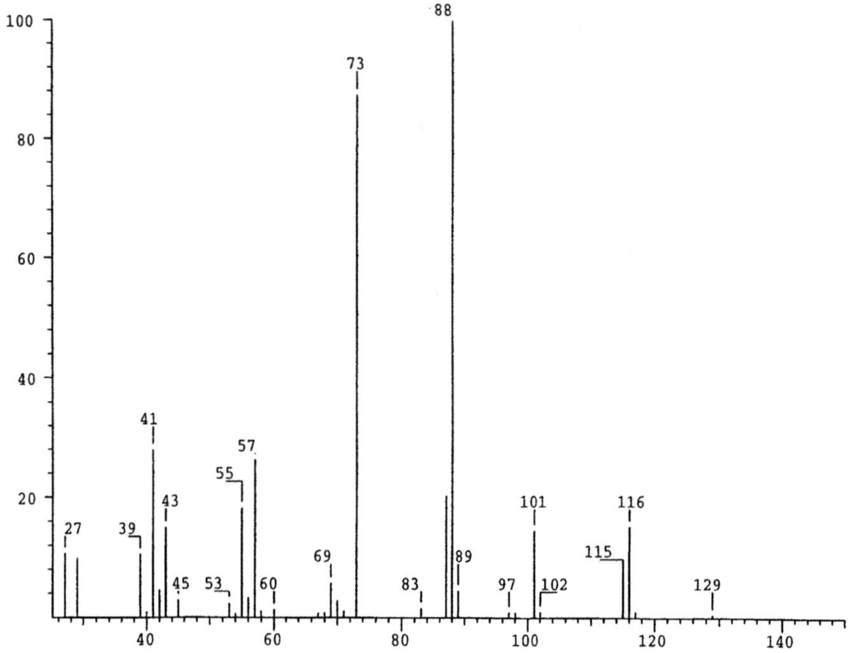

M/Z	Ion Intensity (%)	M/Z	Ion Intensity (%)
27	10.6	73	87.3
29	9.9	87	20.2
39	10.5	88	100.0
41	27.9	89	4.4
42	4.7	101	14.5
43	15.0	115	8.7
55	18.2	116	15.2
57	26.2	129	0.5

Spectrometer	: Finnigan Mat SSQ 700
Inletsystem	: Capillary GC/MS
Source Temperature	: 180°C
Electron Energy	: 70 eV
Scan Range	: 25-400

PM n°
54120

CAS n°
00149-57-5

Name
2-ETHYLHEXANOIC ACID

Solvent	$CDCL_3$
SF	250.13
SI	16384
SW	3125
NS	32

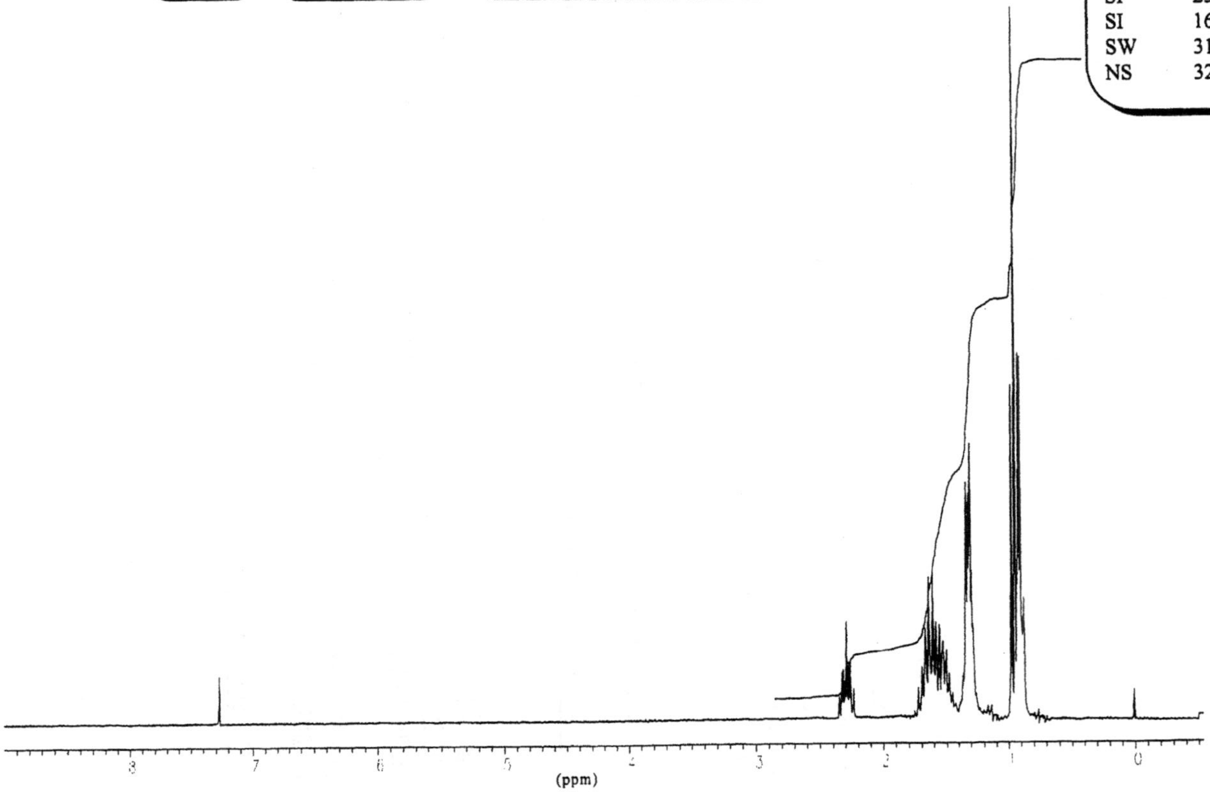

Gallic acid, propyl ester

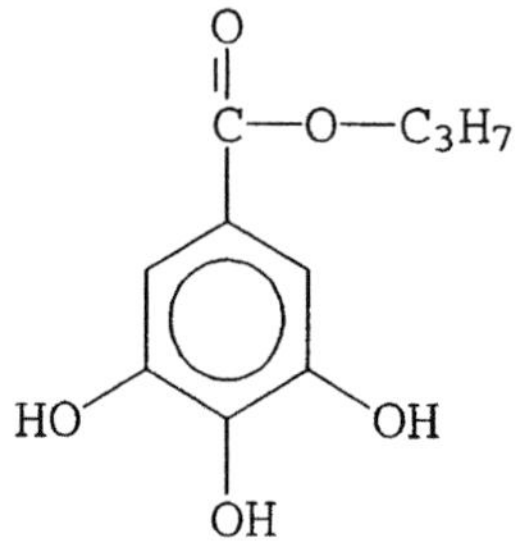

CAS No.	- 00121-79-9
PM Ref. No.	- 55360
Restrictions	- SML = 30 mg/kg as gallic acid
Formula	- $C_{10}H_{12}O_5$
Molecular weight	- 212

Alternative names	- Benzoic acid, 3,4,5-trihydroxy-, propyl ester
Physical Characteristics	- White crystalline powder, mp 150°C. - Soluble in alcohol, ether, water.
Handling	- Store at room temperature.
Safety	- Harmful.
Availability	- Standard sample was supplied.
Current uses	- Antioxidant.
Applications	- Polyolefins.
GC Retention time	- 38.3 min.

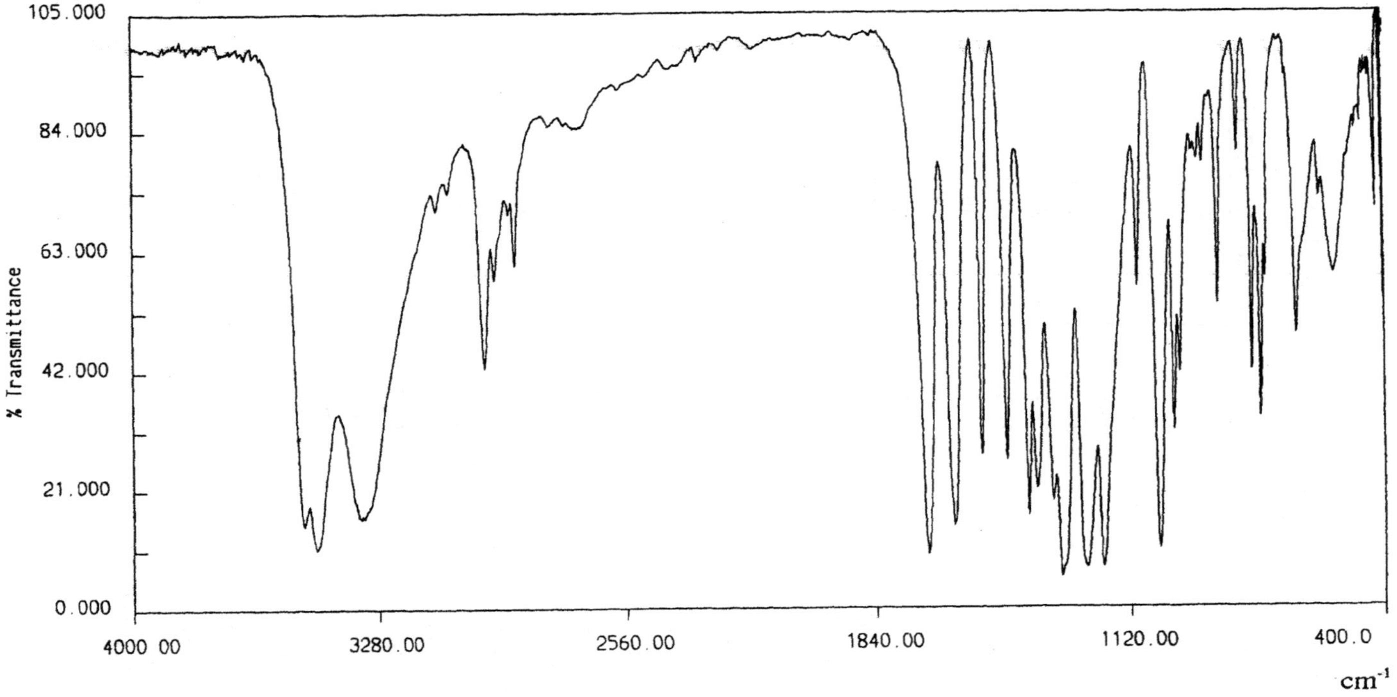

Name : GALLIC ACID, PROPYL ESTER

PM Nr : 55360 CAS Nr : 00121-79-9

KBr pellet

CAS No. 00121-79-9 PM Ref. No. 55360

GALLIC ACID, PROPYL ESTER

Peak wavenumber	% T	Relative intensity
420,5	70,3	31,4
459,1	85,4	15,4
536,3	59,0	43,3
644,3	48,4	54,5
731,1	58,2	44,1
746,5	33,7	70,0
771,6	42,0	61,2
868,1	53,7	48,9
910,5	78,4	22,8
925,9	79,1	22,1
978,0	41,6	61,7
995,4	31,3	72,5
1035,9	10,3	94,7
1099,6	56,8	45,6
1197,9	7,0	98,2
1244,2	7,0	98,2
1315,6	5,3	100,0
1340,7	18,8	85,7
1385,1	20,9	83,5
1408,2	16,1	88,6
1468,0	26,1	78,0
1541,3	27,1	77,0
1618,5	14,6	90,2
1693,7	9,4	95,7
2685,2	84,7	16,2
2880,1	60,5	41,7
2939,9	58,1	44,2
2968,8	42,5	60,7
3071,1	73,1	28,4
3107,7	70,0	31,7
3323,8	16,1	88,6
3468,4	10,6	94,4
3503,2	14,7	90,1

Gallic acid, propyl ester

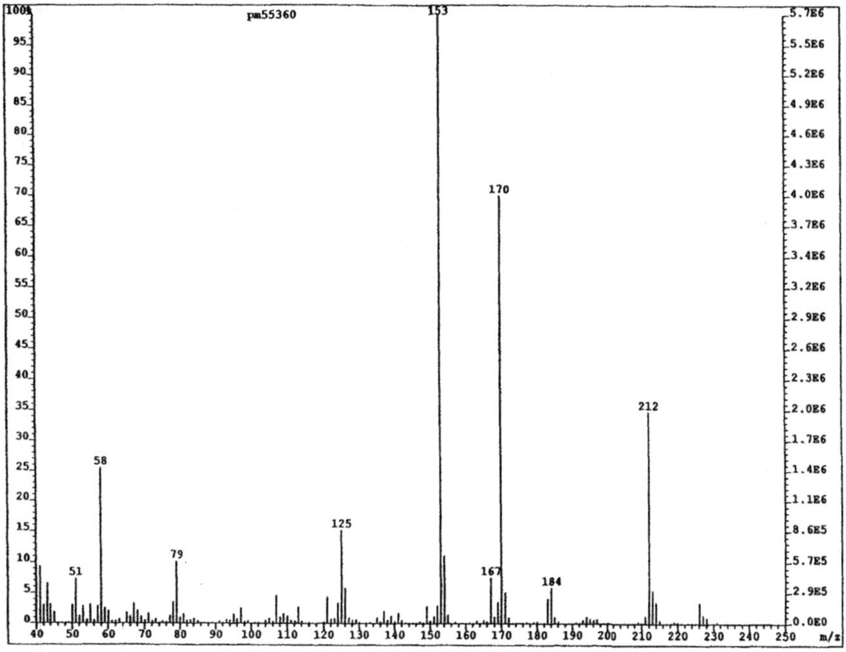

M/Z	Ion Intensity (%)	M/Z	Ion Intensity (%)
41	9.8	153	100.0
43	6.7	154	11.0
51	7.4	167	7.4
58	25.8	170	70.6
79	9.8	184	6.1
125	15.3	212	35.0
126	6.1	213	5.5

Spectrometer : VG Autospec
Inletsystem : Flow injection particle beam MS
Source Temperature : 250°C
Electron Energy : 70 eV
Scan Range : 40-1000

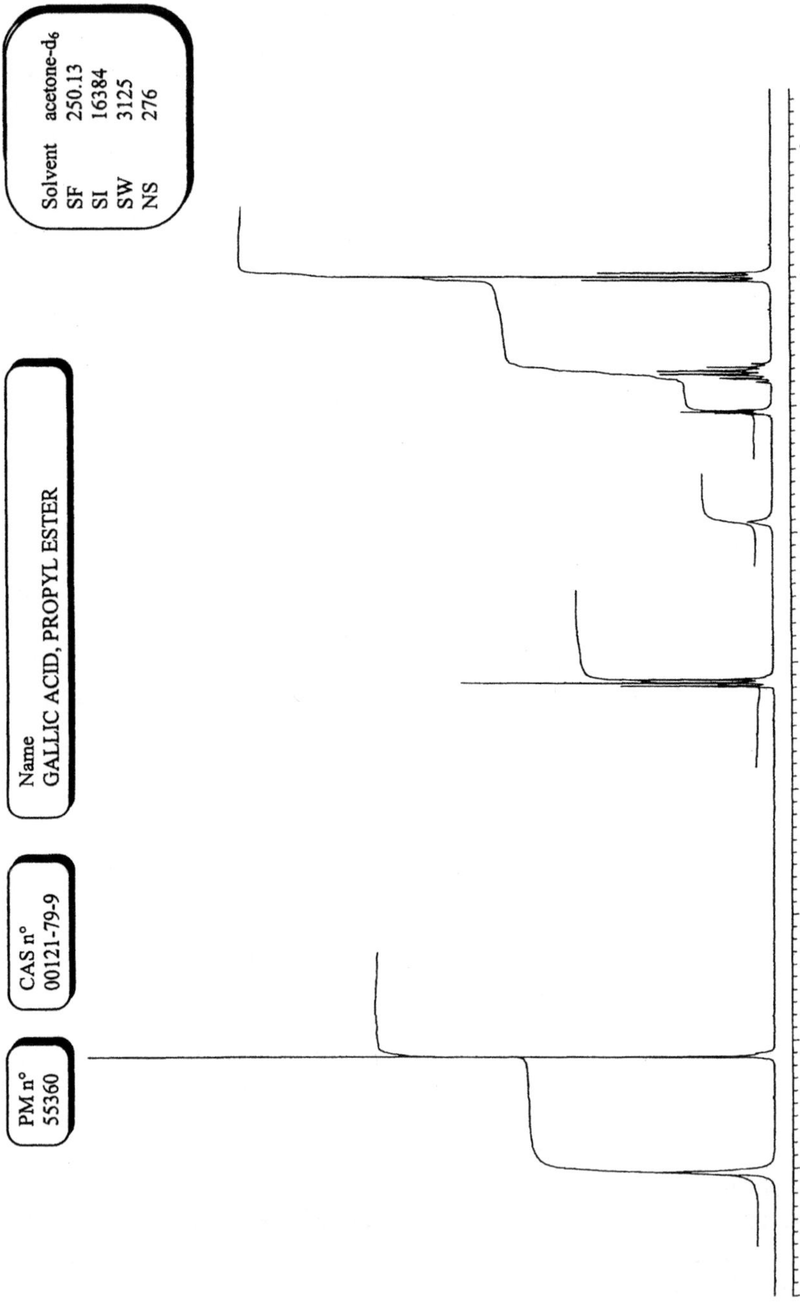
PM n° 55360
CAS n° 00121-79-9
Name GALLIC ACID, PROPYL ESTER
Solvent acetone-d$_6$
SF 250.13
SI 16384
SW 3125
NS 276
0
1
2
3
4
5
6
7
8
(ppm)

Glycerol monooleate

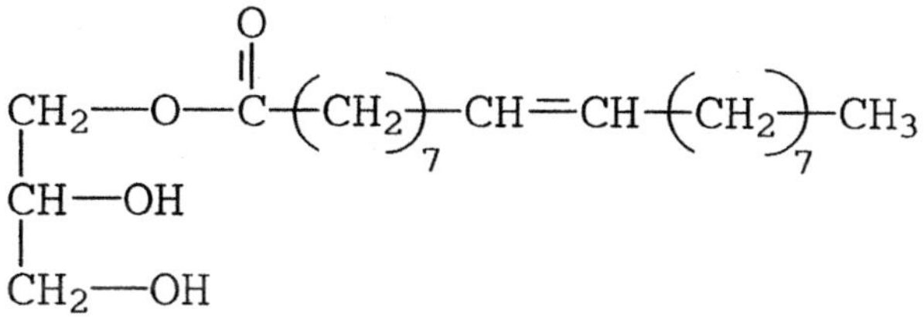

CAS No.	- 25496-72-4
PM Ref. No.	- 56960
Restrictions	- None
Formula	- $C_{21}H_{40}O_4$
Molecular weight	- 356
Alternative names	- Glycerol, 1-mono(9-octadecenoate)
Physical Characteristics	- Yellow liquid, mp 25°C, bp 238 - 240°C. - Soluble in hexane, benzene, alcohol, ether.
Handling	- Store at room temperature.
Safety	-
Availability	- Standard sample supplied.
Current uses	- Lubricant.
Applications	- General use.
GC Retention time	-

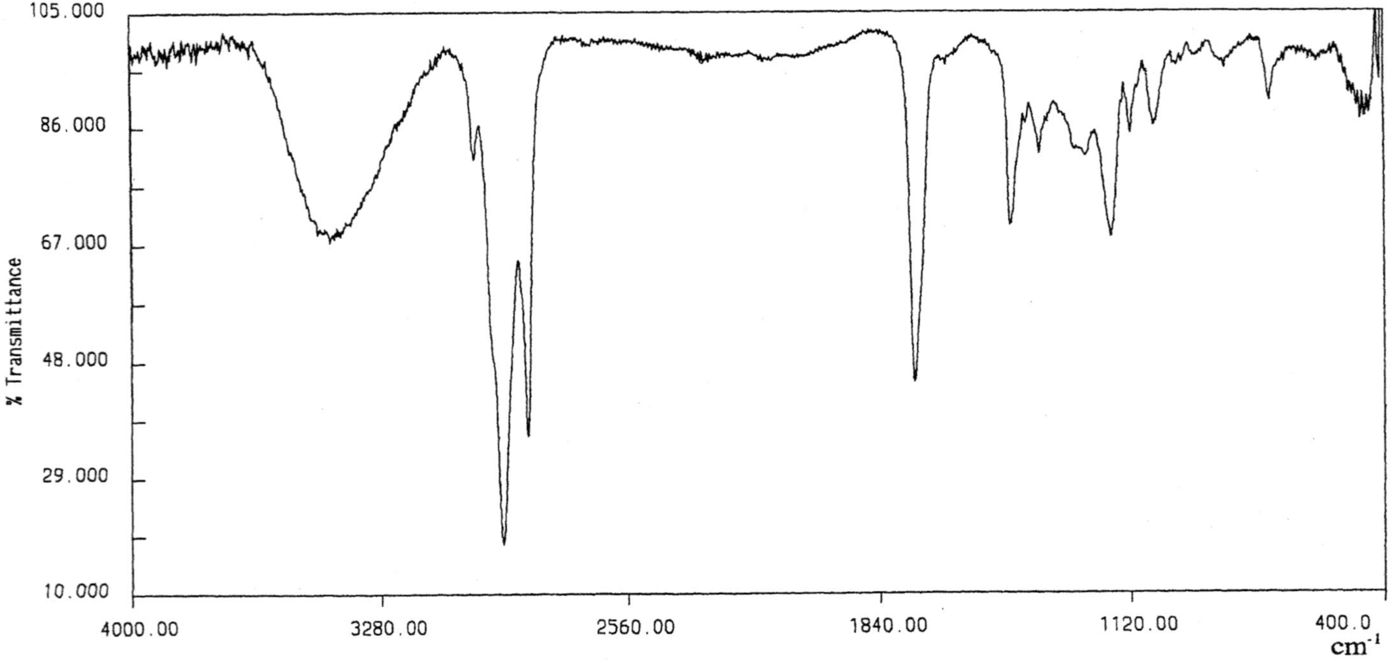

Name : GLYCEROL MONOOLEATE

PM Nr : 56960 CAS Nr : 25496-72-4

KBr pellet

CAS No. 25496-72-4 PM Ref. No. 56960

GLYCEROL MONOOLEATE

Peak wavenumber	% T	Relative intensity
441.8	87.9	14.7
723.4	90.0	12.2
1053.3	85.9	17.2
1118.9	84.6	18.8
1174.8	68.0	39.1
1246.2	80.9	23.3
1379.3	81.4	22.7
1419.8	86.3	16.7
1466.1	69.9	36.7
1741.9	44.6	67.7
2855.0	35.8	78.4
2926.4	18.1	100.0
3007.4	80.6	23.7
3192.6	89.3	13.0
3289.0	77.8	27.1
3372.0	70.2	36.4

Glycerol monooleate

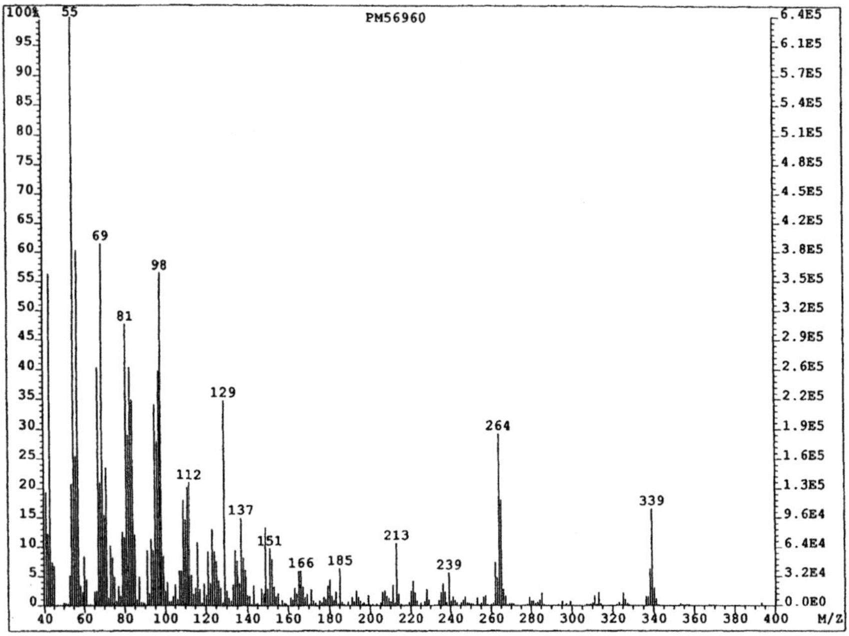

M/Z	Ion Intensity (%)	M/Z	Ion Intensity (%)
43	56.4	98	56.5
55	100.0	109	18.0
56	25.6	111	20.2
57	60.4	112	21.0
69	61.6	129	34.9
71	23.5	137	15.0
81	47.8	213	10.7
83	40.4	264	29.4
96	28.0	264	18.2
97	39.8	339	16.7

Spectrometer	: VG Autospec
Inletsystem	: Flow injection particle beam MS
Source Temperature	: 250°C
Electron Energy	: 70 eV
Scan Range	: 40-1000

PM n°
56960

CAS n°
25496-72-4

Name
GLYCEROL MONOOLEATE

Solvent	$CDCL_3$
SF	250.13
SI	16384
SW	3125
NS	60

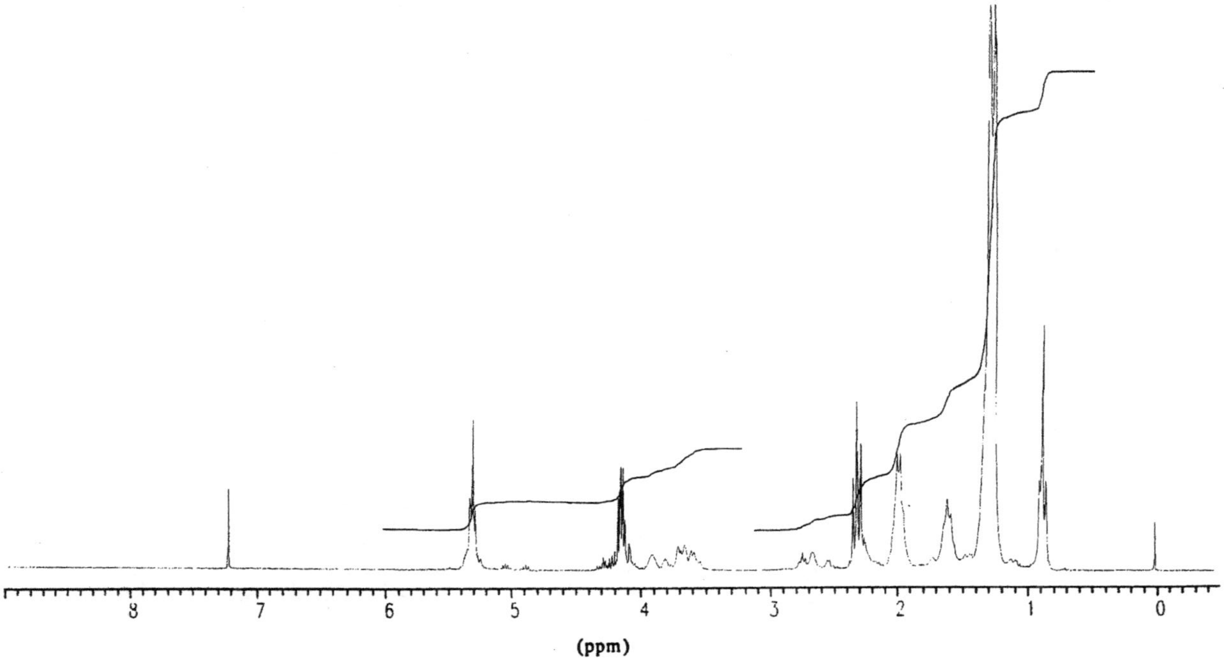

Glycerol monostearate

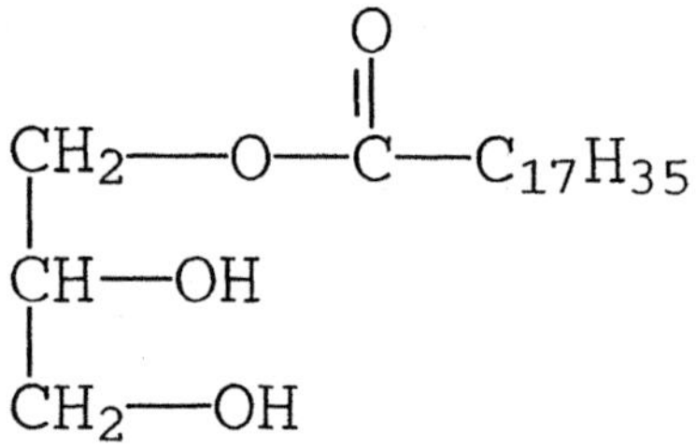

CAS No.	- 31566-31-1
PM Ref. No.	- 57520
Restrictions	- None
Formula	- $C_{21}H_{42}O_4$
Molecular weight	- 358

Alternative names	- Glycerol, 1-monooctadecanoate
Physical Characteristics	- White micro beads, mp approx. 69°C. - Soluble in alcohol, hexane, ether.
Handling	- Store at room temperature.
Safety	- Low toxicity.
Availability	- Standard sample supplied.
Current uses	- Lubricant.
Applications	- General use.
GC Retention time	- 46.5 and 49.6 min.
Note	- Also known as PM Ref. No. 18115.

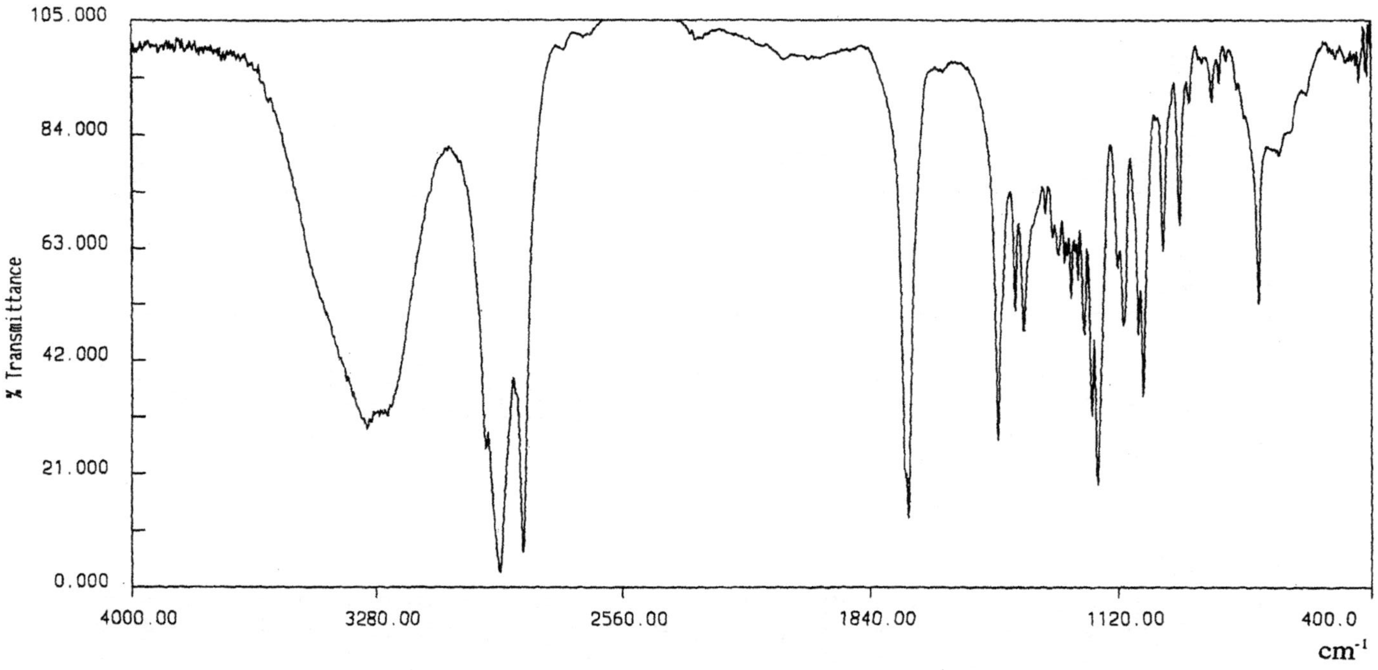

Name : GLYCEROL MONOSTEARATE

PM Nr : 57520 CAS Nr : 31566-31-1

KBr pellet

CAS No. 31566-31-1 PM Ref. No. 57520

GLYCEROL MONOSTEARATE

Peak wavenumber	% T	Relative intensity
434.0	93.4	6.8
582.6	91.3	9.0
632.7	84.3	16.1
661.7	80.2	20.3
721.5	52.6	48.7
943.3	67.4	33.5
991.5	62.5	38.6
1047.5	35.3	66.5
1062.9	46.9	54.5
1105.4	48.6	52.8
1180.6	19.0	83.2
1196.0	31.8	70.1
1215.3	46.9	54.5
1236.5	57.0	44.2
1255.8	53.6	47.7
1267.4	61.6	39.5
1275.1	60.3	40.8
1292.5	61.70	39.3
1331.0	69.71	31.1
1394.7	47.65	53.8
1417.9	51.38	50.0
1469.9	27.25	74.7
1730.4	12.92	89.5
2851.1	6.64	95.9
2916.7	2.67	100.0
3246.6	31.54	70.3

Glycerol monostearate

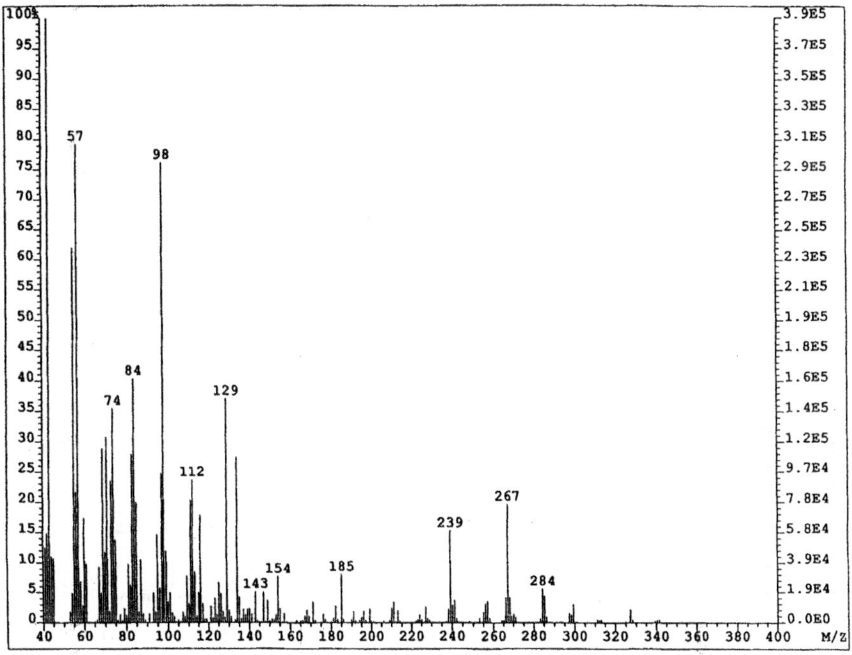

M/Z	Ion Intensity (%)	M/Z	Ion Intensity (%)
43	100.0	97	24.7
44	11.1	98	76.2
55	62.0	99	12.0
57	79.3	111	20.4
69	29.0	112	23.87
71	30.8	116	17.9
73	23.6	129	37.2
74	35.6	134	27.5
84	40.4	239	15.4
95	14.7	267	19.7

Spectrometer : VG Autospec
Inletsystem : Flow injection particle beam MS
Source Temperature : 250°C
Electron Energy : 70 eV
Scan Range : 40-1000

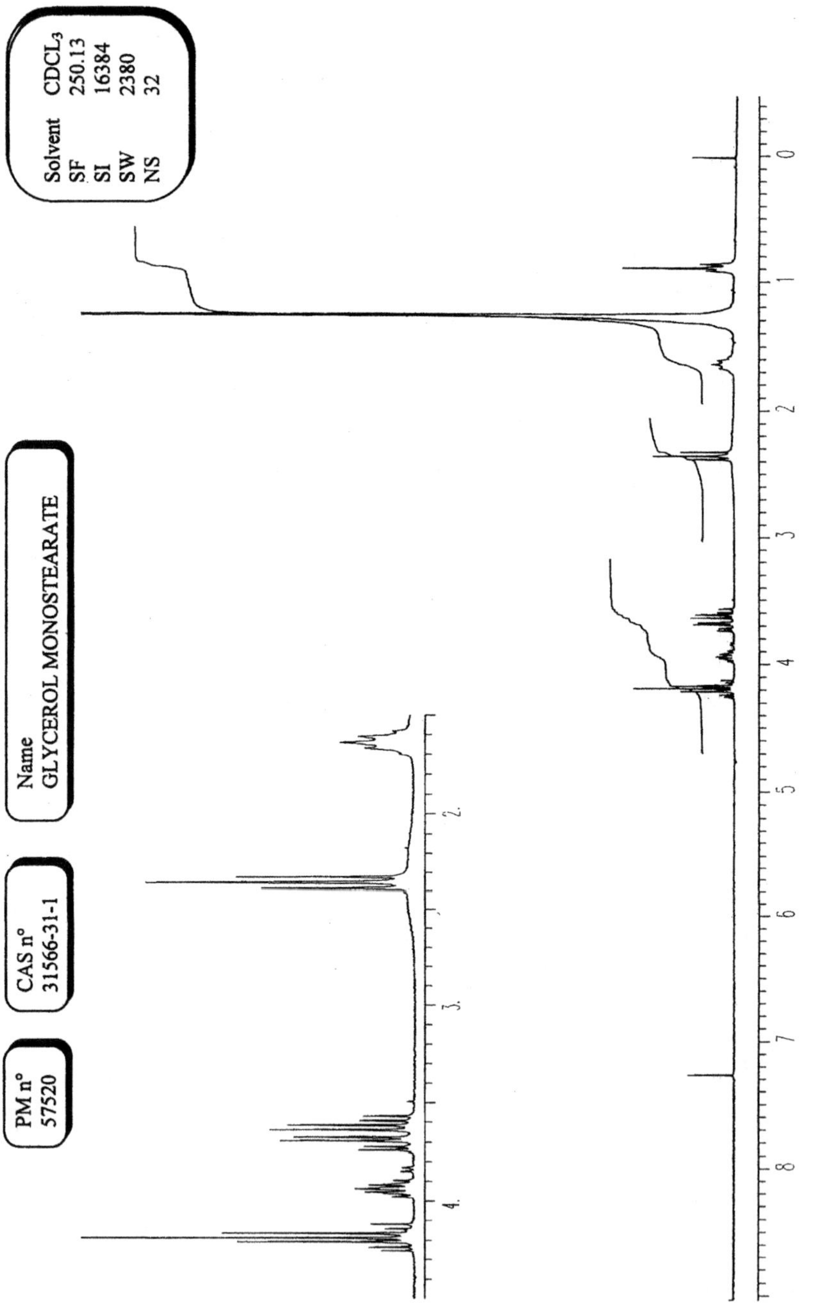
PM n°
57520
CAS n°
31566-31-1
Name
GLYCEROL MONOSTEARATE
Solvent CDCL₃
SF 250.13
SI 16384
SW 2380
NS 32
(ppm)

Glycerol triacetate

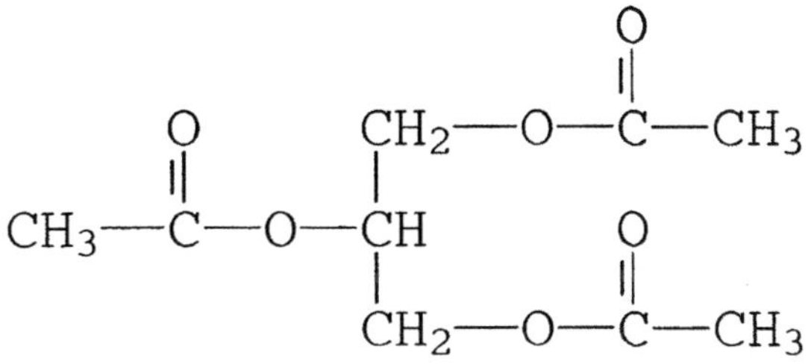

CAS No.	- 00102-76-1
PM Ref. No.	- 57760
Restrictions	- None
Formula	- $C_9H_{14}O_6$
Molecular weight	- 218

Alternative names	- Triacetin
Physical Characteristics	- Colourless liquid, mp -78°C, bp 258 - 260°C. - Soluble in most organic solvents.
Handling	- Store at room temperature.
Safety	-
Availability	- Standard sample was supplied.
Current uses	- Lubricant, carrier.
Applications	- General use.
GC Retention time	- 22.9 min.

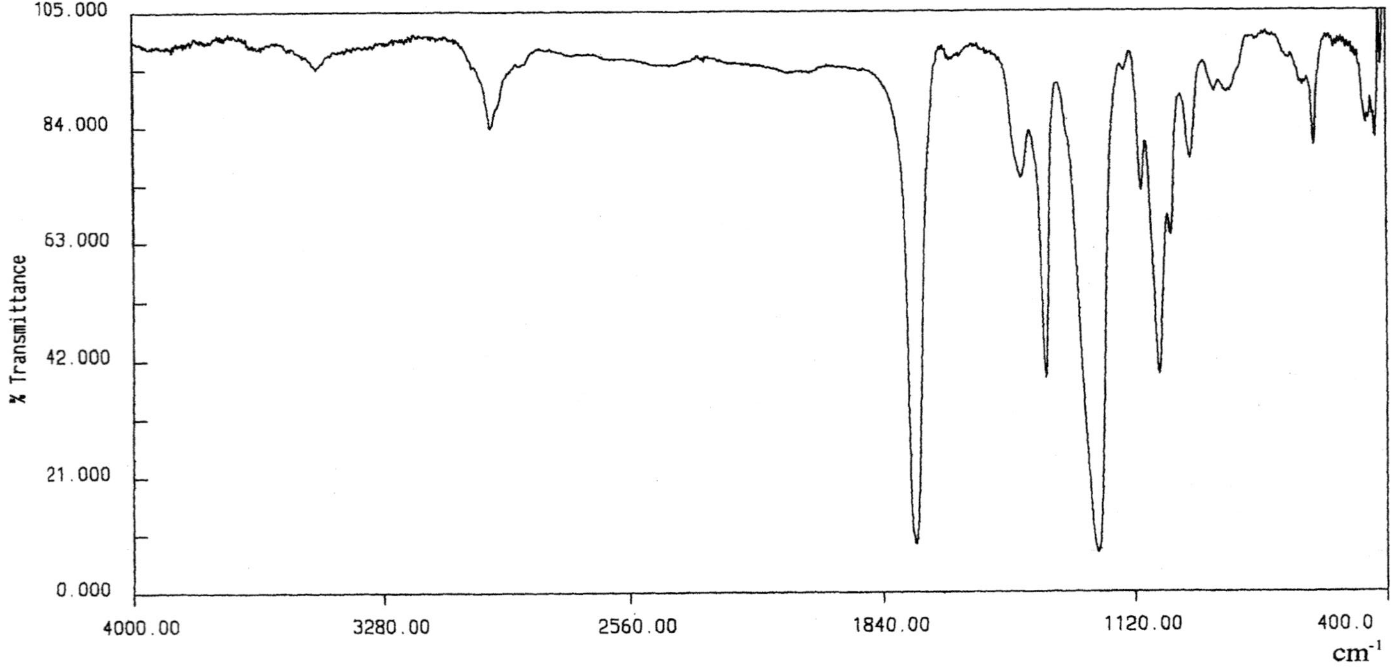

Name : GLYCEROL TRIACETATE

PM Nr : 57760 CAS Nr : 00102-76-1

KBr pellet

CAS No. 00102-76-1 PM Ref. No. 57760

GLYCEROL TRIACETATE

Peak wavenumber	% T	Relative intensity
416.7	84.6	15.7
430.2	70.8	29.8
457.2	75.3	25.2
605.7	66.8	33.9
688.7	82.7	17.7
748.5	87.7	12.5
844.9	76.4	24.1
962.6	68.9	31.7
1018.5	43.7	57.3
1051.3	19.1	82.5
1101.5	61.2	39.6
1221.1	1.8	100.0
1373.5	21.1	80.4
1439.1	69.0	31.6
1477.7	85.2	15.1
1684.1	43.3	57.7
1743.9	3.9	97.9
2843.4	81.5	18.9
2901.3	75.2	25.2
3476.2	88.9	11.3

Glycerol triacetate

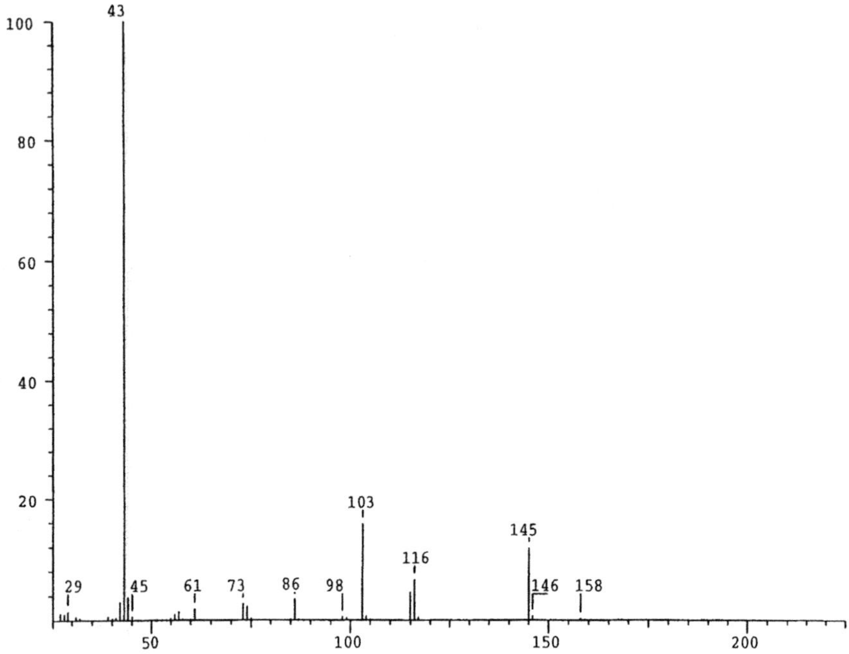

M/Z	Ion Intensity (%)	M/Z	Ion Intensity (%)
27	1.1	73	2.6
29	1.4	74	2.2
42	2.9	86	3.5
43	100.0	103	15.8
44	3.8	115	4.6
56	0.9	116	6.7
57	1.4	145	11.8
61	1.9	146	0.7

Spectrometer	: Finnigan Mat SSQ 700
Inletsystem	: Capillary GC/MS
Source Temperature	: 180°C
Electron Energy	: 70 eV
Scan Range	: 25-250

PM n° 57760

CAS n° 00102-76-1

Name GLYCEROLTRIACETATE

Solvent	$CDCL_3$
SF	250.13
SI	16384
SW	3125
NS	32

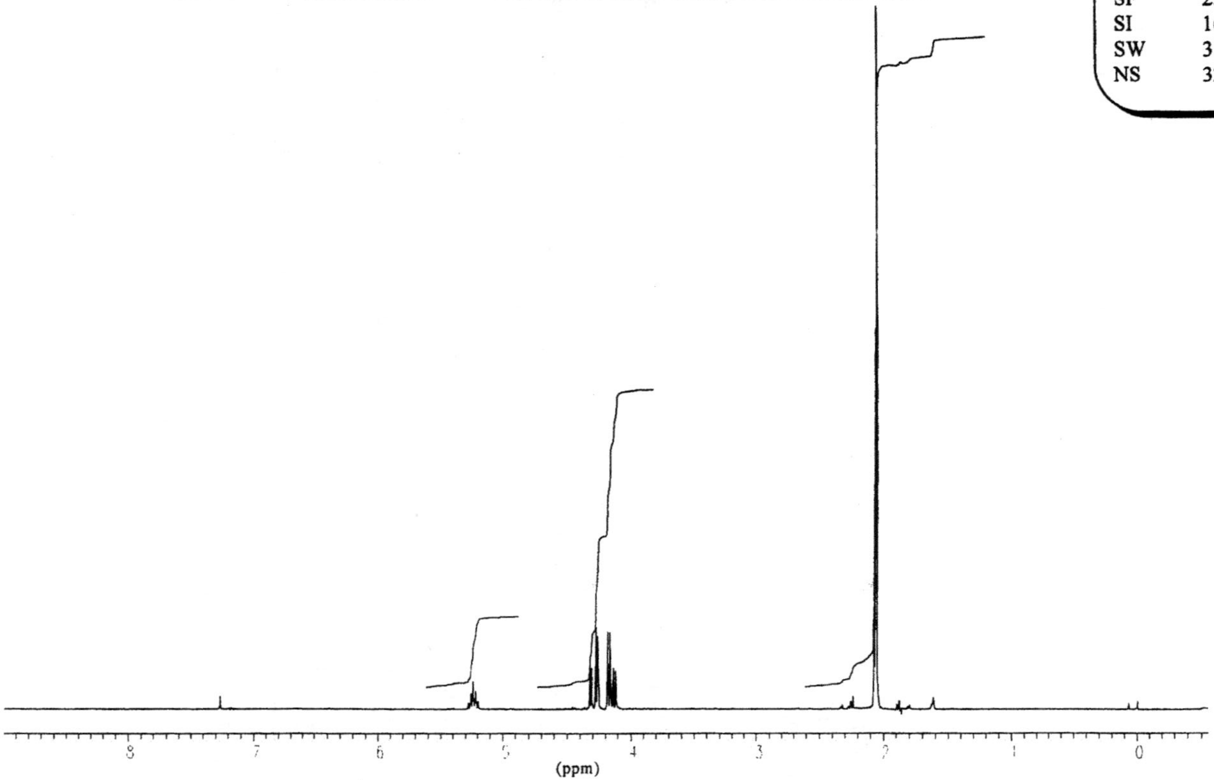

1,6-Hexamethylene bis(3-(3,5-di-tert-butyl-4-hydroxyphenyl)propionamide)

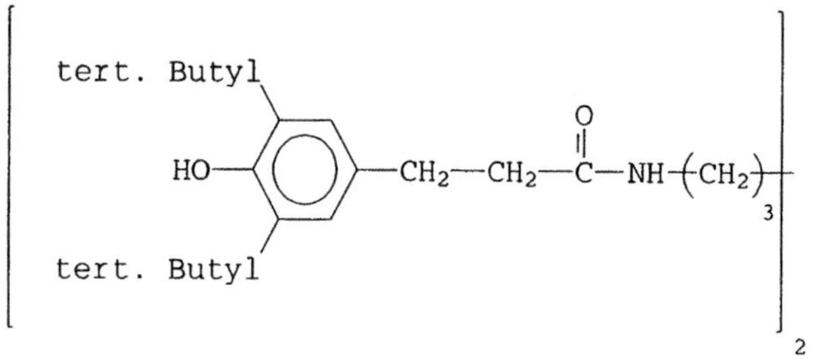

CAS No.	- 23128-74-7
PM Ref. No.	- 59120
Restrictions	- SML = 45 mg/kg
Formula	- $C_{40}H_{64}N_2O_4$
Molecular weight	- 637

Alternative names	- N,N'-hexamethylene bis(3,5-di-tert-butyl-4-hydroxy-hydrocinnamide
Physical Characteristics	- White crystalline powder, mp 156 - 161°C. - Soluble in acetone, methanol, chloroform.
Handling	- Store at room temperature.
Safety	- Harmful.
Availability	- Standard sample was supplied.
Current uses	- Antioxidant.
Applications	- PA, PVA, PBT, polyesters, polyacetals.
GC Retention time	-

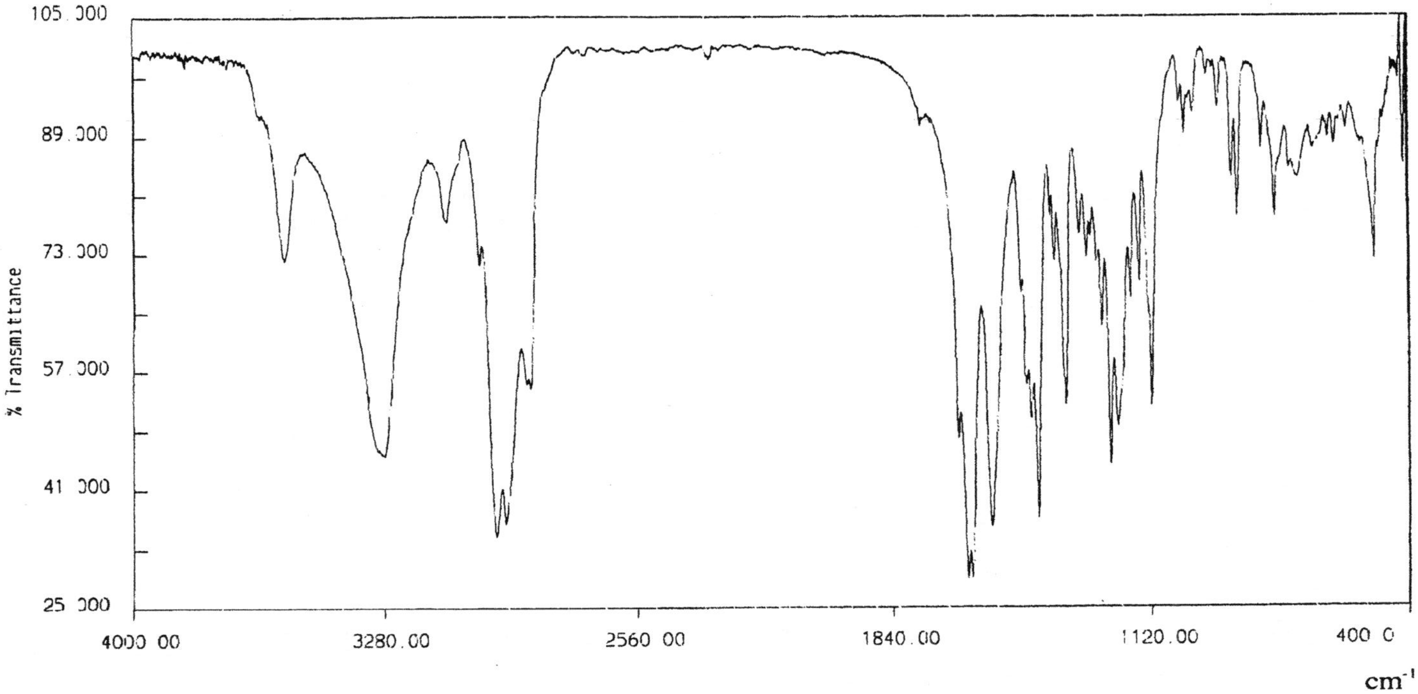

Name : 1,6-HEXAMETHYLENE BIS(3-(3,5-DI-Tert-BUTYL-4-HYDROXYPHENYL)PROPIONAMIDE)

PM Nr : 59120 CAS Nr : 23128-74-7

KBr pellet

CAS No. 23128-74-7 PM Ref. No. 59120

1,6-HEXAMETHYLENE BIS(3-(3,5-DI-tert-BUTYL-4-HYDROXYPHENYL)PROPIONAMIDE)

Peak wavenumber	% T	Relative intensity
410,9	85,2	20,8
468,8	91,3	12,2
493,8	72,2	39,1
569,1	89,9	14,2
603,8	87,8	17,2
621,2	88,7	15,9
661,7	87,3	17,9
704,1	83,3	23,5
769,7	78,1	30,8
806,3	87,3	17,9
875,8	78,1	30,8
889,3	83,3	23,5
1026,3	89,3	15,0
1120,8	52,2	67,2
1155,5	69,2	43,3
1180,6	66,8	46,7
1215,3	49,6	70,9
1234,6	44,2	78,5
1259,7	63,1	51,9
1290,5	75,4	34,6
1302,1	72,4	38,8
1321,4	75,7	34,2
1358,1	52,4	66,9
1390,9	71,9	39,5
1435,2	37,1	88,4
1454,5	50,5	69,6
1564,5	35,9	90,1
1620,4	29,0	99,8
1632,0	28,9	100,0
1657,1	47,7	73,5
2860,8	54,8	63,6
2932,2	36,5	89,3
2957,2	34,7	91,8
3096,1	77,4	31,8
3283,2	45,5	76,6
3570,7	72,1	39,2

1,6-Hexamethylene bis(3-(3,5-di-tert-butyl-4-hydroxyphenyl)propionamide)

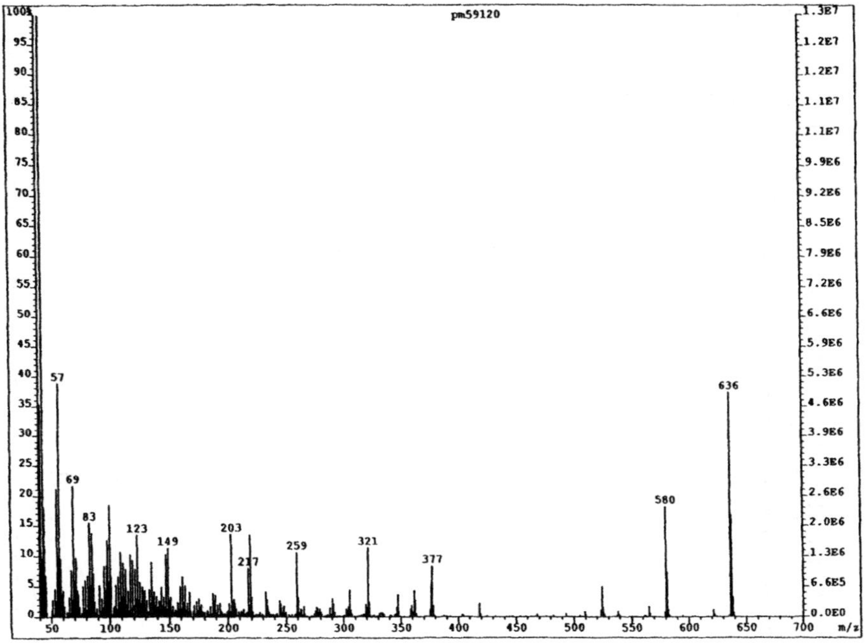

M/Z	Ion Intensity (%)	M/Z	Ion Intensity (%)
41	35.0	109	11.0
43	100.0	123	14.1
44	18.4	149	11.7
55	21.5	203	14.7
57	39.3	219	14.1
69	22.1	259	11.0
83	16.0	321	11.7
85	14.1	580	19.6
98	12.9	636	38.0
100	18.4	637	17.2

Spectrometer : VG Autospec
Inletsystem : Flow injection particle beam MS
Source Temperature : 250°C
Electron Energy : 70 eV
Scan Range : 40-1000

PM n°	CAS n°	Name
59120	23128-74-7	1,6-HEXAMETHYLENE BIS(3-(3,5-DI-TERT-BUTYL-4-HYDROXYPHENYL)PROPIONAMIDE)

Solvent	$CDCL_3$
SF	250.13
SI	16384
SW	3125
NS	32

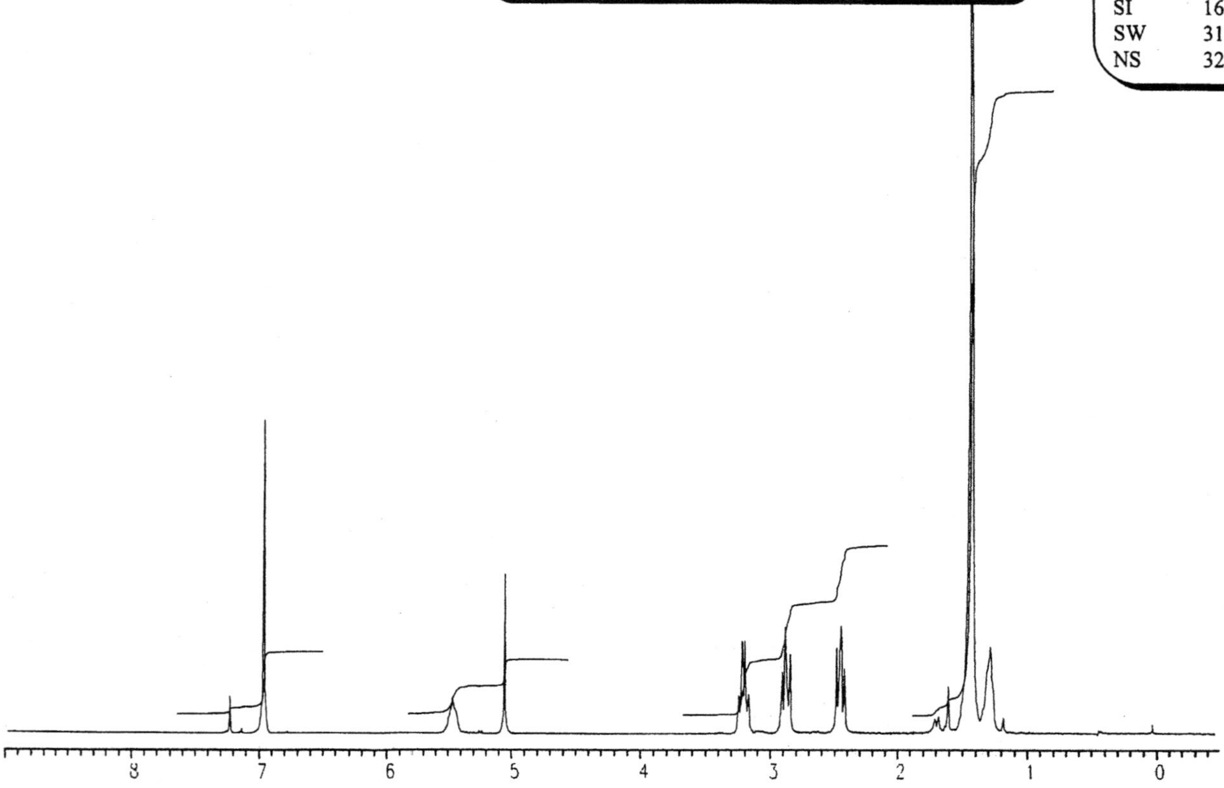

1,6-Hexamethylene bis(3-(3,5-di-tert-butyl-4-hydroxyphenyl)propionate)

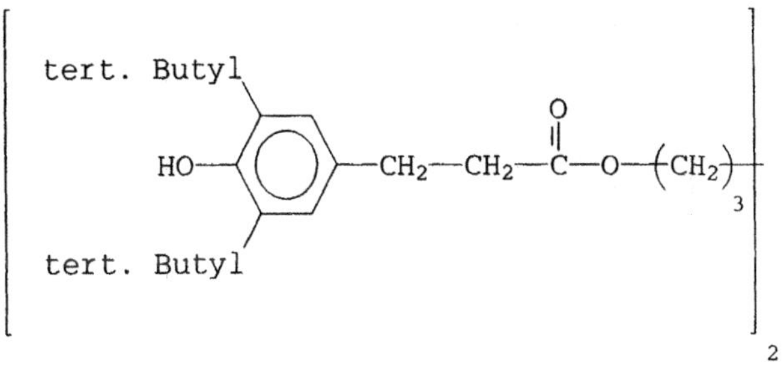

CAS No.	- 35074-77-2
PM Ref. No.	- 59200
Restrictions	- SML = 6.0 mg/kg
Formula	- $C_{40}H_{62}O_6$
Molecular weight	- 639

Alternative names	- Benzenepropanoic acid, 3,5-bis(1,1-di-methylethyl)-4-hydroxy-1,6-hexanediyl ester
Physical Characteristics	- White crystalline powder, mp 104 - 108°C. - Soluble in benzene, chloroform, ethyl acetate.
Handling	- Store at room temperature.
Safety	-
Availability	- Standard sample was supplied.
Current uses	- Antioxidant.
Applications	- Polyacetals.
GC Retention time	- 73.8 min.

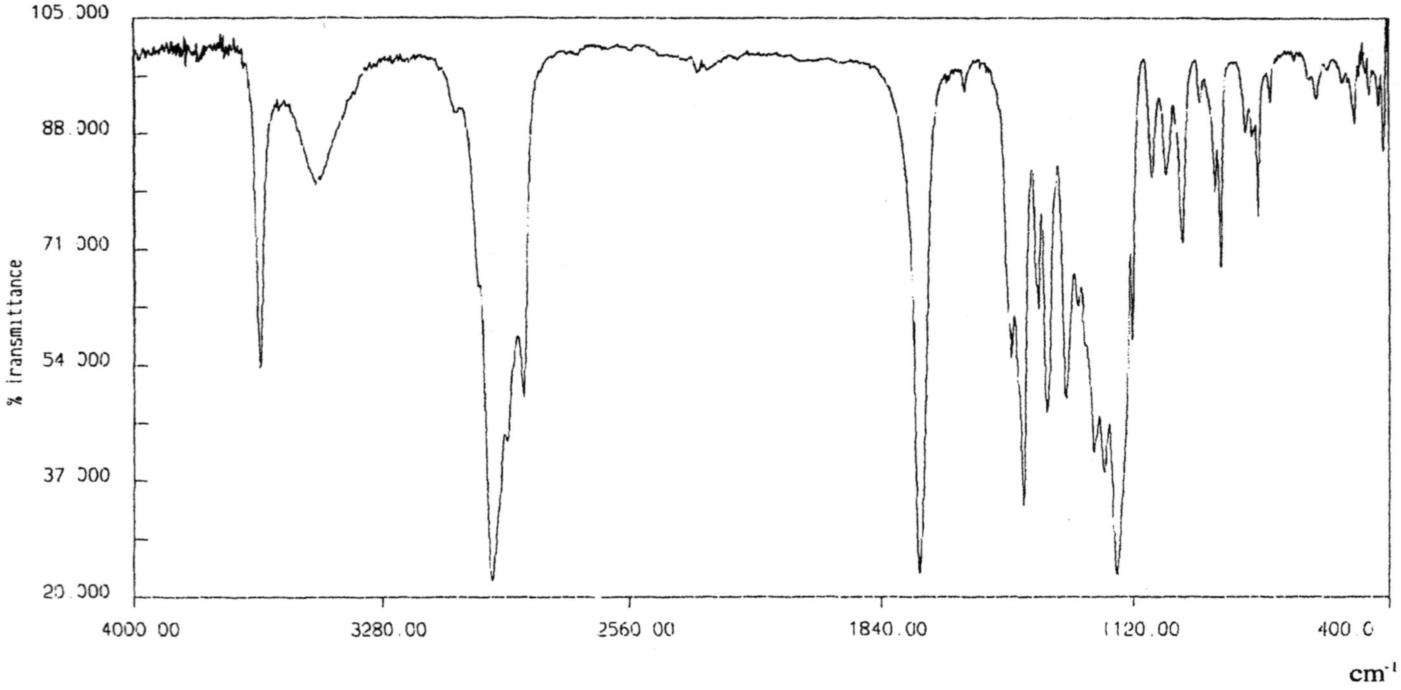

Name : 1,6-HEXAMETHYLENE BIS(3-(3,5-DI-Tert-BUTYL-4-HYDROXYPHENYL)PROPIONATE)

PM Nr : 59200 CAS Nr : 35074-77-2

KBr pellet

1,6-HEXAMETHYLENE BIS(3-(3,5-DI-tert-BUTYL-4-HYDROXYPHENYL)PROPIONATE)

Peak wavenumber	% T	Relative intensity
414,8	85,1	19,2
495,8	89,4	13,7
767,8	75,9	31,1
802,5	88,0	15,5
871,9	68,4	40,7
889,3	79,4	26,6
981,9	72,0	36,1
1028,2	81,8	23,5
1066,8	81,3	24,1
1122,7	57,7	54,5
1167,1	23,3	98,9
1205,7	38,1	79,8
1234,6	41,2	75,8
1313,7	48,9	65,9
1369,6	46,9	68,4
1392,8	62,2	48,7
1437,1	33,2	86,1
1471,9	55,0	58,0
1732,3	23,4	98,7
2868,5	49,3	65,4
2916,7	42,7	73,9
2959,2	22,4	100,0
3464,6	80,9	24,6
3640,1	53,6	59,8

1,6-Hexamethylene bis(3-(3,5-di-tert-butyl-4-hydroxyphenyl)propionate)

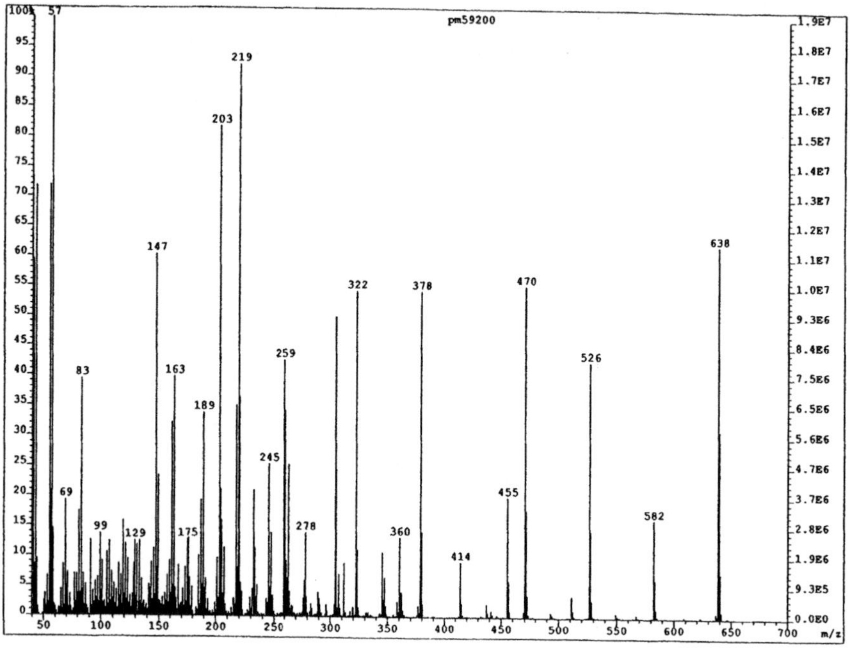

M/Z	Ion Intensity (%)	M/Z	Ion Intensity (%)
41	59.5	259	42.9
43	71.8	260	33.7
55	71.8	305	49.7
57	100.0	322	54.8
83	39.9	378	54.6
147	60.1	470	55.2
163	40.5	526	42.9
189	33.7	638	62.6
203	82.2	639	31.1
219	92.0	640	7.5

Spectrometer	: VG Autospec
Inletsystem	: Flow injection particle beam MS
Source Temperature	: 250°C
Electron Energy	: 70 eV
Scan Range	: 40-1000

PM n°
59200

CAS n°
35074-77-2

Name
1,6-HEXAMETHYLENE BIS(3-(3,5-DI-TERT-BUTYL-4-HYDROXYPHENYL) PROPIONATE)

Solvent	$CDCL_3$
SF	250.13
SI	16384
SW	3125
NS	16

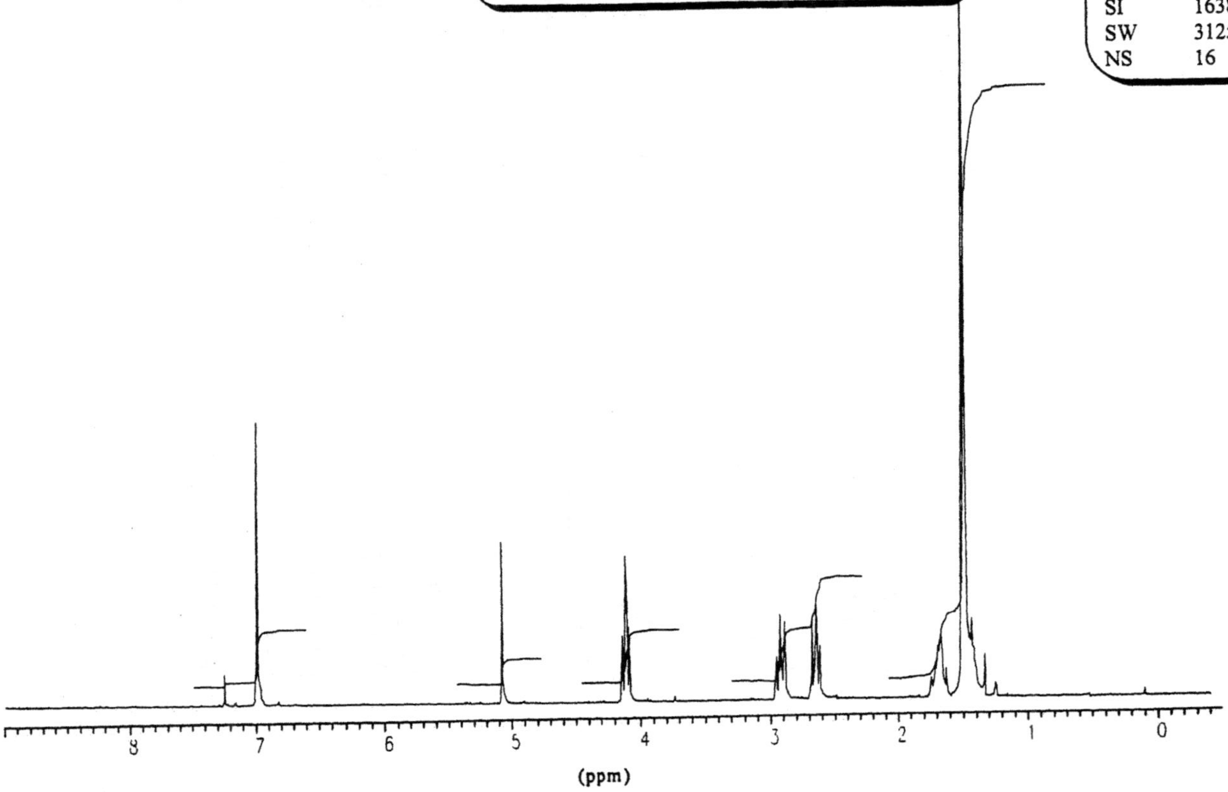

4-Hydroxybenzoic acid, methyl ester

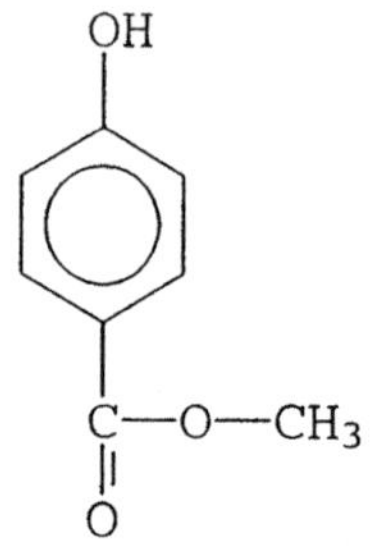

CAS No.	- 00099-76-3
PM Ref. No.	- 60200
Restrictions	- None
Formula	- $C_8H_8O_3$
Molecular weight	- 152

Alternative names	- Methylparaben
Physical Characteristics	- White needles, mp 131°C, bp 270 - 280°C decomposes. - Soluble in alcohol, acetone, ether.
Handling	- Store at room temperature.
Safety	-
Availability	- Standard sample was supplied.
Current uses	- Preservative.
Applications	- Antioxidant, paper.
GC Retention time	- 25.7 min.

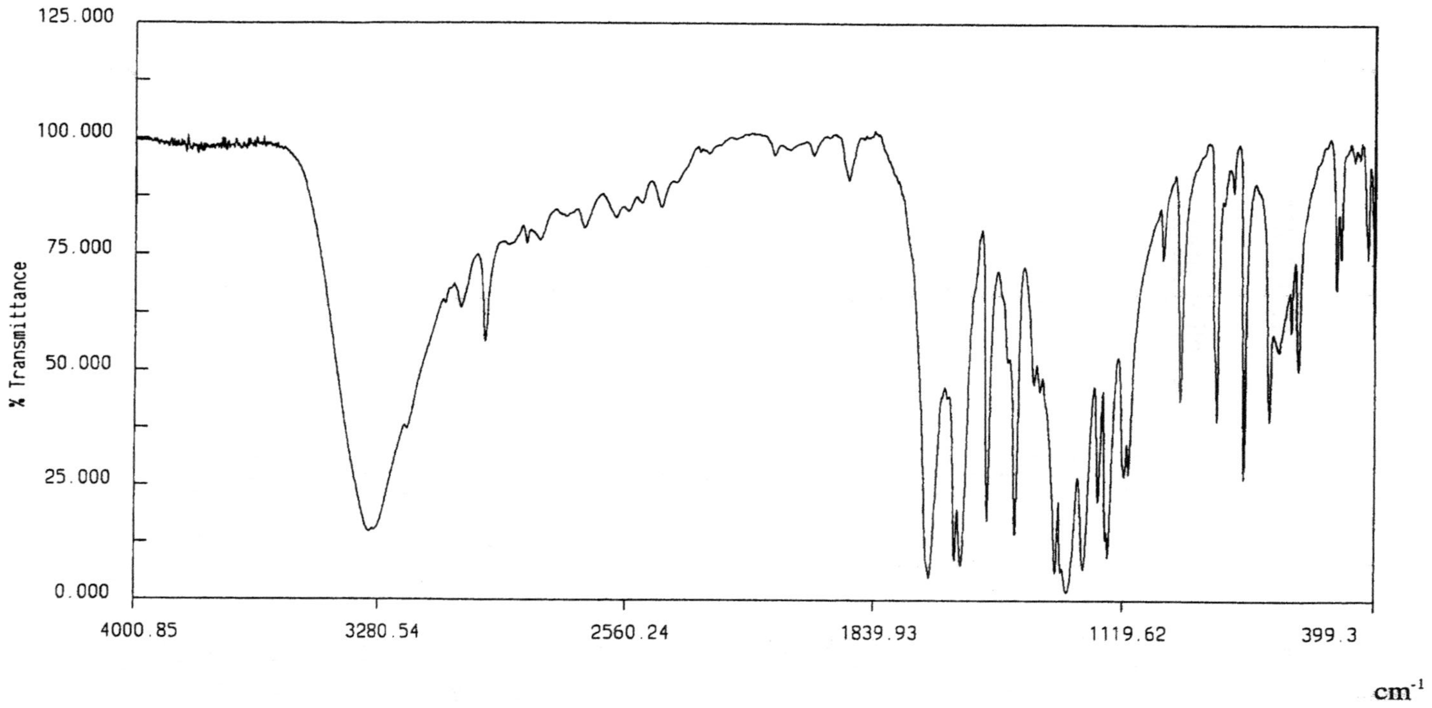

Name : 4-HYDROXYBENZOIC ACID, METHYL ESTER

PM Nr : 60200 CAS Nr : 00099-76-3

KBr pellet

CAS No. 00099-76-3 PM Ref. No. 60200

4-HYDROXYBENZOIC ACID METHYL ESTER

Peak wavenumber	% T	Relative intensity
420.5	74.4	26.1
497.7	74.6	25.8
509.3	67.7	32.8
617.3	50.2	50.7
638.5	58.6	42.1
671.3	54.4	46.4
698.3	39.2	61.8
771.6	26.6	74.7
850.7	39.3	61.8
956.8	43.6	57.3
1008.9	74.1	26.4
1107.3	27.6	73.6
1118.9	27.1	74.2
1163.2	9.4	92.2
1194.1	21.6	79.8
1234.6	6.8	94.8
1279.0	1.7	100.0
1315.6	6.1	95.5
1379.3	47.0	53.9
1435.2	14.5	87.0
1514.3	17.5	83.9
1589.5	7.6	94.0
1606.9	8.7	92.9
1680.2	5.0	96.7
2459.5	85.2	15.0
2459.5	85.2	15.0
2515.5	86.1	14.2
2681.4	80.8	19.6
2810.6	78.1	22.3
2899.4	77.0	23.4
2965.0	56.2	44.6
3034.4	63.5	37.1
3078.8	64.5	36.2
3194.5	37.2	63.9
3292.9	15.1	86.4

4-Hydroxybenzoic acid, methyl ester

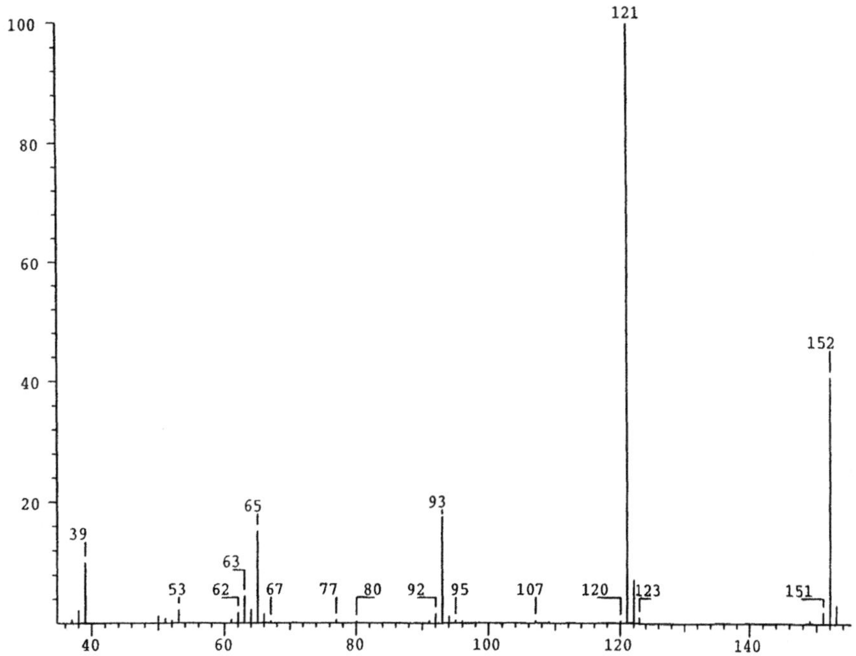

M/Z	Ion Intensity (%)	M/Z	Ion Intensity (%)
38	2.2	92	1.6
39	10.1	93	17.8
50	1.2	94	1.1
53	2.2	121	100.0
62	1.7	122	7.3
63	4.5	123	1.0
64	2.3	151	2.0
65	15.3	152	41.0
66	1.5	153	3.16

Spectrometer	: Finnigan Mat SSQ 700
Inletsystem	: Capillary GC/MS
Source Temperature	: 180°C
Electron Energy	: 70 eV
Scan Range	: 25-450

PM n°	CAS n°	Name
60200	00099-76-3	4-HYDROXYBENZOIC ACID, METHYL ESTER

Solvent	$CDCL_3$
SF	250.13
SI	16384
SW	3125
NS	32

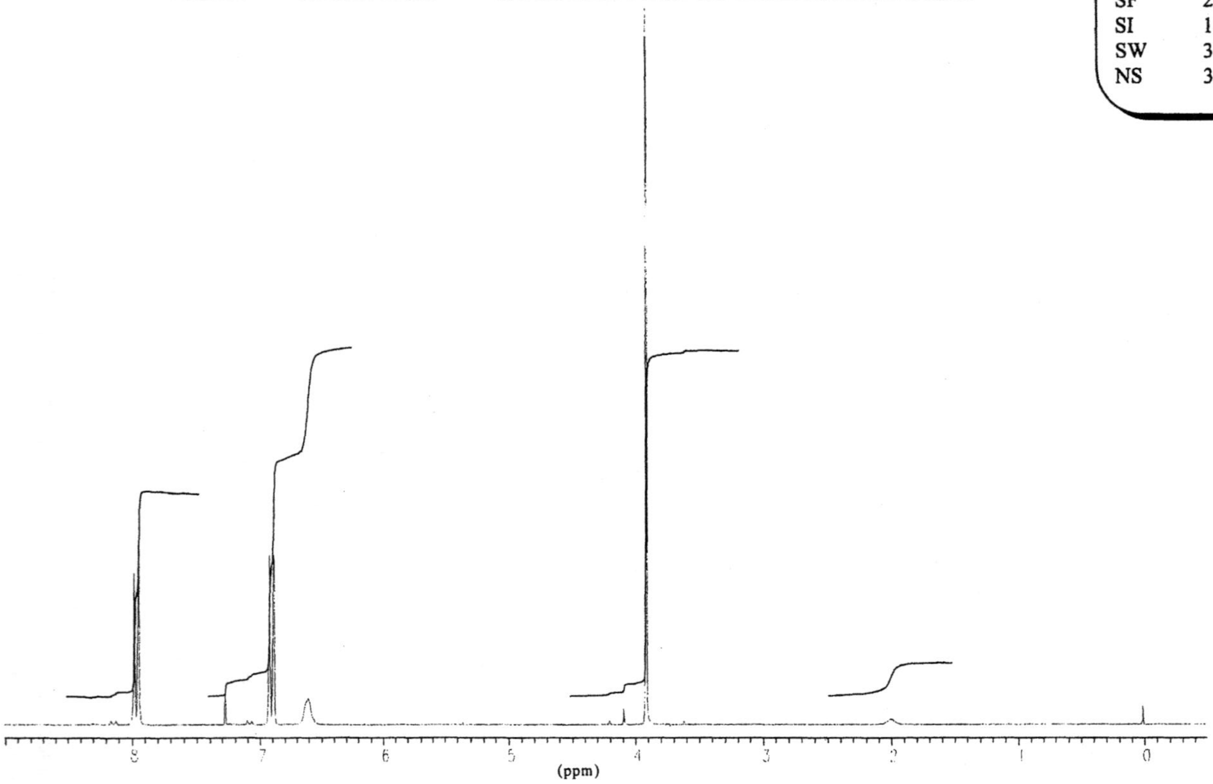

2-(2-Hydroxy-3,5-bis(1,1-dimethylbenzyl)phenyl)benzotriazole

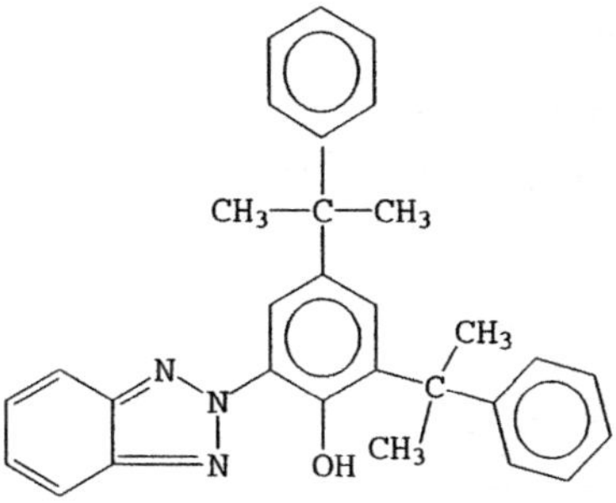

CAS No.	- 70321-86-7
PM Ref. No.	- 60320
Restrictions	- SML = 1.5 mg/kg
Formula	- $C_{30}H_{29}N_3O$
Molecular weight	- 447

Alternative names	- 2-(2H-benzotriazol-2-yl)-4,6-bis(1-methyl-1-phenylethyl)phenol
Physical Characteristics	- Slightly yellow powder, mp 137 - 141°C. - Soluble in benzene, chloroform, dichloromethane.
Handling	- Store at room temperature.
Safety	-
Availability	- Standard sample was supplied.
Current uses	- UV-stabilizer.
Applications	- PC, polyesters.
GC Retention time	- 60.3 min.

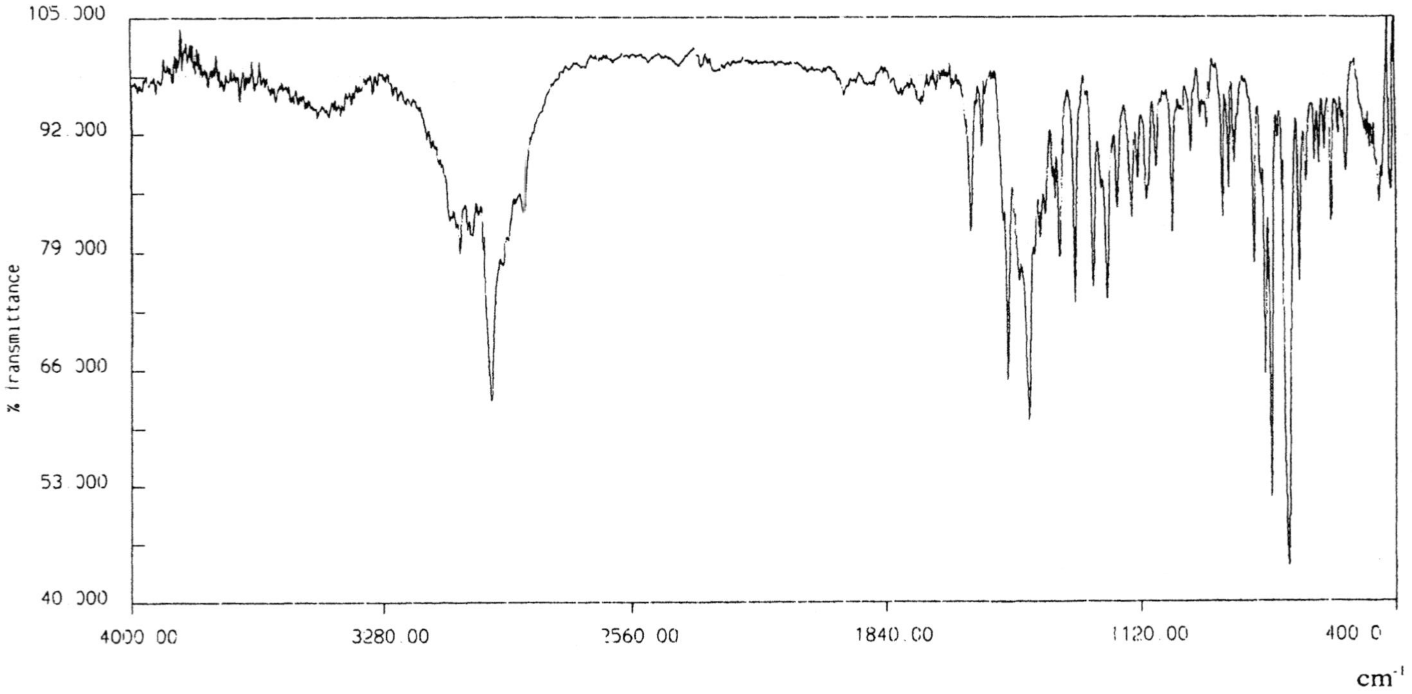

Name : 2-(2-HYDROXY-3,5-BIS(1,1-DIMETHYLBENZYL) PHENYL)BENZOTRIAZOLE

PM Nr : 60320 CAS Nr : 70321-86-7

KBr pellet

CAS No. 70321-86-7 PM Ref. No. 60320

2-(2-HYDROXY-3,5-BIS(1,1-DIMETHYLBENZYL)PHENYL)BENZOTRIAZOLE

Peak wavenumber	% T	Relative intensity
410,9	85,7	25,5
445,6	84,2	28,2
538,2	87,6	22,1
580,6	82,2	31,7
613,4	88,5	20,5
626,9	88,7	20,1
652,0	86,5	24,1
671,3	75,6	43,5
702,2	43,9	100,0
750,4	51,7	86,1
767,8	65,5	61,5
796,7	77,6	39,9
852,6	88,4	20,7
870,0	85,8	25,3
887,4	82,6	31,0
976,1	89,9	18,0
1030,1	81,0	33,9
1074,5	88,2	21,0
1103,4	84,5	27,6
1126,6	86,6	23,9
1145,9	82,6	31,0
1186,4	83,7	29,0
1215,3	73,7	46,9
1253,9	75,0	44,6
1304,0	73,3	47,6
1348,4	78,3	38,7
1387,0	82,9	30,5
1402,4	80,4	34,9
1425,6	78,3	38,7
1435,2	60,3	70,7
1495,0	64,8	62,7
1599,2	81,1	33,7
2916,7	80,2	35,3
2965,0	62,5	66,8
3019,0	80,6	34,6
3055,6	78,8	37,8

2-(2-Hydroxy-3,5-bis(1,1-dimethylbenzyl)phenyl)benzotriazole

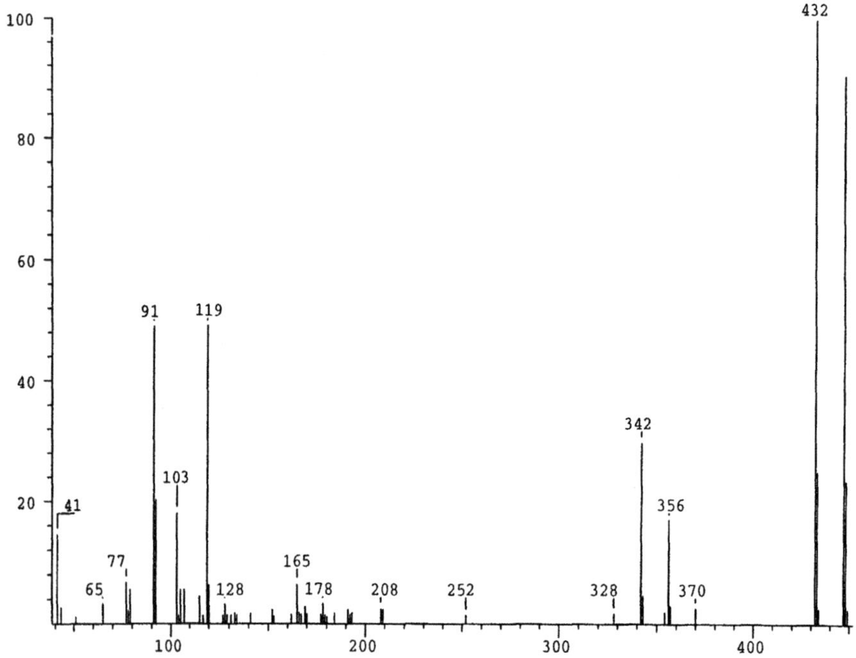

M/Z	Ion Intensity (%)	M/Z	Ion Intensity (%)
41	14.6	120	6.4
65	3.3	342	29.9
77	6.7	356	17.2
91	49.1	432	100.0
92	20.3	433	25.1
103	18.1	447	90.7
105	5.7	448	23.5
119	49.3	449	2.5

Spectrometer	: Finnigan Mat SSQ 700
Inletsystem	: Capillary GC/MS
Source Temperature	: 180°C
Electron Energy	: 70 eV
Scan Range	: 35-500

PM n°	CAS n°	Name
60320	70321-86-7	2-(2-HYDROXY-3,5-BIS(1,1-DIMETHYLBENZYL) PHENYL) BENZOTRIAZOLE

Solvent	$CDCL_3$
SF	250.13
SI	16384
SW	3125
NS	64

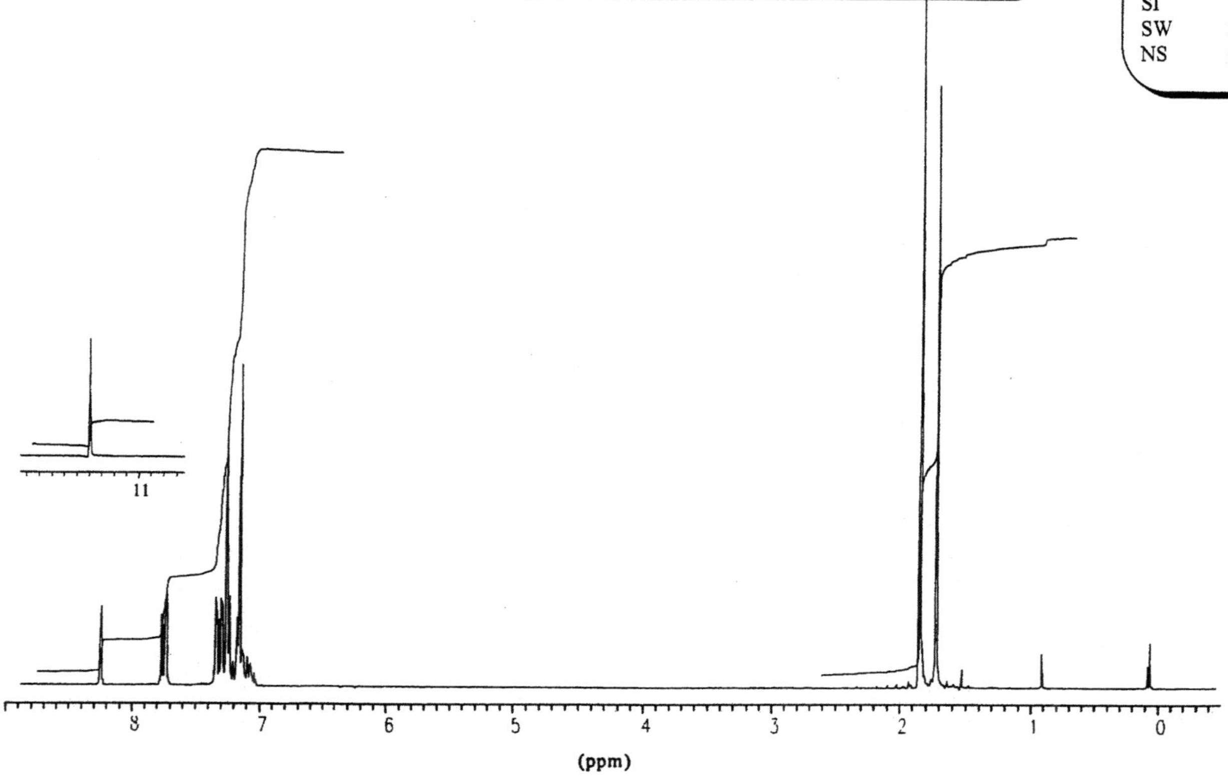

2-(2-Hydroxy-3-tert-butyl-5-methylphenyl)-5-chlorobenzotriazole

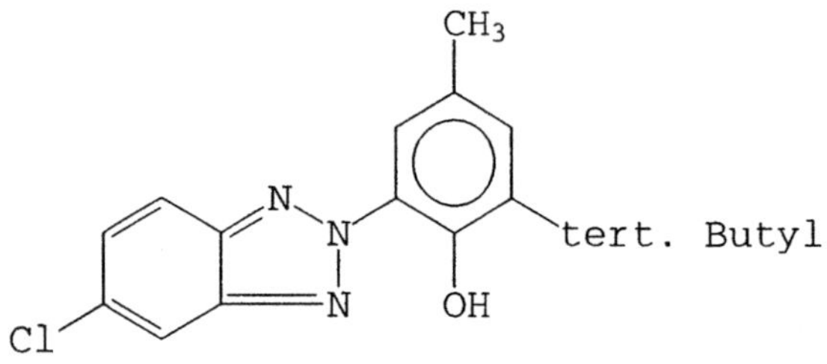

CAS No.	- 03896-11-5
PM Ref. No.	- 60400
Restrictions	- SML = 30 mg/kg with other benzotriazole derivatives
Formula	- $C_{17}H_{18}ClN_3O$
Molecular weight	- 315.5

Alternative names - 2-(3'-tert-Butyl-2'-hydroxy-5'-methylphenol)-5-chlorobenzotriazole

Physical Characteristics - Slightly yellow powder, mp 138 - 141°C.
- Soluble in benzene, toluene, cyclohexane.

Handling - Store at room temperature.

Safety -

Availability - Standard sample was supplied.

Current uses - UV-stabilizer.

Applications - General use.

GC Retention time - 47.2 min.

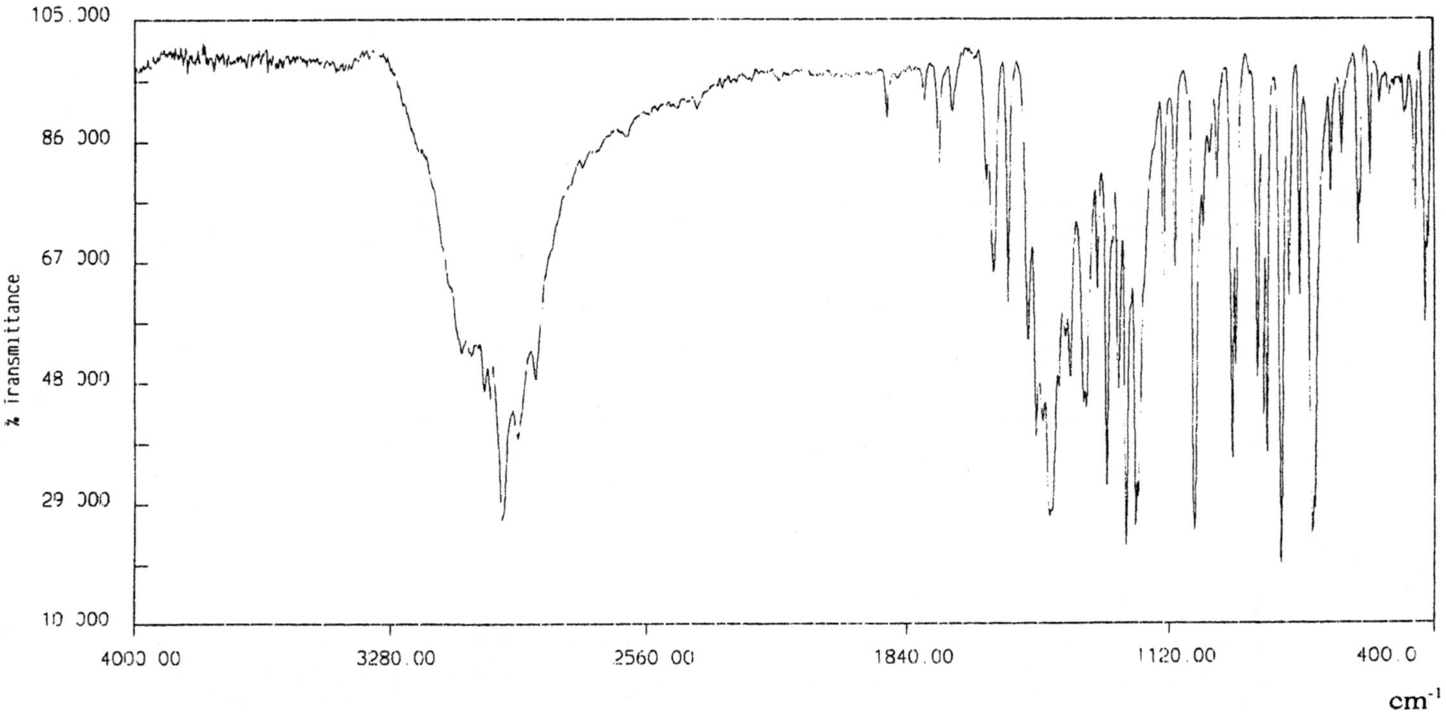

Name : 2-(2-HYDROXY-3-Tert-BUTYL-5-METHYLPHENYL) -5-CHLORO-2-BENZOTRIAZOLE

PM Nr : 60400 CAS Nr : 03896-11-5

KBr pellet

CAS No. 03896-11-5 PM Ref. No. 60400

2-(2-HYDROXY-3-tert-BUTYL-5-METHYLPHENYL)-5-CHLORO -2-BENZOTRIAZOLE

Peak wavenumber	% T	Relative intensity
422,5	57,3	53,2
447,5	71,9	35,0
474,6	90,1	12,3
540,1	91,8	10,2
565,2	80,5	24,3
598,0	69,6	37,8
640,4	83,9	20,0
669,4	78,0	27,4
719,5	24,4	94,1
754,3	61,5	47,9
785,1	60,9	48,7
806,3	19,7	100,0
846,9	36,9	78,6
858,4	42,8	71,2
873,9	48,9	63,6
935,6	50,6	61,5
945,2	35,9	79,8
1008,9	84,0	19,9
1026,3	72,3	34,5
1049,4	24,7	93,7
1103,4	66,1	42,2
1136,2	59,7	50,2
1215,3	25,5	92,8
1240,4	22,3	96,7
1261,6	46,8	66,2
1292,5	31,9	84,8
1317,5	62,6	46,6
1348,4	43,8	70,0
1392,8	48,8	63,7
1448,7	27,1	90,8
1485,4	39,3	75,6
1504,7	54,5	56,6
1560,6	60,1	49,7
1601,1	65,2	43,3
1620,4	79,8	25,1
2424,8	91,0	11,2
2619,7	86,6	16,7
2702,6	84,3	19,5
2868,5	48,5	64,1
2918,7	39,2	75,7
2961,1	26,4	91,6
2993,9	45,3	68,1

2-(2-Hydroxy-3-tert-butyl-5-methylphenyl)-5-chlorobenzotriazole

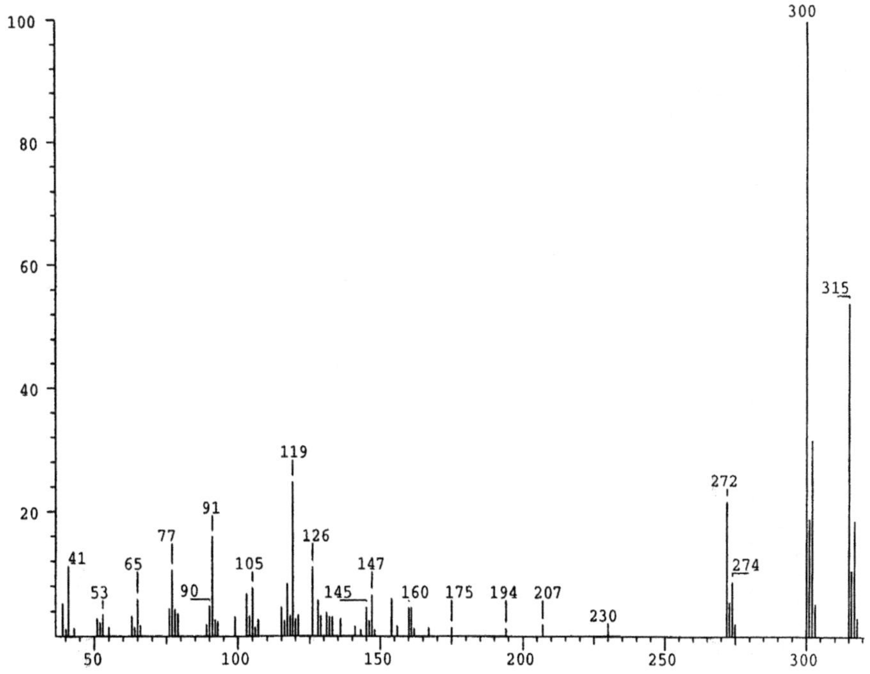

M/Z	Ion Intensity (%)	M/Z	Ion Intensity (%)
41	10.1	272	21.7
53	3.5	274	8.7
65	6.0	300	100.0
77	10.6	301	18.9
91	15.9	302	31.7
105	7.8	315	54.0
117	8.3	316	10.6
119	24.8	317	18.6
126	11.1	318	2.9

Spectrometer	: Finnigan Mat SSQ 700
Inletsystem	: Capillary GC/MS
Source Temperature	: 180°C
Electron Energy	: 70 eV
Scan Range	: 35-400

PM n°	CAS n°	Name
60400	03896-11-5	2-(2-HYDROXY-3-TERT-BUTYL-5- METHYLPHENYL) -5-CHLOROBENZOTRIAZOLE

Solvent	$CDCL_3$
SF	250.13
SI	16384
SW	3125
NS	16

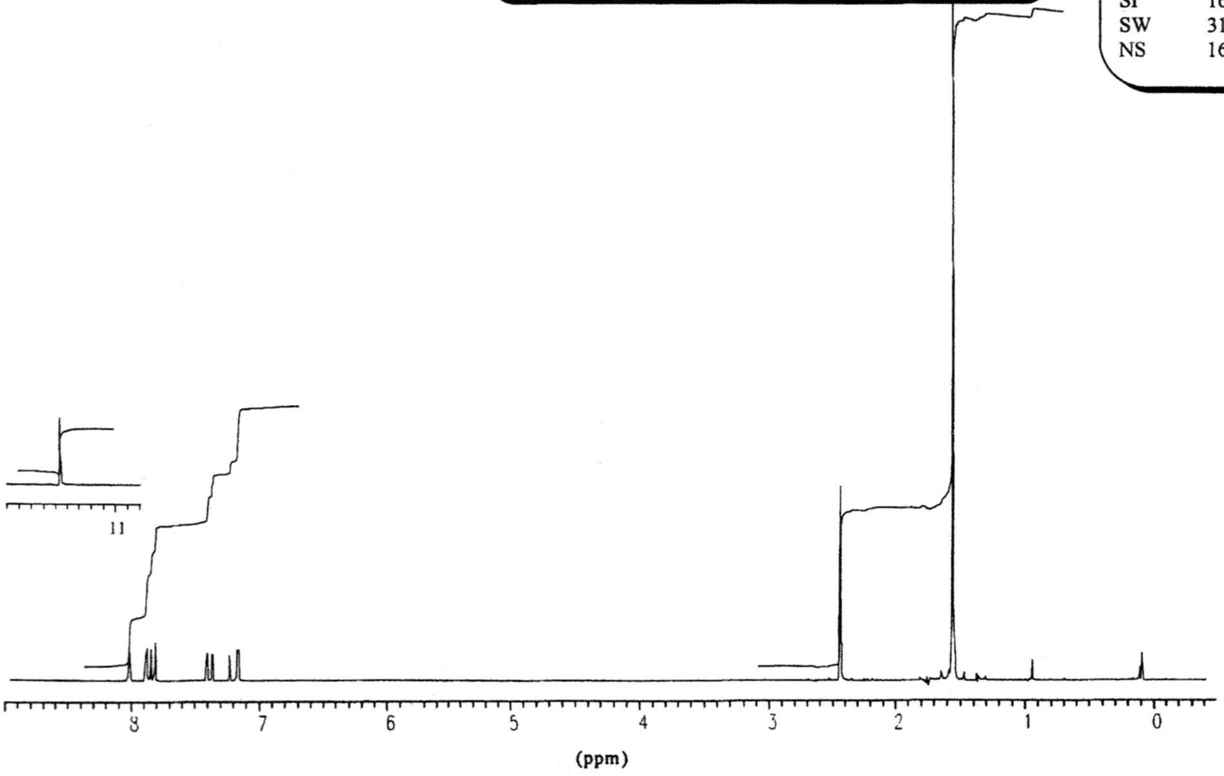

1-(2-Hydroxyethyl)-4-hydroxy-2,2,6,6-tetramethyl piperidine-succinic acid, dimethyl ester, copolymer

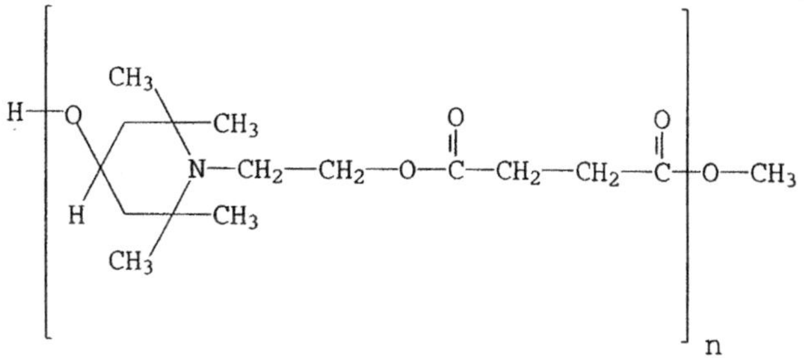

CAS No.	- 65447-77-0
PM Ref. No.	- 60800
Restrictions	- SML = 30 mg/kg Specifications to be added.
Formula	- $[C_{31}H_{54}N_2O_9]_n$
Molecular weight	- 1500 - 5000

Alternative names - Butanedionic acid, polymer with 4-hydroxy-2,2,6,6-tertamethyl-1-piperidineethanol

Physical Characteristics - Slightly yellow powder, mp 100 - 120°C.
- Soluble in benzene, chloroform, toluene, dichloromethane.

Handling - Store at room temperature.
Safety - Harmful.
Availability - Standard sample was supplied.

Current uses - UV-stabilizer.
Applications - EVA, PP, PE.

GC Retention time -

Note - Also known as CAS No. 70198-29-7, and PM Ref. No. 81215, and 91150.

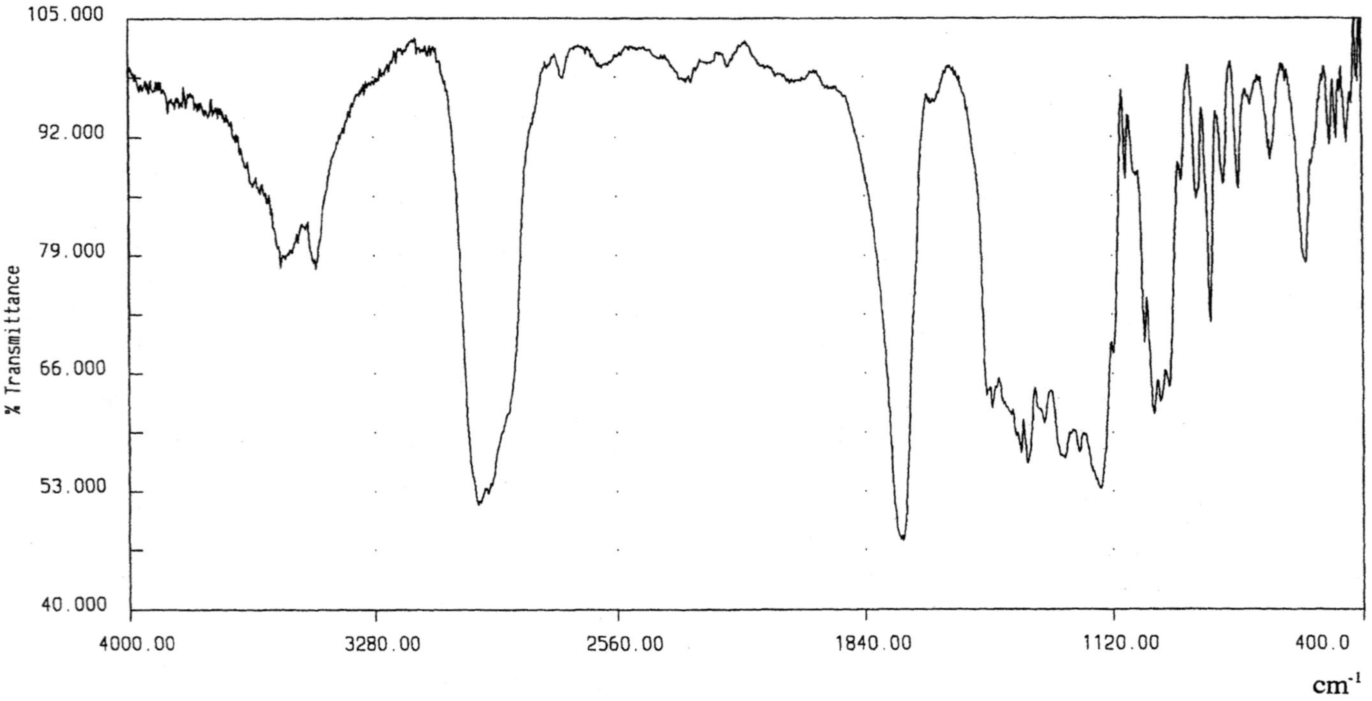

Name : 1-(2-HYDROXYETHYL)-4-HYDROXY-2,2,6,6-TETRAMETHYLPIPERIDINE-SUCCINIC ACID, DIMETHYLESTER, COPOLYMER

PM Nr : 60800 CAS Nr : 65447-77-0

KBr pellet

CAS No. 65447-77-0 PM Ref. No. 60800

1-(2-HYDROXYETHYL)-4-HYDROXY-2,2,6,6-TETRAMETHYLPIPERIDINE-SUCCINIC ACID, DIMETHYL ESTER, COPOLYMER (M.W.1500-5000)

Peak wavenumber	% T	Relative intensity
445,6	90,6	18,3
474,6	92,0	15,5
493,8	87,7	23,9
563,3	75,5	47,6
665,5	87,8	23,7
758,1	88,8	21,8
839,1	74,8	49,0
877,7	87,1	25,1
920,2	88,4	22,5
956,8	65,5	67,0
981,9	63,9	70,1
999,3	62,7	72,5
1028,2	70,8	56,7
1116,9	69,9	58,5
1155,5	54,5	88,4
1257,7	57,9	81,8
1317,5	61,9	74,0
1365,8	57,2	83,2
1385,1	58,4	80,8
1468,0	63,6	70,7
1732,3	48,5	100,0
2945,7	48,8	99,5
3453,0	72,8	52,8

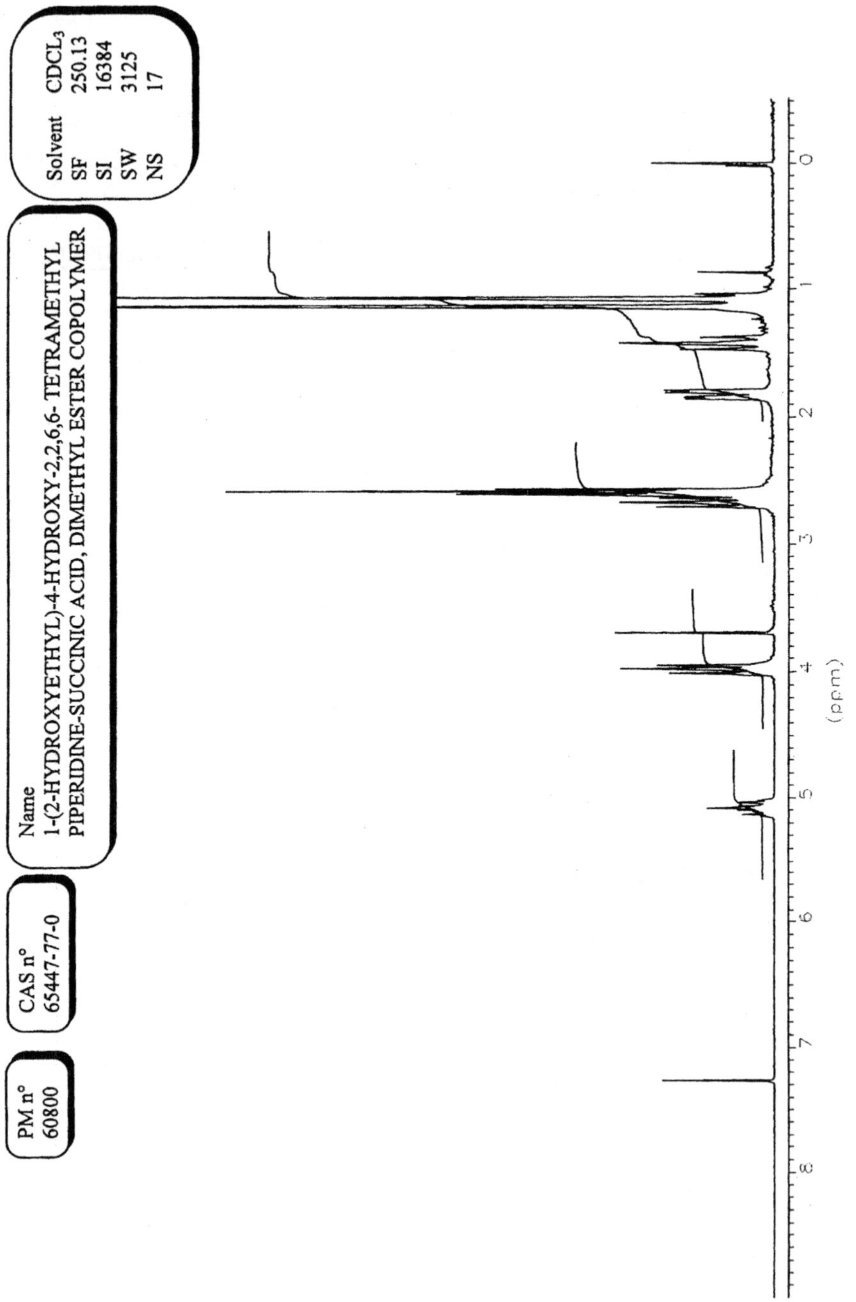
PM n°
60800
CAS n°
65447-77-0
Name
1-(2-HYDROXYETHYL)-4-HYDROXY-2,2,6,6- TETRAMETHYL PIPERIDINE-SUCCINIC ACID, DIMETHYL ESTER COPOLYMER
Solvent CDCL3
SF 250.13
SI 16384
SW 3125
NS 17
8
7
6
5
4
3
2
1
0
(ppm)

2-Hydroxy-4-methoxybenzophenone

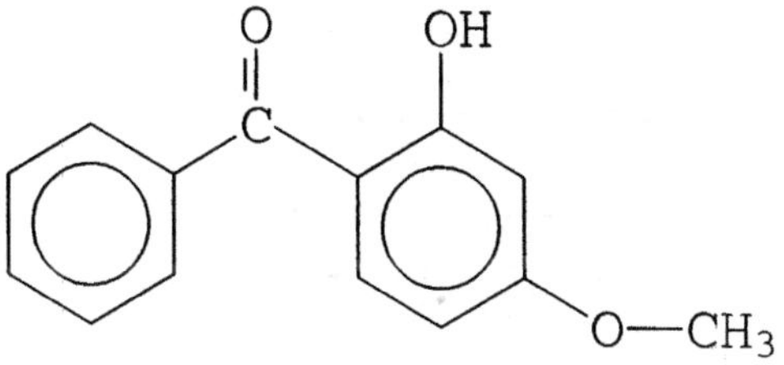

CAS No.	- 00131-57-7
PM Ref. No.	- 61360
Restrictions	- SML = 6.0 mg/kg
Formula	- $C_{14}H_{12}O_3$
Molecular weight	- 228

Alternative names	- Oxybenzone
Physical Characteristics	- White crystalline powder, mp 63 - 65°C, bp 150 - 160°C (5 mm Hg). - Soluble in most organic solvents.
Handling	- Store at room temperature.
Safety	-
Availability	- Standard sample was supplied.
Current uses	- UV-stabilizer.
Applications	- General use.
GC Retention time	- 38.2 min.

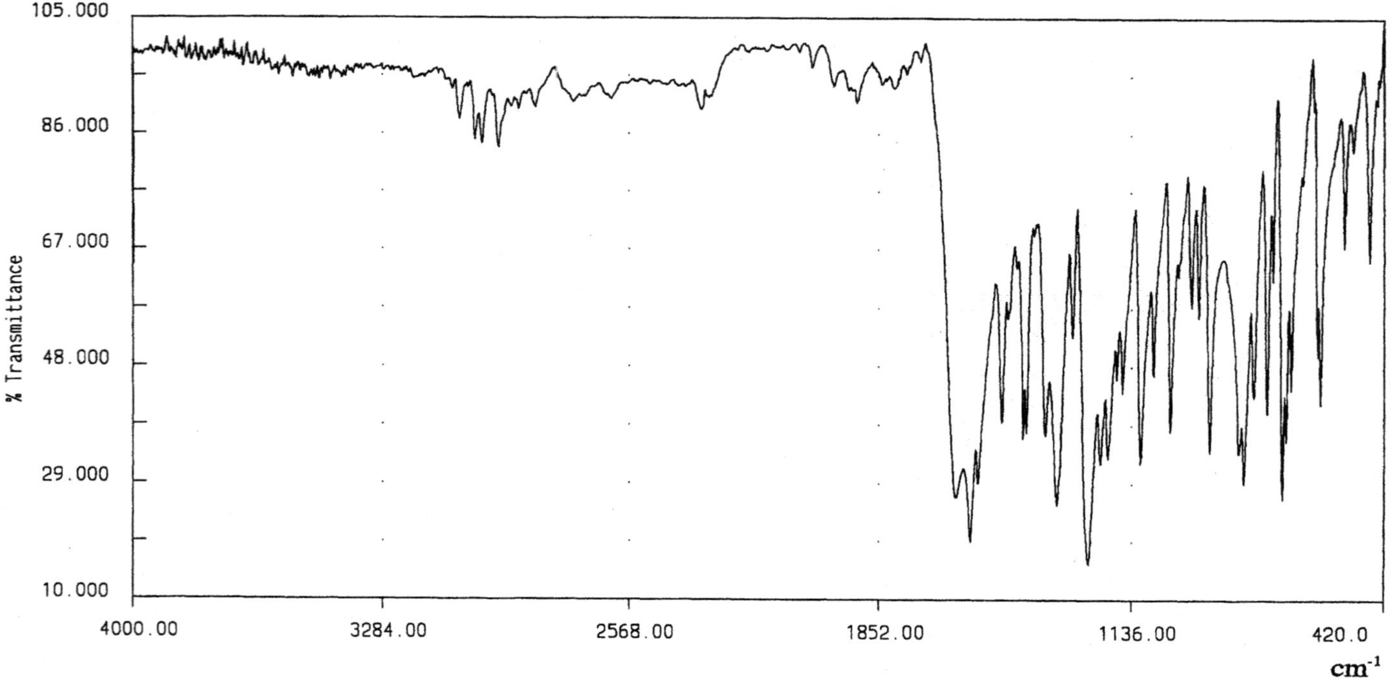

Name : 2-HYDROXY-4-METHOXYBENZOPHENONE

PM Nr : 61360 CAS Nr : 00131-57-7

KBr pellet

CAS No. 00131-57-7 PM Ref. No. 61360

2-HYDROXY-4-METHOXYBENZOPHENONE

Peak wavenumber	% T	Relative intensity
461.0	69.2	36.2
532.4	69.0	36.4
599.9	43.0	67.1
682.9	43.1	66.9
698.3	36.0	75.3
708.0	28.0	84.8
733.0	58.4	48.9
750.4	40.2	70.4
789.0	41.6	68.8
817.9	29.0	83.5
914.4	34.5	77.0
945.2	51.7	56.8
964.5	51.5	57.1
1026.3	34.6	76.9
1074.5	40.6	69.9
1113.1	30.3	82.0
1163.2	37.4	73.6
1178.7	38.8	72.0
1205.7	27.3	85.5
1225.0	21.7	92.1
1259.7	15.0	100.0
1304.0	43.3	66.7
1348.4	23.4	90.1
1381.2	30.6	81.7
1435.2	30.5	81.8
1444.9	29.5	83.0
1489.2	43.0	67.0
1506.6	31.1	81.0
1574.1	23.7	89.8
1595.3	17.5	97.0
1633.9	85.8	16.7
2947.6	84.4	18.3
2995.8	84.4	18.4
3017.0	85.6	17.0
3061.4	89.0	12.9

2-Hydroxy-4-methoxybenzophenone

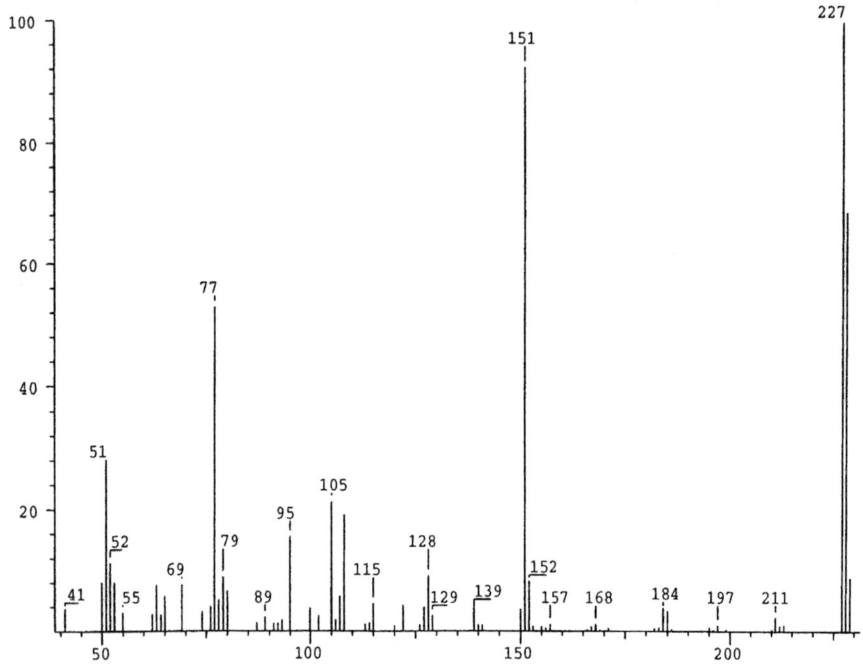

M/Z	Ion Intensity (%)	M/Z	Ion Intensity (%)
41	3.8	105	21.3
50	8.1	108	19.2
51	27.2	128	9.2
52	11.3	151	92.5
53	8.1	152	8.3
77	52.9	227	100.0
79	9.0	228	68.8
95	15.5	229	8.8

Spectrometer	: Finnigan Mat SSQ 700
Inletsystem	: Capillary GC/MS
Source Temperature	: 180°C
Electron Energy	: 70 eV
Scan Range	: 40-500

PM n°	CAS n°	Name
61360	00131-57-7	2-HYDROXY-4-METHOXYBENZOPHENONE

SF	250.13
SI	8192
SW	2380
NS	32

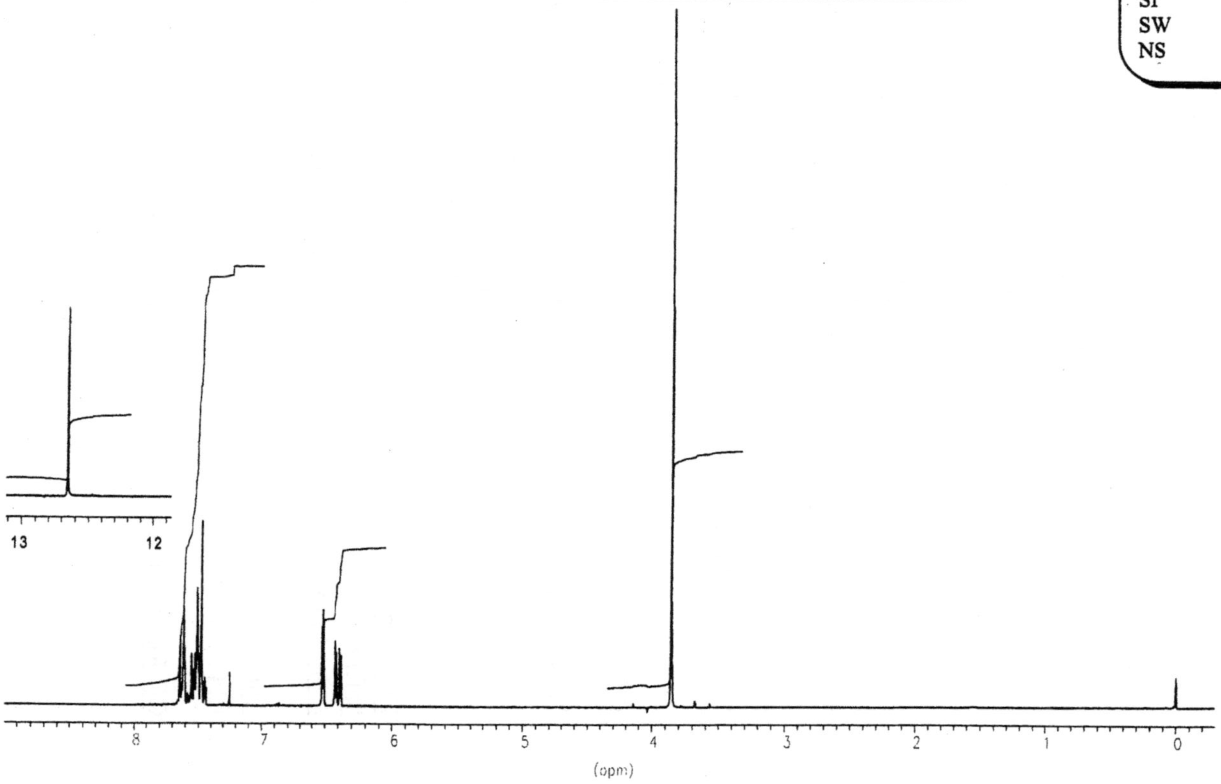

2-(2-Hydroxy-5-methylphenyl)benzotriazole

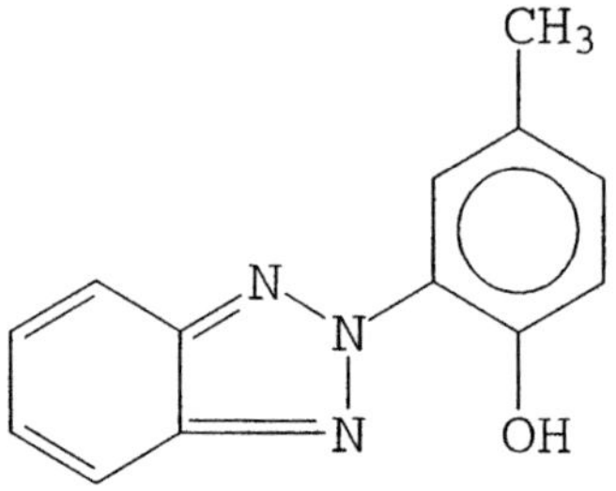

CAS No.	- 02440-22-4
PM Ref. No.	- 61440
Restrictions	- SML = 30 mg/kg with other benzotriazole derivatives
Formula	- $C_{13}H_{11}N_3O$
Molecular weight	- 225

Alternative names -

Physical Characteristics
- Light yellow crystalline powder, mp 128 - 132°C.
- Soluble in benzene, toluene, acetone, chloroform.

Handling - Store at room temperature.

Safety -

Availability - Standard sample was supplied.

Current uses - UV-absorber.

Applications - General use.

GC Retention time - 38.9 min.

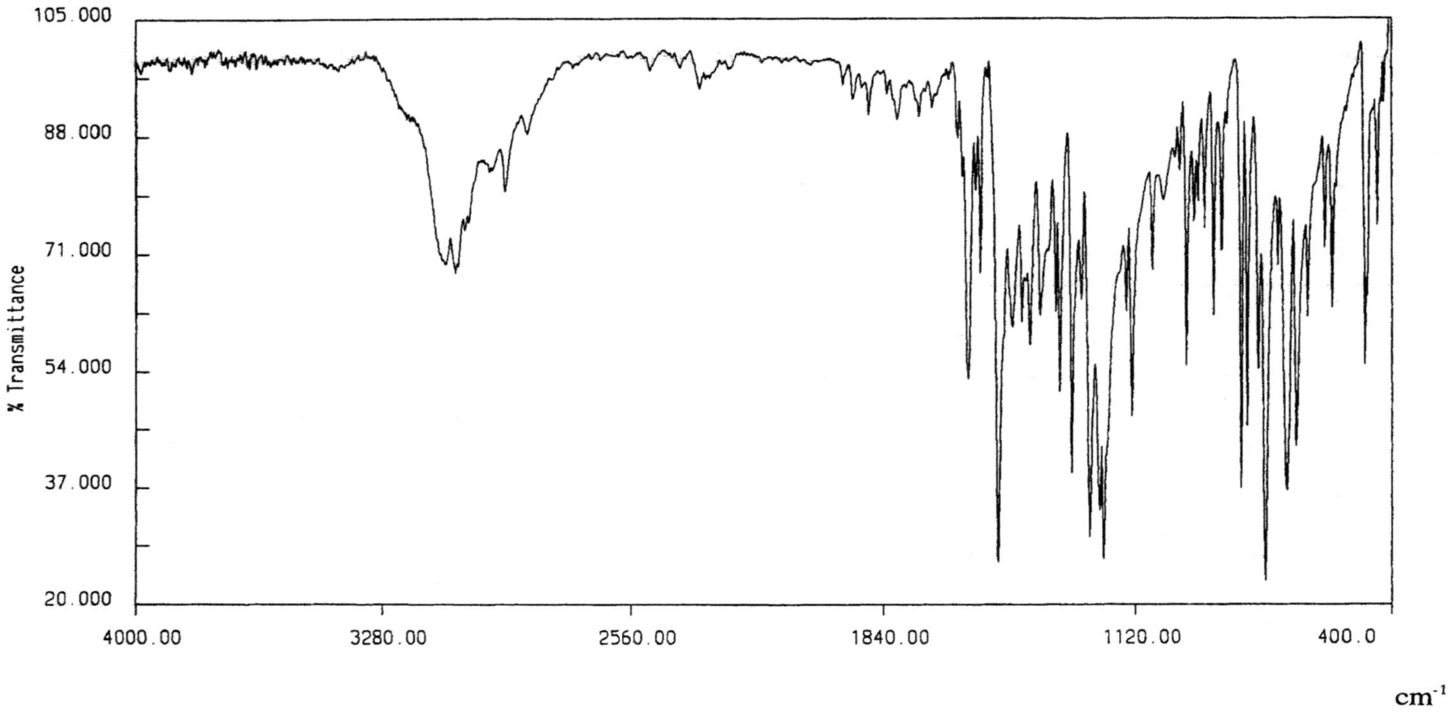

Name : 2-(2-HYDROXY-5-METHYLPHENYL)BENZOTRIAZOLE

PM Nr : 61440 CAS Nr : 02440-22-4

KBr pellet

CAS No. 02440-22-4 PM Ref. No. 61440

2-(2-HYDROXY-5-METHYLPHENYL)BENZOTRIAZOLE

Peak wavenumber	% T	Relative intensity
439.8	75.3	32.3
472.6	54.9	58.8
561.4	63.3	47.9
580.6	72.0	36.5
628.9	62.0	49.6
661.7	43.0	74.4
688.7	36.6	82.8
713.8	69.4	40.0
748.5	23.4	100.0
769.7	54.4	59.6
800.6	45.9	70.6
817.9	36.9	82.4
873.9	71.6	37.1
897.0	62.2	49.3
922.1	74.9	32.8
939.5	78.7	27.8
953.0	75.8	31.7
974.2	54.8	59.0
991.5	83.1	22.1
1039.8	78.8	27.6
1070.6	68.7	40.8
1132.4	47.3	68.8
1145.9	62.8	48.6
1211.4	26.6	95.9
1223.0	33.6	86.8
1252.0	29.7	91.8
1275.1	64.5	46.4
1304.0	39.0	79.6
1338.8	50.9	64.1
1348.4	62.6	48.8
1392.8	62.2	49.3
1421.7	57.9	55.0
1446.8	61.2	50.7
1473.8	60.3	51.8
1514.3	26.1	96.5
1564.5	68.3	41.4
1578.0	80.2	25.8
1599.2	52.8	61.6
1628.1	87.8	15.9
1738.1	90.9	11.9
1799.8	90.4	12.5
1880.8	91.0	11.7
2858.9	88.2	15.3
2922.5	80.1	26.0
3024.8	75.5	32.0
3065.3	68.3	41.4
3096.1	69.5	39.8

2-(2-Hydroxy-5-methylphenyl)benzotriazole

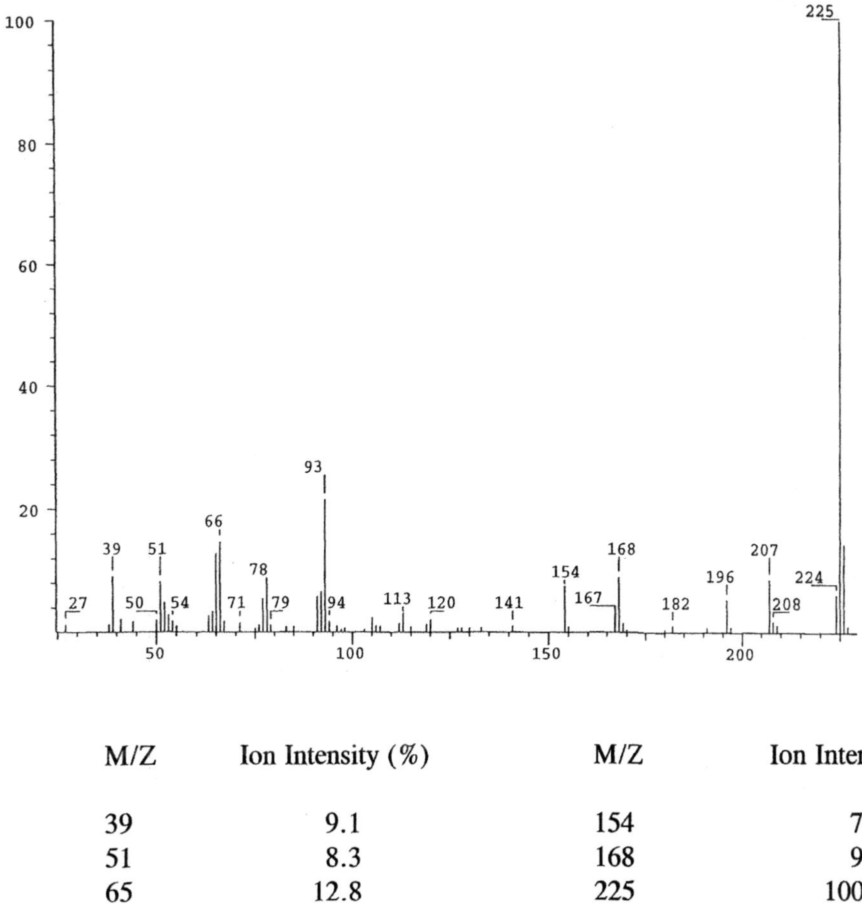

M/Z	Ion Intensity (%)	M/Z	Ion Intensity (%)
39	9.1	154	7.6
51	8.3	168	9.0
65	12.8	225	100.0
66	14.7	226	14.4
93	21.5	227	1.0

Spectrometer	: Finnigan Mat SSQ 700
Inletsystem	: Capillary GC/MS
Source Temperature	: 180°C
Electron Energy	: 70 eV
Scan Range	: 25-300

PM n°
61440

CAS n°
02440-22-4

Name
2-(2-HYDROXY-5-METHYLPHENYL)BENZOTRIAZOL

Solvent	$CDCL_3$
SF	250.13
SI	16384
SW	3125
NS	32

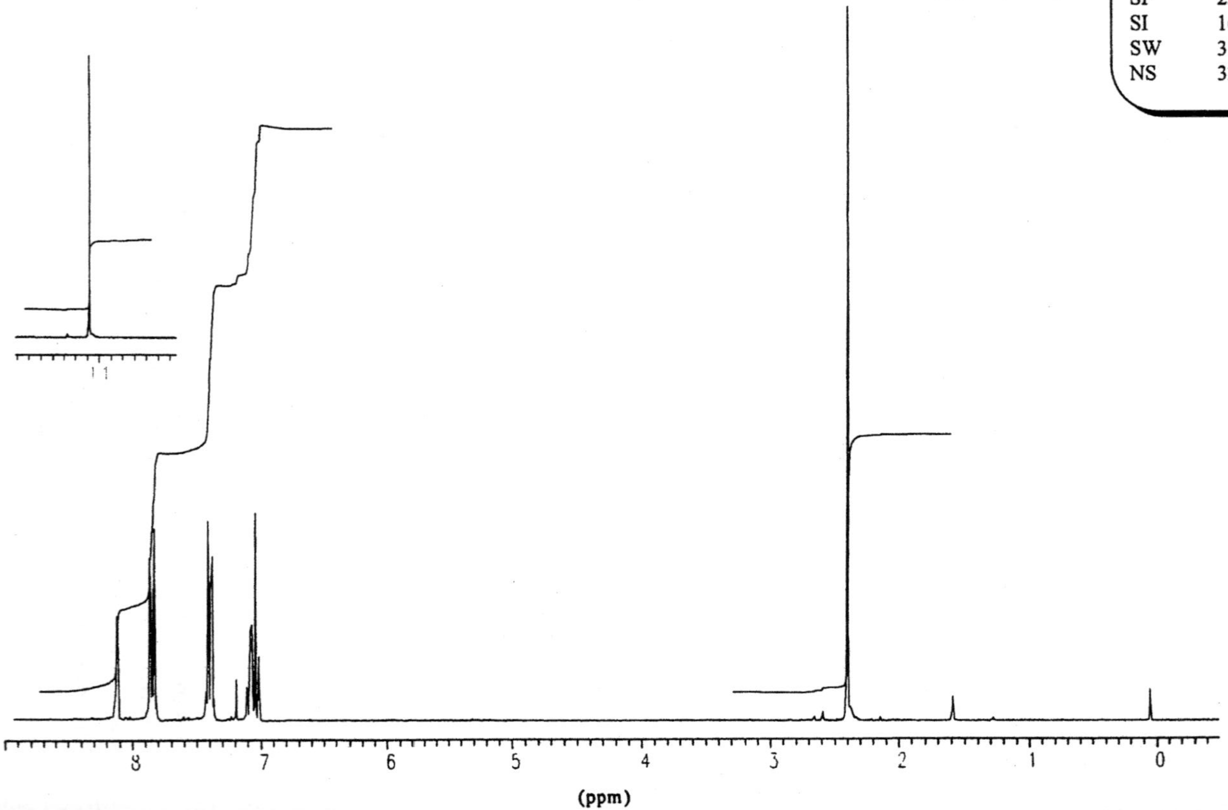

2-Hydroxy-4-n-octyloxybenzophenone

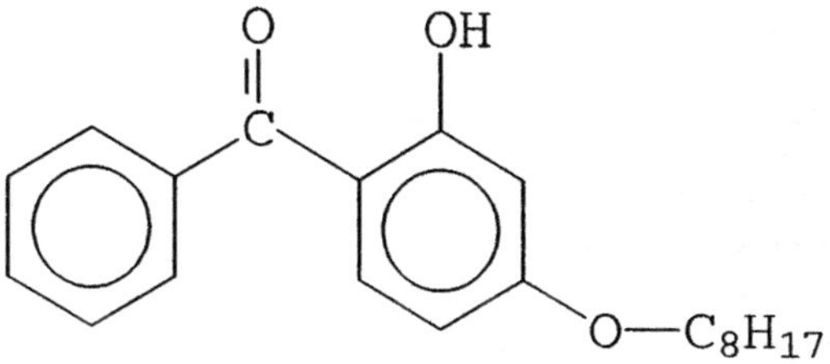

CAS No.	- 01843-05-6
PM Ref. No.	- 61600
Restrictions	- SML = 6.0 mg/kg
Formula	- $C_{21}H_{26}O_3$
Molecular weight	- 326
Alternative names	- Octabenzone Methanone, (2-hydroxy-4-(octyloxy)-phenyl) phenyl-
Physical Characteristics	- Pale yellow powder, mp 45 - 46°C. - Soluble in alcohol, benzene.
Handling	- Store at room temperature.
Safety	-
Availability	- Standard sample was supplied.
Current uses	- UV-stabilizer.
Applications	- General use.
GC Retention time	- 50.3 min.

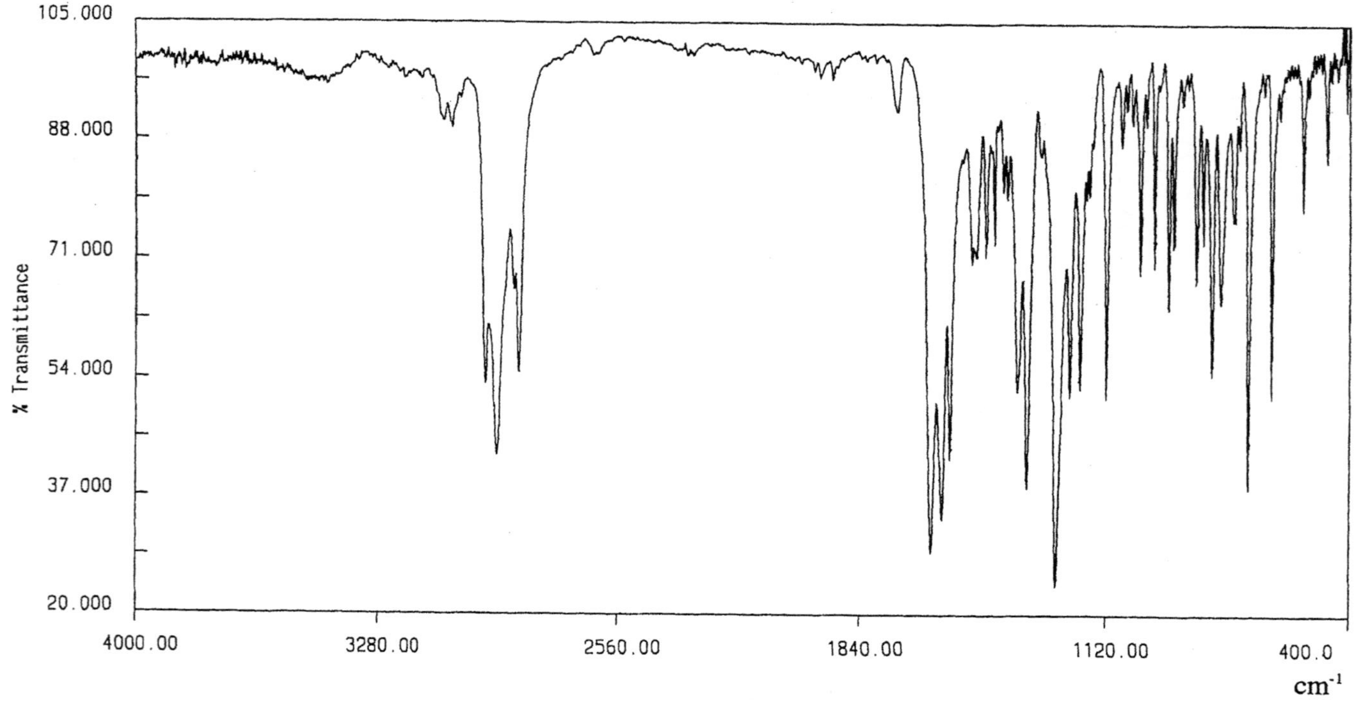

Name : 2-HYDROXY-4-n-OCTYLOXYBENZOPHENONE

PM Nr : 61600 CAS Nr : 01843-05-6

KBr pellet

CAS No. 01843-05-6 PM Ref. No. 61600

2-HYDROXY-4-n-OCTYLOXYBENZOPHENONE

Peak wavenumber	% T	Relative intensity
407.0	92.5	9.9
464.9	85.0	19.8
518.9	94.6	7.1
536.3	78.0	29.0
603.8	91.2	11.7
628.9	51.4	64.1
652.0	94.8	6.8
698.3	38.2	81.5
725.3	87.0	17.1
738.8	76.5	31.1
781.3	64.8	46.4
806.3	54.6	59.9
833.4	73.2	35.3
854.6	67.6	42.8
920.2	72.8	36.0
933.7	63.9	47.6
978.0	69.9	39.8
1020.5	69.0	41.0
1041.7	90.3	12.8
1074.5	87.1	17.1
1120.8	51.3	64.3
1169.0	79.9	26.5
1197.9	52.6	62.6
1226.9	51.5	64.0
1267.4	24.2	100.0
1352.3	38.3	81.5
1379.3	52.2	63.1
1406.3	79.5	27.1
1419.8	80.5	25.7
1444.9	73.1	35.5
1471.9	71.3	37.9
1496.9	71.3	37.8
1576.0	42.3	76.1
1597.3	33.8	87.4
1630.1	28.9	93.9
2920.6	42.9	75.4
2955.3	53.2	61.8
3055.6	89.6	13.7

2-Hydroxy-4-n-octyloxybenzophenone

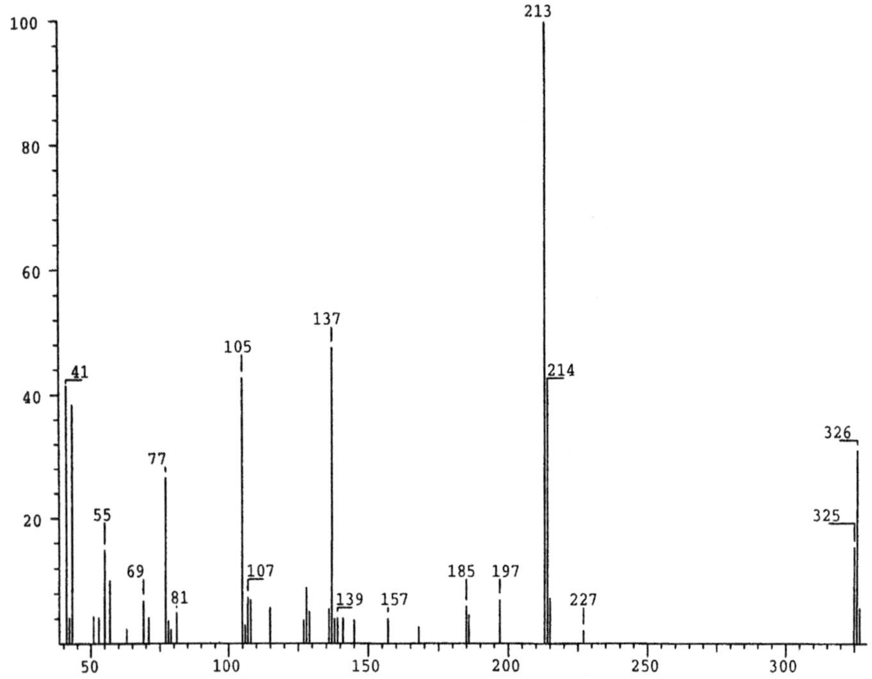

M/Z	Ion Intensity (%)	M/Z	Ion Intensity (%)
41	41.4	136	5.6
43	38.4	137	47.5
55	14.9	185	6.0
57	10.1	197	7.0
69	6.9	213	100.0
77	26.5	214	41.7
105	42.7	215	7.3
107	7.4	325	15.5
108	7.0	326	31.1
128	8.9	327	5.8

Spectrometer	: Finnigan Mat SSQ 700
Inletsystem	: Capillary GC/MS
Source Temperature	: 180°C
Electron Energy	: 70 eV
Scan Range	: 40-500

PM n° 61600

CAS n° 01843-05-6

Name 2-HYDROXY-4-n-OCTYLOXYBENZOPHENONE

Solvent	$CDCL_3$
SF	250.13
SI	16384
SW	4504
NS	32

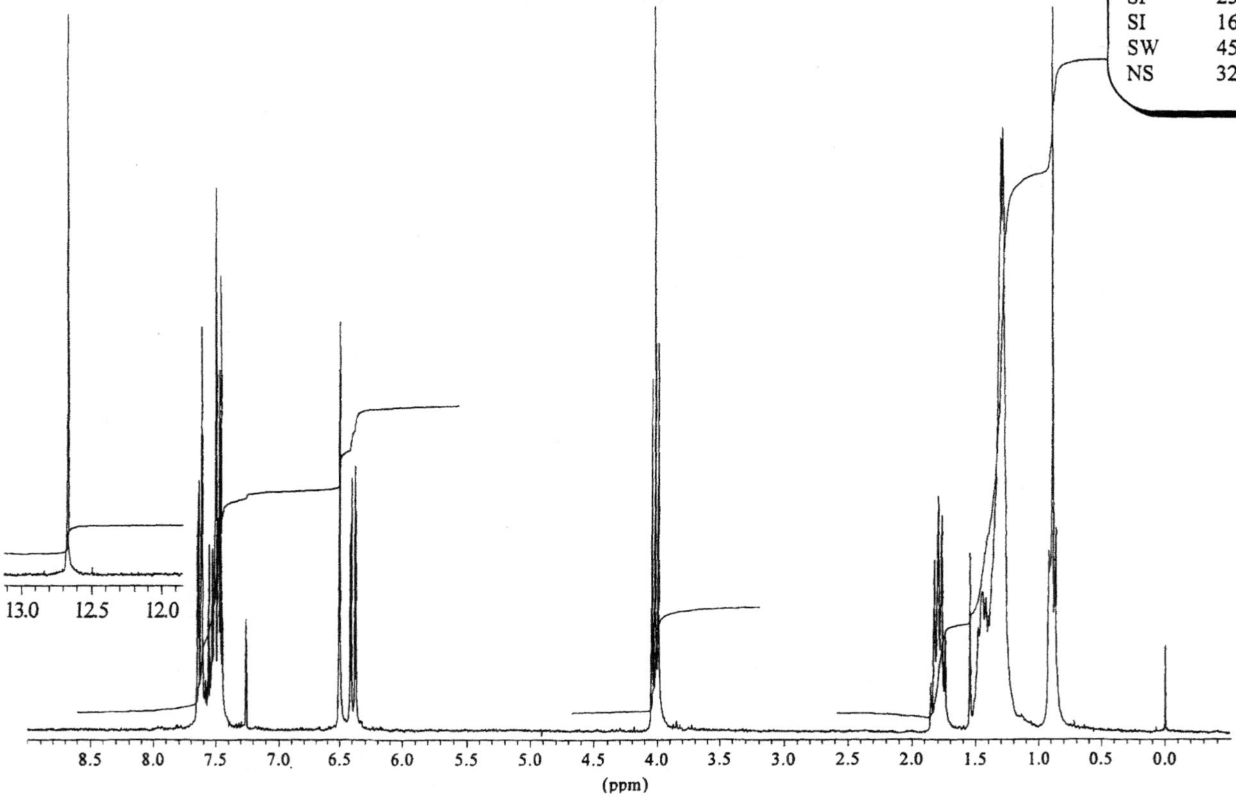

2,2'-Methylenebis(4-ethyl-6-tert-butylphenol)

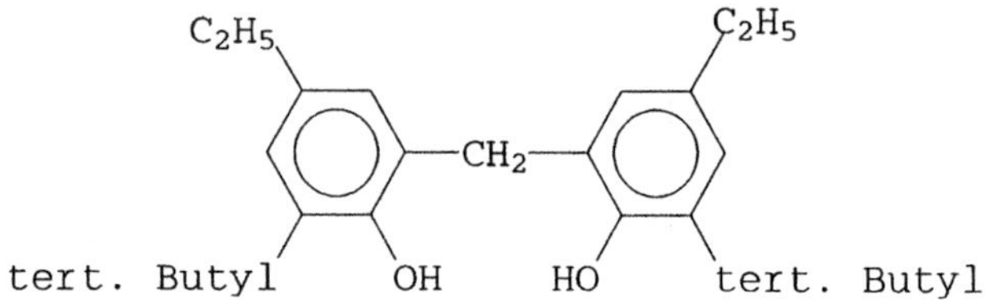

CAS No.	- 00088-24-4
PM Ref. No.	- 66400
Restrictions	- SML = 1.5 mg/kg With PM Ref. No. 66480
Formula	- $C_{25}H_{36}O_2$
Molecular weight	- 368
Alternative names	- Bis(2-hydroxy-3-tert-butyl-5-ethylphenyl) methane
Physical Characteristics	- White powder, mp 119 - 122°C. - Soluble in acetone, benzene, alcohol, chloroform.
Handling	- Store at room temperature.
Safety	-
Availability	- Standard sample was supplied.
Current uses	- Antioxidant.
Applications	- General use.
GC Retention time	- 46.4 min.

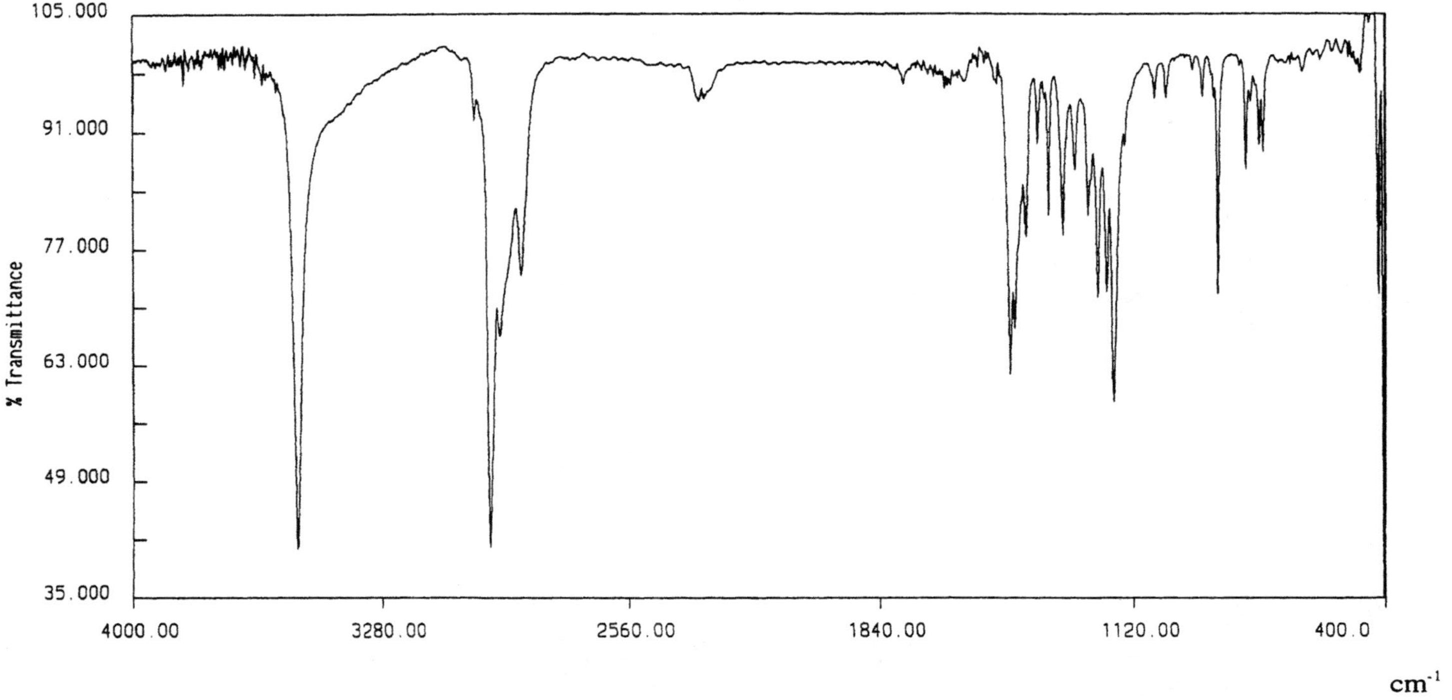

Name : 2,2'-METHYLENEBIS(4-ETHYL-6-TERT-BUTYLPHENOL)

PM Nr : 66400 CAS Nr : 0088-24-4

KBr pellet

CAS No. 00088-24-4 PM Ref. No. 6640

2,2'-METHYLENEBIS(4-ETHYL-6-TERT-BUTYLPHENOL)

Peak wavenumber	% T	Relative intensity
418.6	71.7	48.0
760.1	89.5	17.8
798.6	86.5	22.9
881.6	71.6	48.1
1147.8	89.4	17.9
1178.7	58.5	70.2
1199.9	71.8	47.7
1225.0	71.2	48.8
1252.0	81.1	32.0
1288.6	86.5	22.9
1323.3	78.7	36.1
1363.8	81.1	32.1
1394.7	89.6	17.6
1427.5	78.6	36.3
1458.4	67.4	55.1
1471.9	61.8	64.6
2930.2	66.5	56.8
2961.1	41.1	99.7
3530.2	40.9	100.0

2,2'-Methylenebis(4-ethyl-6-tert-butylphenol)

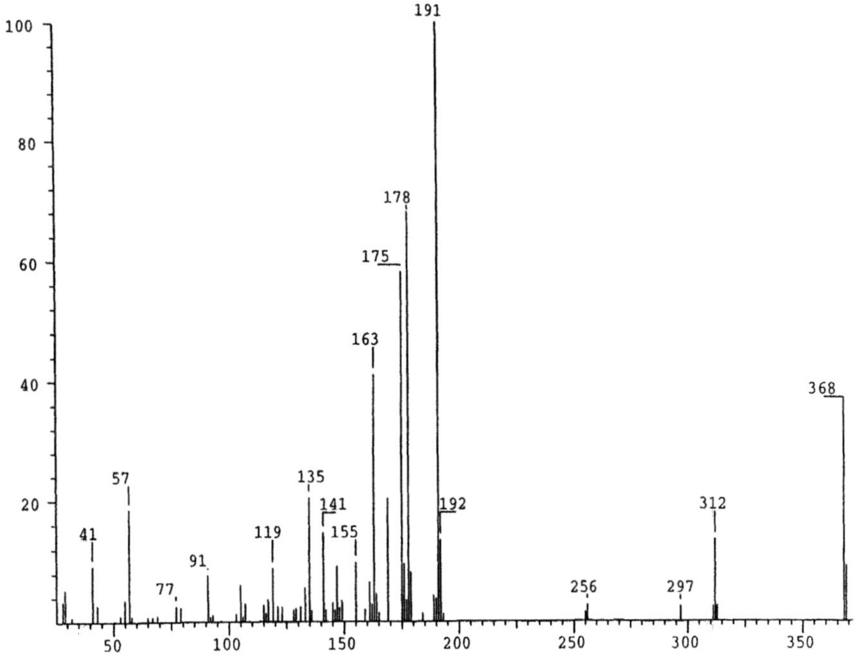

M/Z	Ion Intensity (%)	M/Z	Ion Intensity (%)
57	18.6	191	100.0
135	20.5	192	13.5
141	14.7	256	2.6
163	41.1	312	13.6
175	58.1	368	36.2
178	68.0	369	9.3

Spectrometer	: Finnigan Mat SSQ 700
Inletsystem	: Capillary GC/MS
Source Temperature	: 180°C
Electron Energy	: 70 eV
Scan Range	: 25-400

PM n°	CAS n°	Name
66400	00088-24-4	2,2'-METHYLENEBIS(4-ETHYL-6-TERT-BUTYLPHENOL)

Solvent	$CDCL_3$
SF	250.13
SI	16384
SW	3125
NS	32

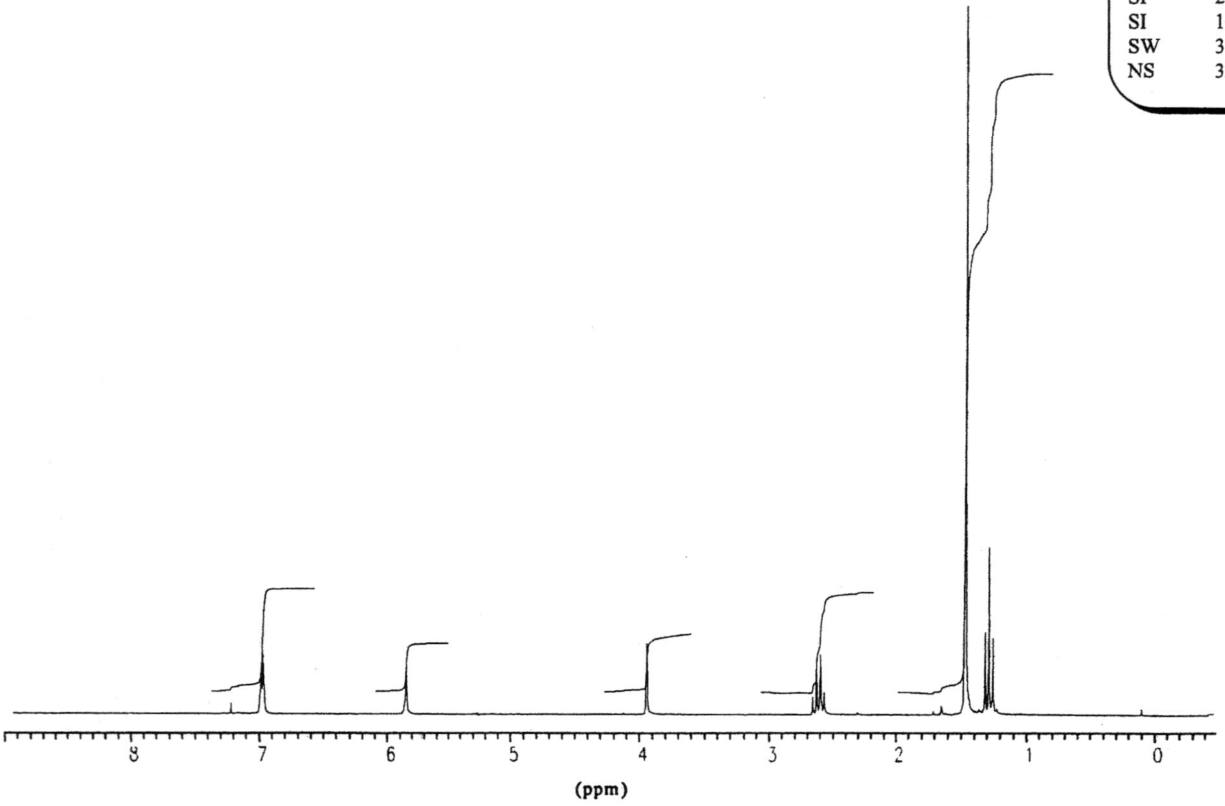

2,2'-Methylenebis(4-methyl-6-tert-butylphenol)

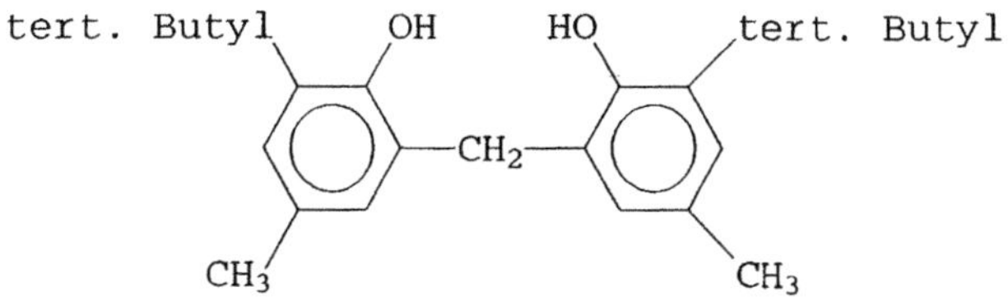

CAS No.	- 00119-47-1
PM Ref. No.	- 66480
Restrictions	- SML = 1.5 mg/kg With PM Ref. No. 66400
Formula	- $C_{23}H_{32}O_2$
Molecular weight	- 340
Alternative names	- Di(2-hydroxy-3-tert-butyl-5-methylphenyl) methane
Physical Characteristics	- White powder, mp approx. 125°C. - Soluble in most organic solvents.
Handling	- Store at room temperature.
Safety	-
Availability	- Standard sample was supplied.
Current uses	- Antioxidant.
Applications	- General use.
GC Retention time	- 44.9 min.

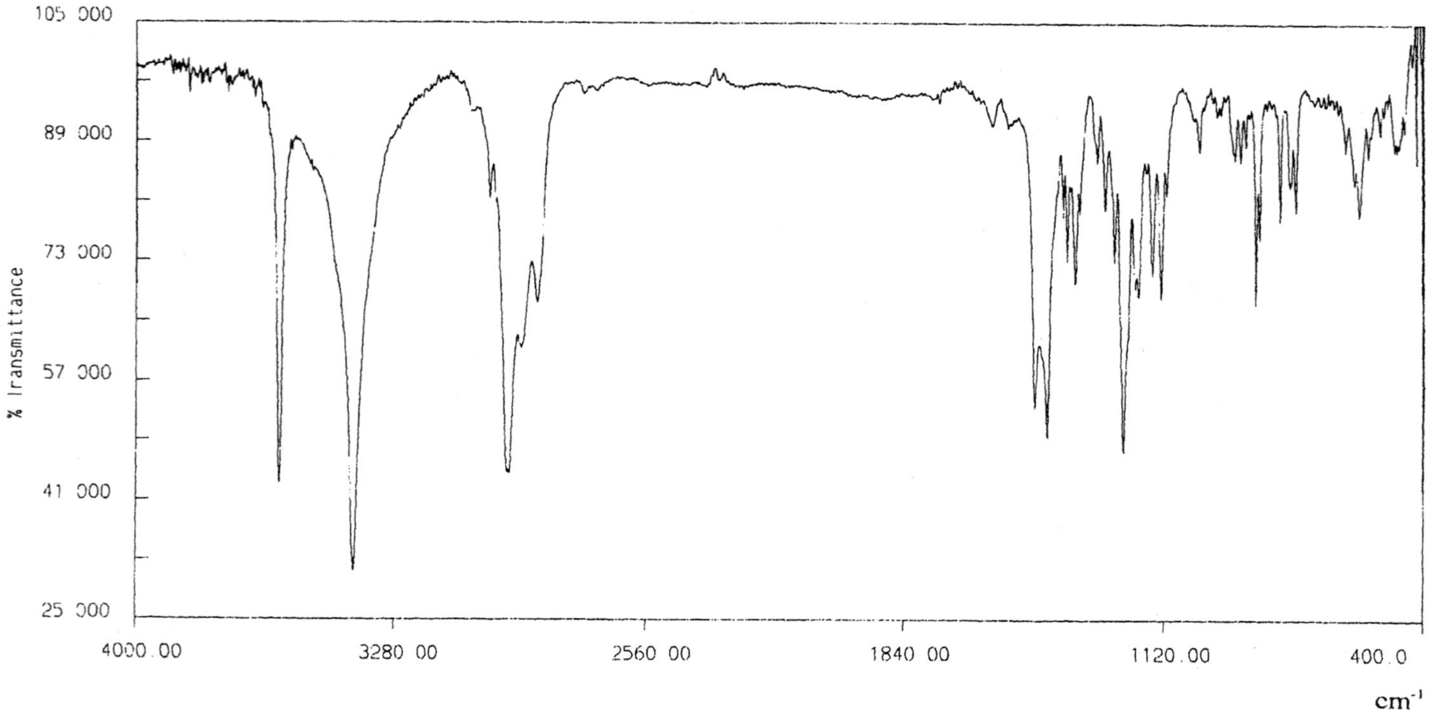

Name : 2,2'-METHYLENEBIS(4-METHYL-6-Tert-BUTYLPHENOL)

PM Nr : 66480 CAS Nr : 00119-47-1

KBr pellet

CAS No. 00119-47-1 PM Ref. No. 66480

2,2'-METHYLENEBIS(4-METHYL-6-tert-BUTYLPHENOL)

Peak wavenumber	% T	Relative intensity
418,6	86,3	19,9
453,3	90,6	13,7
474,6	87,7	17,9
520,8	90,1	14,4
555,6	87,0	18,9
582,6	79,1	30,4
619,2	87,7	17,9
760,1	79,8	29,4
775,5	82,9	24,9
802,5	78,6	31,2
858,4	76,1	34,8
868,1	67,4	47,5
897,0	88,4	16,9
910,5	86,4	19,8
927,9	86,6	19,5
1024,3	87,7	17,9
1115,0	81,8	26,5
1130,4	68,0	46,6
1155,5	71,1	42,1
1192,2	68,3	46,2
1232,7	47,5	76,4
1257,7	72,7	39,7
1284,8	79,8	29,4
1306,0	86,2	20,1
1367,7	70,1	43,5
1388,9	72,9	39,5
1400,5	79,1	30,4
1442,9	49,6	73,4
1479,6	53,3	68,0
1597,3	91,2	12,8
2872,4	67,4	47,5
2918,7	61,4	56,2
2957,2	44,5	80,8
3007,4	81,4	27,1
3265,9	90,6	13,7
3393,2	31,3	100,0
3603,5	43,3	82,6

2,2'-Methylenebis(4-methyl-6-tert-butylphenol)

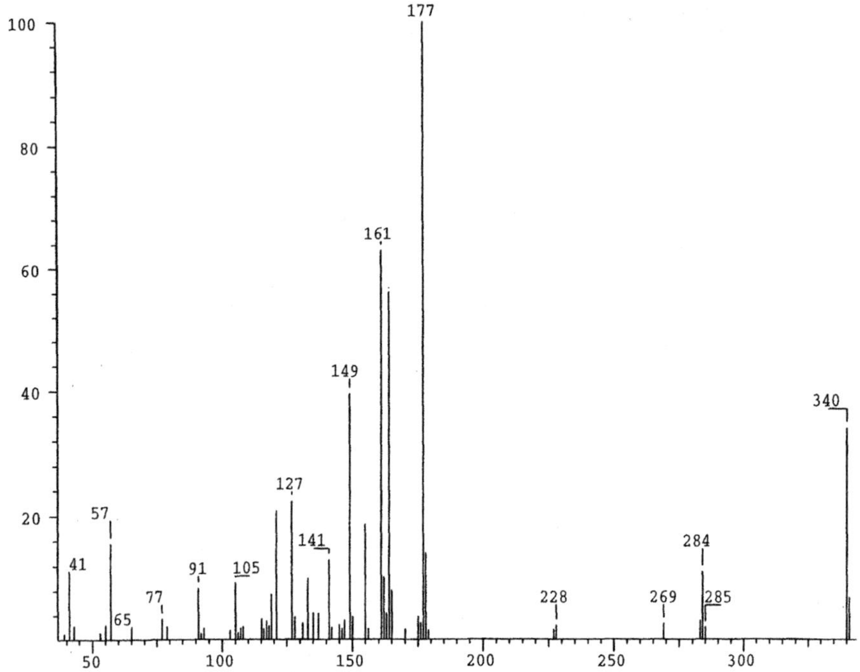

M/Z	Ion Intensity (%)	M/Z	Ion Intensity (%
41	10.1	155	18.6
57	15.5	161	63.0
91	8.4	162	10.2
105	9.3	164	56.2
121	20.7	177	100.0
127	22.3	178	14.1
133	10.0	284	11.2
141	13.0	340	34.3
149	39.5	341	7.1

Spectrometer	: Finnigan Mat SSQ 700
Inletsystem	: Capillary GC/MS
Source Temperature	: 180°C
Electron Energy	: 70 eV
Scan Range	: 35-450

PM n°	CAS n°	Name
66480	00119-47-1	2,2'-METHYLENEBIS(4-METHYL-6-TERT-BUTYLPHENOL)

Solvent	$CDCL_3$
SF	250.13
SI	16384
SW	3125
NS	16

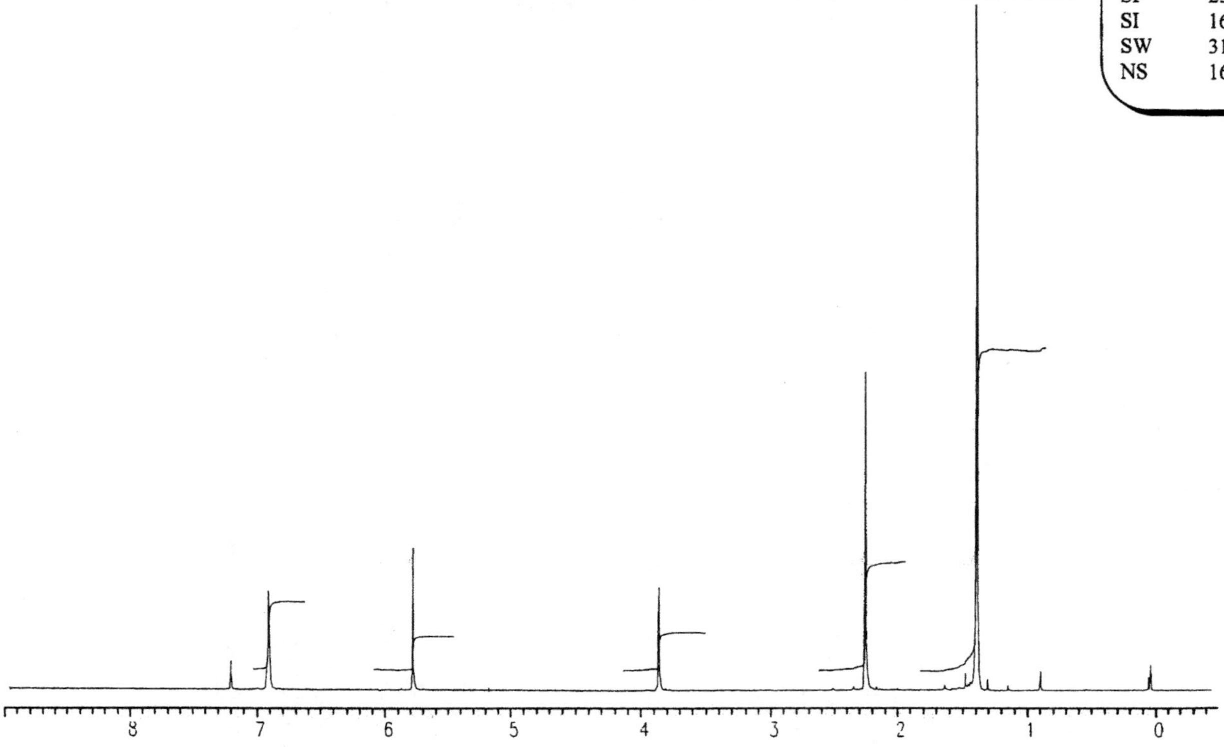

2,2'-Methylenebis(4-methyl-6-cyclohexylphenol)

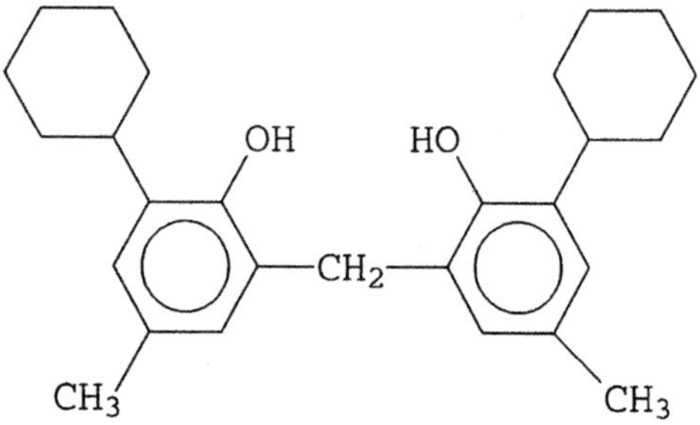

CAS No.	- 04066-02-8
PM Ref. No.	- 66560
Restrictions	- SML = 3.0 mg/kg With PM Ref. No. 66580
Formula	- $C_{27}H_{36}O_2$
Molecular weight	- 392

Alternative names - Bis(2-hydroxy-3-cyclohexyl-5-methyl-phenyl)methane

Physical Characteristics - White powder, mp approx. 125°C.
- Soluble in acetone, alcohol, ethyl acetate.

Handling - Store at room temperature.

Safety -

Availability - Standard sample was supplied.

Current uses - Antioxidant.

Applications - Styrene co-polymers, rubber.

GC Retention time - 56.3 min.

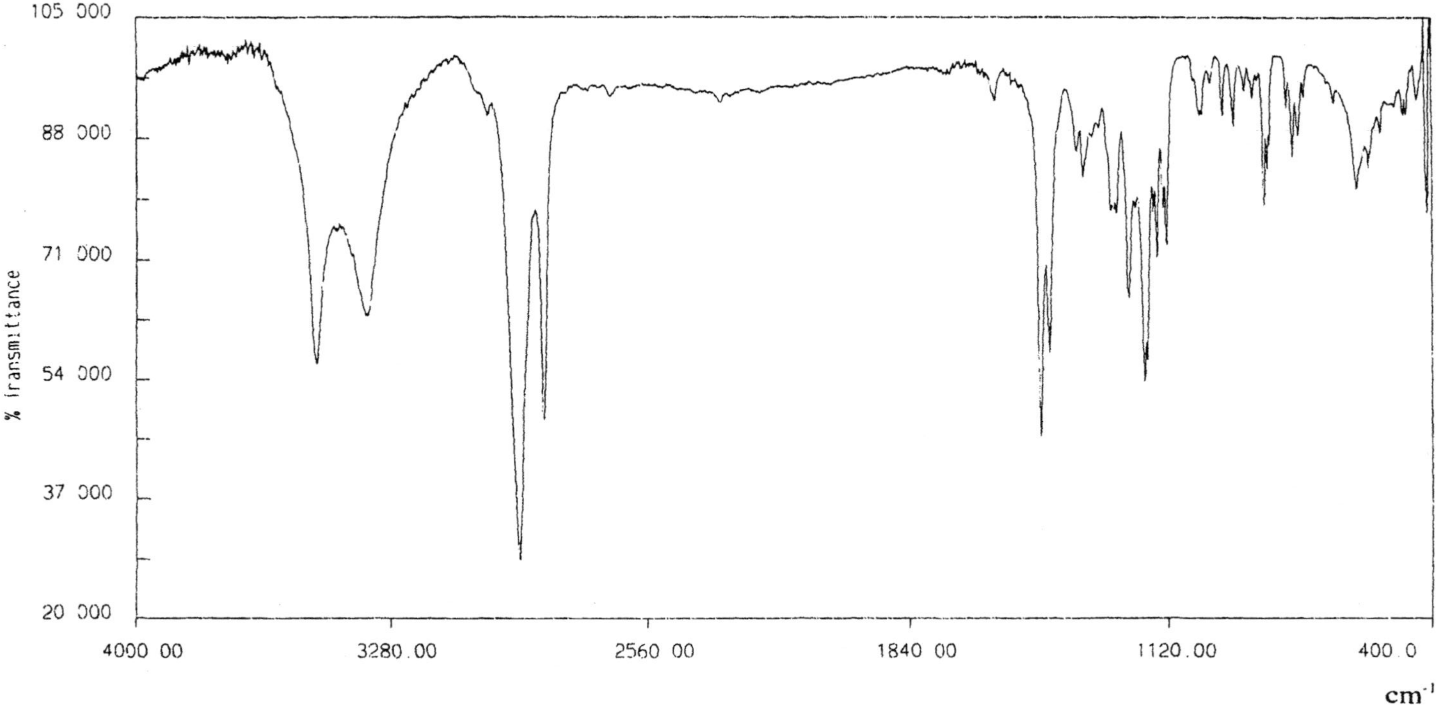

Name : 2,2'-METHYLENEBIS(4-METHYL-6-CYCLOHEXYLPHENOL)

PM Nr : 66560 CAS Nr : 04066-02-8

KBr pellet

CAS No. 04066-02-8 PM Ref. No. 66560

2,2'-METHYLENEBIS(4-METHYL-6-CYCLOHEXYLPHENOL)

Peak wavenumber	% T	Relative intensity
410,9	77,7	31,1
439,8	93,5	9,1
468,8	91,4	12,0
540,1	88,7	15,8
572,9	84,0	22,3
605,7	81,0	26,5
767,8	88,5	16,1
783,2	85,4	20,4
852,6	83,9	22,5
860,4	78,7	29,7
1124,6	73,2	37,4
1153,6	71,4	39,9
1188,3	53,6	64,8
1230,7	65,6	48,0
1265,5	77,5	31,4
1358,1	82,6	24,3
1450,7	57,7	59,1
1475,7	45,9	75,5
2853,1	48,2	72,3
2922,5	28,4	100,0
3350,8	63,0	51,7
3491,6	56,1	61,3

2,2'-Methylenebis(4-methyl-6-cyclohexylphenol)

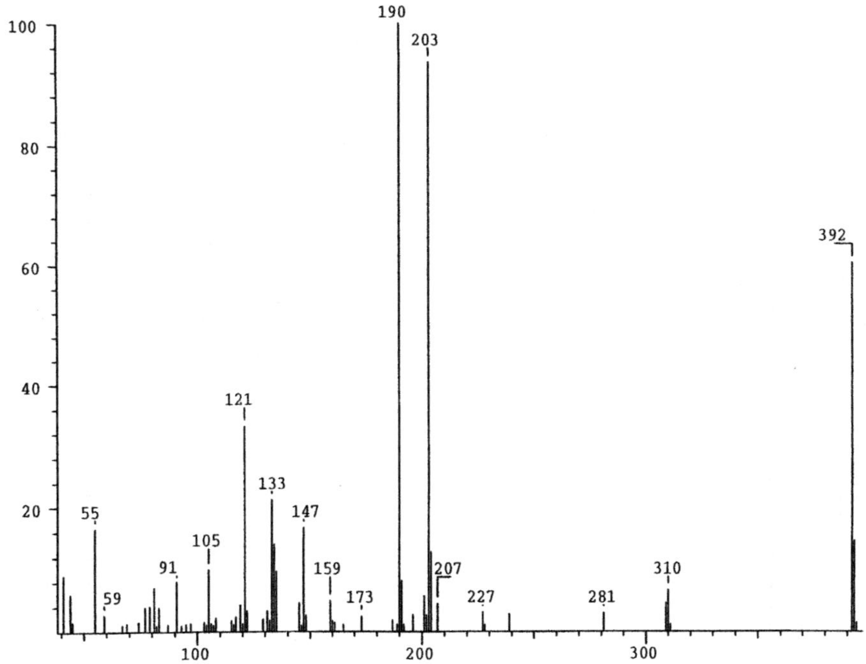

M/Z	Ion Intensity (%)	M/Z	Ion Intensity (%)
55	16.6	190	100.0
91	8.1	191	8.3
105	10.2	203	93.6
121	33.3	204	12.9
133	21.3	310	6.8
134	14.1	392	60.6
135	9.7	393	14.8
147	16.7		

Spectrometer	: Finnigan Mat SSQ 700
Inletsystem	: Capillary GC/MS
Source Temperature	: 180°C
Electron Energy	: 70 eV
Scan Range	: 35-500

PM n°	CAS n°	Name	Solvent	$CDCL_3$
66560	04066-02-8	2,2'-METHYLENEBIS(4-METHYL-6-CYCLOHEXYLPHENOL)	SF	250.13
			SI	16384
			SW	3125
			NS	12

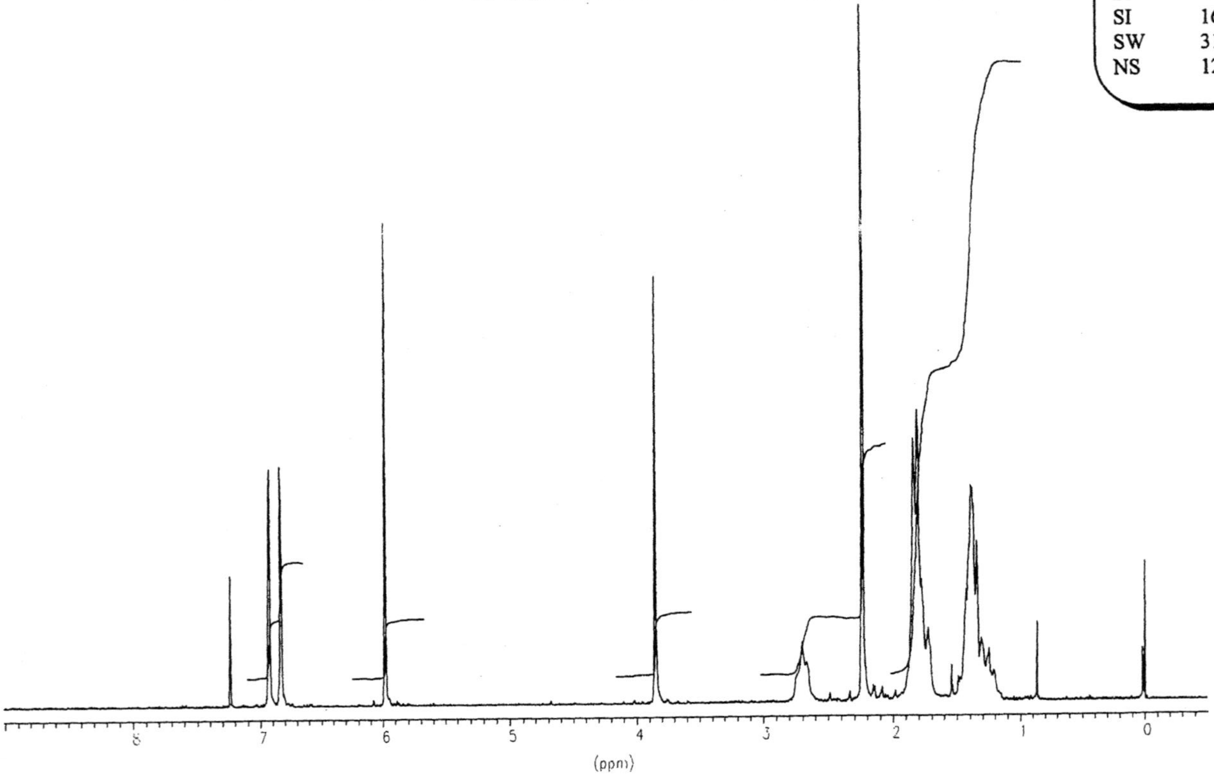

Monochlorobenzene

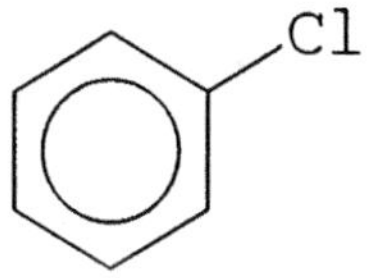

CAS No.	- 00108-90-7
PM Ref. No.	- 67280
Restrictions	- SML = 36 mg/kg
Formula	- C_6H_5Cl
Molecular weight	- 112
Alternative names	- Chlorobenzene
Physical Characteristics	- Colourless liquid, mp - 45°C, bp 131 - 132°C. - Soluble in alcohol, benzene, ether, chloroform.
Handling	- Store at room temperature.
Safety	- Harmful.
Availability	- Standard sample was supplied.
Current uses	- Solvent.
Applications	- PC, PSO.
GC Retention time	- 7.5 min.

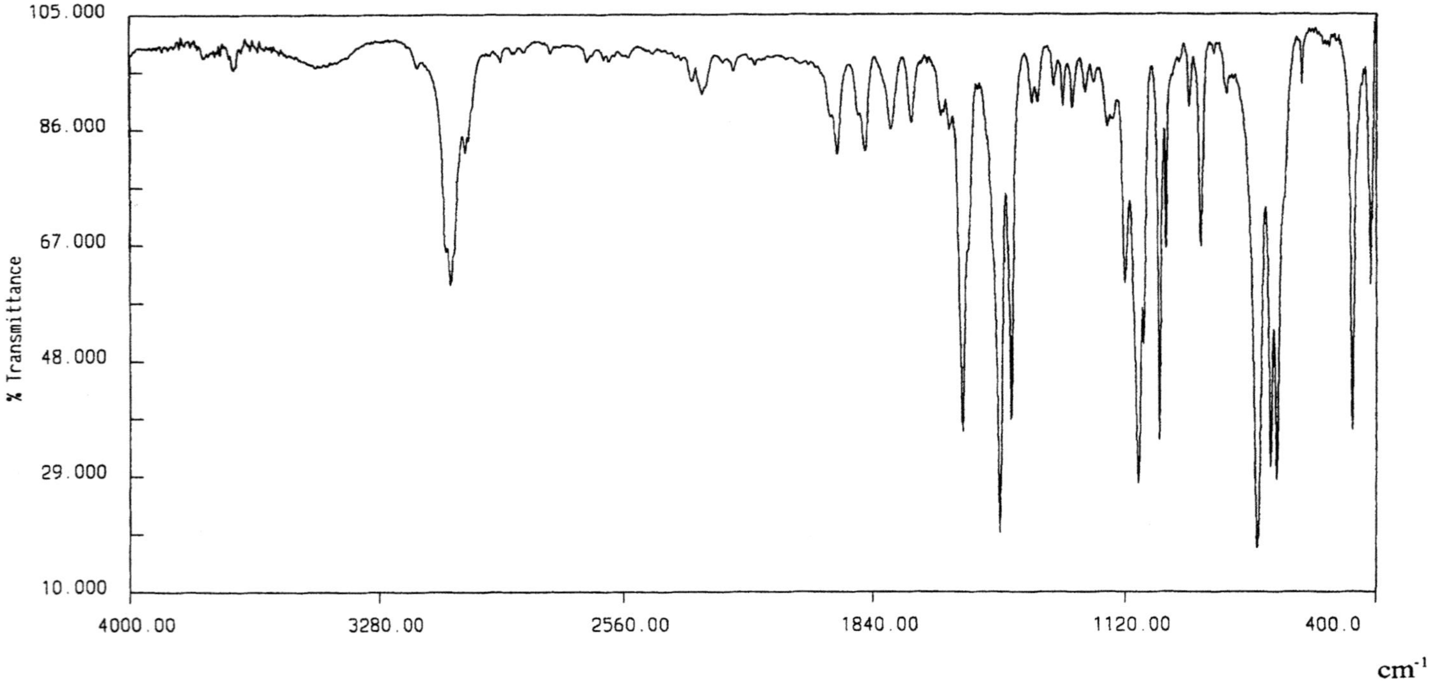

Name : MONOCHLOROBENZENE

PM Nr : 67280 CAS Nr : 00108-90-7

KBr pellet

CAS No. 00108-90-7 PM Ref. No. 67280

MONOCHLOROBENZENE

Peak wavenumber	% T	Relative intensity
416.7	60.5	47.7
468.8	36.4	76.7
613.4	93.3	8.1
684.8	28.2	86.6
702.2	30.4	84.0
740.8	17.1	100.0
902.8	66.7	40.1
935.6	89.6	12.5
1003.1	66.5	40.4
1022.4	34.8	78.6
1068.7	50.6	59.6
1084.1	27.7	87.2
1122.7	60.8	47.3
1446.8	38.2	74.5
1477.7	19.5	97.0
1583.8	36.0	77.1
1730.4	86.9	15.7
1788.2	85.9	17.0
1863.5	82.3	21.4
1944.5	81.7	22.0
3026.7	82.1	21.6
3071.1	60.3	47.8

Monochlorobenzene

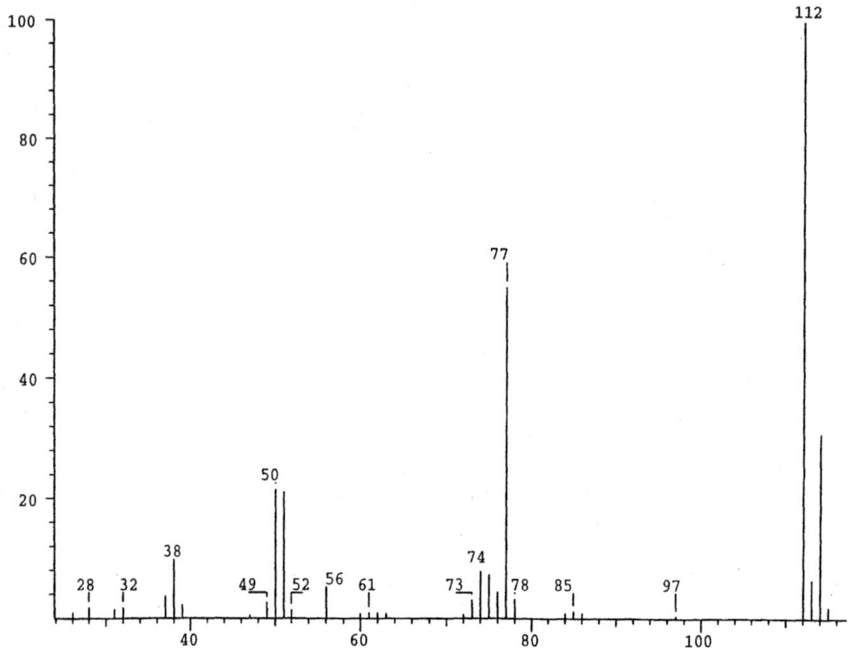

M/Z	Ion Intensity (%)	M/Z	Ion Intensity (%)
37	3.7	75	7.2
38	8.8	76	4.4
39	2.1	77	55.1
49	2.6	78	3.2
50	21.5	97	0.4
51	21.1	112	100.0
56	4.2	113	6.4
73	3.1	114	30.9
74	7.8	115	1.9

Spectrometer : Finnigan Mat SSQ 700
Inletsystem : Capillary GC/MS
Source Temperature : 180°C
Electron Energy : 70 eV
Scan Range : 25-500

PM n°	CAS n°	Name
67280	00108-90-7	MONOCHLOROBENZENE

Solvent	$CDCL_3$
SF	250.13
SI	16384
SW	3125
NS	32

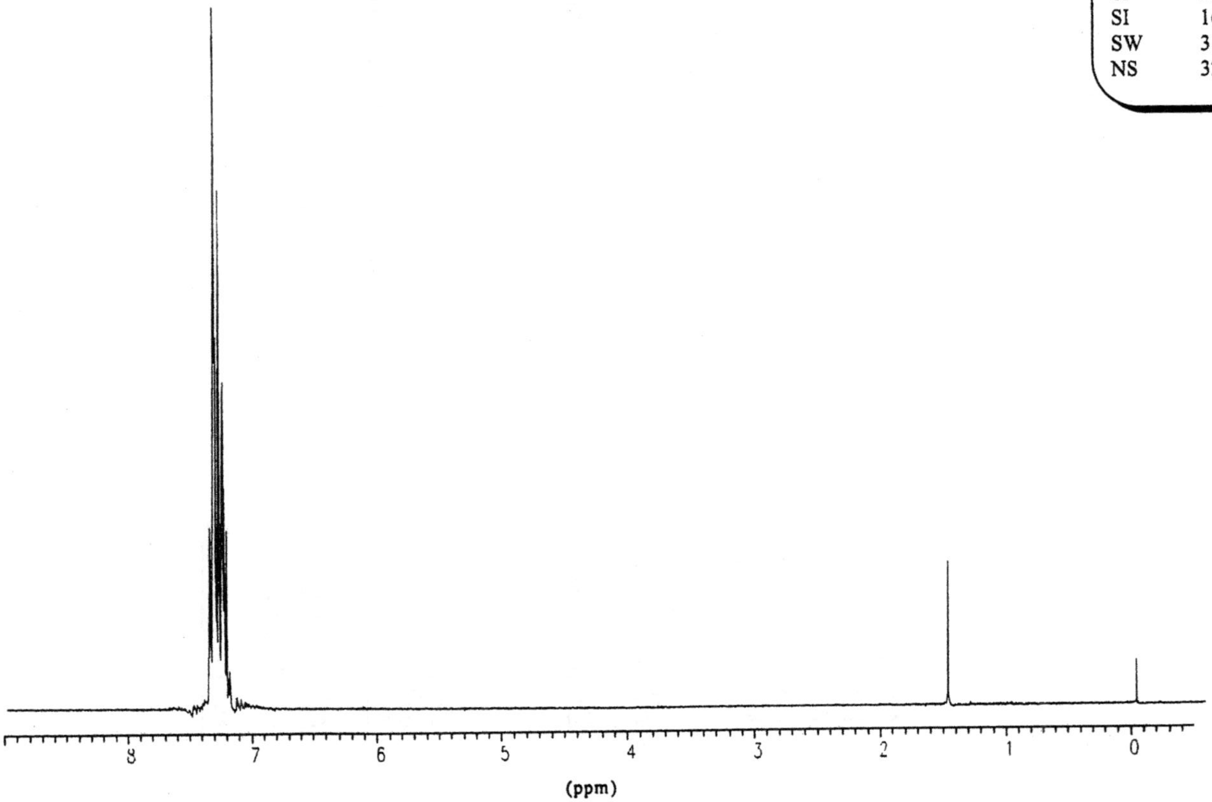

1-Octadecanol

$$C_{18}H_{37}—OH$$

CAS No.	- 00112-92-5
PM Ref. No.	- 68225
Restrictions	- None
Formula	- $C_{18}H_{38}O$
Molecular weight	- 270

Alternative names	- Stearyl alcohol
Physical Characteristics	- White flakes or granules, mp 56 - 60°C, bp 210°C (15 mm Hg). - Soluble in most organic solvents.
Handling	- Store at room temperature.
Safety	-
Availability	- Standard sample was supplied.
Current uses	- Lubricant.
Applications	- General use.
GC Retention time	- 39.7 min.

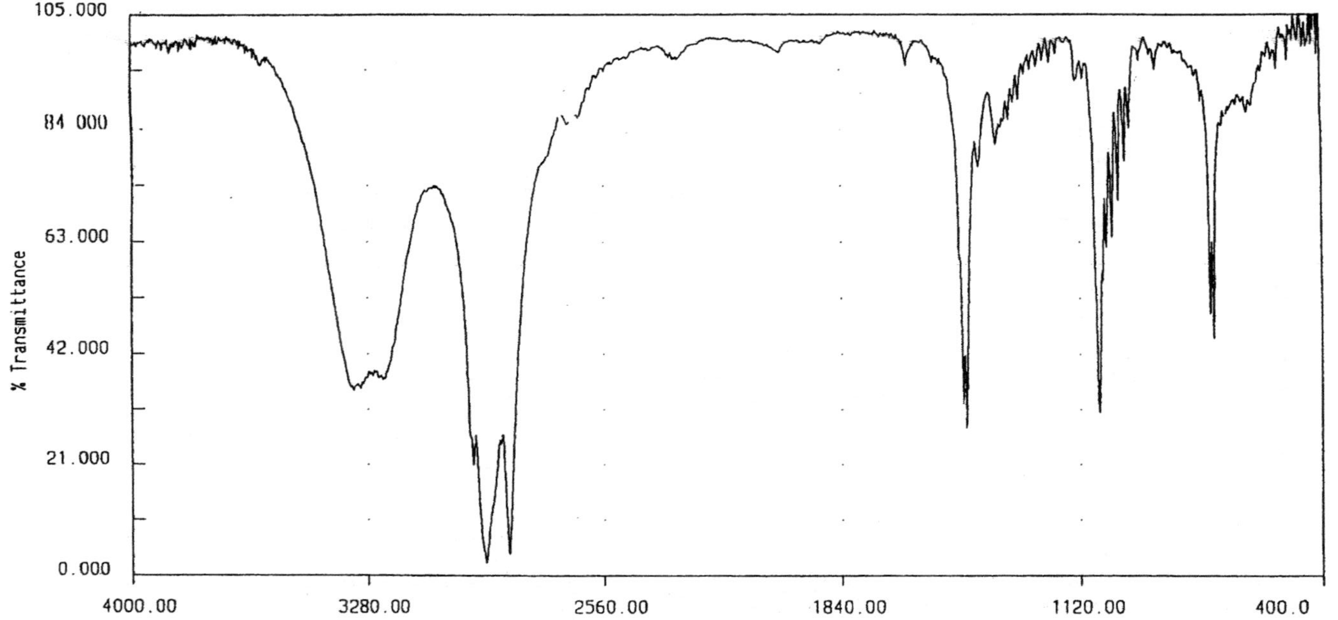

Name : 1-OCTADECANOL

PM Nr : 68225 CAS Nr : 00112-92-5

KBr pellet

CAS No. 00112-92-5 PM Ref. No. 68225

1-Octadecanol

Peak wavenumber	% T	Relative intensity
526.6	94.6	5.5
572.9	94.1	6.0
603.8	87.9	12.4
719.5	45.1	56.2
731.1	49.6	51.6
968.4	84.3	16.1
983.8	78.1	22.4
1005.0	70.6	30.1
1024.3	63.9	36.9
1039.7	62.0	38.9
1062.9	30.9	70.7
1317.5	88.5	11.8
1333.0	85.6	14.7
1363.8	84.1	16.3
1373.5	81.1	19.3
1425.6	76.9	23.6
1464.2	27.9	73.8
1473.8	32.5	69.1
2633.2	85.9	14.4
2849.2	4.0	98.3
2918.7	2.3	100.0
2955.3	20.7	81.2
3233.1	37.1	64.4

1-Octadecanol

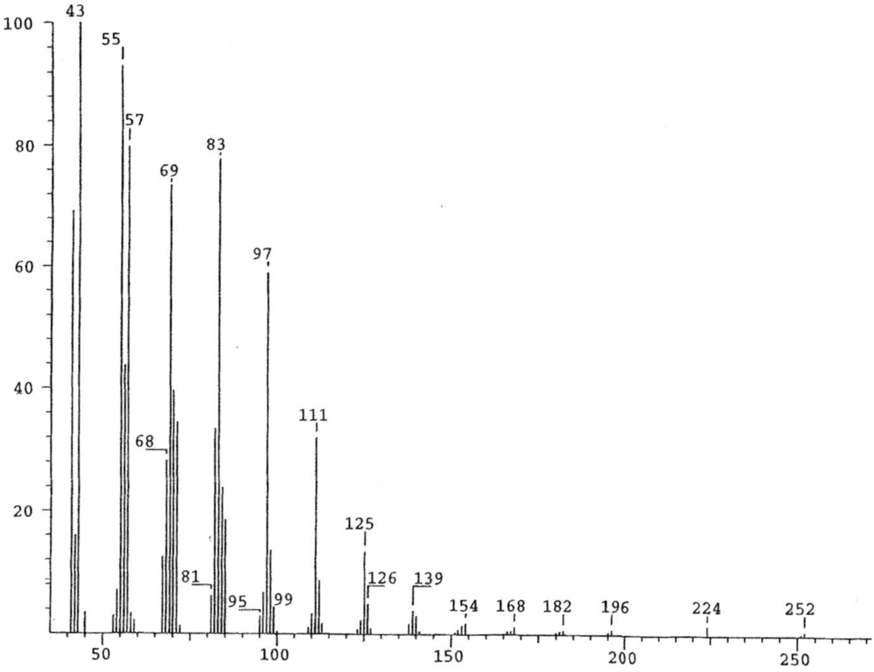

M/Z	Ion Intensity (%)	M/Z	Ion Intensity (%)
41	69.2	83	77.9
43	100.0	84	23.9
55	93.1	85	18.6
56	43.7	97	59.0
57	79.9	98	13.6
67	12.6	111	32.0
68	28.2	112	8.7
69	73.6	125	13.4
71	34.5	224	1.5
82	33.4	225	0.7

Spectrometer	: Finnigan Mat SSQ 700
Inletsystem	: Capillary GC/MS
Source Temperature	: 180°C
Electron Energy	: 70 eV
Scan Range	: 40-500

PM n°	CAS n°	Name
68225	00112-92-5	1-OCTADECANOL

Solvent	$CDCL_3$
SF	250.13
SI	16384
SW	6494
NS	8

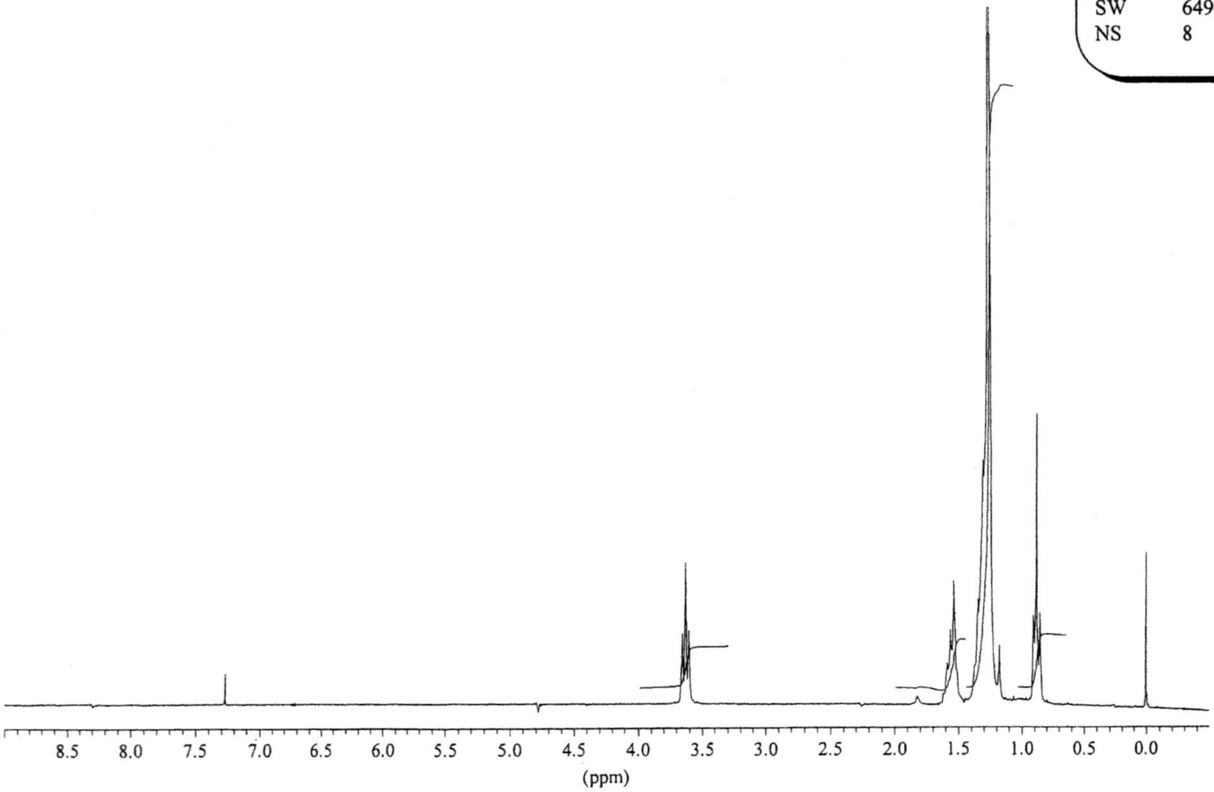

Octadecyl 3-(3,5-di-tert-butyl-4-hydroxyphenyl) propionate

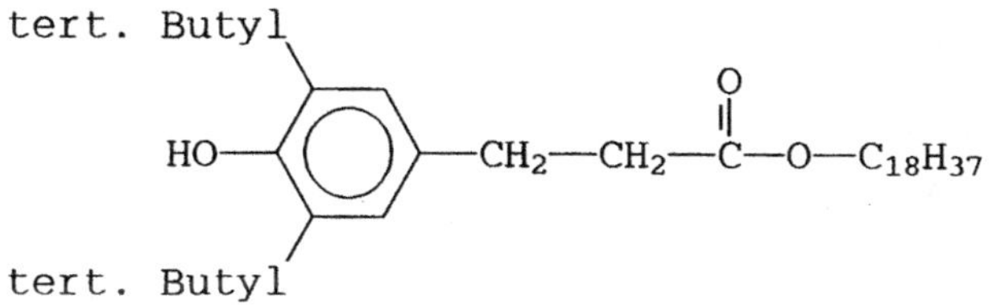

CAS No.	- 02082-79-3
PM Ref. No.	- 68320
Restrictions	- SML = 6.0 mg/kg
Formula	- $C_{35}H_{62}O_3$
Molecular weight	- 530

Alternative names - Benzenepropanoic acid, 3,5-bis(1,1-dimethylethyl)-4-hydroxy-, octadecyl ester

Physical Characteristics - White powder or flakes, mp 51.5 - 52.5°C.
- Soluble in acetone, acetic acid, benzene, chloroform, hexane.

Handling - Store at room temperature.

Safety -

Availability - Standard sample was supplied.

Current uses - Antioxidant.

Applications - General use.

GC Retention time - 61.1 min.

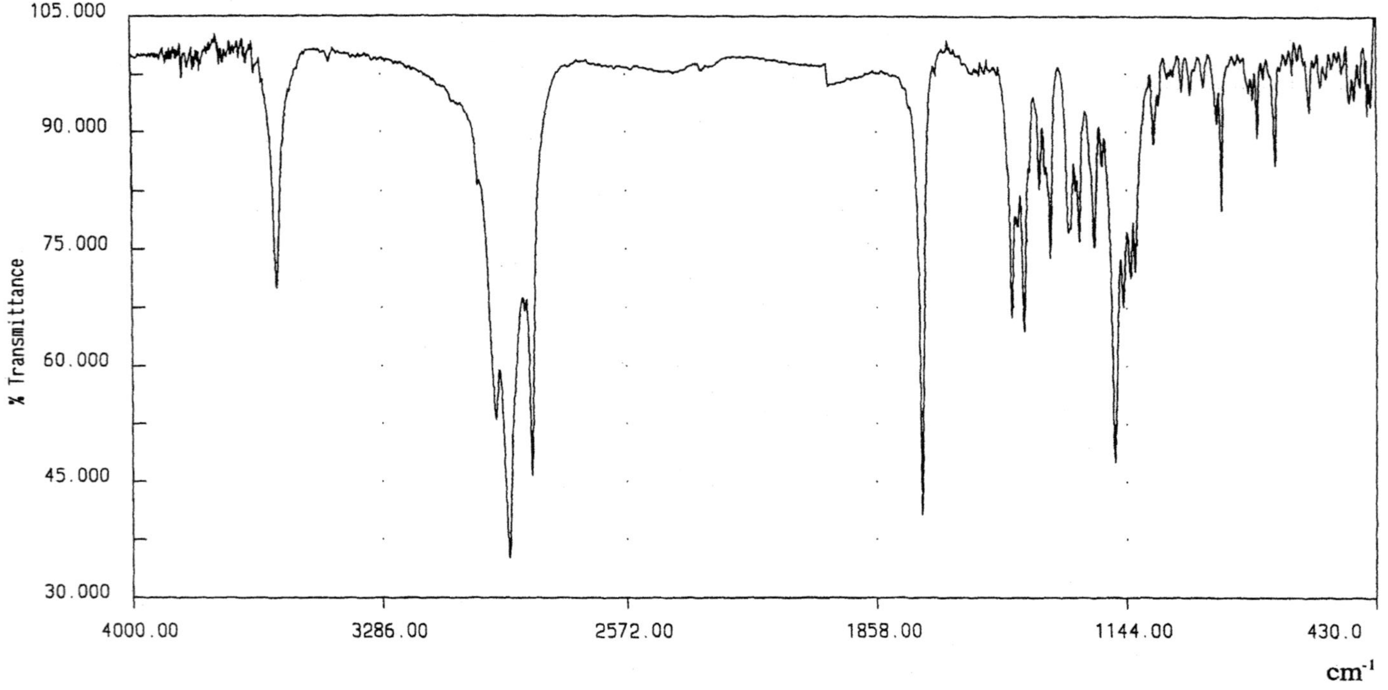

Name : OCTADECYL 3-(3,5-DI-TERT-BUTYL-4-HYDROXYPHENYL) PROPIONATE

PM Nr : 68320 CAS Nr : 02082-79-3

KBr pellet

CAS No. 02082-79-3 PM Ref. No. 68320

OCTADECYL 3-(3,5-DI-TERT-BUTYL-4-HYDROXYPHENYL) PROPIONATE

Peak wavenumber	% T	Relative intensity
721.5	90.4	16.1
744.6	91.3	14.7
771.6	92.5	12.7
785.1	94.4	9.5
796.7	94.5	9.3
875.8	89.1	18.4
1070.6	82.6	29.3
1122.7	71.5	48.1
1153.6	78.7	35.8
1178.7	64.3	60.2
1236.5	81.9	30.5
1277.0	63.0	62.4
1435.2	77.1	38.5
1469.9	74.3	43.3
1728.4	40.7	100.0
2851.1	58.0	70.8
2918.7	43.6	95.1
2957.2	54.73	76.3

Octadecyl 3-(3,5-di-tert-butyl-4-hydroxyphenyl) propionate

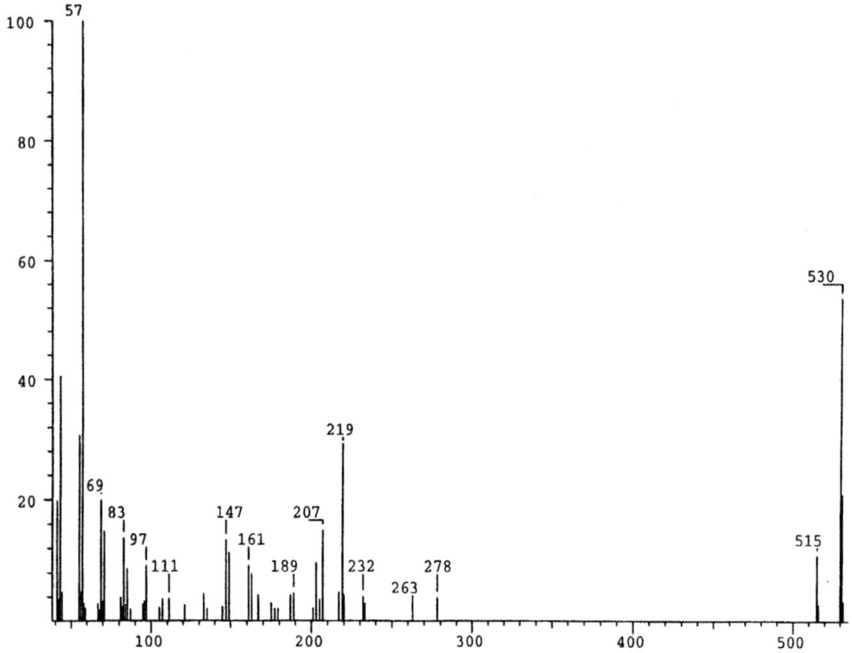

M/Z	Ion Intensity (%)	M/Z	Ion Intensity (%)
43	40.5	161	9.2
55	30.7	203	9.7
57	100.0	207	15.0
69	19.9	219	29.3
71	14.8	263	3.2
83	13.7	278	3.9
97	9.2	515	10.9
111	3.8	530	53.8
147	13.4	531	20.9
149	11.3	532	3.4

Spectrometer	: Finnigan Mat SSQ 700
Inletsystem	: Capillary GC/MS
Source Temperature	: 180°C
Electron Energy	: 70 eV
Scan Range	: 40-600

PM n°	CAS n°	Name
68320	02082-79-3	OCTADECYL 3-(3,5-DI-tert-BUTYL-4-HYDROXYPHENYL)PROPIONATE

Solvent	$CDCL_3$
SF	250.13
SI	16384
SW	2380
NS	32

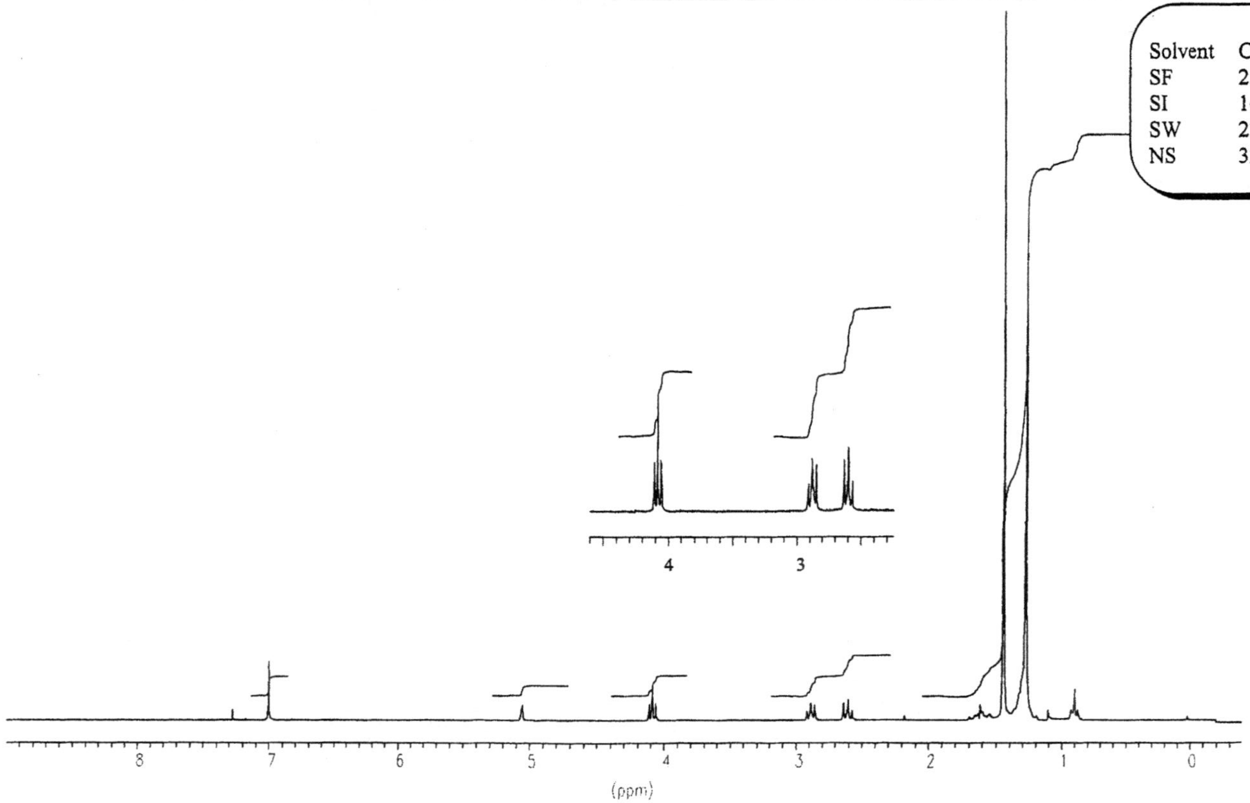

Oleamide

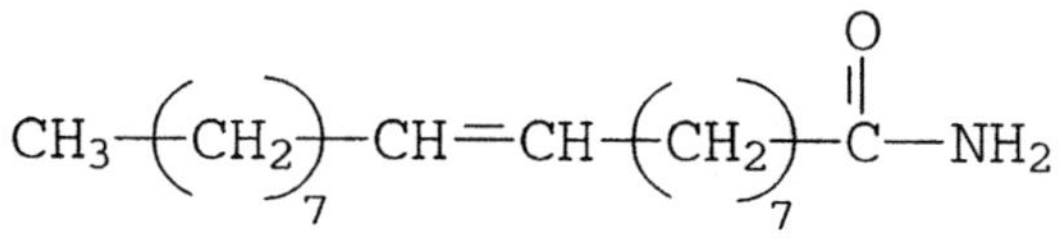

CAS No.	- 00301-02-0
PM Ref. No.	- 68960
Restrictions	- None
Formula	- $C_{18}H_{35}NO$
Molecular weight	- 281

Alternative names	- 9-Octadecenoic acid, amide
Physical Characteristics	- Yellow pearls, mp 76°C, bp decomposes. - Soluble in alcohol, ether, toluene.
Handling	- Store at room temperature.
Safety	-
Availability	- Standard sample was supplied.
Current uses	- Antistatic agent, antiblocking agent.
Applications	- Polyolefins.
GC Retention time	- 44.1 min.

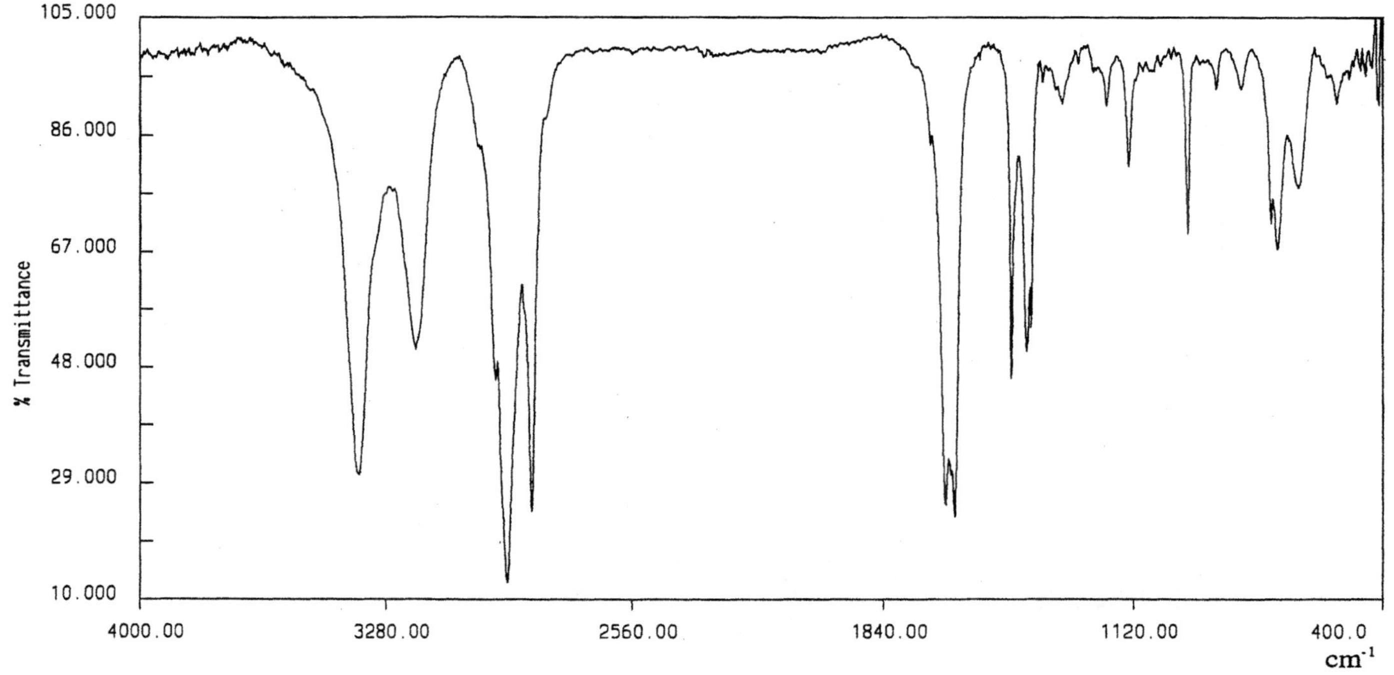

Name : OLEAMIDE

PM Nr : 68960 CAS Nr : 00301-02-0

KBr pellet

CAS No. 00301-02-0 PM Ref. No. 68960

OLEAMIDE

Peak wavenumber	% T	Relative intensity
409.0	90.7	10.6
493.8	94.9	5.8
528.6	90.9	10.4
640.4	77.2	26.1
702.2	67.3	37.4
721.5	71.5	32.6
964.5	69.9	34.4
1412.1	54.4	52.2
1425.6	50.5	56.7
1469.9	45.9	61.9
1633.9	23.4	87.7
1659.0	25.3	85.4
2851.1	24.2	86.7
2920.6	12.6	100.0
2957.2	45.7	62.1
3362.3	30.2	79.9

Oleamide

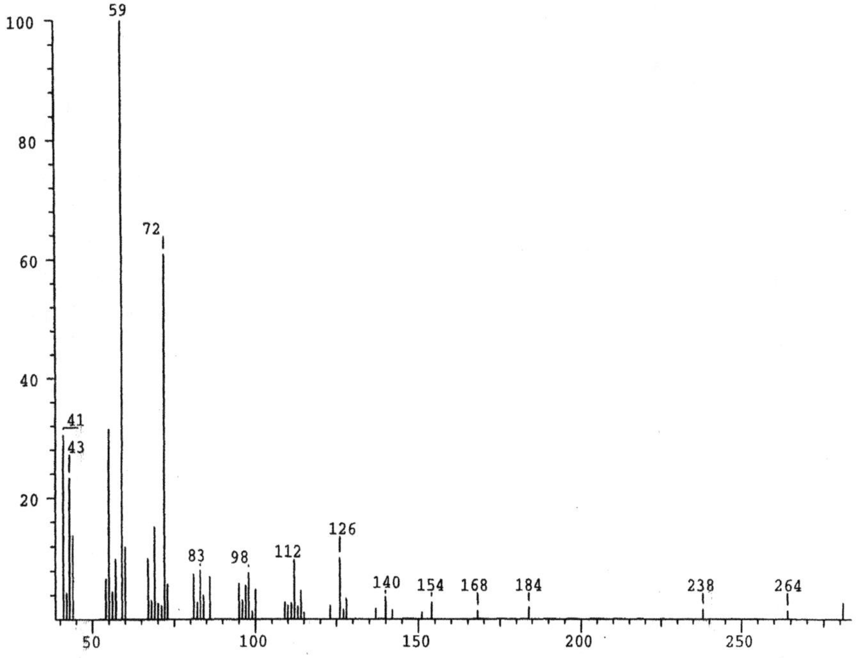

M/Z	Ion Intensity (%)	M/Z	Ion Intensity (%)
41	30.5	98	7.6
43	23.4	112	8.7
44	13.8	126	10.0
55	31.5	140	3.68
59	100.0	154	2.8
60	11.9	168	1.3
67	10.0	184	2.0
69	15.2	238	1.7
72	60.8	264	1.4
83	8.1	281	2.7

Spectrometer : Finnigan Mat SSQ 700
Inletsystem : Capillary GC/MS
Source Temperature : 180°C
Electron Energy : 70 eV
Scan Range : 35-500

PM n°	CAS n°	Name
68960	00301-02-0	OLEAMIDE

Solvent	$CDCL_3$
SF	250.13
SI	16384
SW	3125
NS	32

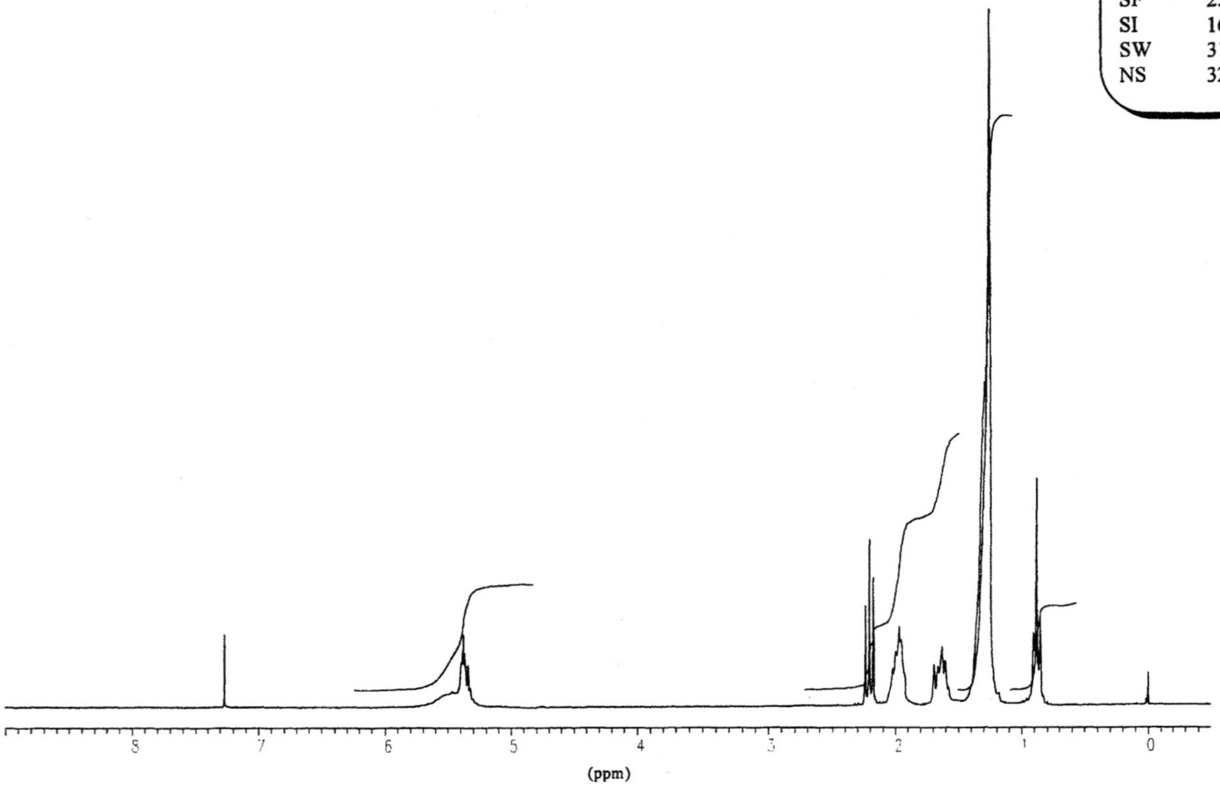

Hydrocarbon waxes, paraffin and microcrystalline (hydrogenated)

CAS No.	- 08002-74-2 and 63231-60-7
PM Ref. No.	- 71281
Restrictions	- SML = 3.0 mg/kg for all hydrogenated products
Formula	- C_nH_{2n+2}
Molecular weight	-
Alternative names	-
Physical Characteristics	- White crystalline powder, mp 50 - 57°C. - Soluble in toluene, benzene, chloroform, ether.
Handling	- Store below 4°C.
Safety	-
Availability	- Standard sample was supplied.
Current uses	- Lubricant.
Applications	- PS, PVC, PP, PE.
GC Retention time	-

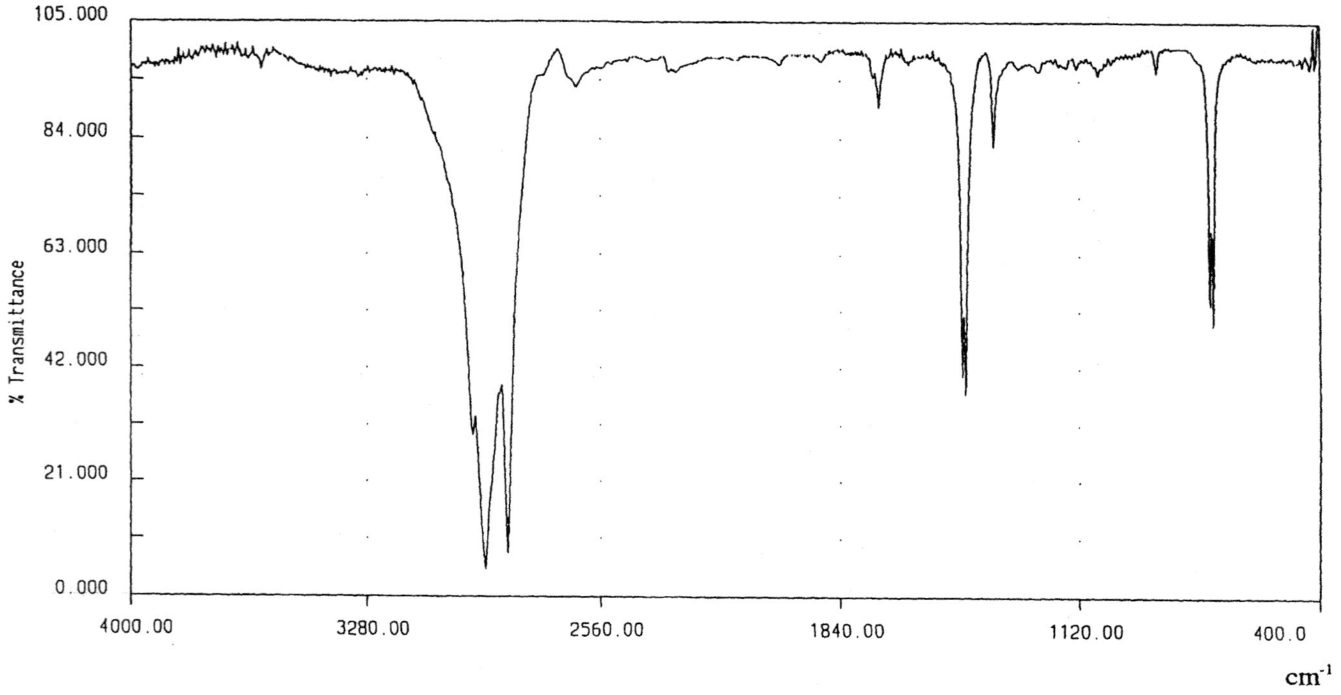

Name : PARAFFIN WAXES AND HYDROCARBON WAXES MICROCRYSTALLINE

PM Nr : 71281 CAS Nr : 08002-74-2
63231-60-7

KBr pellet

CAS No. 08002-74-2 and 63231-60-7 PM Ref. No. 71281

PARAFFIN WAXES AND HYDROCARBON WAXES MICROCRYSTALLINE

Peak wavenumber	% T	Relative intensity
719.5	50.0	52.6
729.2	53.7	48.7
1464.2	37.2	66.0
1473.8	40.6	62.5
2849.2	7.7	97.1
2916.7	4.9	100.0
2955.3	29.5	74.1

Hydrocarbon waxes, paraffin and microcrystalline (hydrogenated)

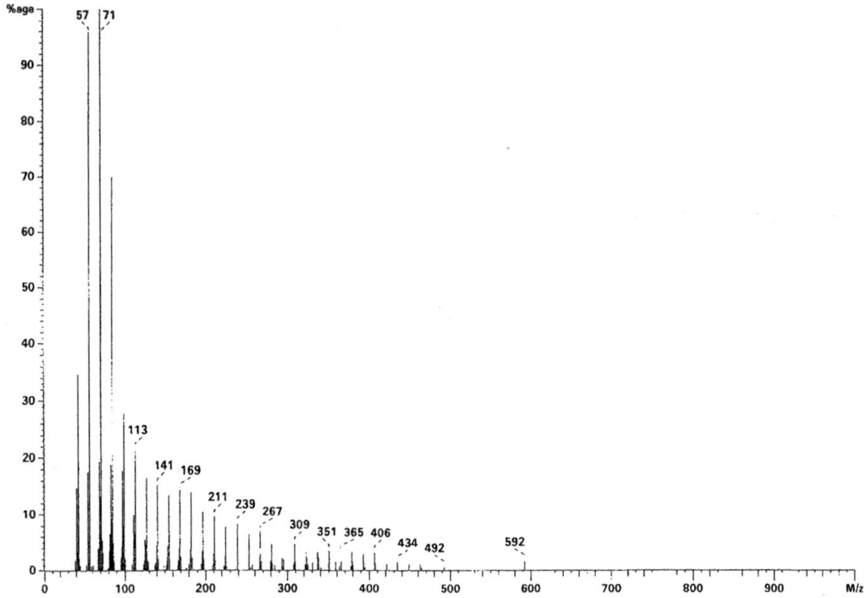

M/Z	Ion Intensity (%)	M/Z	Ion Intensity (%)
41	14.7	97	19.3
43	34.7	99	28.0
55	18.0	111	9.3
56	12.7	113	21.3
57	96.7	127	16.7
69	19.3	141	15.3
70	13.3	155	13.3
71	100.0	169	14.7
83	18.7	183	13.3
85	70.0	197	10.0

Spectrometer	: VG Autospec
Inletsystem	: Probe inlet
Source Temperature	: 350-400°C
Electron Energy	: 70 eV
Scan Range	: 40-1000

PM n°	CAS n°	Name
71281	8002-74-2 63231-60-7	PARAFFIN WAXES AND HYDROCARBON WAXES MICROCRYSTALLINE

Solvent	$CDCL_3$
SF	250.13
SI	16384
SW	6494
NS	24

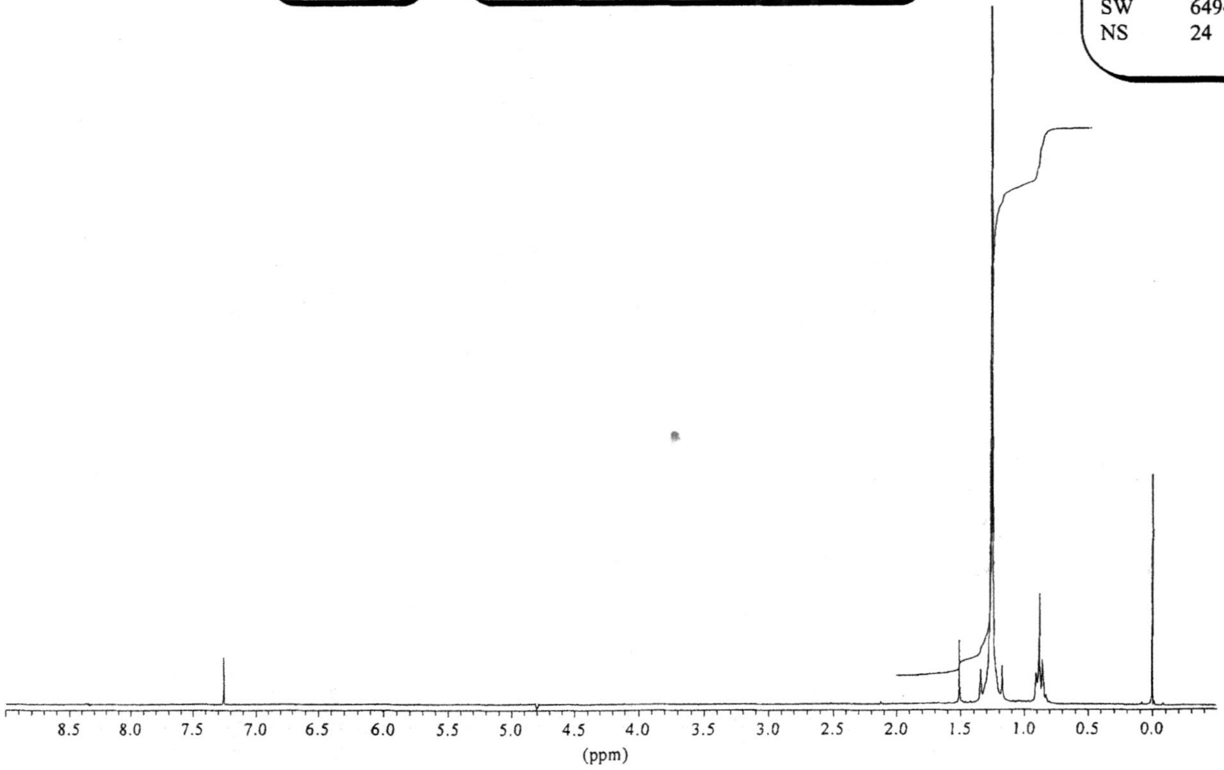

Pentaerythritol tetrakis(3-(3,5-di-tert-butyl-4-hydroxyphenyl)propionate)

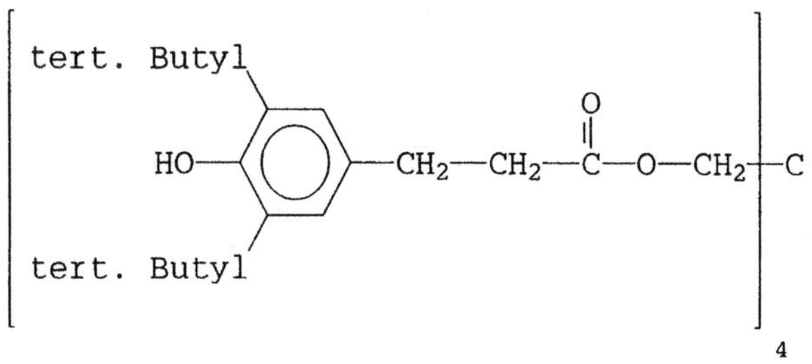

CAS No.	- 06683-19-8
PM Ref. No.	- 71680
Restrictions	- None
Formula	- $C_{73}H_{108}O_{12}$
Molecular weight	- 1178

Alternative names	- Tetrakis(methylene(3,5-di-tert-butyl-4-hydroxy-hydrocinnate))methane
Physical Characteristics	- White powder, mp 110 - 125°C. - Soluble in acetone, benzene, acetic acid, chloroform.
Handling	- Store at room temperature.
Safety	-
Availability	- Standard sample was supplied.
Current uses	- Antioxidant and heat stabilizer.
Applications	- General use.
GC Retention time	-

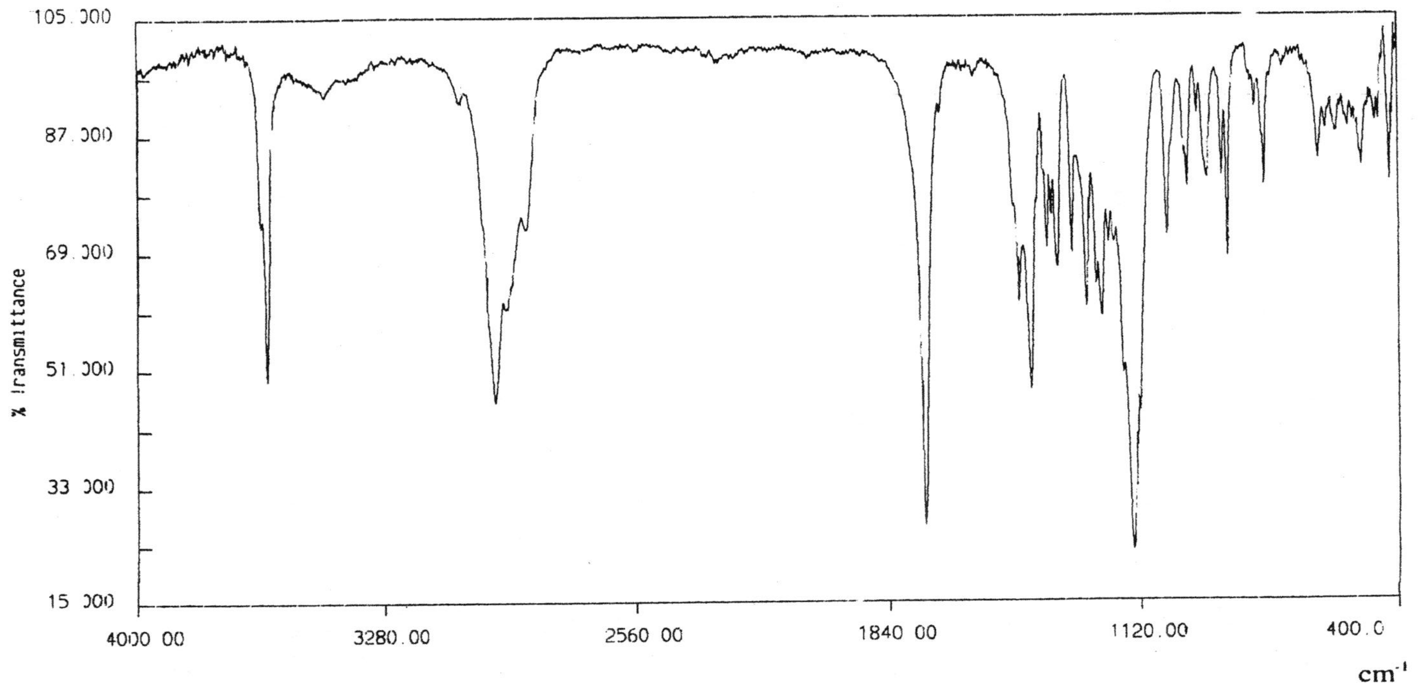

Name : PENTAERYTHRITOL TETRAKIS(3-(3,5-DI-Tert-BUTYL-4-HYDROXYPHENYL) PROPIONATE)

PM Nr : 71680 CAS Nr : 06683-19-8

KBr pellet

CAS No. 06683-19-8 PM Ref. No. 71680

PENTAERYTHRITOL TETRAKIS(3-(3,5-DI-tert-BUTYL-4-HYDROXYPHENYL)PROPIONATE

Peak wavenumber	% T	Relative intensity
422,5	79,7	26,3
453,3	89,1	14,1
499,6	81,9	23,5
538,2	88,1	15,4
572,9	87,1	16,7
599,9	87,6	16,1
619,2	83,0	22,0
771,6	79,0	27,2
873,9	68,0	41,5
889,3	80,5	25,3
931,7	80,2	25,7
960,7	90,1	12,8
987,7	78,8	27,5
1045,6	71,4	37,1
1142,0	22,8	100,0
1215,3	70,3	38,5
1232,7	59,0	53,1
1277,0	60,4	51,3
1317,5	68,6	40,7
1360,0	66,4	43,5
1377,3	74,4	33,2
1388,9	69,4	39,7
1437,1	47,7	67,8
1469,9	61,2	50,3
1740,0	26,8	94,9
2878,1	72,4	35,8
2930,2	60,2	51,6
2963,0	45,9	70,1
3626,6	49,4	65,6

PM n°	CAS n°	Name
71680	06683-19-8	PENTAERYTHRITOL TETRAKIS (3-(3,5-DI-TERT-BUTYL-4-HYDROXYPHENYL) PROPIONATE)

Solvent	$CDCL_3$
SF	250.13
SI	16384
SW	3125
NS	16

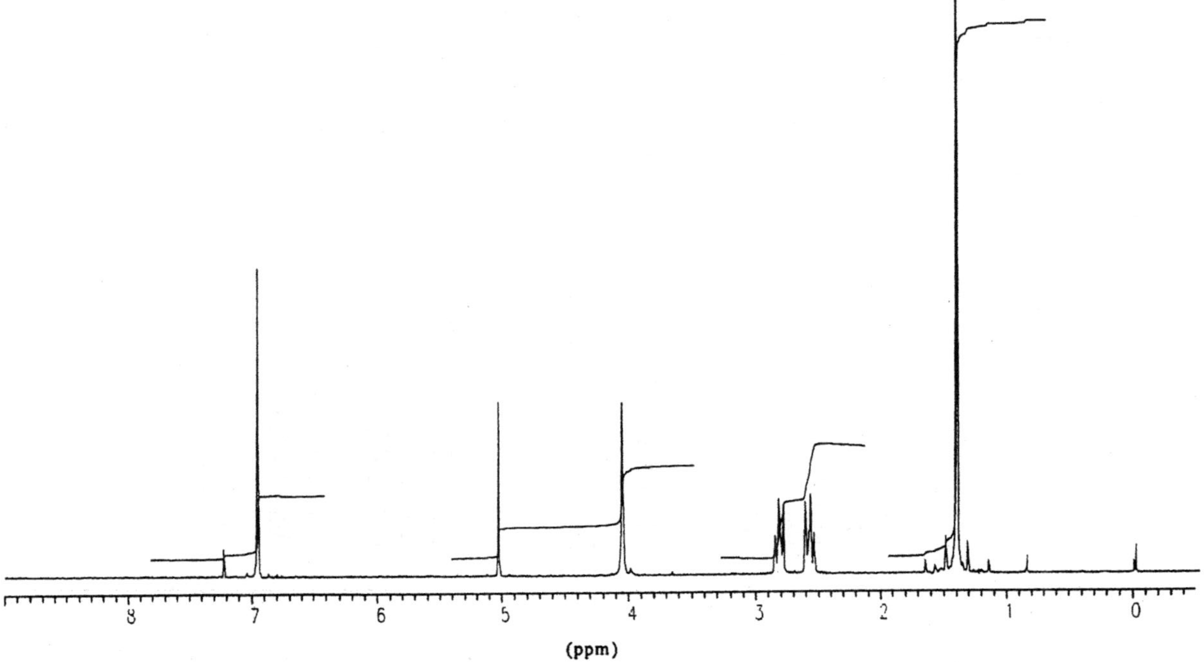

2-Phenylindole

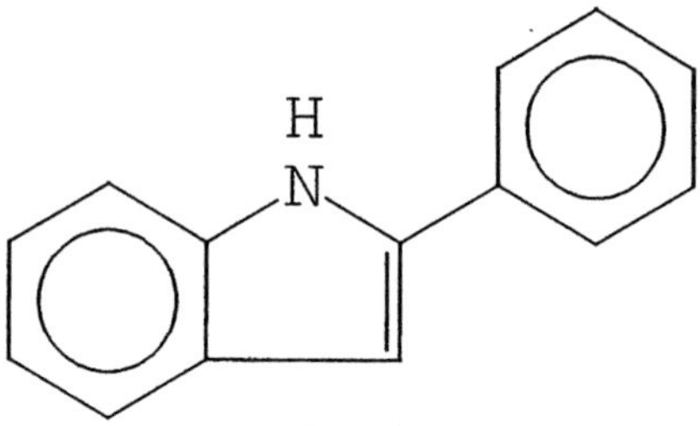

CAS No.	- 00948-65-2
PM Ref. No.	- 72160
Restrictions	- SML = 15 mg/kg
Formula	- $C_{14}H_{11}N$
Molecular weight	- 193

Alternative names	-
Physical Characteristics	- White powder, mp 188 - 190°C, bp 250°C (10 mm Hg). - Soluble in most organic solvents.
Handling	- Store at room temperature.
Safcty	-
Availability	- Standard sample was supplied.
Current uses	- Antioxidant.
Applications	- PVC, PVDC.
GC Retention time	- 38.5 min.

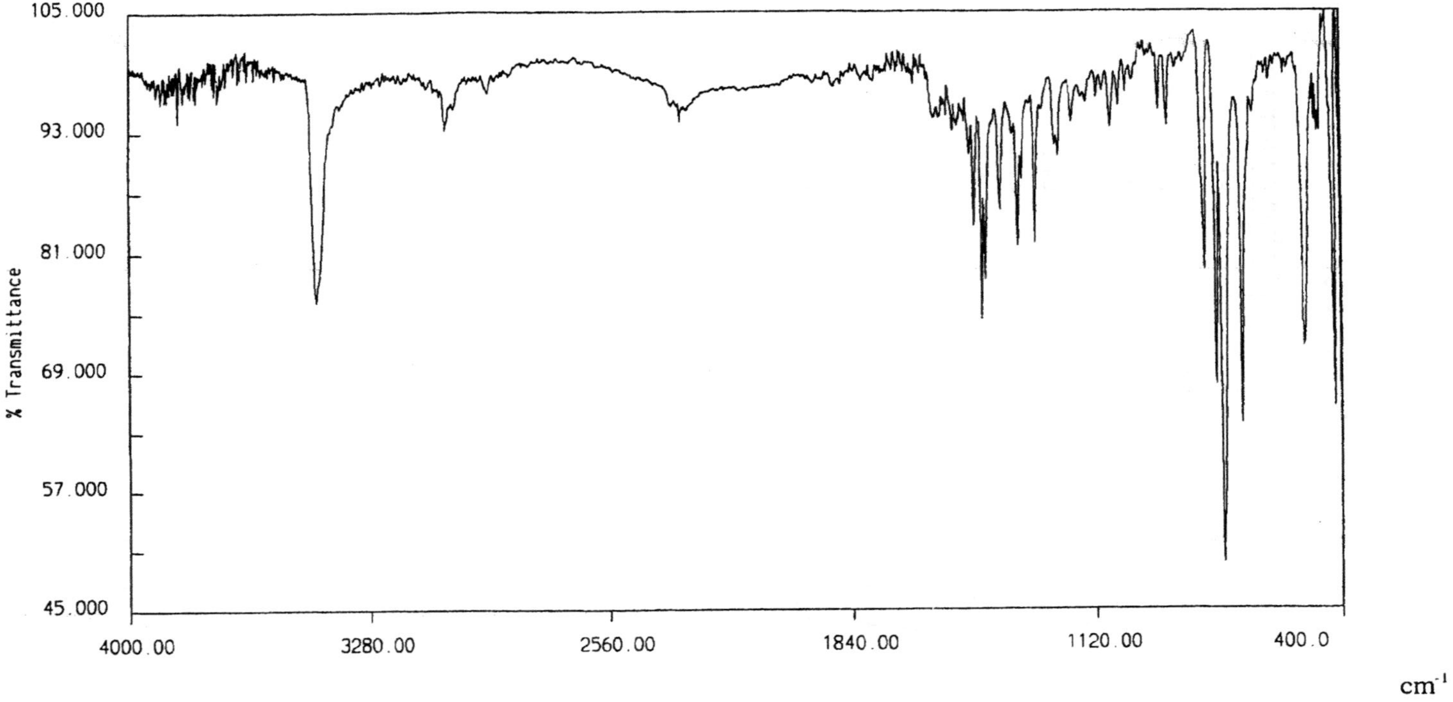

Name : 2-PHENYLINDOLE

PM Nr : 72160 CAS Nr : 00948-65-2

KBr pellet

CAS No. 00948-65-2 PM Ref. No. 72160

2-PHENYLINDOLE

Peak wavenumber	% T	Relative intensity
420,5	65,4	68,6
466,8	93,0	13,9
509,3	71,5	56,5
688,7	63,7	71,9
742,7	49,6	99,9
763,9	67,7	64,0
796,7	79,1	41,4
1300,2	81,9	35,9
1338,8	88,2	23,4
1352,3	81,5	36,7
1402,4	85,1	29,5
1446,8	78,2	43,2
1458,4	74,3	50,9
1481,5	83,7	32,3
3443,4	76,1	47,4

2-Phenylindole

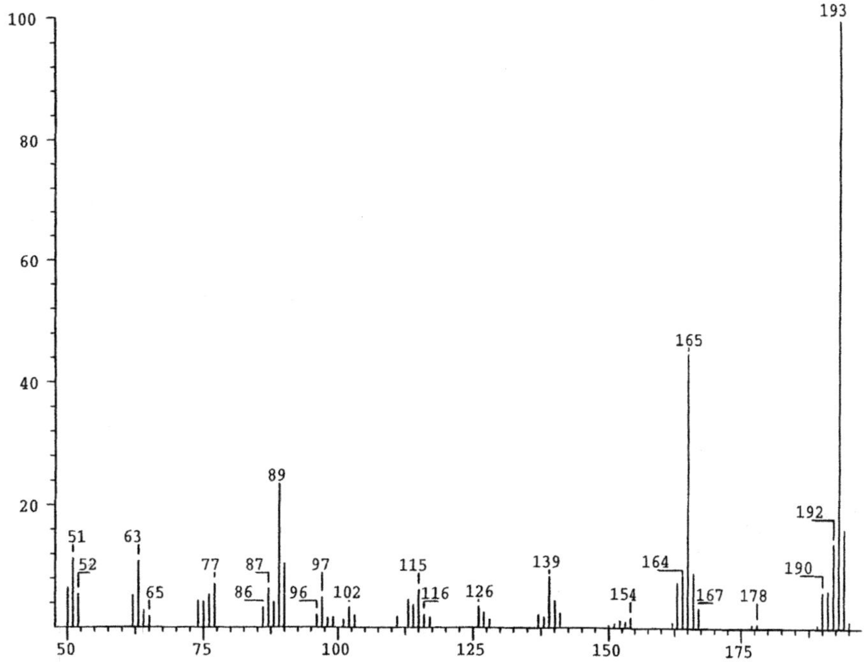

M/Z	Ion Intensity (%)	M/Z	Ion Intensity (%)
51	11.3	165	44.9
63	10.9	166	9.0
89	22.4	178	0.7
90	10.3	192	14.0
115	6.2	193	100.0
139	8.5	194	16.2
164	8.6	195	1.1

Spectrometer	: Finnigan Mat SSQ 700
Inletsystem	: Capillary GC/MS
Source Temperature	: 180°C
Electron Energy	: 70 eV
Scan Range	: 40-500

PM n°	CAS n°	Name
72160	00948-65-2	2-PHENYLINDOLE

Solvent	$CDCl_3$
SF	250.13
SI	16384
SW	3125
NS	65

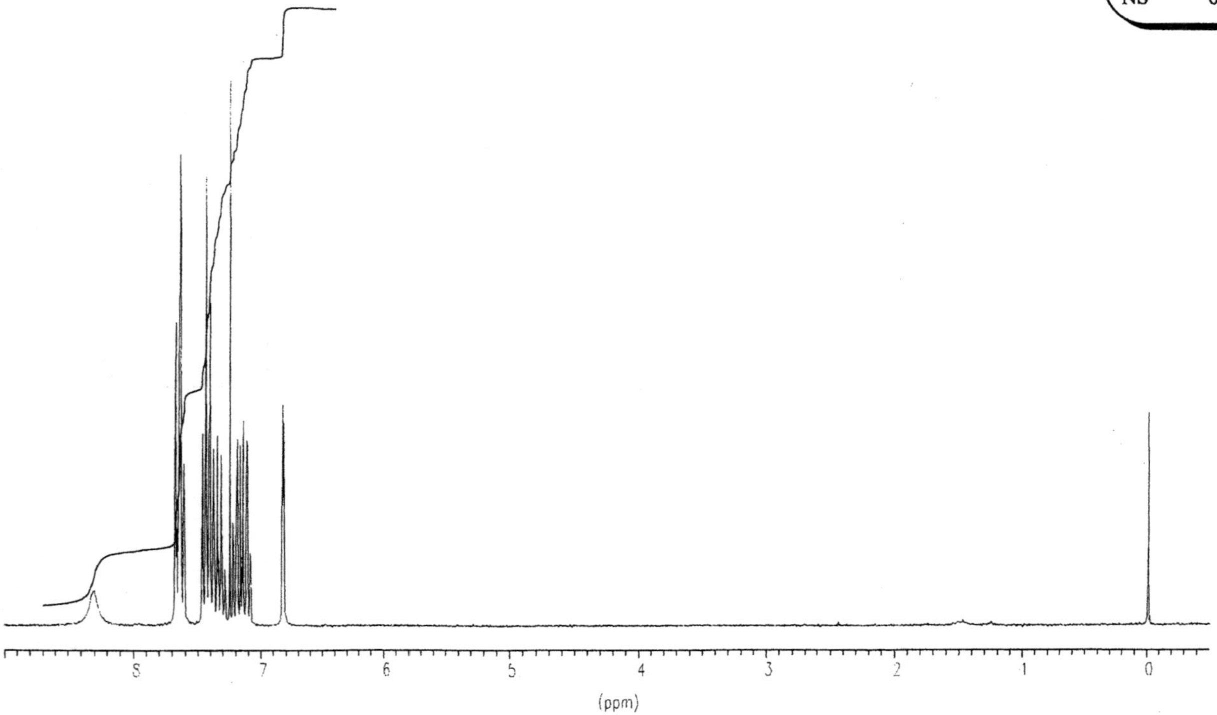

Phosphoric acid, diphenyl 2-ethylhexyl ester

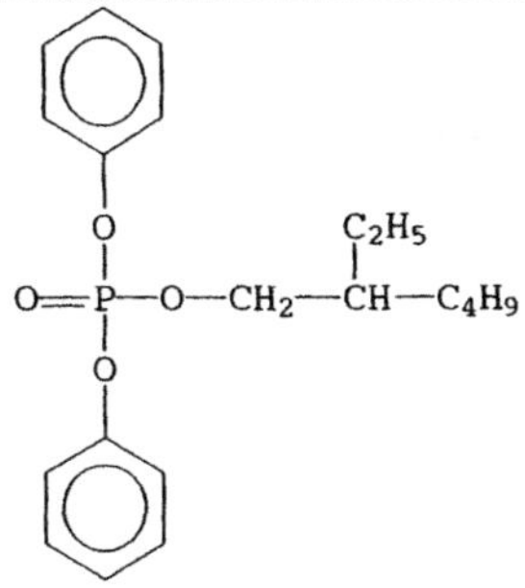

CAS No.	- 01241-94-7
PM Ref. No.	- 72800
Restrictions	- SML = 3.0 mg/kg
Formula	- $C_{20}H_{27}O_4P$
Molecular weight	- 362

Alternative names	- 2-Ethylhexyl diphenyl phosphate
Physical Characteristics	- Colourless liquid, mp -30°C, bp 375°C. - Soluble in alcohol, ether, benzene, chloroform.
Handling	- Store at room temperature.
Safety	-
Availability	- Standard sample was supplied.
Current uses	- Plastizicer.
Applications	- Inks, PVC.
GC Retention time	- 45.1 min.

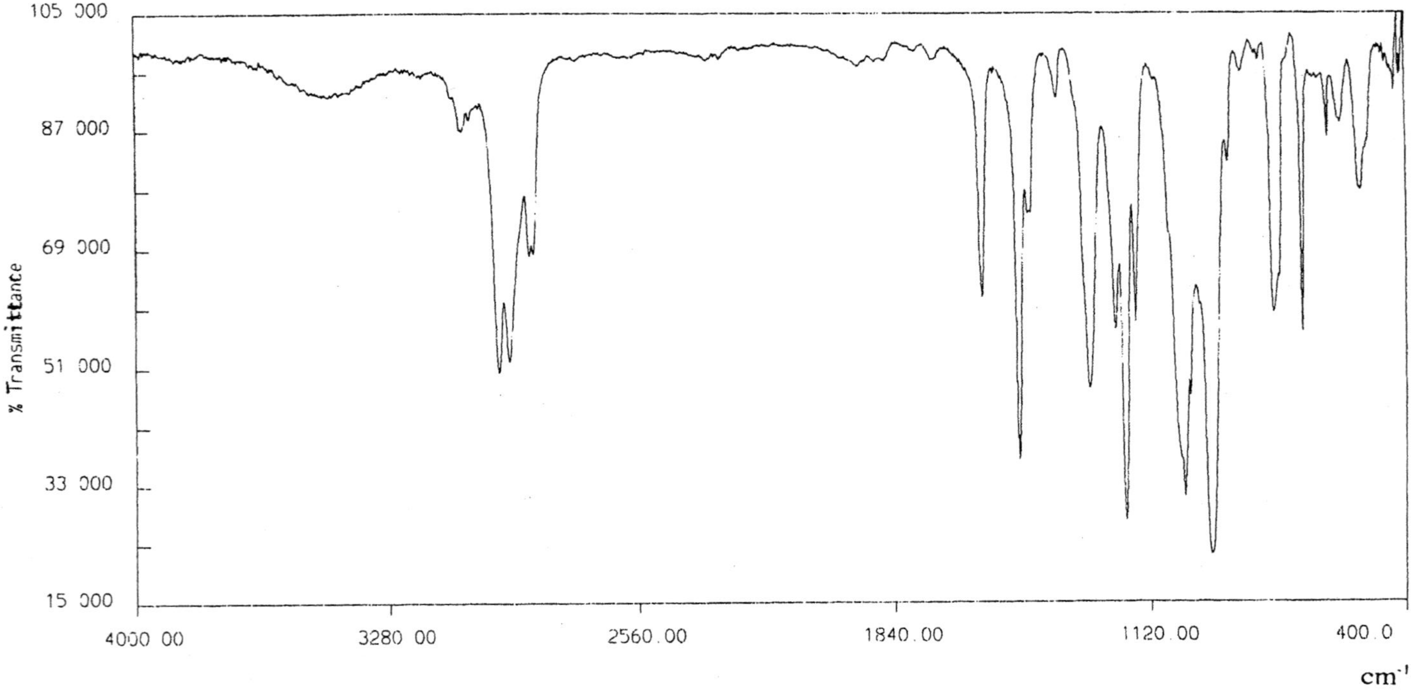

Name : PHOSPHORIC ACID, DIPHENYL 2-ETHYLHEXYL ESTER

PM Nr : 72800 CAS Nr : 01241-94-7

KBr pellet

CAS No. 01241-94-7 PM Ref. No. 72800

PHOSPHORIC ACID, DIPHENYL 2-ETHYLHEXYL ESTER

Peak wavenumber	% T	Relative intensity
428,3	92,9	9,1
526,6	77,8	28,5
580,6	88,1	15,3
617,3	85,7	18,4
644,3	94,8	6,7
688,7	56,3	56,2
771,6	59,2	52,5
904,7	82,0	23,1
951,0	22,2	100,0
1024,3	31,1	88,6
1163,2	57,7	54,4
1192,2	27,6	93,1
1221,1	56,7	55,7
1294,4	47,7	67,3
1466,1	74,3	33,1
1491,2	36,8	81,3
1593,4	61,4	49,6
2862,7	68,2	40,9
2932,2	52,1	61,6
2961,1	50,3	63,9

Phosphoric acid, diphenyl 2-ethylhexyl ester

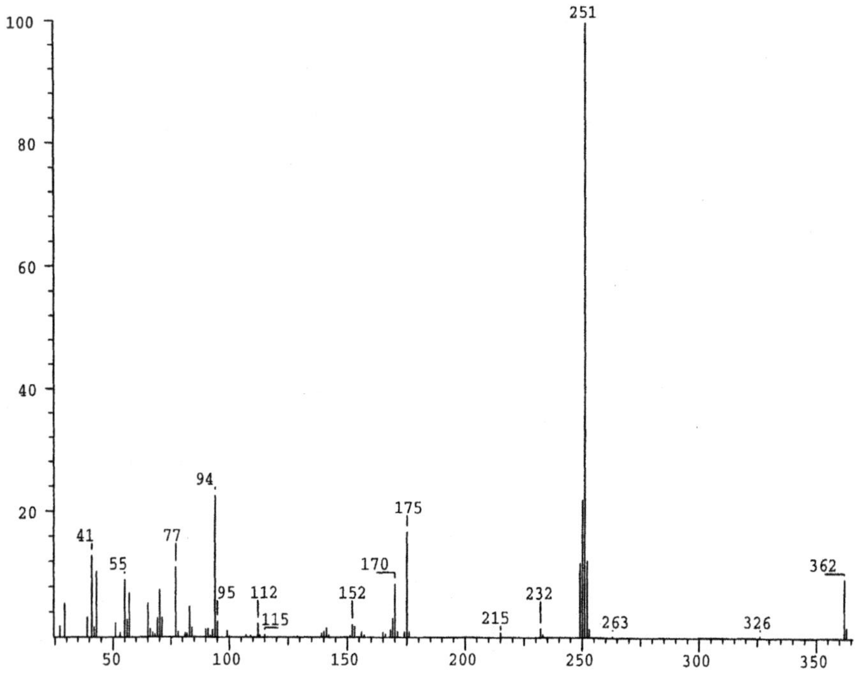

M/Z	Ion Intensity (%)	M/Z	Ion Intensity (%)
41	12.9	249	11.7
43	10.4	250	21.9
55	9.1	251	100.0
77	11.1	252	12.1
94	2.5	362	9.4
175	16.7	363	1.7

Spectrometer	: Finnigan Mat SSQ 700
Inletsystem	: Capillary GC/MS
Source Temperature	: 180°C
Electron Energy	: 70 eV
Scan Range	: 25-450

PM n°
72800

CAS n°
01241-94-7

Name
PHOSPHORIC ACID, DIPHENYL 2-ETHYLHEXYL ESTER

Solvent	$CDCL_3$
SF	250.13
SI	16384
SW	3125
NS	25

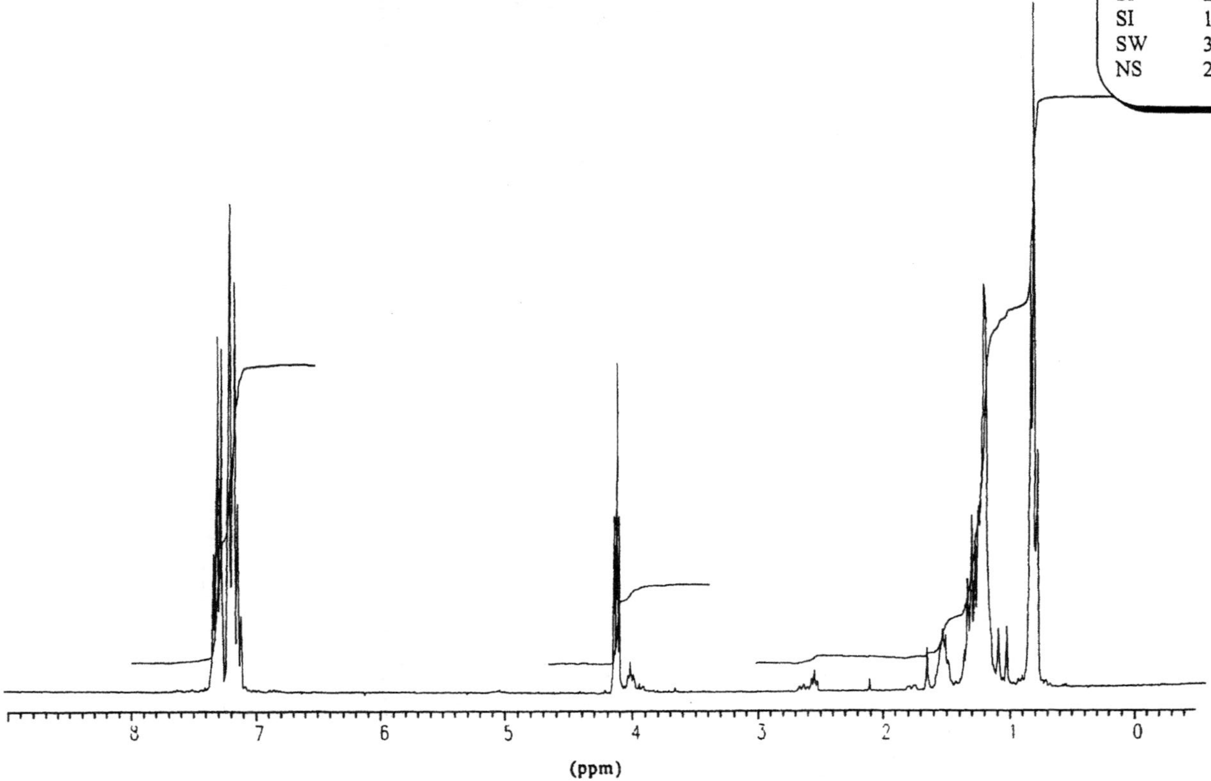

Phosphoric acid, tributoxyethyl ester

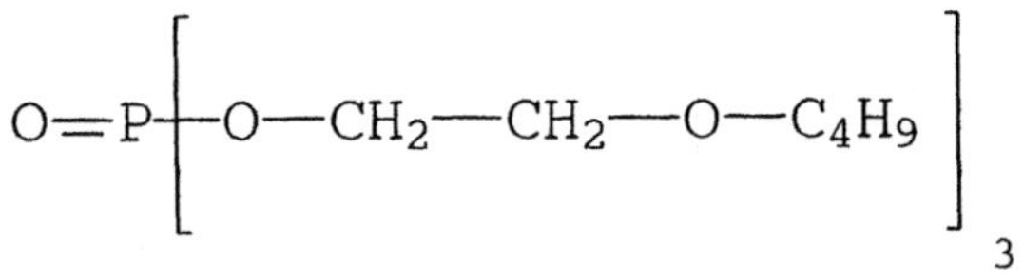

CAS No.	- 00078-51-3
PM Ref. No.	- 73600
Restrictions	- SML = 3.0 mg/kg
Formula	- $C_{18}H_{39}O_7P$
Molecular weight	- 398

Alternative names	- 2-Butoxyethanol, phosphate
Physical Characteristics	- Colourless liquid, mp - 70°C, bp 215 - 228°C (4 mm Hg). - Soluble in most organic solvents.
Handling	- Store at room temperature.
Safety	-
Availability	- Standard sample was supplied.
Current uses	- Plasticizer, flame retardant.
Applications	- Paper.
GC Retention time	- 44.9 min.
Note	- Not widely used for plastic food contact materials.

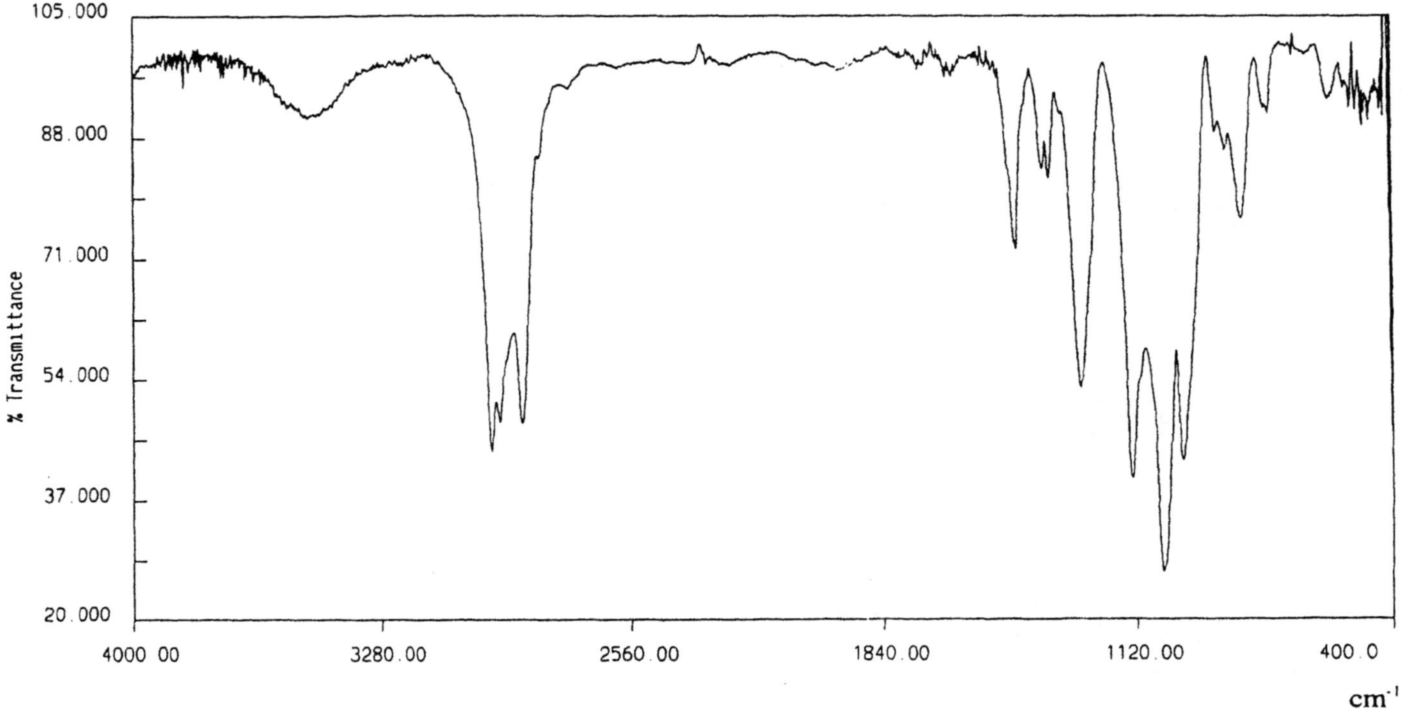

Name : PHOSPHORIC ACID, TRIBUTOXYETHYL ESTER

PM Nr : 73600 CAS Nr : 00078-51-3

KBr pellet

CAS No. 00078-51-3 PM Ref. No. 73600

PHOSPHORIC ACID, TRIBUTOXYETHYL ESTER

Peak wavenumber	% T	Relative intensity
428,3	91,9	11,0
476,5	89,6	14,2
817,9	76,6	31,9
987,7	42,5	78,4
1043,6	26,7	100,0
1134,3	39,9	82,0
1279,0	52,9	64,2
1365,8	82,2	24,3
1381,2	83,6	22,4
1460,3	72,3	37,8
2872,4	47,7	71,3
2936,0	48,0	70,9
2959,2	43,8	76,7

Phosphoric acid, tributoxyethyl ester

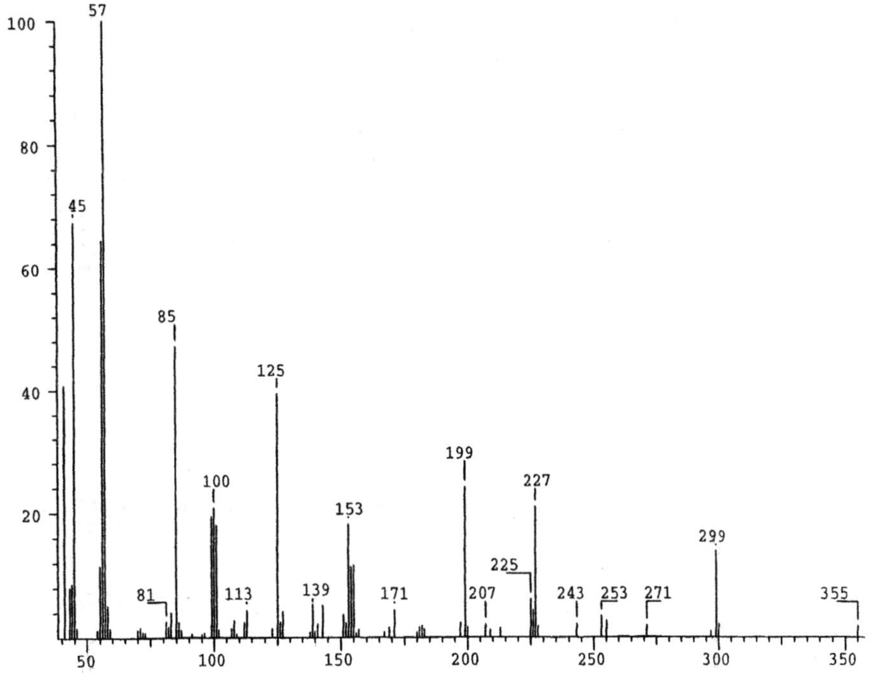

M/Z	Ion Intensity (%)	M/Z	Ion Intensity (%)
41	40.7	125	39.4
45	67.4	153	18.2
55	11.5	154	11.3
56	64.5	155	11.6
57	100.0	199	24.1
85	47.2	227	21.0
99	19.4	299	13.8
100	20.8	355	1.8
101	18.0		

Spectrometer : Finnigan Mat SSQ 700
Inletsystem : Capillary GC/MS
Source Temperature : 180°C
Electron Energy : 70 eV
Scan Range : 40-500

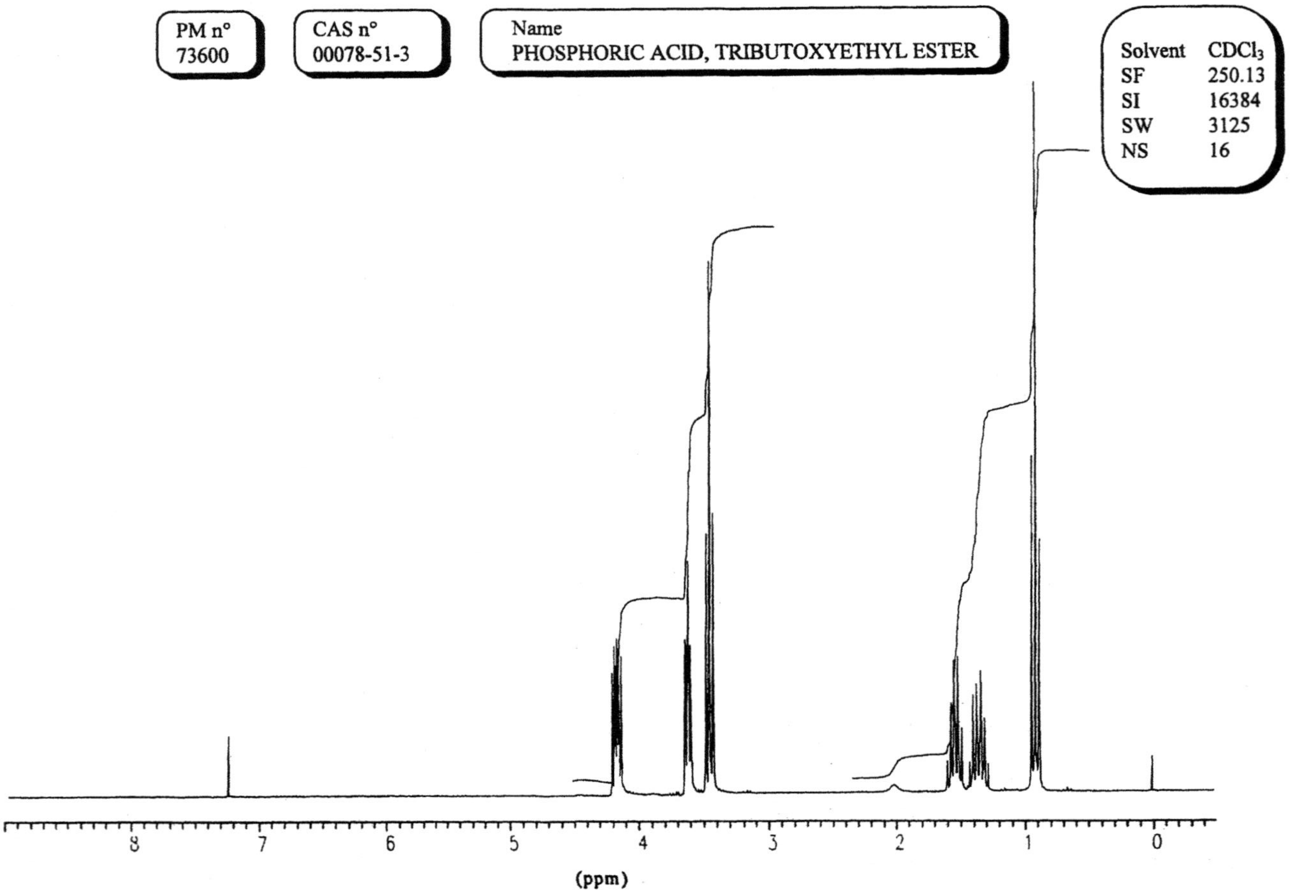
PM n°
73600
CAS n°
00078-51-3
Name
PHOSPHORIC ACID, TRIBUTOXYETHYL ESTER
Solvent CDCl3
SF 250.13
SI 16384
SW 3125
NS 16
8
7
6
5
4
3
2
1
0
(ppm)

Phosphoric acid, tributyl ester

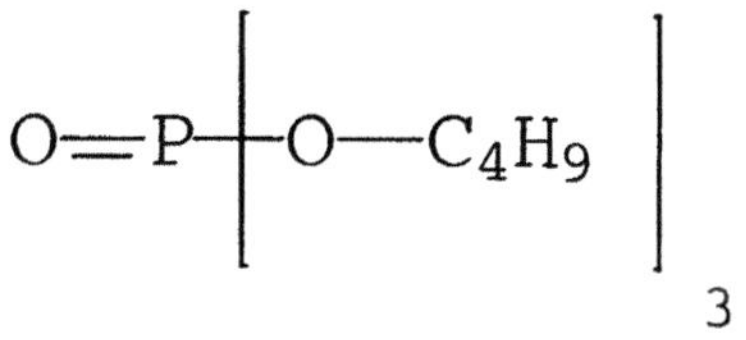

CAS No.	- 00126-73-8
PM Ref. No.	- 73680
Restrictions	- SML = 3.0 mg/kg
Formula	- $C_{12}H_{27}O_4P$
Molecular weight	- 266

Alternative names	- Tributyl phosphate
Physical Characteristics	- Colourless liquid, mp < -80°C, bp 289°C with decomposition. - Soluble in most organic solvents.
Handling	- Store at room temperature.
Safety	- Irritant.
Availability	- Standard sample was supplied.
Current uses	- Plasticizer.
Applications	- Cellulose esters.
GC Retention time	- 30.5 min.

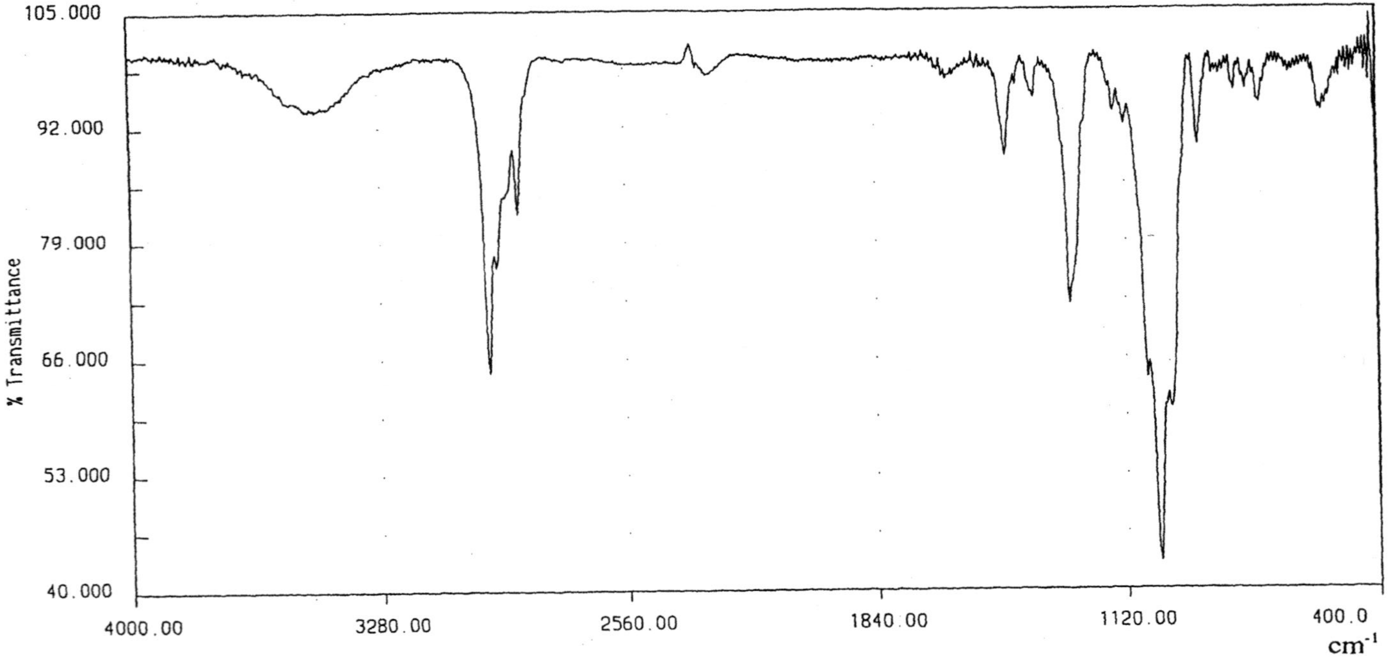

Name : PHOSPHORIC ACID, TRIBUTYL ESTER

PM Nr : 73680 CAS Nr : 00126-73-8

KBr pellet

CAS No. 00126-73-8 PM Ref. No. 73680

PHOSPHORIC ACID, TRIBUTYL ESTER

Peak wavenumber	% T	Relative intensity
544.0	94.4	9.8
555.6	93.8	11.0
993.5	61.0	69.0
1028.2	43.4	100.0
1282.8	71.5	50.4
1466.1	88.4	20.5
2876.2	81.9	32.0
2936.0	75.9	42.6
2961.1	64.3	63.1

Phosphoric acid, tributyl ester

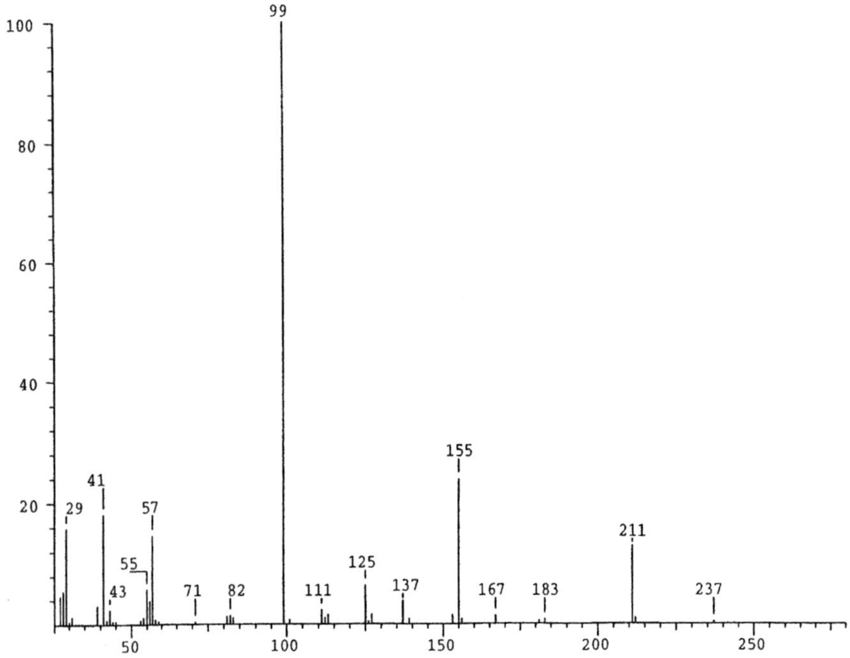

M/Z	Ion Intensity (%)	M/Z	Ion Intensity (%)
27	4.7	57	14.6
28	5.6	99	100.0
29	15.9	111	2.4
39	3.1	125	6.5
41	18.2	137	4.0
43	2.4	155	23.9
55	5.9	211	13.0
56	3.9	237	0.4

Spectrometer : Finnigan Mat SSQ 700
Inletsystem : Capillary GC/MS
Source Temperature : 180°C
Electron Energy : 70 eV
Scan Range : 25-500

PM n°	CAS n°	Name
73680	00126-73-8	PHOSPHORIC ACID, TRIBUTYL ESTER

Solvent	$CDCL_3$
SF	250.13
SI	16384
SW	3125
NS	32

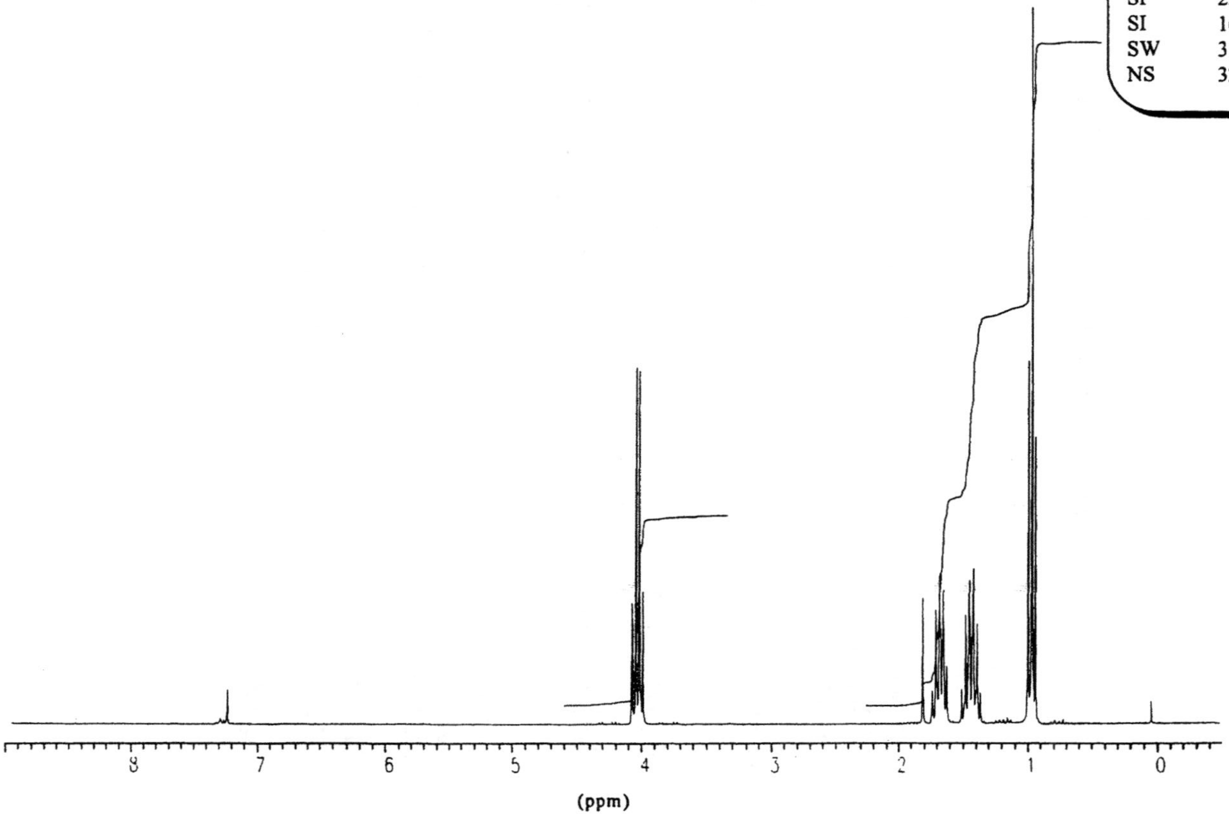

Phosphoric acid, triisobutyl ester

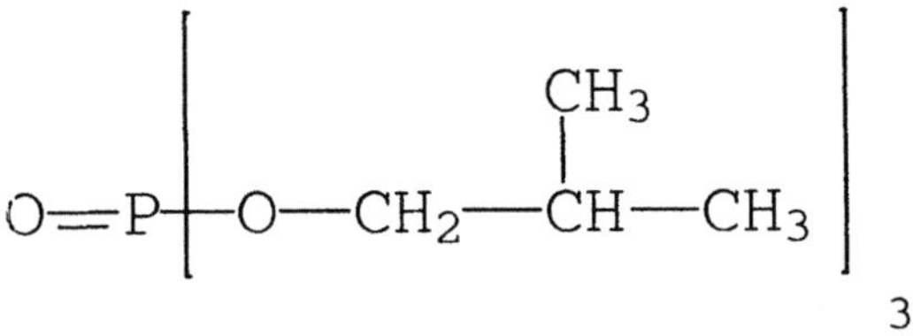

CAS No.	- 00126-71-6
PM Ref. No.	- 73840
Restrictions	- SML = 3.0 mg/kg
Formula	- $C_{12}H_{27}O_4P$
Molecular weight	- 266

Alternative names	- Triisobutyl phosphate
Physical Characteristics	- Colourless liquid, mp abt. -80°C, bp 264°C. - Soluble in alcohol, ether, benzene, water.
Handling	- Store at room temperature.
Safety	-
Availability	- Standard sample was supplied.
Current uses	- Plasticizer.
Applications	- Cellulose esters.
GC Retention time	- 27.5 min.

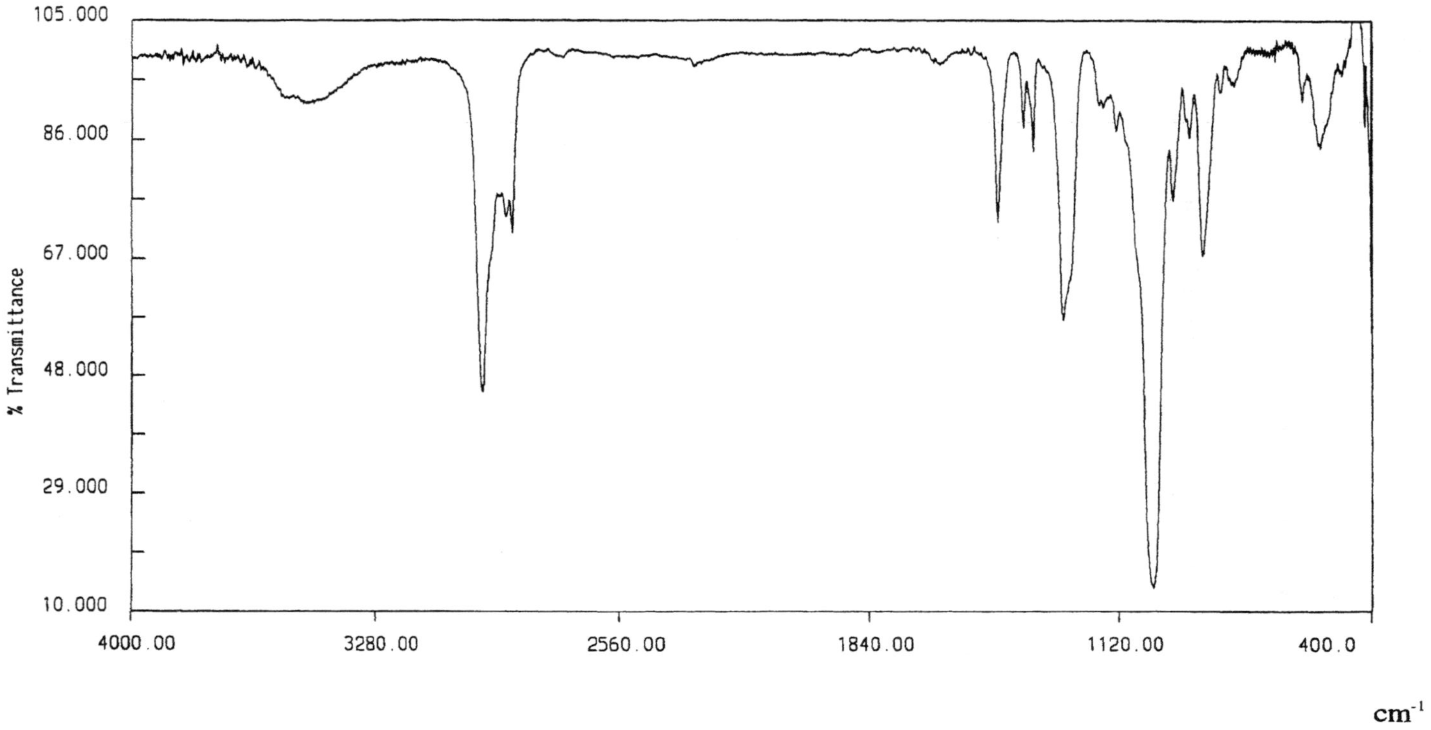

Name : PHOSPHORIC ACID, TRIISOBUTYL ESTER

PM Nr : 73840 CAS Nr : 00126-71-6

KBr pellet

CAS No. 00126-71-6 PM Ref. No. 73840

PHOSPHORIC ACID, TRIISOBUTYL ESTER

Peak wavenumber	% T	Relative intensity
418.6	88.1	13.8
542.1	84.6	17.8
879.7	67.4	37.8
918.2	86.2	16.0
964.5	76.1	27.7
1022.4	13.7	100.0
1284.8	57.0	49.9
1369.6	84.1	18.5
1396.6	87.8	14.1
1471.9	72.8	31.6
2876.2	71.1	33.5
2963.0	45.3	63.5

Phosphoric acid, triisobutyl ester

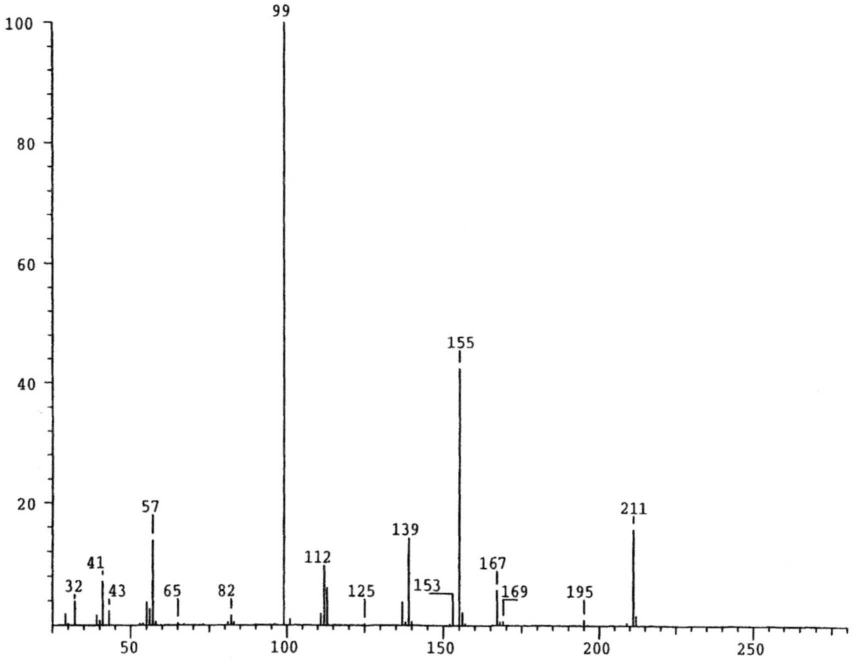

M/Z	Ion Intensity (%)	M/Z	Ion Intensity (%)
32	4.0	113	6.2
41	7.2	137	3.9
43	2.49	139	13.3
55	3.8	153	4.2
56	2.8	155	42.5
57	13.8	156	2.19
99	100.0	167	5.8
112	8.7	211	15.8

Spectrometer : Finnigan Mat SSQ 700
Inletsystem : Capillary GC/MS
Source Temperature : 180°C
Electron Energy : 70 eV
Scan Range : 25-500

PM n°	CAS n°	Name
73840	00126-71-6	PHOSPHORIC ACID, TRIISOBUTYL ESTER

Solvent	$CDCL_3$
SF	250.13
SI	16384
SW	3125
NS	32

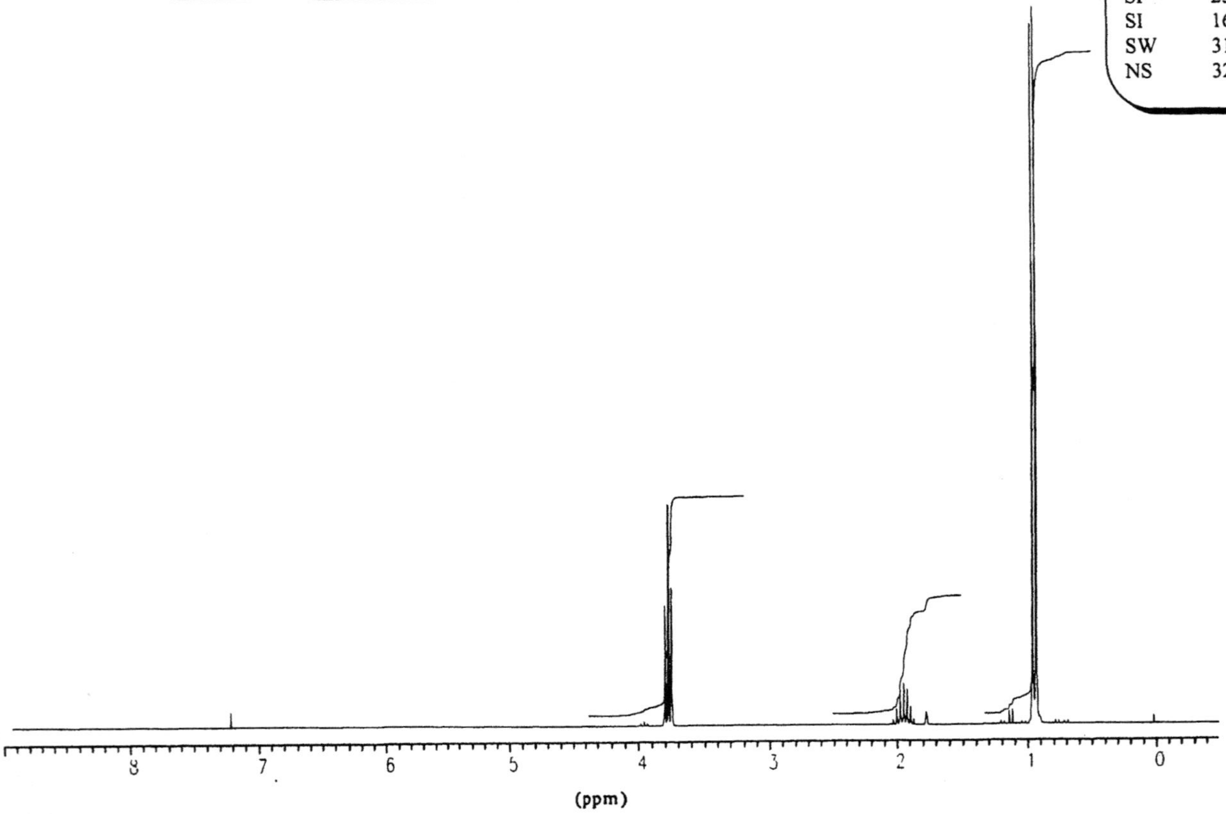

Phosphoric acid, triphenyl ester

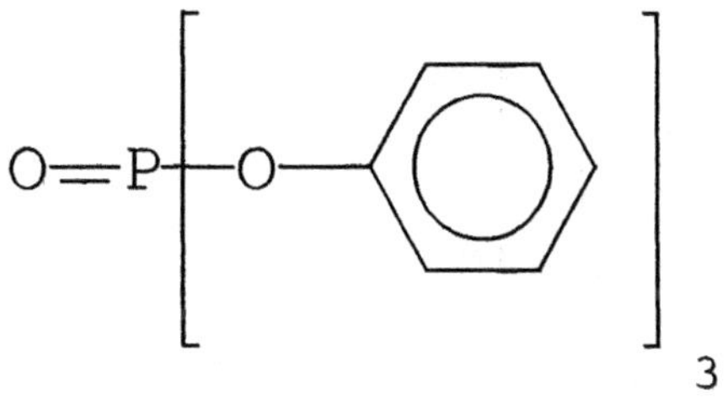

CAS No.	- 00115-86-6
PM Ref. No.	- 73920
Restrictions	- SML = 3.0 mg/kg
Formula	- $C_{18}H_{15}O_4P$
Molecular weight	- 326

Alternative names	- Triphenyl phosphate
Physical Characteristics	- White crystalline powder, mp 50 - 51°C, bp 245°C (11 mm Hg). - Soluble in most organic solvents.
Handling	- Store at room temperature.
Safety	-
Availability	- Standard sample was supplied.
Current uses	- Plasticizer, flame retardant.
Applications	- Paper.
GC Retention time	- 44.4 min.
Note	- Not widely used for plastic food contact materials.

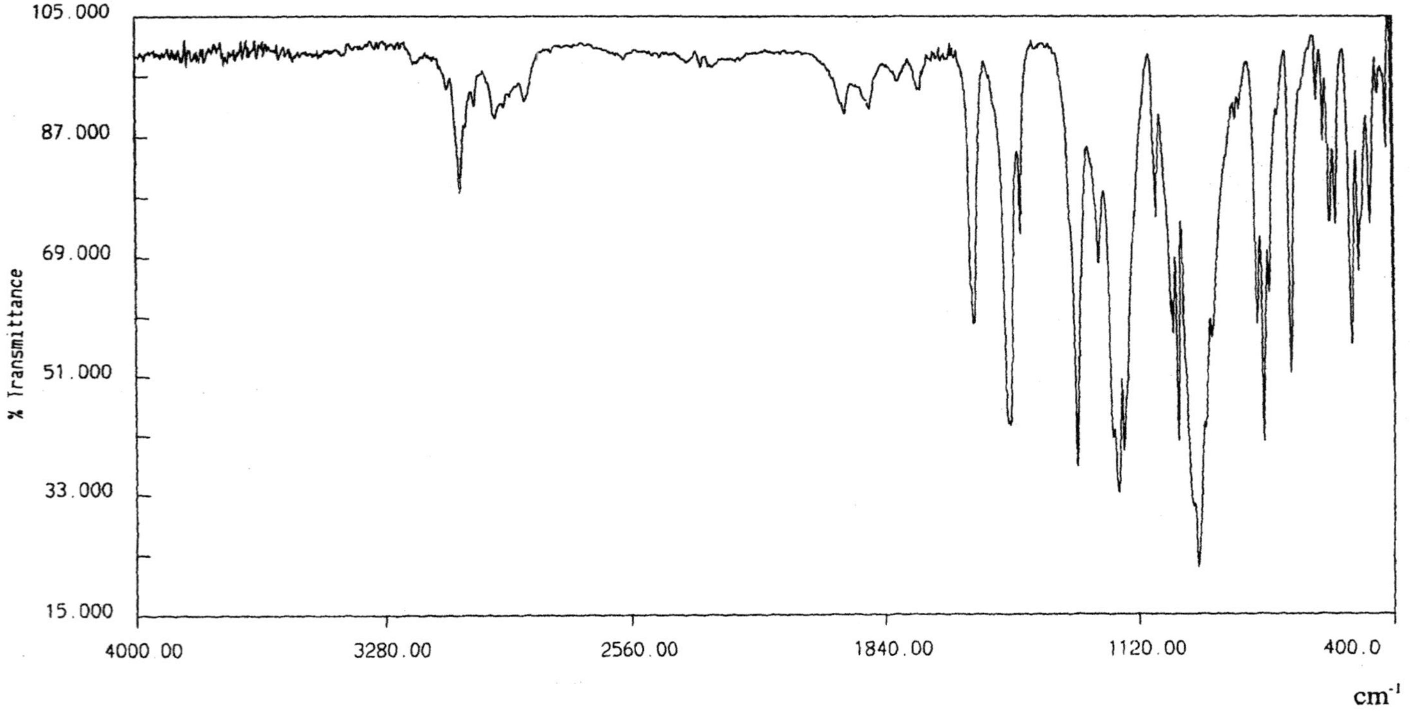

Name : PHOSPHORIC ACID, TRIPHENYL ESTER

PM Nr : 73920 CAS Nr : 00115-86-6

KBr pellet

CAS No. 00115-86-6 PM Ref. No. 73920

PHOSPHORIC ACID,TRIPHENYL ESTER

Peak wavenumber	% T	Relative intensity
418.6	85.1	19.0
441.8	93.2	8.7
466.8	73.7	33.6
499.6	66.6	42.7
518.9	55.8	56.5
565.2	73.7	33.6
580.6	73.9	33.4
596.1	86.1	17.8
690.6	51.4	62.1
727.3	90.0	12.8
752.3	63.6	46.5
769.7	40.9	75.5
787.1	58.9	52.5
846.9	89.2	13.8
914.4	56.7	55.3
953.0	21.7	100.0
1008.9	40.9	75.5
1022.4	57.3	54.6
1070.6	74.7	32.3
1163.2	39.3	77.6
1176.7	33.0	85.6
1234.6	67.8	41.2
1294.4	36.9	80.6
1454.5	72.1	35.7
1485.4	43.1	72.7
1589.5	58.7	52.8
1952.2	90.1	12.7
3061.4	78.4	27.6

Phosphoric acid, triphenyl ester

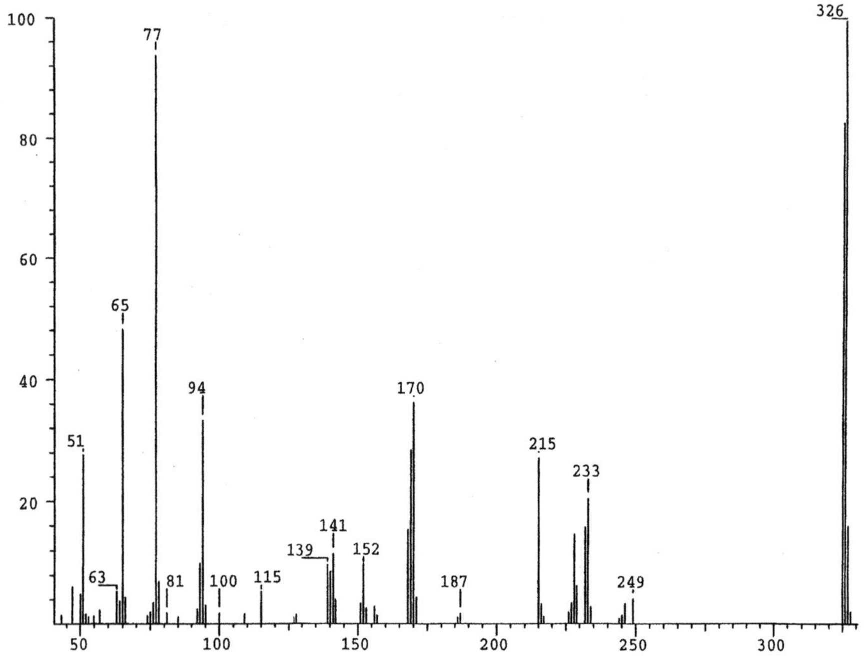

M/Z	Ion Intensity (%)	M/Z	Ion Intensity (%)
51	27.6	169	28.5
65	48.1	170	36.3
77	93.8	215	27.2
78	7.0	228	14.8
93	10.0	232	16.0
94	33.3	233	20.7
139	9.8	325	83.0
141	11.6	326	100.0
152	10.0	327	16.3
168	15.5	328	2.2

Spectrometer : Finnigan Mat SSQ 700
Inletsystem : Capillary GC/MS
Source Temperature : 180°C
Electron Energy : 70 eV
Scan Range : 40-500

PM n°	CAS n°	Name
73920	00115-86-6	PHOSPHORIC ACID, TRIPHENYL ESTER

Solvent	$CDCl_3$
SF	250.13
SI	16384
SW	3125
NS	46

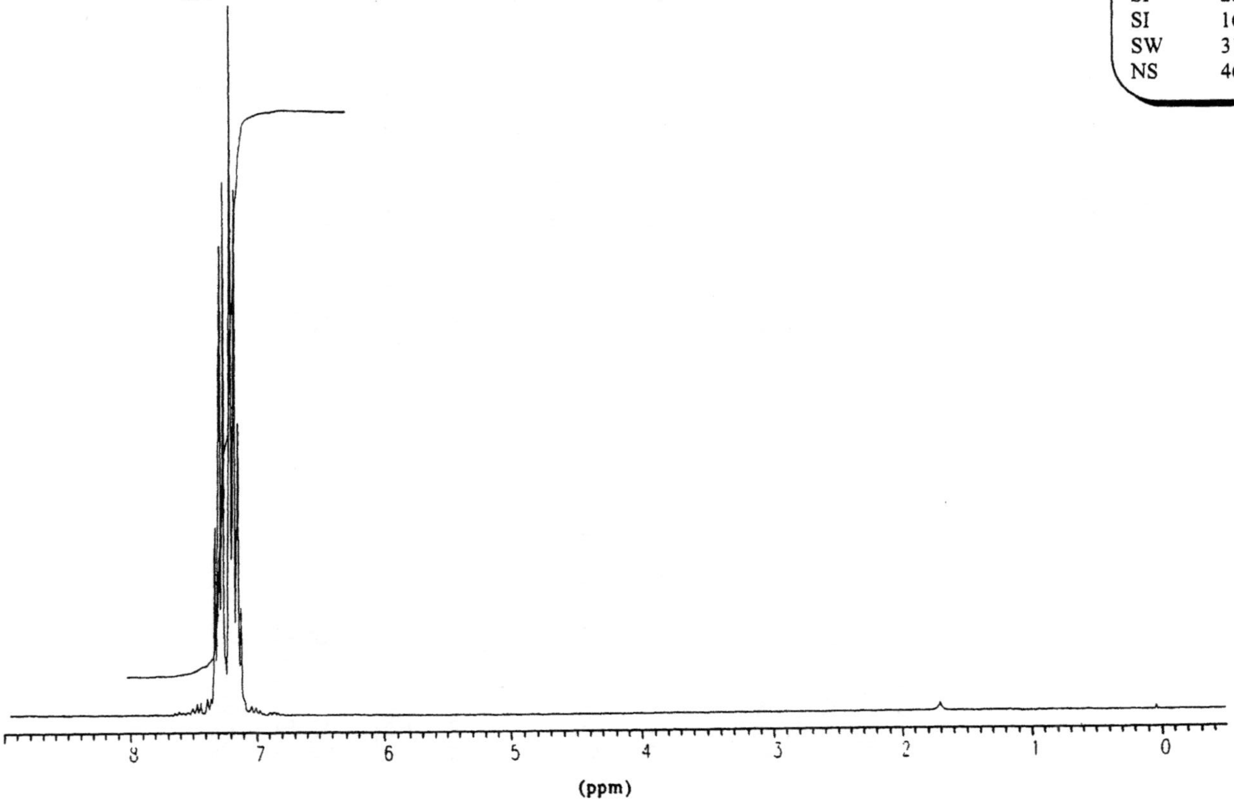

Phosphoric acid, tris(2-ethylhexyl) ester

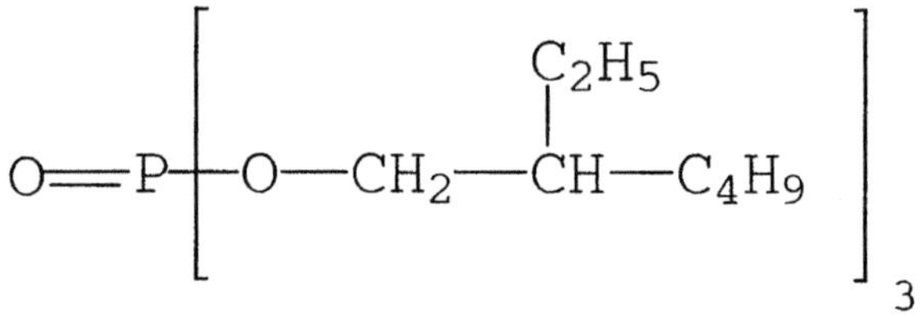

CAS No.	- 00078-42-2
PM Ref. No.	- 74000
Restrictions	- SML = 3.0 mg/kg
Formula	- $C_{24}H_{51}O_4P$
Molecular weight	- 434

Alternative names	- Trioctyl phosphate, Tri(2-ethylhexyl) phosphate
Physical Characteristics	- Colourless liquid, mp -74°C, bp 215°C (4 mm Hg). - Soluble in alcohol, acetone, ether.
Handling	- Store at room temperature.
Safety	-
Availability	- Standard sample was supplied.
Current uses	- Lubricant, plasticizer.
Applications	- Inks, PVC.
GC Retention time	- 64.2 min.

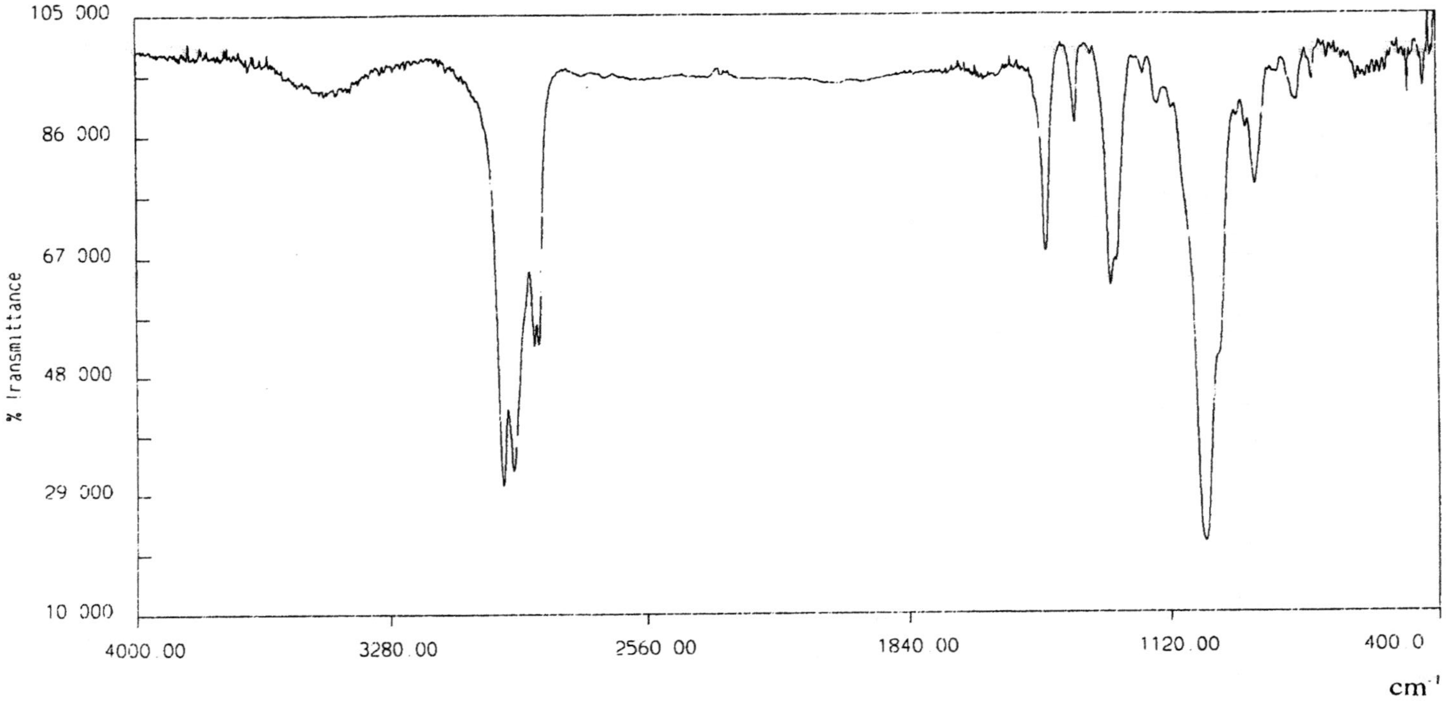

Name : PHOSPHORIC ACID, TRIS(2-ETHYLHEXYL)ESTER

PM Nr : 74000 CAS Nr : 00078-42-2

KBr pellet

CAS No. 00078-42-2 PM Ref. No. 74000

PHOSPHORIC ACID, TRIS(2-ETHYLHEXYL) ESTER

Peak wavenumber	% T	Relative intensity
434,0	93,4	8,4
474,6	92,3	9,8
586,4	94,7	6,7
611,5	94,3	7,2
727,3	94,4	7,1
769,7	91,3	11,0
885,4	78,0	27,9
1022,4	21,2	100,0
1269,3	66,1	43,0
1284,8	62,2	48,0
2862,7	53,0	59,6
2932,2	32,8	85,3
2961,1	30,5	88,2

Phosphoric acid, tris(2-ethylhexyl) ester

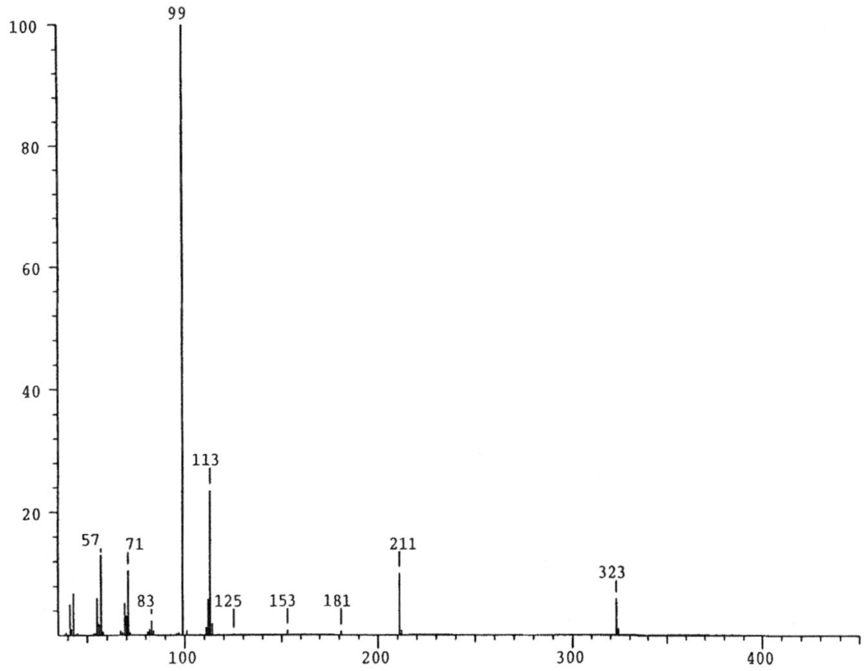

M/Z	Ion Intensity (%)	M/Z	Ion Intensity (%)
41	5.1	99	100.0
43	6.9	112	5.9
55	6.1	113	23.5
57	13.1	211	10.0
69	5.3	212	0.7
71	10.5	323	6.0
83	2.4	324	1.0

Spectrometer	: Finnigan Mat SSQ 700
Inletsystem	: Capillary GC/MS
Source Temperature	: 180°C
Electron Energy	: 70 eV
Scan Range	: 25-700

PM n° 74000

CAS n° 00078-42-2

Name
PHOSPHORIC ACID, TRIS(2-ETHYLHEXYL)ESTER

Solvent	$CDCL_3$
SF	250.13
SI	16384
SW	3125
NS	16

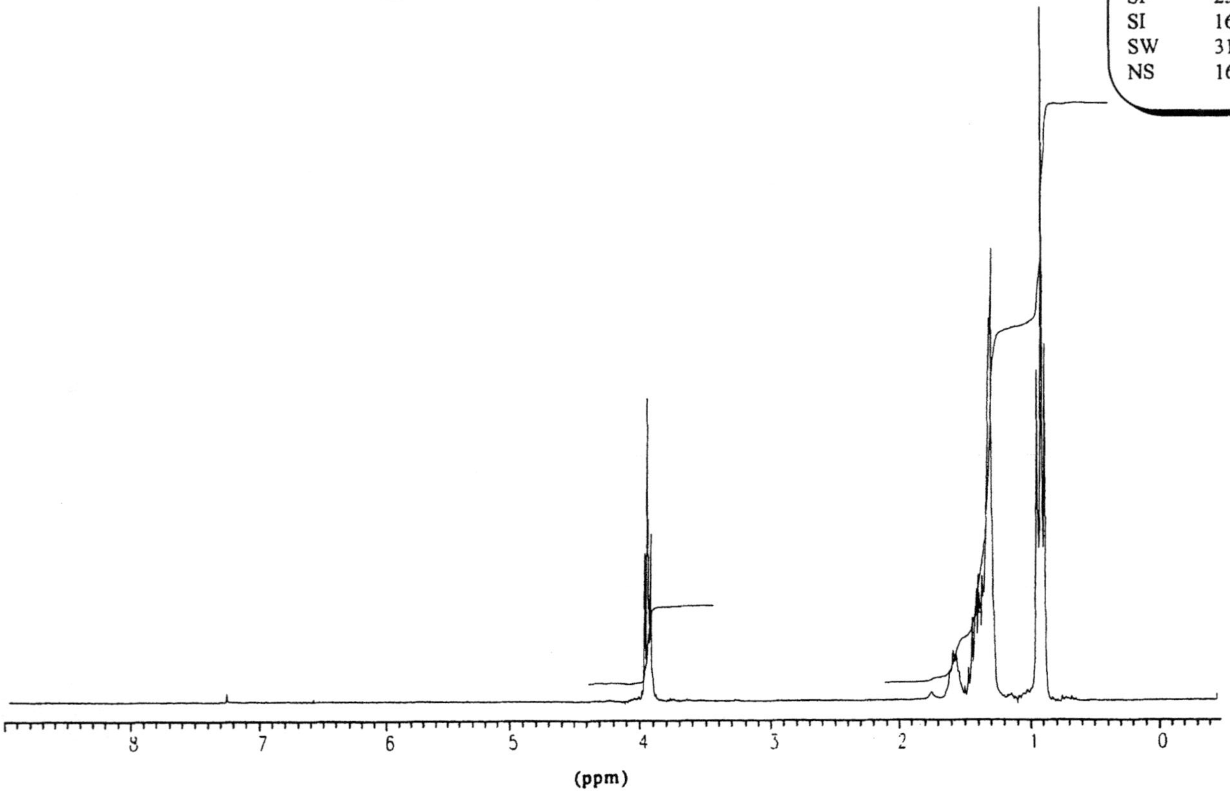

Phosphorous acid, tris(2,4-di-tert-butylphenyl) ester

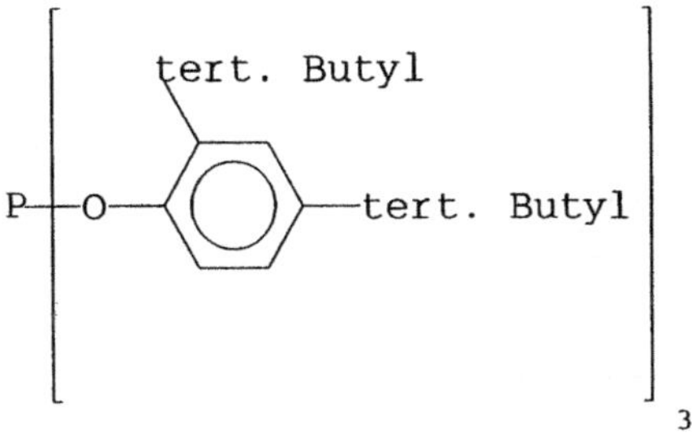

CAS No.	- 31570-04-4
PM Ref. No.	- 74240
Restrictions	- None
Formula	- $C_{42}H_{63}O_3P$
Molecular weight	- 646.94

Alternative names	- Tris(2,4-di-tert-butylphenyl)phosphite
Physical Characteristics	- White powder, mp 180 - 185°C. - Soluble in most organic solvents.
Handling	- Store at room temperature.
Safety	-
Availability	- Standard sample supplied.
Current uses	- Stabilizer, co-stabilizer.
Applications	- General use.
GC Retention time	- 58.9 min. - Di-tert-butylphenol 27.7 min. - Tris(2,4-di-tert-butylphenyl)phosphate 49.2 min.

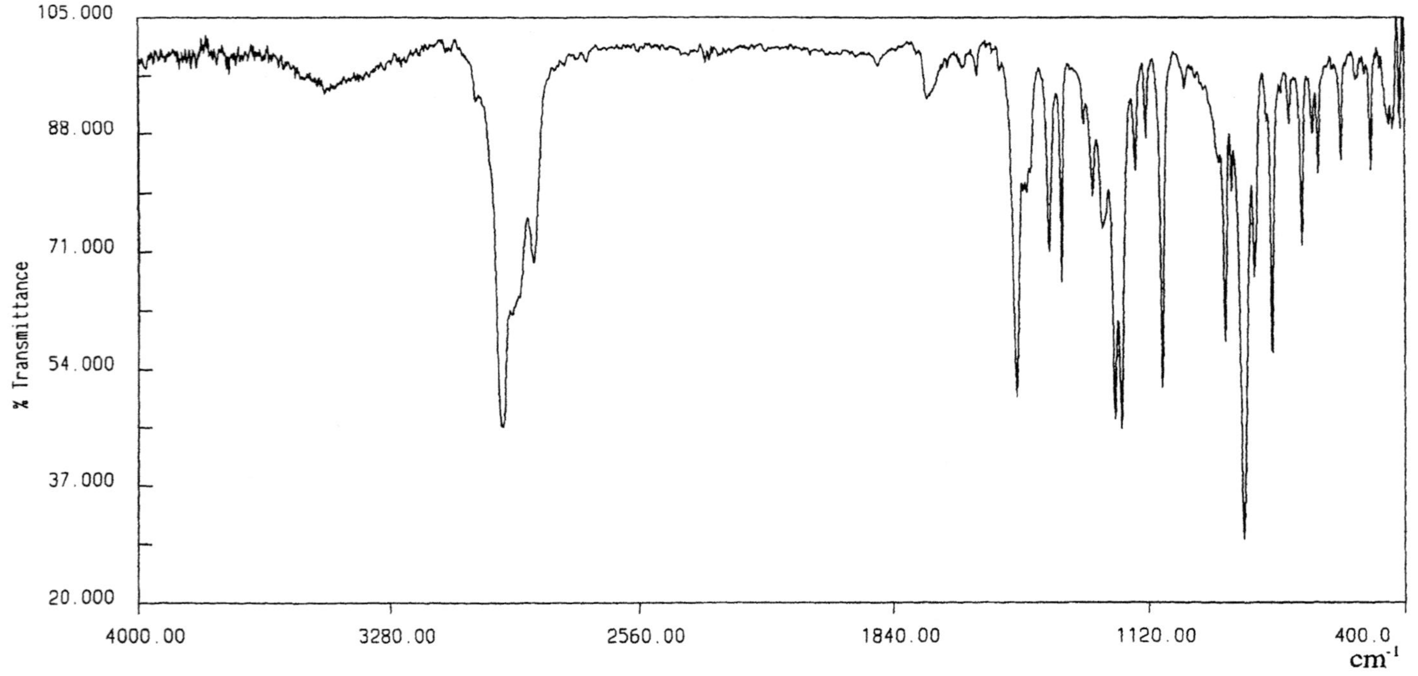

Name : PHOSPHOROUS ACID, TRIS(2,4-DI-TERT-BUTYLPHENYL) ESTER

PM Nr : 74240 CAS Nr : 31570-04-4

KBr pellet

CAS No. 31570-04-4 PM Ref. No. 74240

PHOSPHOROUS ACID, TRIS(2,4 DI-TERT-BUTYLPHENYL) ESTER

Peak wavenumber	% T	Relative intensity
410.9	88.6	16.1
434.0	88.5	16.2
495.8	82.5	24.7
580.6	84.0	22.5
646.2	82.2	25.1
661.7	87.8	17.2
692.5	71.8	39.8
727.3	89.2	15.2
775.5	56.4	61.5
825.6	67.2	46.2
854.6	29.1	100.0
889.3	79.5	28.9
906.7	58.0	59.3
1082.2	51.1	68.9
1126.6	87.2	18.0
1155.5	82.4	24.8
1196.0	45.2	77.3
1213.4	46.62	75.3
1248.1	74.25	36.3
1275.1	78.85	29.8
1361.9	66.44	47.3
1398.6	70.91	41.0
1462.2	79.29	29.2
1491.2	49.82	70.8
2868.5	69.26	43.3
2932.2	61.79	53.9
2961.1	45.48	76.9

Phosphorous acid, tris(2,4-di-tert-butylphenyl) ester

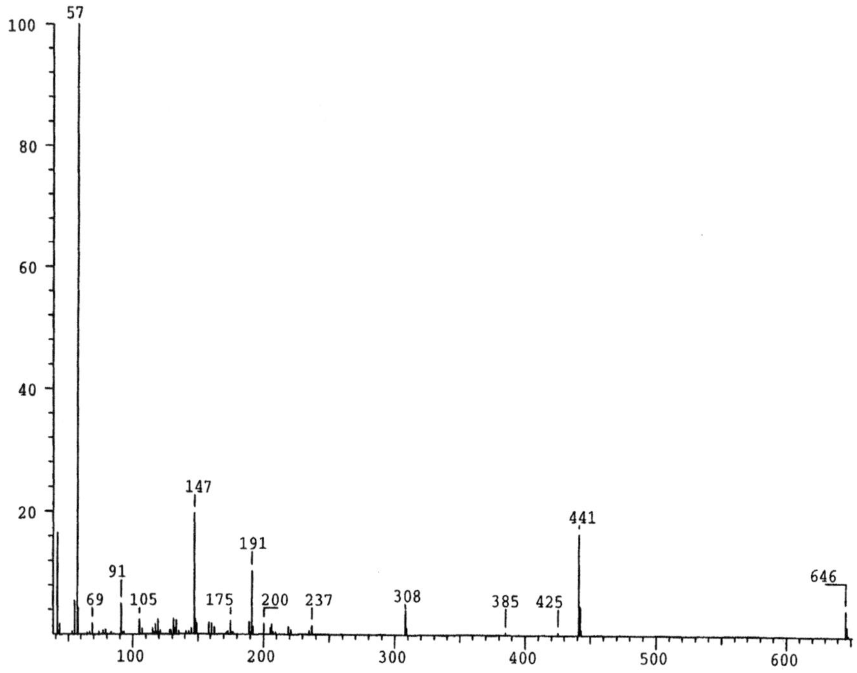

M/Z	Ion Intensity (%)	M/Z	Ion Intensity (%)
41	16.6	191	10.3
55	5.5	308	4.0
57	100.0	441	16.5
58	4.3	646	4.4
91	5.0	647	1.7
105	2.5	648	0.3
147	19.7		

Spectrometer	: Finnigan Mat SSQ 700
Inletsystem	: Capillary GC/MS
Source Temperature	: 180°C
Electron Energy	: 70 eV
Scan Range	: 40-700

Di-tert-butylphenol

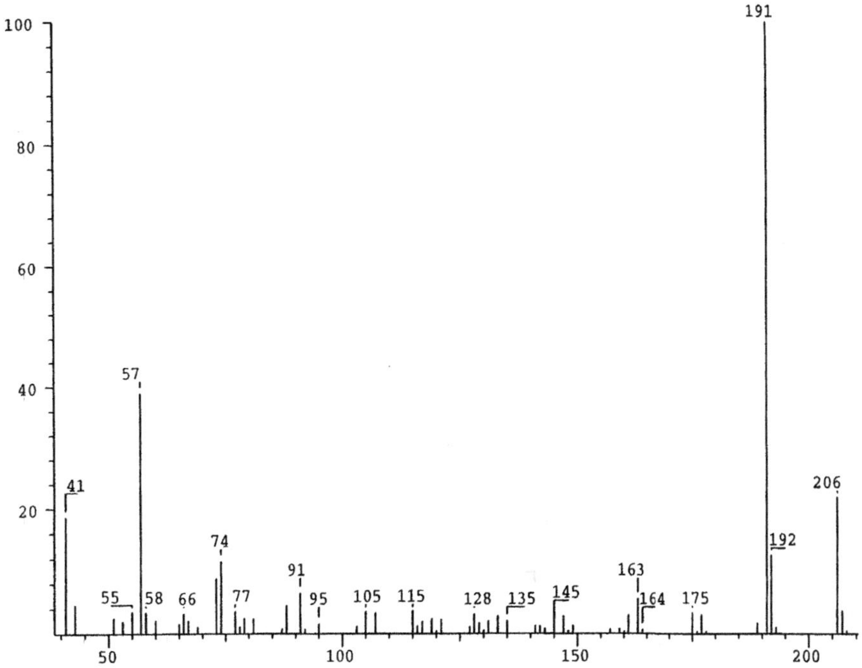

M/Z	Ion Intensity (%)	M/Z	Ion Intensity (%)
41	18.6	145	4.3
43	4.6	163	5.6
57	39.0	175	3.4
58	3.4	191	100
73	8.9	192	12.7
74	11.6	206	21.9
88	4.6	207	3.7
91	6.5	208	0.5

Spectrometer	: Finnigan Mat SSQ 700
Inletsystem	: Capillary GC/MS
Source Temperature	: 180°C
Electron Energy	: 70 eV
Scan Range	: 40-700

Tris(2,4-di-tert-butylphenyl)phosphate

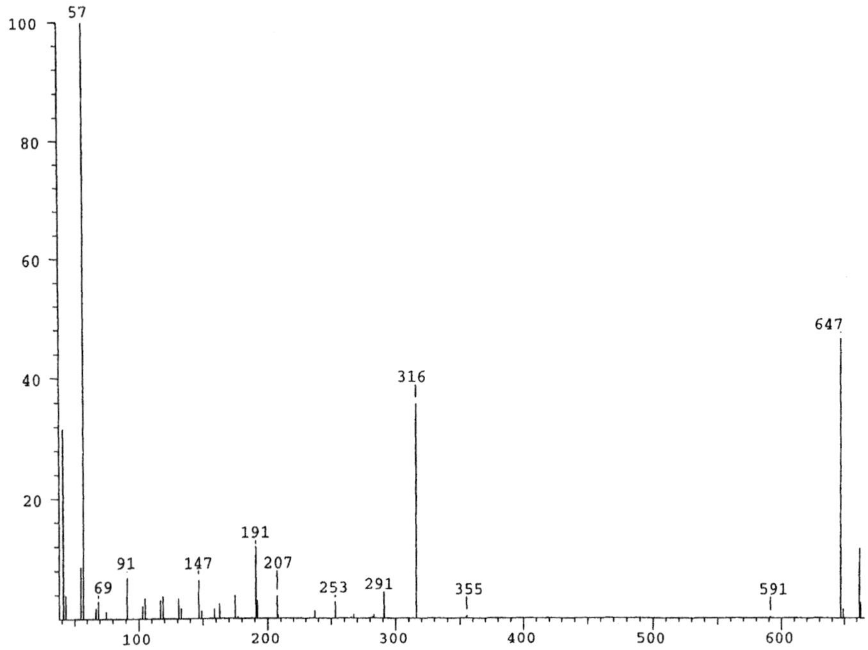

M/Z	Ion Intensity (%)	M/Z	Ion Intensity (%)
41	31.7	175	3.8
55	8.6	191	11.9
57	100.0	316	35.6
91	7.0	647	46.5
119	3.7	662	11.6
147	6.4	663	2.7

Spectrometer	: Finnigan Mat SSQ 700
Inletsystem	: Capillary GC/MS
Source Temperature	: 180°C
Electron Energy	: 70 eV
Scan Range	: 40-700

PM n°	CAS n°	Name
74240	31570-04-4	PHOSPHOROUS ACID, TRIS(2,4-DI-tert-BUTYLPHENYL) ESTER

Solvent	$CDCL_3$
SF	250.13
SI	16384
SW	2380
NS	32

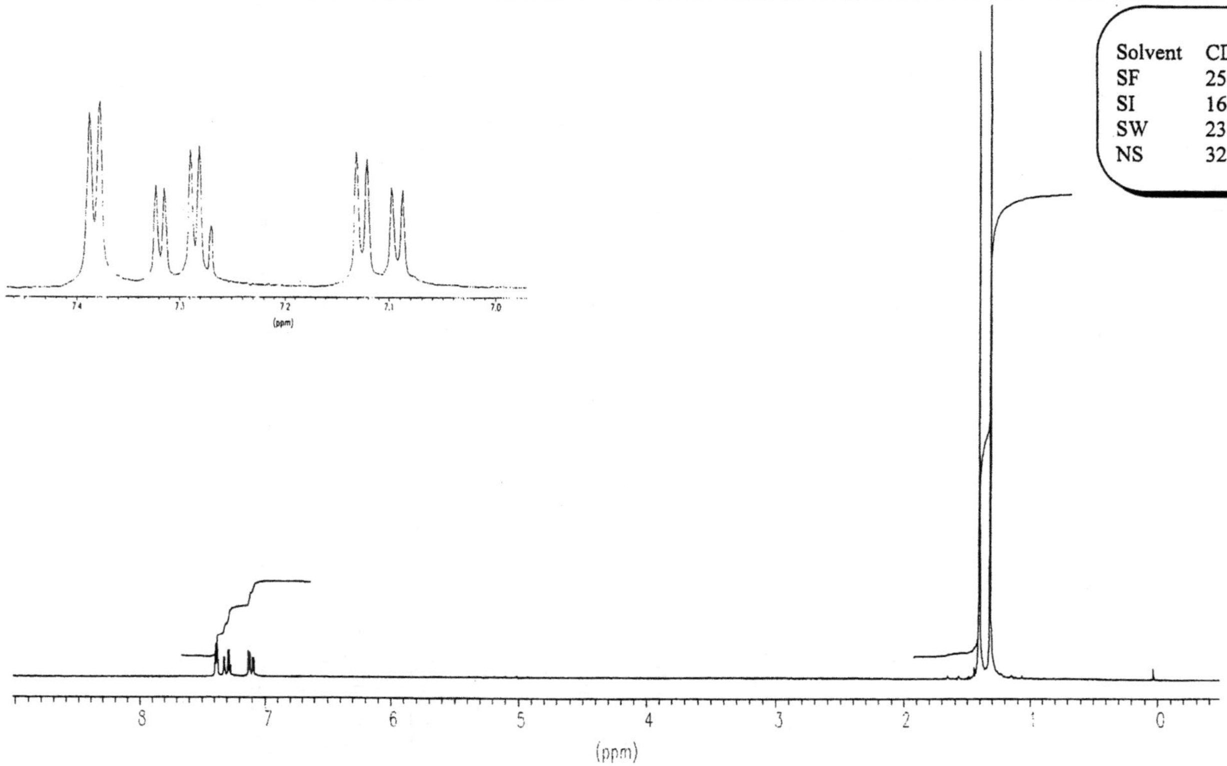

Phosphorous acid, tri(nonyl-and/or dinonylphenyl) ester

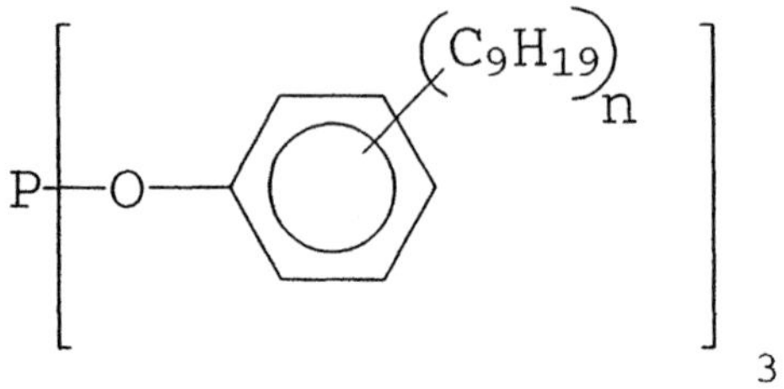

CAS No.	- 26523-78-4 and 01333-21-7
PM Ref. No.	- 74400
Restrictions	- SML = 30 mg/kg
Formula	- n = 1 and/or 2
Molecular weight	-

Alternative names	- Tris(nonylphenyl)phosphite
Physical Characteristics	- Slightly yellow viscous liquid. - Soluble in most organic solvents.
Handling	- Store at room temperature.
Safety	-
Availability	- Standard sample was supplied.
Current uses	- Stabilizer, antioxidant.
Applications	- General use.
GC Retention time	- Degradation product, cluster 31.8 - 33.3 min.

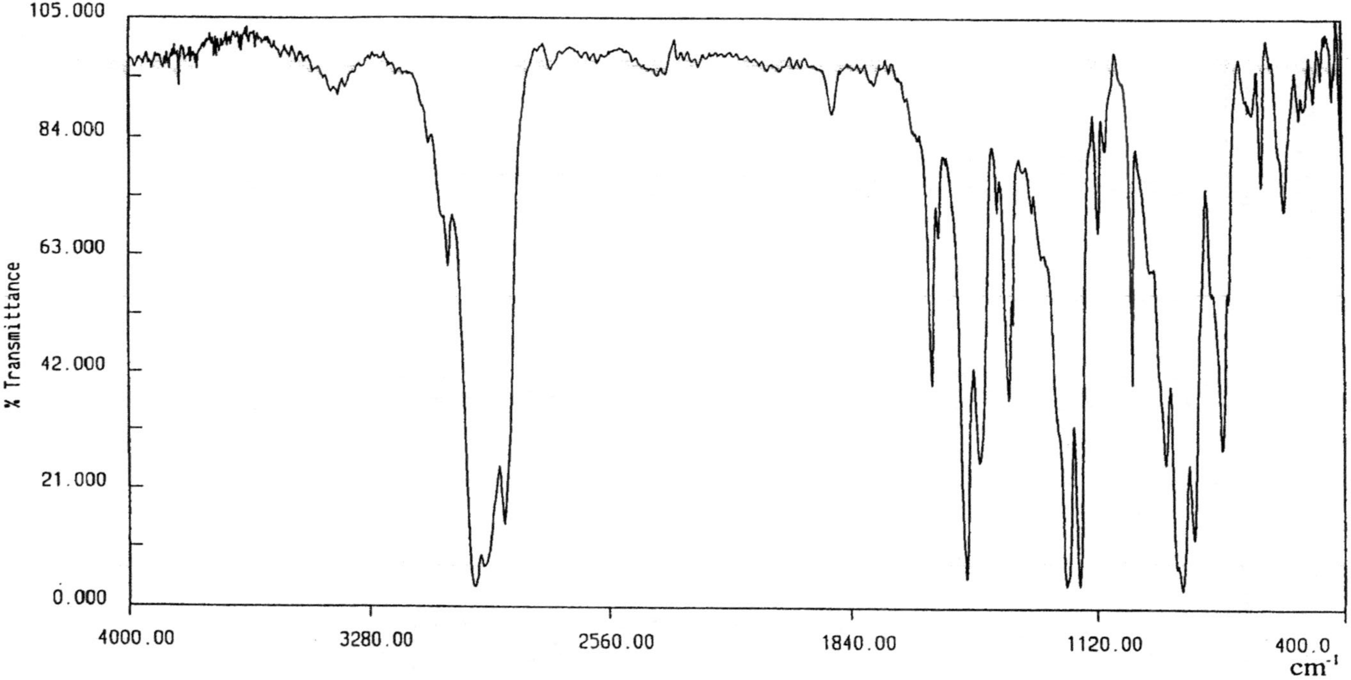

Name : PHOSPHOROUS ACID, TRIS(NONYL-AND/OR DINONYLPHENYL) ESTER

PM Nr : 74400 CAS Nr : 26523-78-4
01333-21-7

KBr pellet

CAS No. 26523-78-4 and 00133-21-7 PM Ref. No. 74400

PHOSPHOROUS ACID, TRIS(NONYL and/or DINONYLPHENYL) ESTER

Peak wavenumber	% T	Relative intensity
405,1	87,3	13,1
428,3	79,3	21,3
509,3	83,9	16,6
522,8	82,4	18,1
569,1	65,1	35,9
632,7	67,7	33,3
657,8	79,5	21,1
731,1	49,5	52,0
752,3	25,6	76,6
835,3	11,1	91,6
870,0	2,9	100,0
916,3	23,4	78,9
1016,6	36,6	65,3
1093,8	74,0	26,8
1113,1	60,7	40,5
1172,9	3,5	99,4
1211,4	3,4	99,5
1365,8	45,3	56,3
1379,3	33,0	69,0
1410,1	62,8	38,3
1466,1	22,8	79,5
1504,7	4,6	98,3
1583,8	58,2	43,1
1603,0	34,8	67,2
1892,4	80,2	20,4
2386,2	86,6	13,8
2727,7	86,8	13,6
2872,4	13,0	89,6
2932,2	6,2	96,6
2963,0	3,0	99,9
3038,3	53,6	47,8
3752,0	89,9	10,4
3854,2	86,0	14,4

Phosphorous acid, tri(nonyl-and/or dinonylphenyl) ester

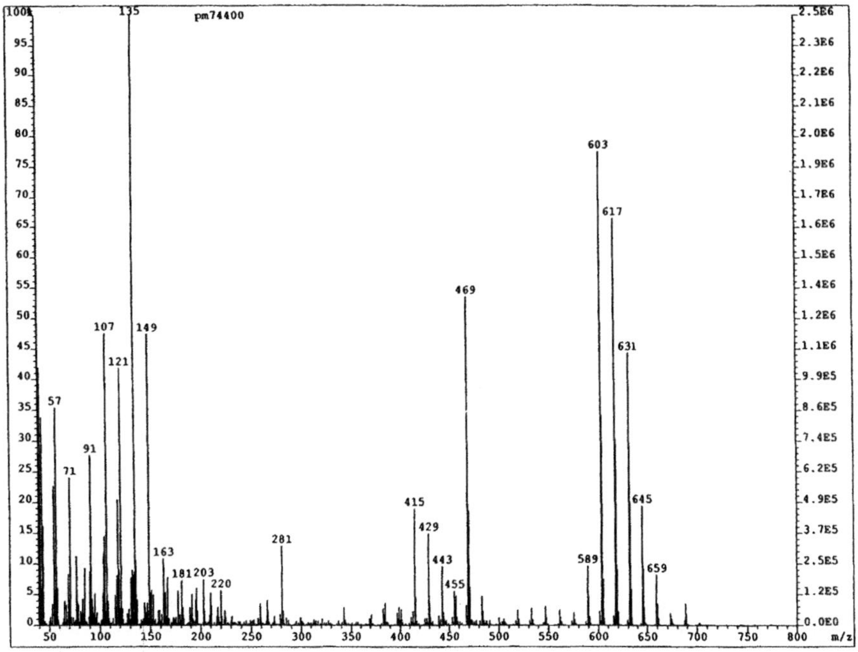

M/Z	Ion Intensity (%)	M/Z	Ion Intensity (%)
41	42.3	149	46.6
43	34.4	469	53.4
55	22.7	470	18.4
57	35.0	603	77.9
71	24.5	604	34.4
91	28.2	617	67.5
107	47.9	618	30.7
118	20.2	631	44.8
121	42.3	632	20.2
135	100.0	645	19.6

Spectrometer : VG Autospec
Inletsystem : Flow injection particle beam MS
Source Temperature : 250°C
Electron Energy : 70 eV
Scan Range : 40-1000

Degradation product

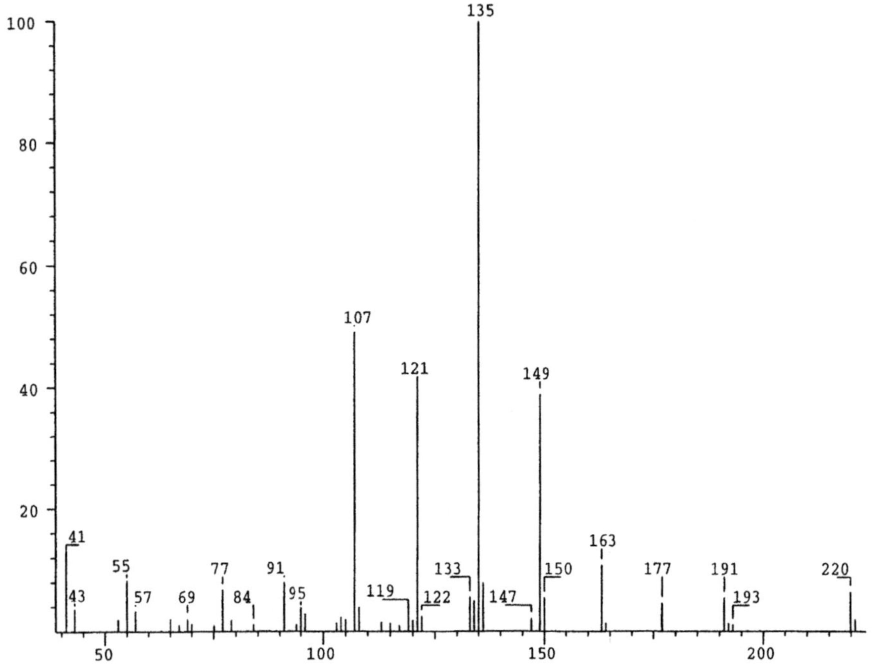

M/Z	Ion Intensity (%)	M/Z	Ion Intensity (%)
41	13.2	149	38.9
55	8.3	163	10.8
77	6.7	177	4.7
91	7.9	191	5.5
107	49.1	220	6.5
121	40.7	221	1.9
135	100.0		

Spectrometer	: Finnigan Mat SSQ 700
Inletsystem	: Capillary GC/MS
Source Temperature	: 180°C
Electron Energy	: 70 eV
Scan Range	: 40-700

PM n°	CAS n°	Name
74400	26523-78-4 01333-21-7	PHOSPHOROUS ACID, TRIS(NONYL-AND/OR DINONYLPHENYL) ESTER

Solvent	$CDCl_3$
SF	250.13
SI	16384
SW	3125
NS	110

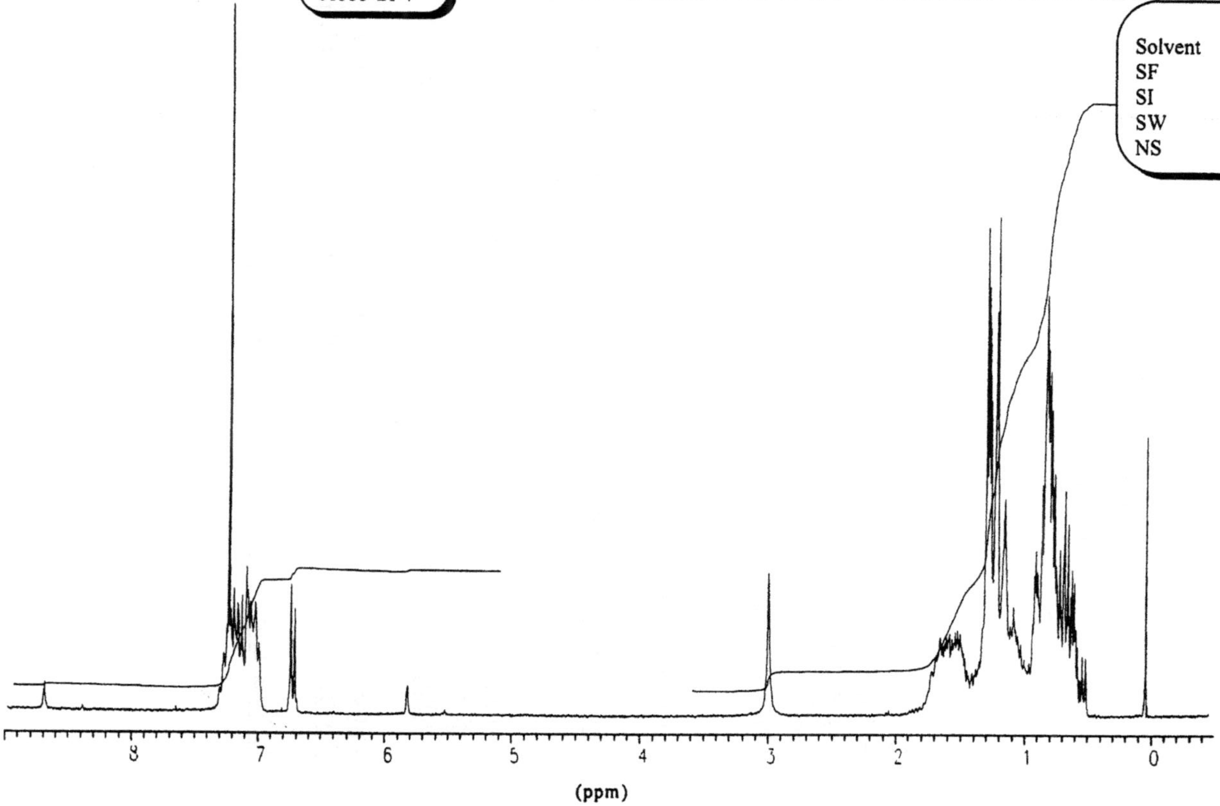

Phthalic acid, benzyl butyl ester

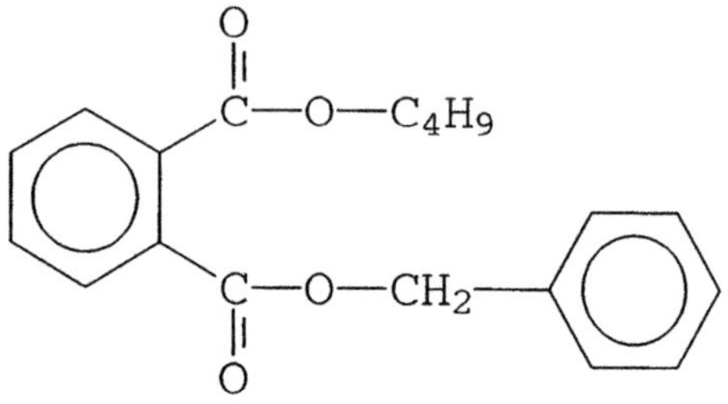

CAS No.	- 00085-68-7
PM Ref. No.	- 74560
Restrictions	- SML = 6.0 mg/kg
Formula	- $C_{19}H_{20}O_4$
Molecular weight	- 312

Alternative names	- Benzylbutyl phthalate
Physical Characteristics	- Colourless liquid, mp -35°C, bp 370°C. - Soluble in most organic solvents.
Handling	- Store at room temperature.
Safety	-
Availability	- Standard sample was supplied.
Current uses	- Plasticizer, lubricant, carrier.
Applications	- PVC, PVDC, styrene co-polymers, inks.
GC Retention time	- 43.7 min.

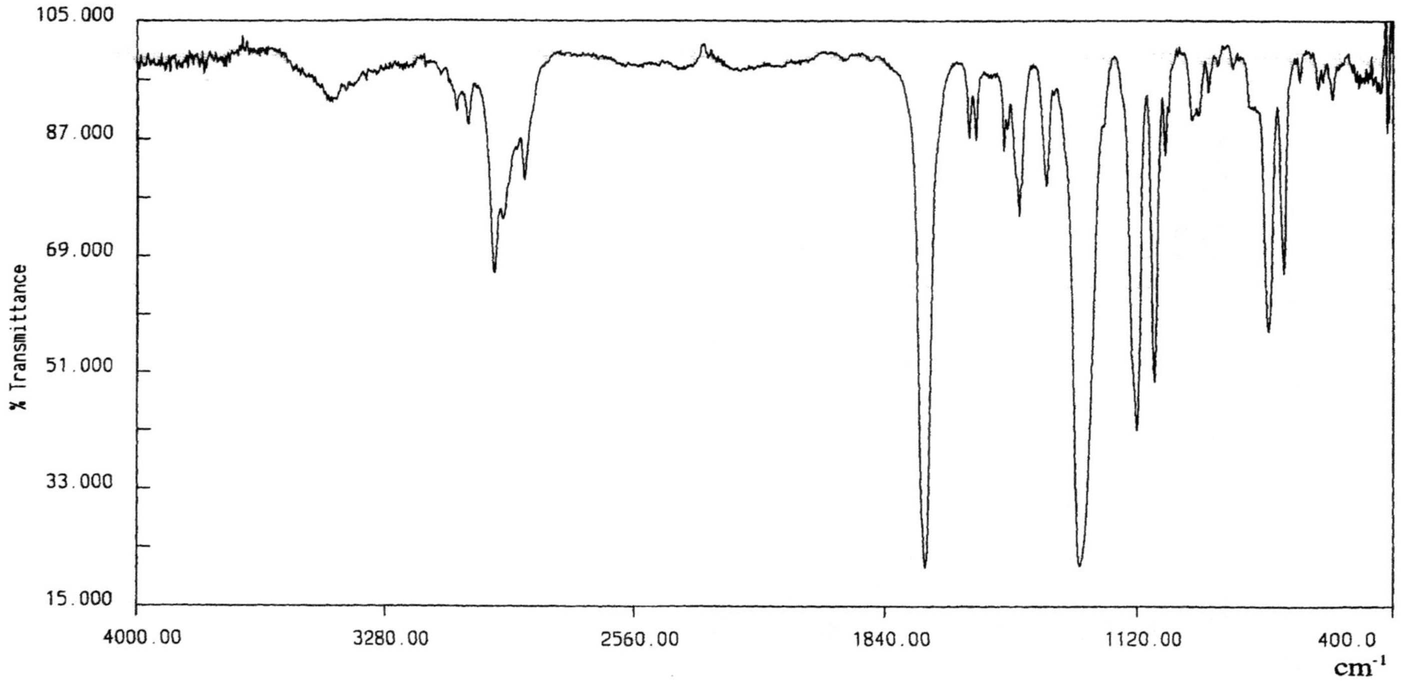

Name : PHTHALIC ACID, BENZYL BUTYL ESTER

PM Nr : 74560 CAS Nr : 00085-68-7

KBr pellet

CAS No. 00085-68-7 PM Ref. No. 74560

PHTHALIC ACID, BENZYL BUTYL ESTER

Peak wavenumber	% T	Relative intensity
416.7	87.9	15.3
432.1	94.0	7.6
698.3	66.2	42.8
742.7	57.2	54.1
1039.8	84.4	19.8
1070.6	49.5	63.9
1122.7	42.0	73.4
1284.8	21.1	99.9
1377.3	79.7	25.7
1456.4	75.1	31.5
1489.2	88.4	14.7
1498.9	85.1	18.8
1599.2	87.1	16.4
1726.5	21.0	100.0
2874.3	80.7	24.4
2934.1	74.5	32.2
2959.2	66.4	42.6

Phthalic acid, benzyl butyl ester

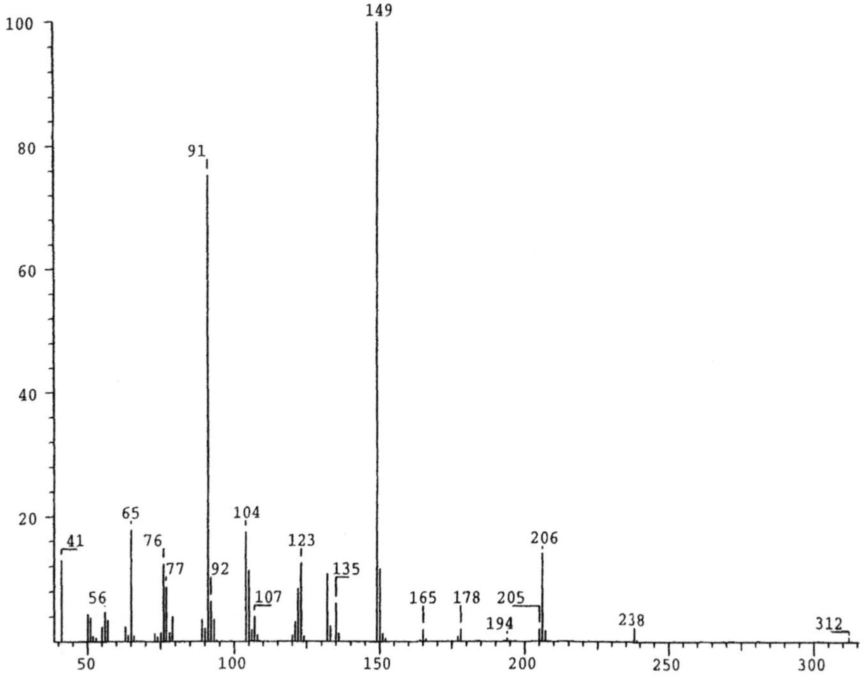

M/Z	Ion Intensity (%)	M/Z	Ion Intensity (%)
41	12.9	132	10.8
65	17.8	133	2.5
76	12.4	135	6.1
77	8.7	149	100.0
91	75.4	150	11.6
104	17.5	206	14.2
105	11.3	238	1.3
122	8.4	312	0.8
123	12.5	313	0.1

Spectrometer	: Finnigan Mat SSQ 700
Inletsystem	: Capillary GC/MS
Source Temperature	: 180°C
Electron Energy	: 70 eV
Scan Range	: 40-500

PM n°
74560

CAS n°
00085-68-7

Name
PHTHALIC ACID, BENZYL BUTYL ESTER

Solvent	$CDCL_3$
SF	250.13
SI	16384
SW	2380
NS	32

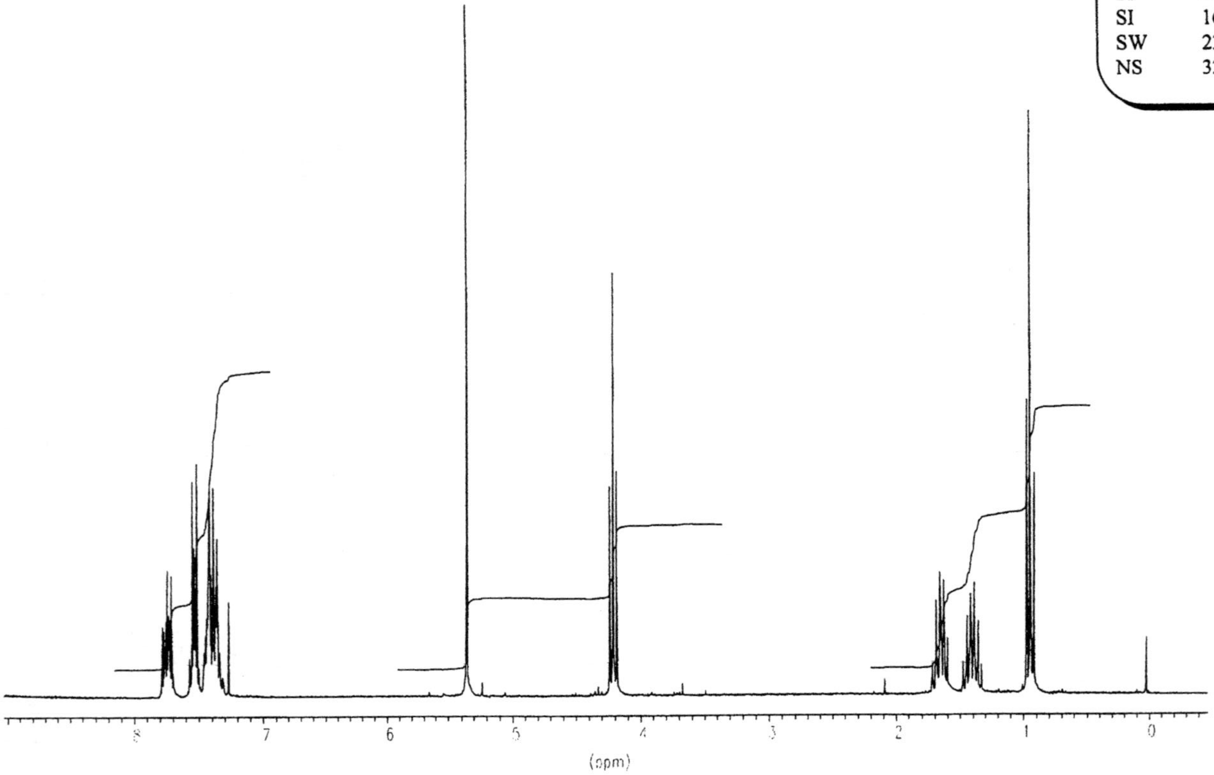

Phthalic acid, bis(2-ethylhexyl) ester

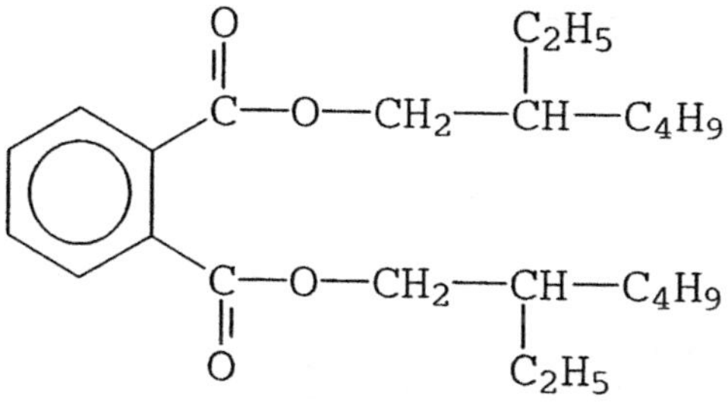

CAS No.	- 00117-81-7
PM Ref. No.	- 74640
Restrictions	- SML = 3.0 mg/kg
Formula	- $C_{24}H_{38}O_4$
Molecular weight	- 390

Alternative names - Dioctyl phthalate, Di(2-ethylhexyl) phthalate

Physical Characteristics - Light coloured liquid, mp -50°C, bp 384°C.
- Soluble in most organic solvents.

Handling - Store at room temperature.
Safety -
Availability - Standard sample was supplied.

Current uses - Plasticizer, lubricant, carrier, release agent.
Applications - PVC, PVDC, styrene co-polymers, inks.

GC Retention time - 47.2 min.

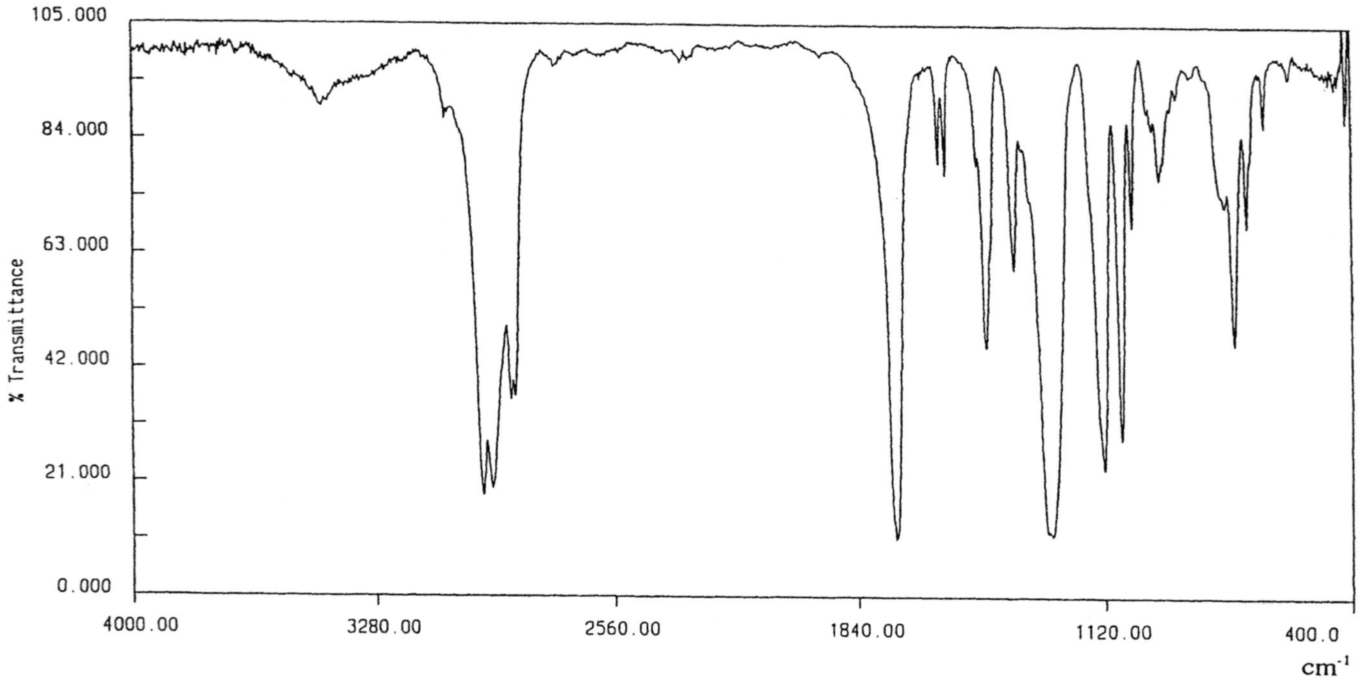

Name : PHTHALIC ACID, BIS(2-ETHYLHEXYL) ESTER

PM Nr : 74640 CAS Nr : 00117-81-7

KBr pellet

CAS No. 00117-81-7 PM Ref. No. 74640

PHTHALIC ACID, BIS(2-ETHYLHEXYL) ESTER

Peak wavenumber	% T	Relative intensity
414.8	87.5	14.0
437.9	94.3	6.4
447.5	93.6	7.2
652.0	86.6	15.0
706.0	68.3	35.5
742.7	46.5	59.8
958.7	76.9	25.9
1039.8	68.3	35.5
1072.6	29.1	79.4
1122.7	23.6	85.5
1275.1	11.2	99.3
1381.2	60.2	44.5
1464.2	45.8	60.7
1581.8	77.5	25.1
1601.1	79.6	22.8
1730.4	10.6	100.0
2862.7	36.7	70.8
2932.2	19.88	89.6
2961.1	18.50	91.2
3427.9	90.61	10.5

Phthalic acid, bis(2-ethylhexyl) ester

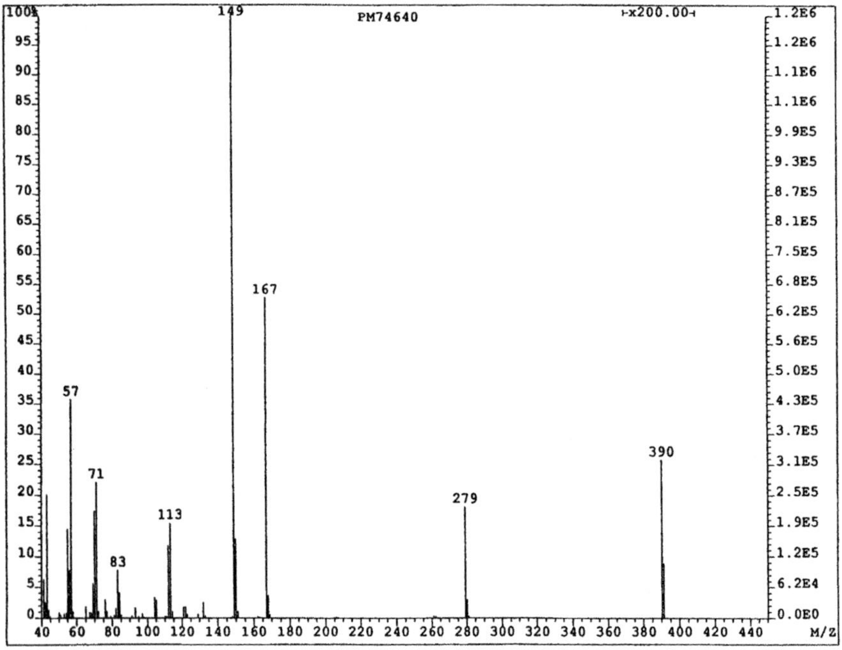

M/Z	Ion Intensity (%)	M/Z	Ion Intensity (%)
41	6.4	83	7.9
43	20.2	112	11.8
55	14.6	113	15.6
56	8.0	149	100.0
57	35.9	150	13.0
69	5.7	167	52.9
70	17.5	279	18.2
71	22.2		

Spectrometer	: VG Autospec
Inletsystem	: Flow injection particle beam MS
Source Temperature	: 250°C
Electron Energy	: 70 eV
Scan Range	: 40-1000

PM n°	CAS n°	Name
74640	00117-81-7	PHTHALIC ACID, BIS(2-ETHYLHEXYL) ESTER

Solvent	$CDCl_3$
SF	250.13
SI	16384
SW	2380
NS	32

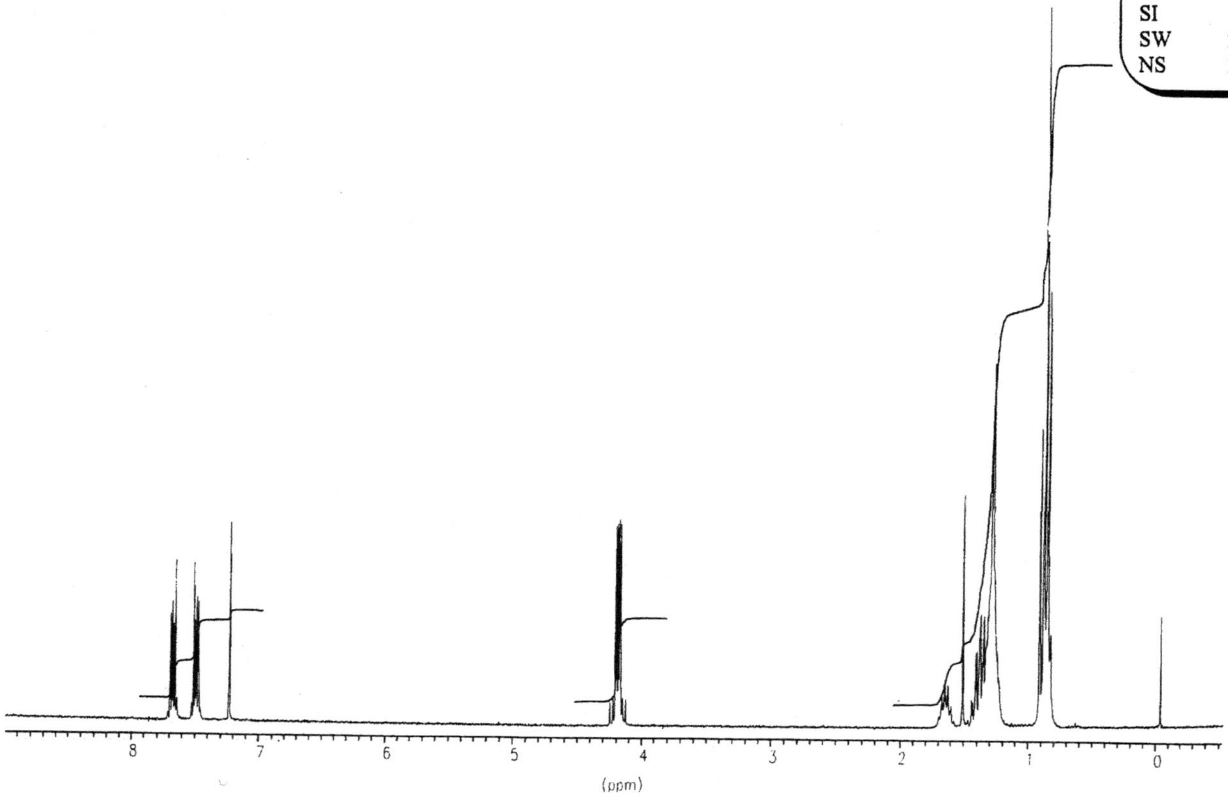

Phthalic acid, dibutyl ester

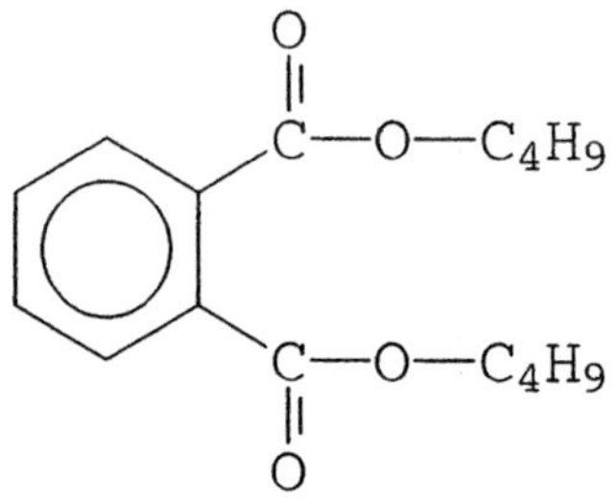

CAS No.	- 00084-74-2
PM Ref. No.	- 74880
Restrictions	- SML = 3.0 mg/kg
Formula	- $C_{16}H_{22}O_4$
Molecular weight	- 278

Alternative names	- Dibutyl phthalate
Physical Characteristics	- Colourless liquid, mp -35°C, bp 340°C. - Soluble in alcohol, ether, benzene.
Handling	- Store at room temperature.
Safety	-
Availability	- Standard sample was supplied.
Current uses	- Plasticizer.
Applications	- PVC.
GC Retention time	- 36.7 min.

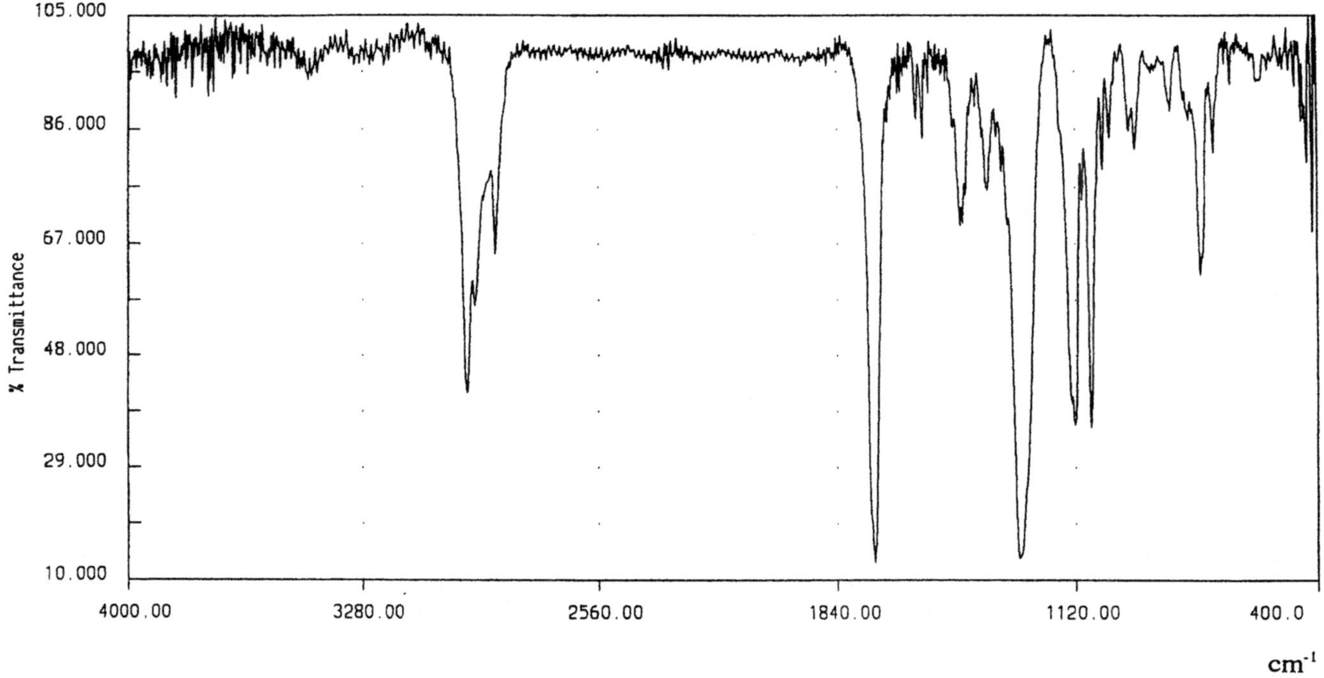

Name : PHTHALIC ACID, DIBUTYL ESTER

PM Nr : 74880 CAS Nr : 00084-74-2

KBr pellet

CAS No. 00084-74-2 PM Ref. No. 74880

PHTHALIC ACID, DIBUTYL ESTER

Peak wavenumber	% T	RELATIVE INTENSITY
412.8	69.0	35.6
430.2	80.5	22.4
445.6	87.2	14.7
467.1	93.9	7.0
704.1	82.0	20.7
746.5	61.8	43.9
943.3	82.6	20.0
1018.5	84.4	17.9
1039.8	79.3	23.8
1074.5	35.6	73.9
1099.6	74.2	29.6
1122.7	36.2	73.2
1288.6	13.5	99.3
1340.7	78.8	24.3
1385.1	75.8	27.8
1450.7	75.1	28.6
1466.1	70.0	34.4
1726.5	12.9	100.0
2876.2	65.1	40.1
2937.9	56.3	50.2
2961.9	41.5	67.2

Phthalic acid, dibutyl ester

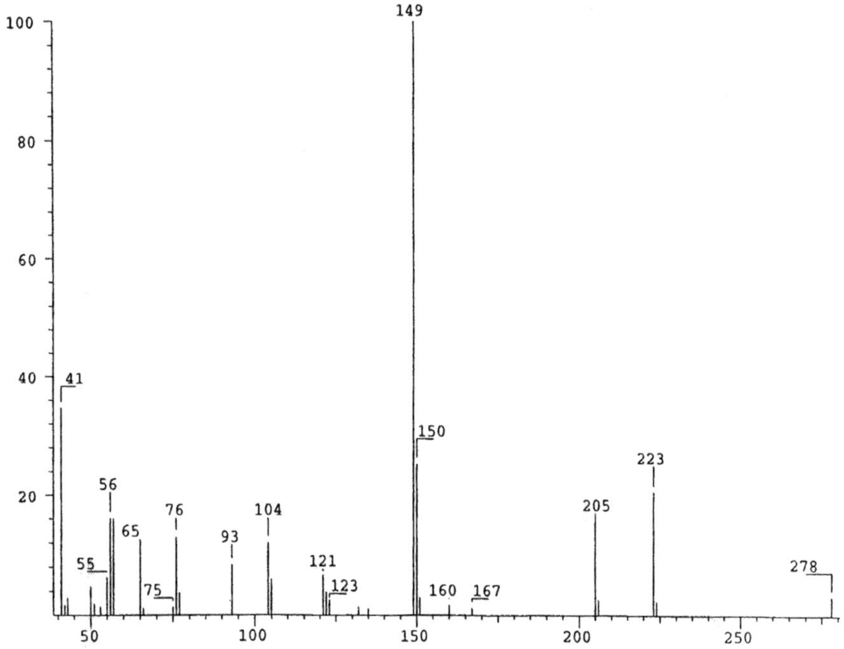

M/Z	Ion Intensity (%)	M/Z	Ion Intensity (%)
41	34.7	104	12.1
56	16.2	105	6.0
57	16.0	149	100.0
65	11.6	150	25.4
76	13.0	205	16.1
77	3.7	223	20.7
93	8.5	278	3.2

Spectrometer	: Finnigan Mat SSQ 700
Inletsystem	: Capillary GC/MS
Source Temperature	: 180°C
Electron Energy	: 70 eV
Scan Range	: 40-300

PM n°	CAS n°	Name
74880	00084-74-2	PHTHALIC ACID, DIBUTYL ESTER

Solvent	$CDCl_3$
SF	250.13
SI	16384
SW	3125
NS	64

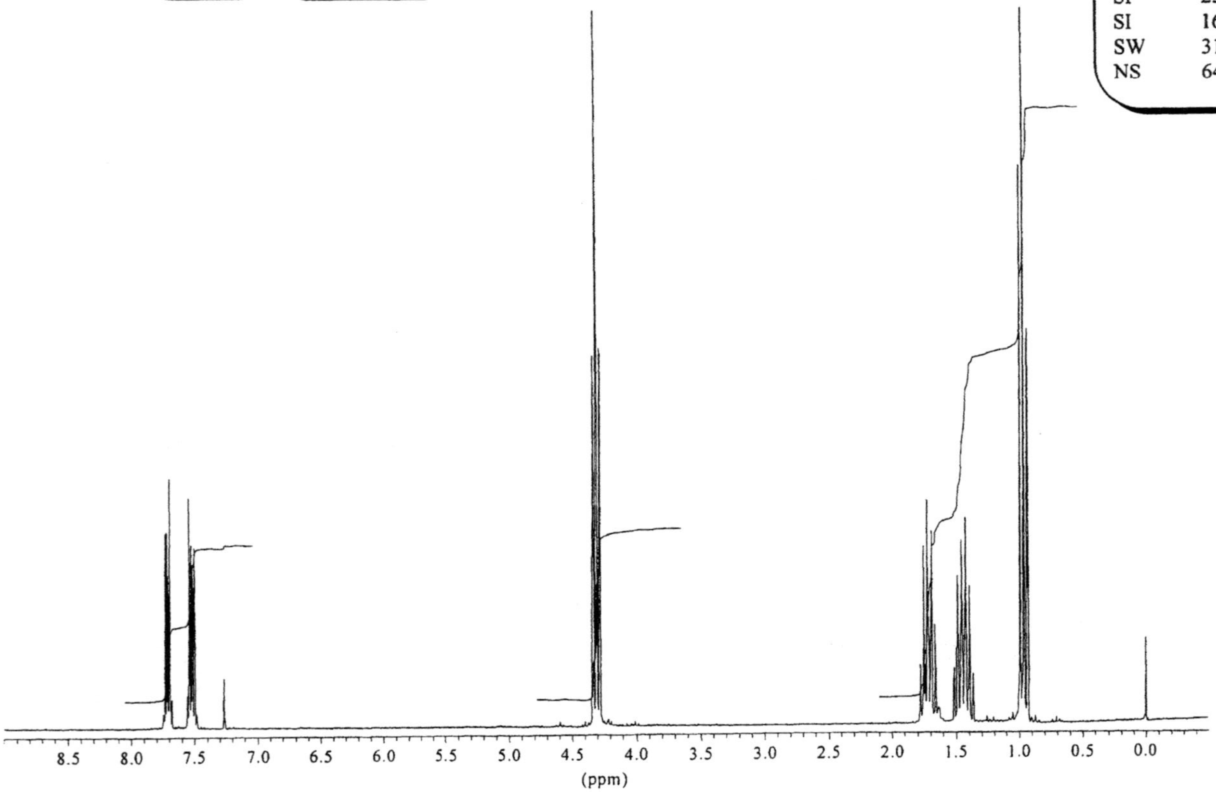

Phthalic acid, dicyclohexyl ester

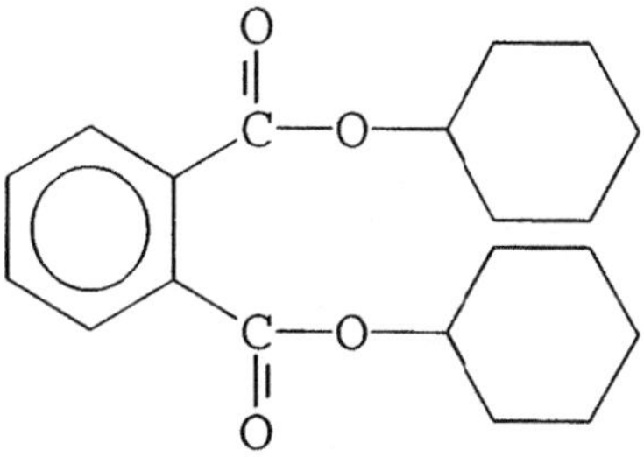

CAS No.	- 00084-61-7
PM Ref. No.	- 74960
Restrictions	- SML = 6.0 mg/kg
Formula	- $C_{20}H_{26}O_4$
Molecular weight	- 330.43

Alternative names	- Dicyclohexyl phthalate
Physical Characteristics	- White powder, mp 63 - 65°C, bp 220 - 230°C. - Soluble in most organic solvents.
Handling	- Store at room temperature.
Safety	-
Availability	- Standard sample was supplied.
Current uses	- Plasticizer.
Applications	- PVC, PVDC.
GC Retention time	- 46.3 min.

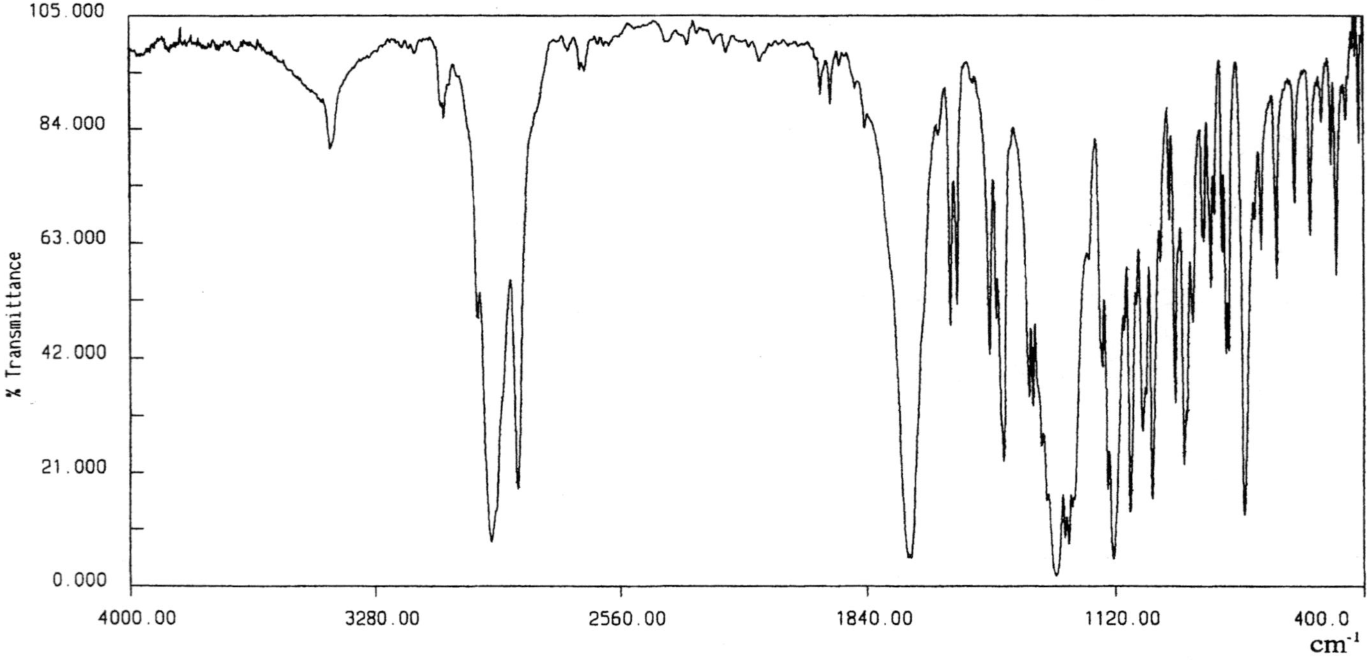

Name : PHTHALIC ACID, DICYCLOHEXYL ESTER

PM Nr : 74960 CAS Nr : 00084-61-7

KBr pellet

CAS No. 00084-61-7 PM Ref. No. 74960

PHTHALIC ACID, DICYCLOHEXYLESTER

Peak wavenumber	% T	Relative intensity
410.9	79.8	20.6
449.5	86.6	13.7
476.5	58.6	42.2
490.0	79.9	20.5
549.8	66.5	34.2
594.2	77.4	23.1
648.2	61.4	39.4
692.5	66.1	34.6
711.8	71.8	28.8
740.8	13.8	87.9
794.8	42.5	58.7
804.4	60.1	40.7
827.6	67.3	33.4
839.1	54.0	46.9
858.4	62.3	38.5
891.2	48.1	52.9
916.3	22.3	79.3
943.3	33.5	67.8
958.7	66.6	34.1
987.7	59.1	41.7
1010.8	15.9	85.8
1039.8	28.6	72.8
1076.4	13.6	88.1
1126.6	5.0	96.9
1142.0	17.9	83.7
1157.4	40.4	60.8
1196.0	60.0	40.8
1246.2	14.7	87.0
1257.7	7.8	94.0
1267.4	9.0	92.8
1294.4	2.0	100.0
1360.0	33.7	67.6
1373.5	35.4	65.9
1446.8	23.4	78.1
1468.0	50.2	50.8
1487.3	43.5	57.6
1579.9	52.9	48.0
1599.2	49.0	52.0
1632.0	84.5	15.8
1713.0	5.3	96.6
1944.5	89.3	10.9
1973.4	90.8	9.4
2662.1	88.8	11.4
2860.8	17.1	84.6
2937.9	7.8	94.0
2976.5	47.5	53.6
3074.9	84.1	16.2
3416.4	79.2	21.2

Phthalic acid, dicyclohexyl ester

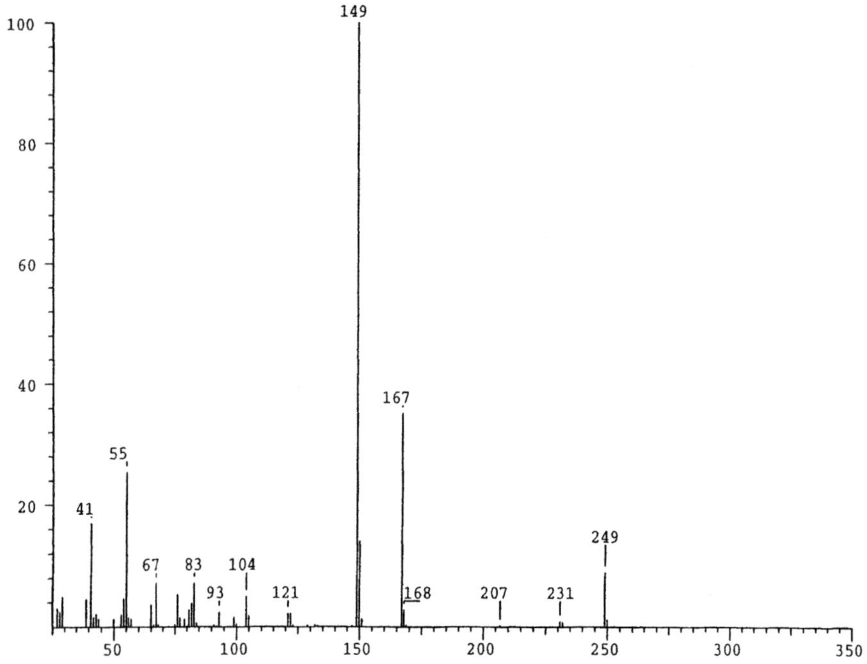

M/Z	Ion Intensity (%)	M/Z	Ion Intensity (%)
41	16.9	149	100.0
55	25.3	150	14.1
67	7.2	167	35.1
77	1.6	249	9.0
104	5.1		

Spectrometer	: Finnigan Mat SSQ 700
Inletsystem	: Capillary GC/MS
Source Temperature	: 180°C
Electron Energy	: 70 eV
Scan Range	: 25-450

PM n°	CAS n°	Name
74960	00084-61-7	PHTHALIC ACID, DICYCLOHEXYL ESTER

Solvent	$CDCL_3$
SF	250.13
SI	16384
SW	3125
NS	72

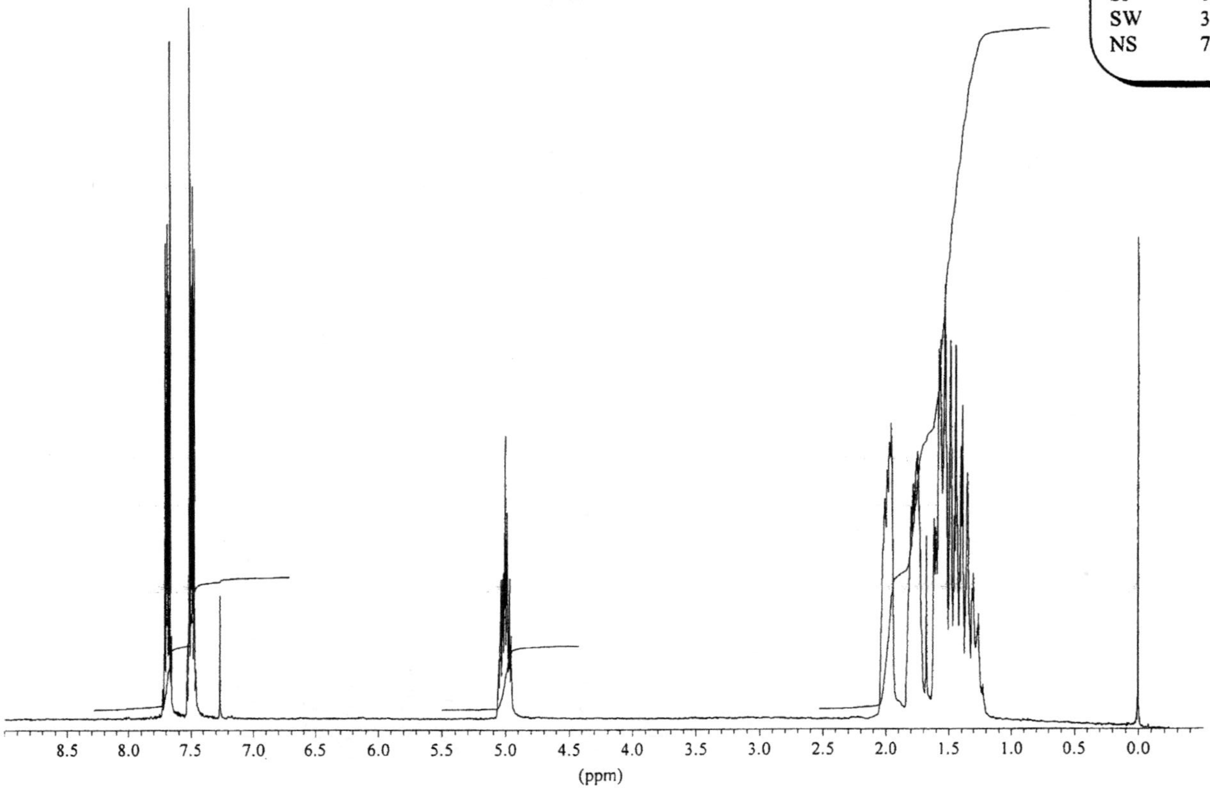

Phthalic acid, diethyl ester

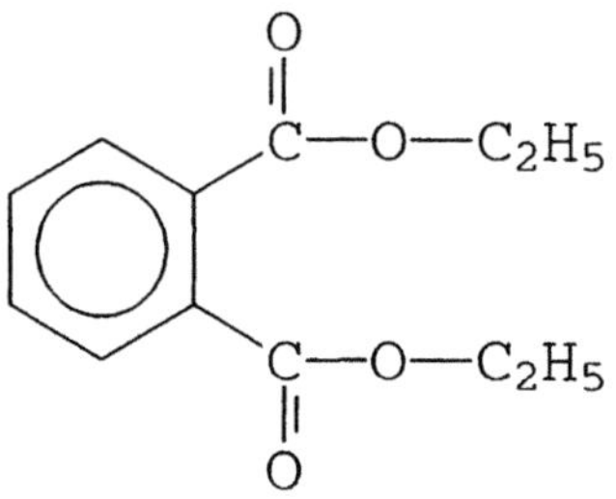

CAS No.	- 00084-66-2
PM Ref. No.	- 75120
Restrictions	- SML = 12 mg/kg
Formula	- $C_{12}H_{14}O_4$
Molecular weight	- 222

Alternative names	- Diethyl phthalate
Physical Characteristics	- Colourless liquid, mp -40°C, bp 298°C. - Soluble in most organic solvents.
Handling	- Store at room temperature.
Safety	-
Availability	- Standard sample was supplied.
Current uses	- Plasticizer.
Applications	- General use.
GC Retention time	- 29.1 min.

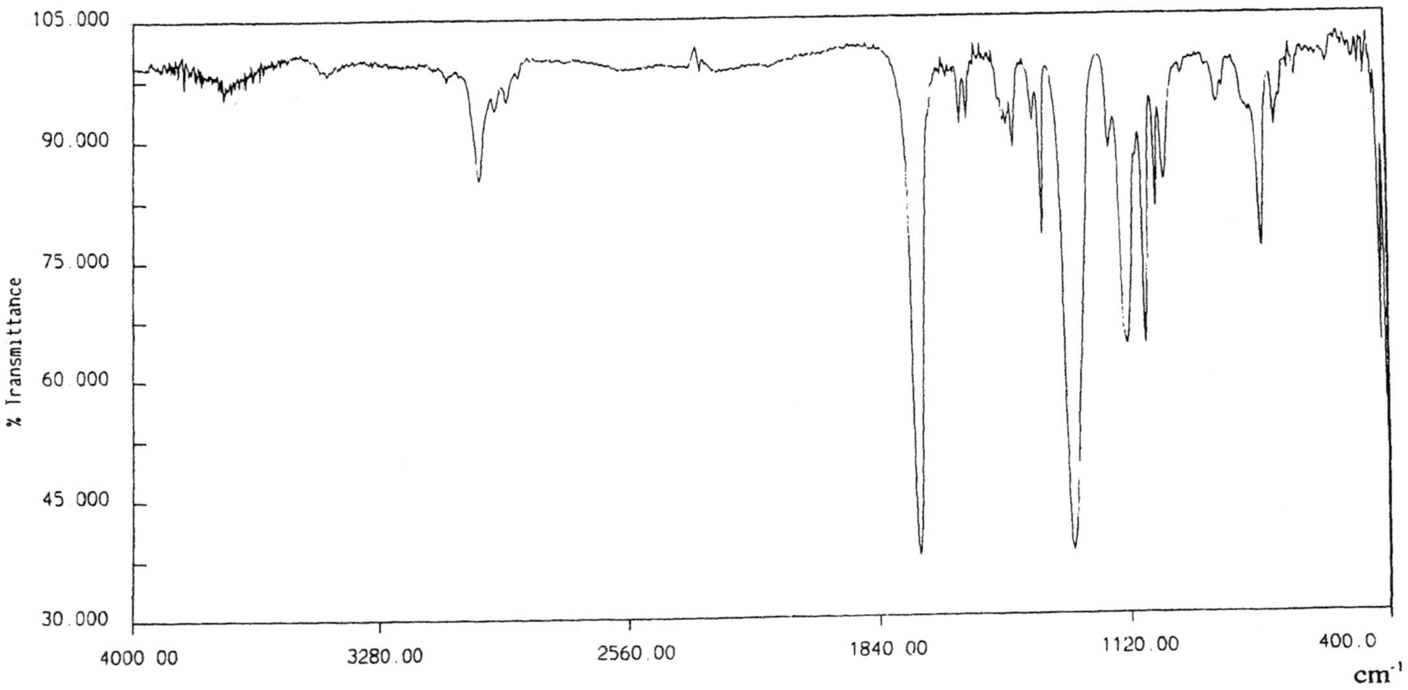

Name : PHTHALIC ACID, DIETHYL ESTER

PM Nr : 75120 CAS Nr : 00084-66-2

KBr pellet

CAS No. 00084-66-2 PM Ref. No. 75120

PHTHALIC ACID, DIETHYL ESTER

Peak wavenumber	% T	Relative intensity
418.6	63.5	58.4
744.6	75.8	38.7
1016.6	84.4	24.9
1041.7	81.0	30.4
1072.6	63.7	58.0
1126.6	63.6	58.2
1174.8	88.3	18.7
1286.7	37.8	99.5
1367.7	77.5	36.0
1448.7	88.5	18.4
1726.5	37.5	100.0
2984.2	85.0	24.0

Phthalic acid, diethyl ester

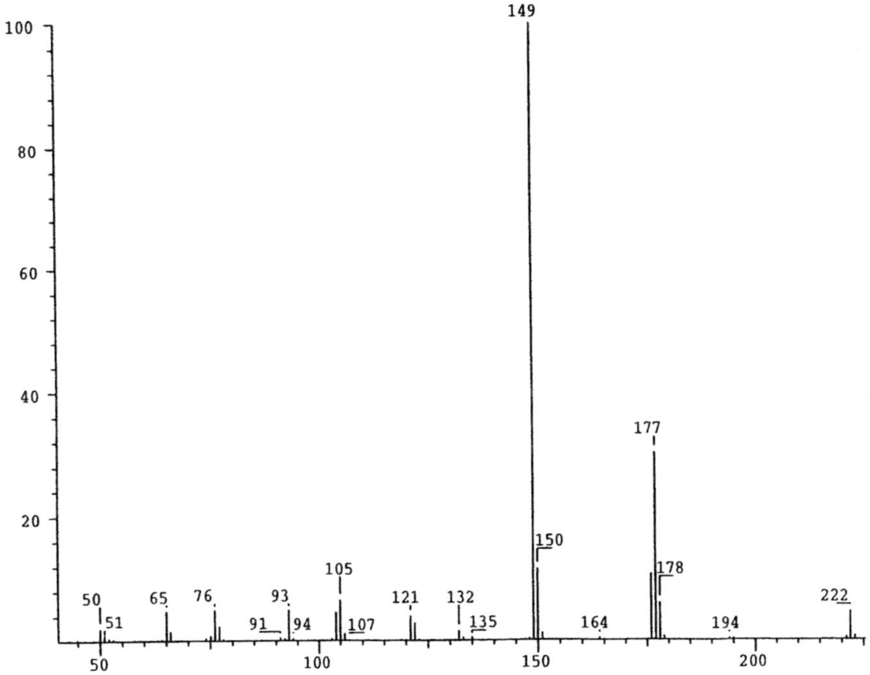

M/Z	Ion Intensity (%)	M/Z	Ion Intensity (%)
50	2.0	150	11.6
65	4.8	176	10.8
76	5.1	177	30.3
93	5.5	178	6.1
105	6.6	222	4.6
149	100.0	223	0.6

Spectrometer	: Finnigan Mat SSQ 700
Inletsystem	: Capillary GC/MS
Source Temperature	: 180°C
Electron Energy	: 70 eV
Scan Range	: 40-500

PM n°	CAS n°	Name
75120	00084-66-2	PHTHALIC ACID, DIETHYL ESTER

Solvent	$CDCl_3$
SF	250.13
SI	16384
SW	3125
NS	16

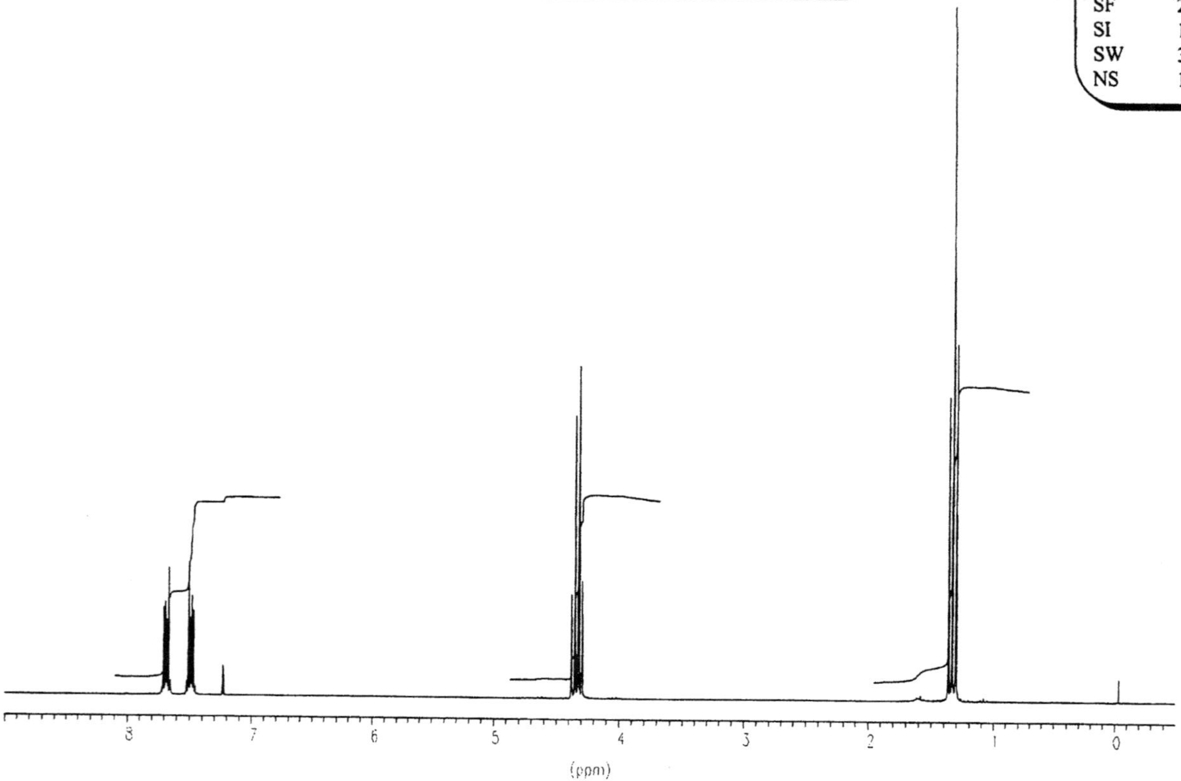

Phthalic acid, diisobutyl ester

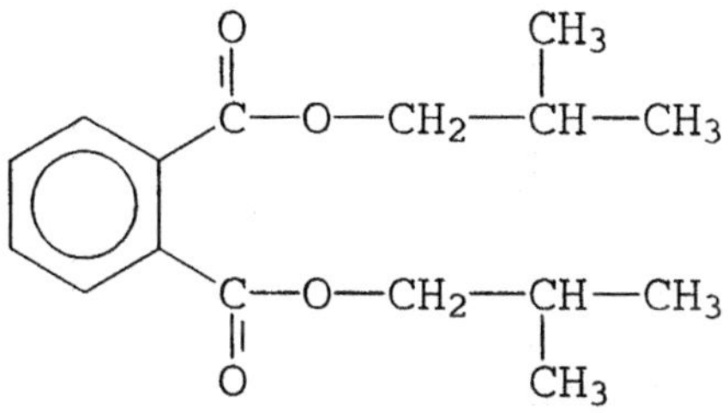

CAS No.	- 00084-69-5
PM Ref. No.	- 75280
Restrictions	- SML = 3.0 mg/kg
Formula	- $C_{16}H_{22}O_4$
Molecular weight	- 278

Alternative names	- Diisobutyl phthalate
Physical Characteristics	- Colourless liquid, mp -50°C, bp 295 - 298°C. - Soluble in most organic solvents.
Handling	- Store at room temperature.
Safety	-
Availability	- Standard sample was supplied.
Current uses	- Plasticizer.
Applications	- General use.
GC Retention time	- 35.2 min.

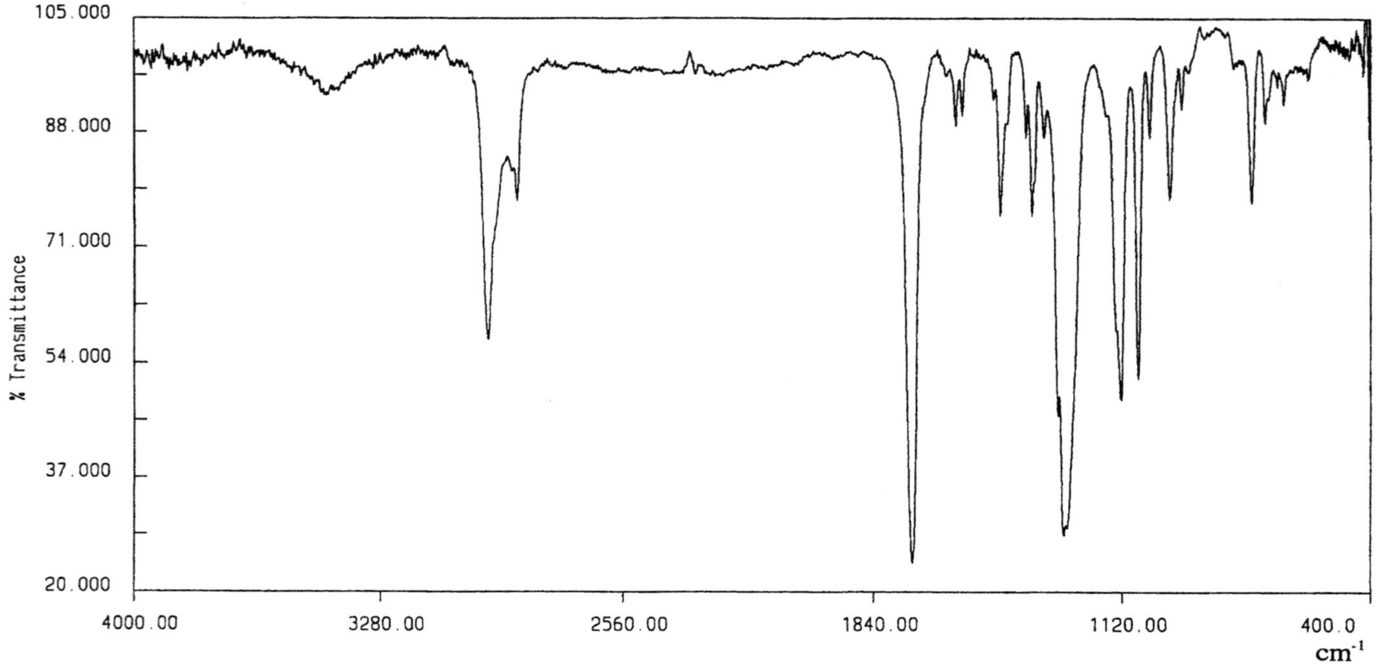

Name : PHTHALIC ACID, DIISOBUTYL ESTER

PM Nr : 75280 CAS Nr : 84-69-5

KBr pellet

CAS No. 00084-69-5 PM Ref. No. 75280

PHTHALIC ACID, DIISOBUTYL ESTER

Peak wavenumber	% T	Relative intensity
652.0	90.2	13.0
706.0	82.2	23.5
744.6	69.7	40.0
947.2	87.8	16.1
981.9	76.1	31.6
1039.8	86.4	17.9
1072.6	51.3	64.3
1122.7	48.4	68.1
1286.7	28.4	94.5
1342.6	87.5	16.4
1377.3	75.8	31.9
1394.7	87.7	16.2
1469.9	74.1	34.2
1581.8	88.4	15.3
1601.1	87.7	16.2
1726.5	24.2	100.0
2876.2	78.2	28.8
2963.0	57.99	55.4

Phthalic acid, diisobutyl ester

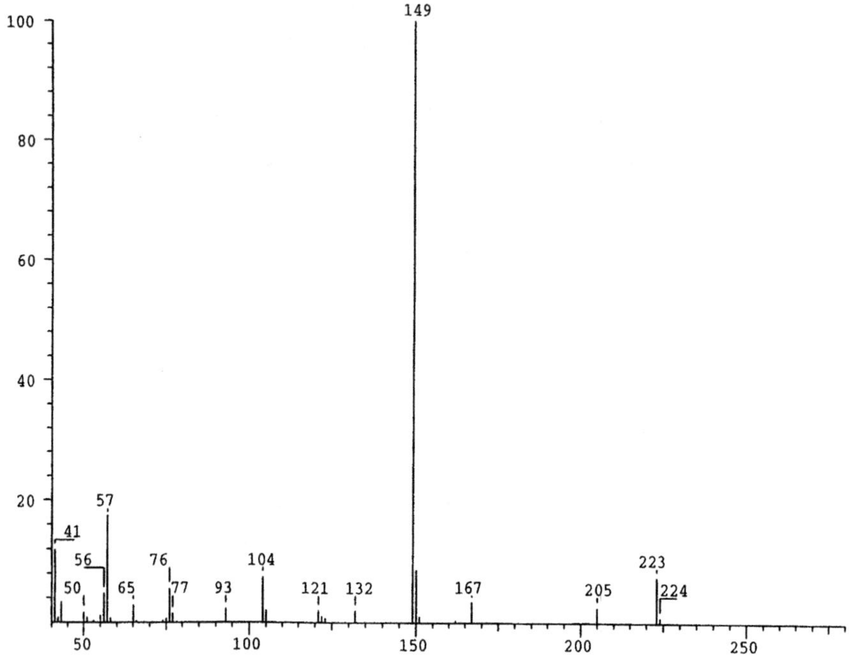

M/Z	Ion Intensity (%)	M/Z	Ion Intensity (%)
41	11.8	105	2.0
43	3.3	121	1.9
56	4.7	149	100.0
57	17.5	150	8.6
65	2.8	167	3.4
76	5.6	205	2.5
77	1.4	223	7.4
93	2.2	224	0.8
104	7.4		

Spectrometer : Finnigan Mat SSQ 700
Inletsystem : Capillary GC/MS
Source Temperature : 180°C
Electron Energy : 70 eV
Scan Range : 40-500

PM n°	CAS n°	Name
75280	0084-69-5	PHTHALIC ACID, DIISOBUTYL ESTER

Solvent	$CDCL_3$
SF	250.13
SI	16384
SW	2380
NS	16

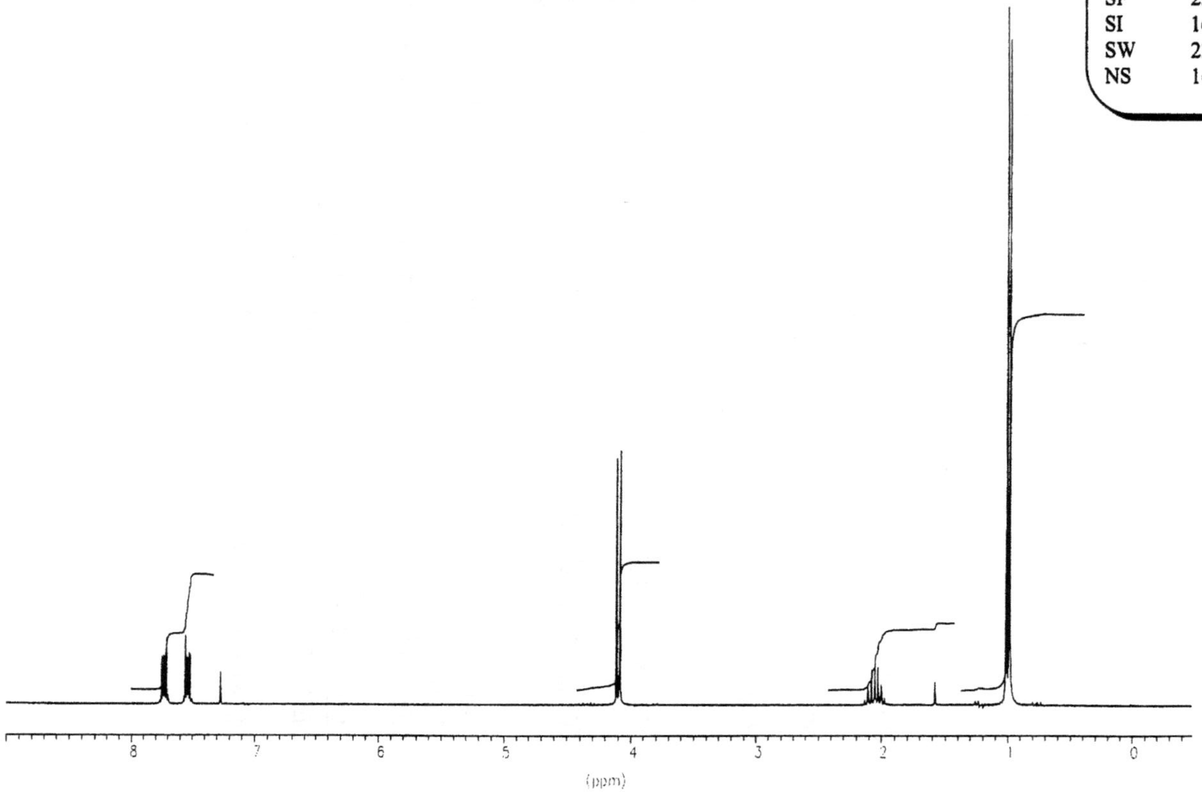

Phthalic acid, diisodecyl ester

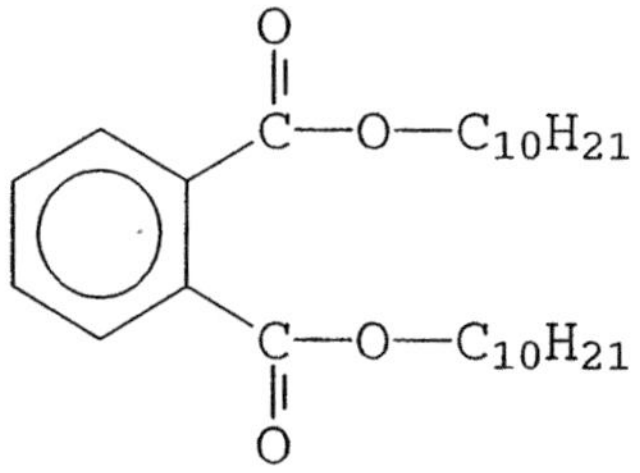

CAS No. - 26761-40-0
PM Ref. No. - 75360
Restrictions - Re-examined
Formula - $C_{28}H_{46}O_4$
Molecular weight - 446

Alternative names - Diisodecyl phthalate

Physical Characteristics - Colourless liquid, mp -50°C, bp 450°C.
- Soluble in most organic solvents.

Handling - Store at room temperature.
Safety -
Availability - Standard sample was supplied.

Current uses - Plasticizer.
Applications - PVC.

GC Retention time - Cluster 48.3 - 54.3 min.

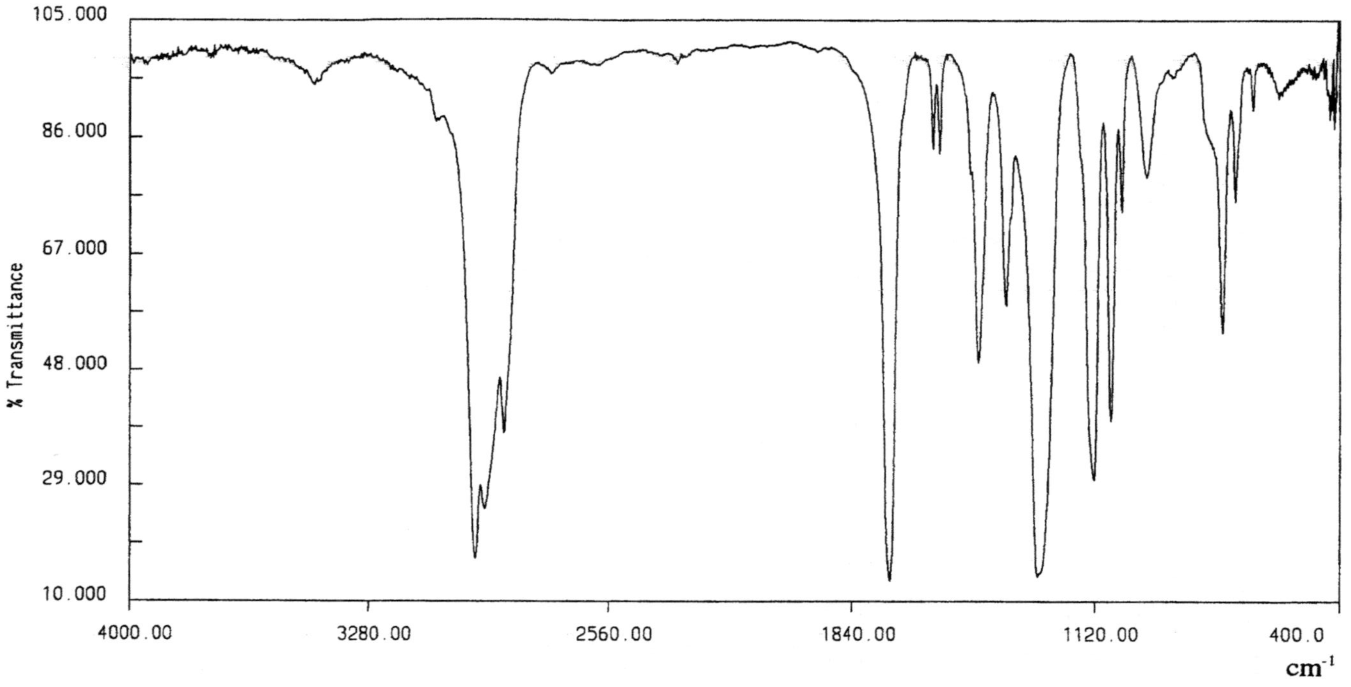

Name : PHTHALIC ACID, DIISODECYL ESTER

PM Nr : 75360 CAS Nr : 26761-40-0

KBr pellet

CAS No. 26761-40-0 PM Ref. No. 75360

PHTHALIC ACID, DIISODECYL ESTER

Peak wavenumber	% T	Relative intensity
414.8	87.4	14.6
428.3	88.8	12.9
542.1	94.4	6.5
565.2	92.7	8.5
706.0	75.4	28.4
742.7	54.0	53.0
966.5	79.4	23.7
1039.8	73.7	30.3
1072.6	39.4	69.9
1122.7	29.9	80.9
1286.7	13.9	99.3
1381.2	58.5	47.9
1466.1	49.1	58.6
1581.8	83.3	19.3
1601.1	84.0	18.4
1728.4	13.3	100.0
2874.3	37.4	72.2
2930.2	24.91	86.6
2959.2	16.89	95.8

Phthalic acid, diisodecyl ester

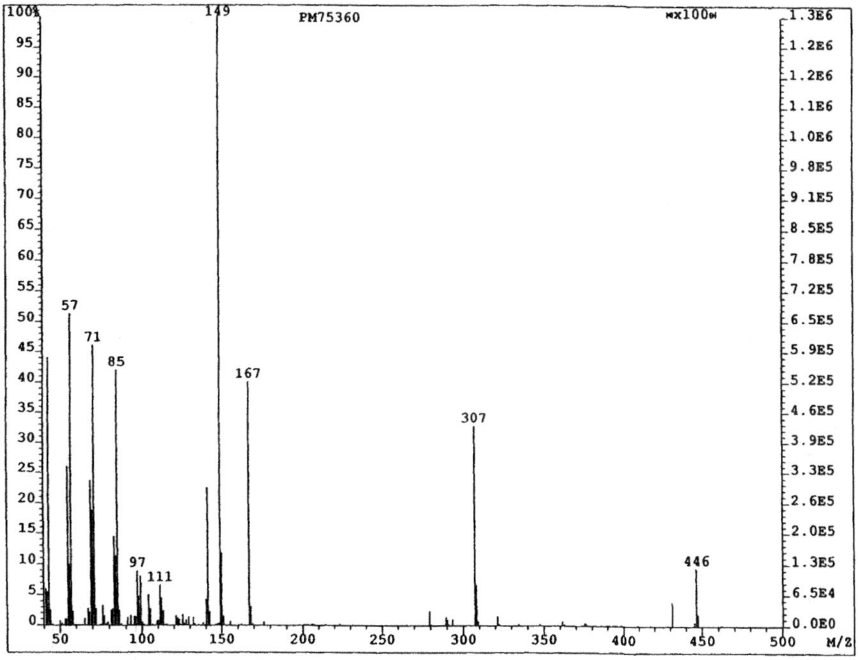

M/Z	Ion Intensity (%)	M/Z	Ion Intensity (%)
43	44.1	97	8.9
55	26.1	99	8.1
56	10.1	104	5.1
57	51.4	111	6.6
69	23.8	141	22.6
70	18.9	149	100.0
71	46.2	150	11.8
83	14.5	167	40.1
84	11.3	307	32.8
85	42.0	308	6.7

Spectrometer	: VG Autospec
Inletsystem	: Flow injection particle beam MS
Source Temperature	: 250°C
Electron Energy	: 70 eV
Scan Range	: 40-1000

PM n°	CAS n°	Name
75360	26761-40-0	PHTHALIC ACID, DIISODECYL ESTER

Solvent	$CDCL_3$
SF	250.13
SI	32768
SW	2380
NS	32

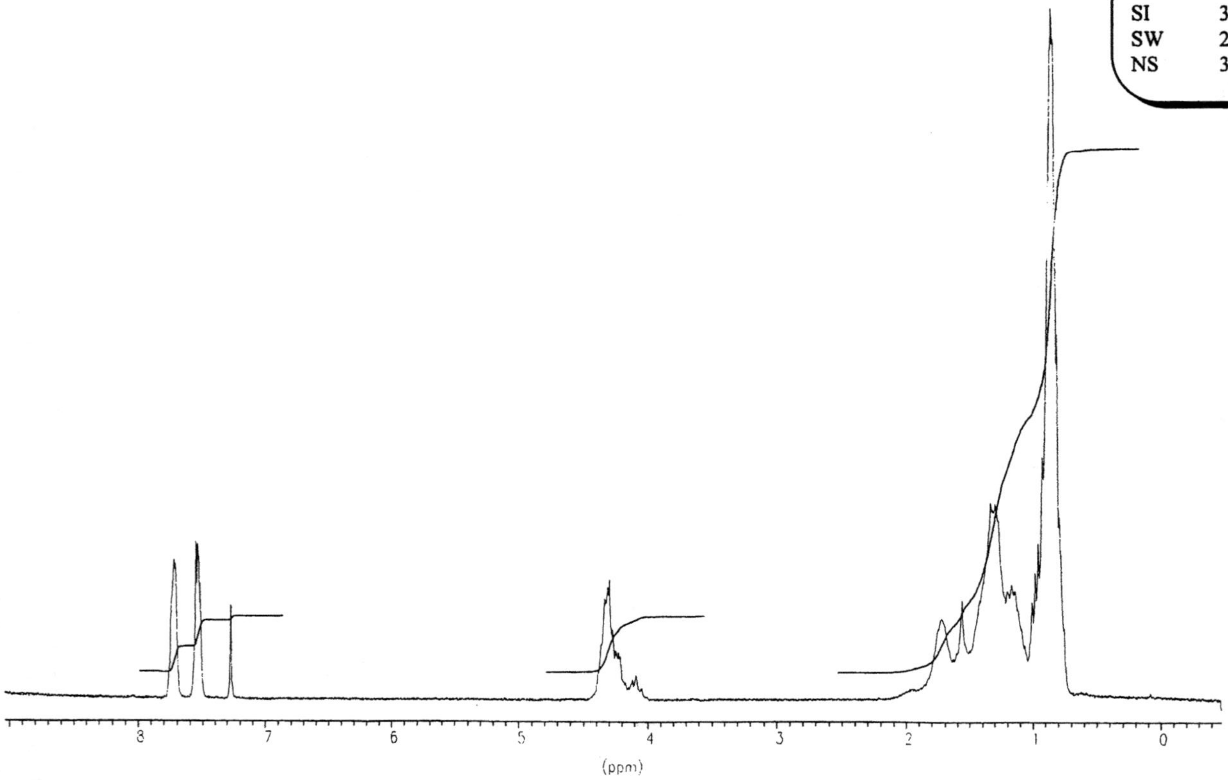

Phthalic acid, diisononyl ester

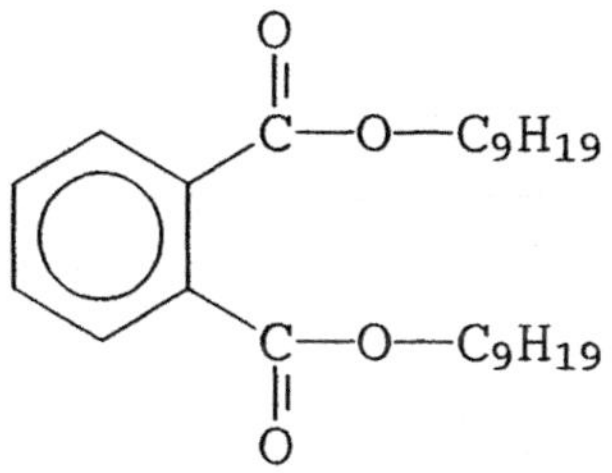

CAS No.	- 28533-12-0
PM Ref. No.	- 75440
Restrictions	- Re-examined
Formula	- $C_{26}H_{42}O_4$
Molecular weight	- 418

Alternative names	- Diisononyl phthalate
Physical Characteristics	- Colourless liquid, mp -48°C. - Soluble in most organic solvents.
Handling	- Store at room temperature.
Safety	-
Availability	- Standard sample was supplied.
Current uses	- Plasticizer.
Applications	- PVC.
GC Retention time	- Cluster 47.9 - 52.1 min.

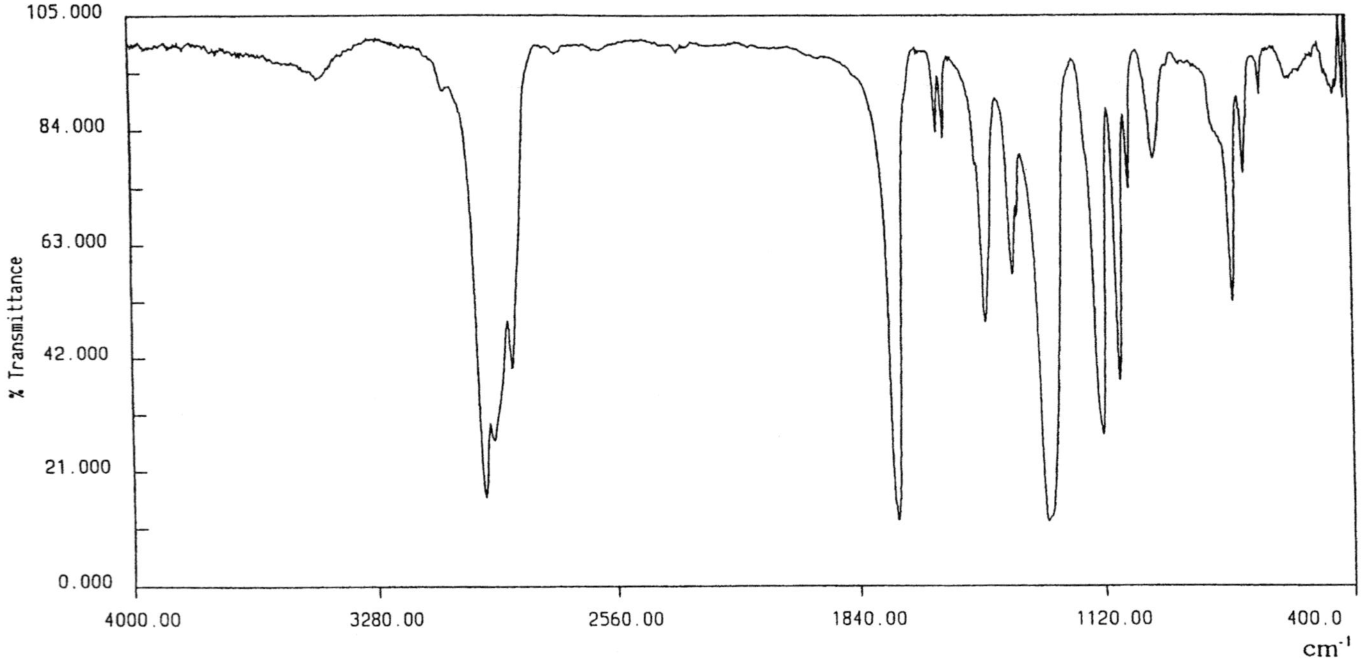

Name : PHTHALIC ACID, DIISONONYL ESTER

PM Nr : 75440 CAS Nr : 28553-12-0

KBr pellet

CAS No. 28533-12-0 PM Ref. No. 75440

PHTHALIC ACID, DIISONONYL ESTER

Peak wavenumber	% T	Relative intensity
410.9	89.8	11.6
441.8	90.5	10.8
536.3	94.4	6.3
706.0	76.0	27.1
742.7	52.7	53.6
966.5	78.7	24.1
1039.8	73.3	30.2
1074.5	37.8	70.4
1124.6	27.8	81.8
1286.7	11.6	100.0
1367.7	68.2	36.0
1383.1	57.4	48.2
1466.1	48.6	58.2
1581.8	82.4	19.9
1601.1	83.6	18.6
1728.4	11.9	99.7
2874.3	40.1	67.8
2928.3	26.63	83.0
2959.2	16.35	94.7

Phthalic acid, diisononyl ester

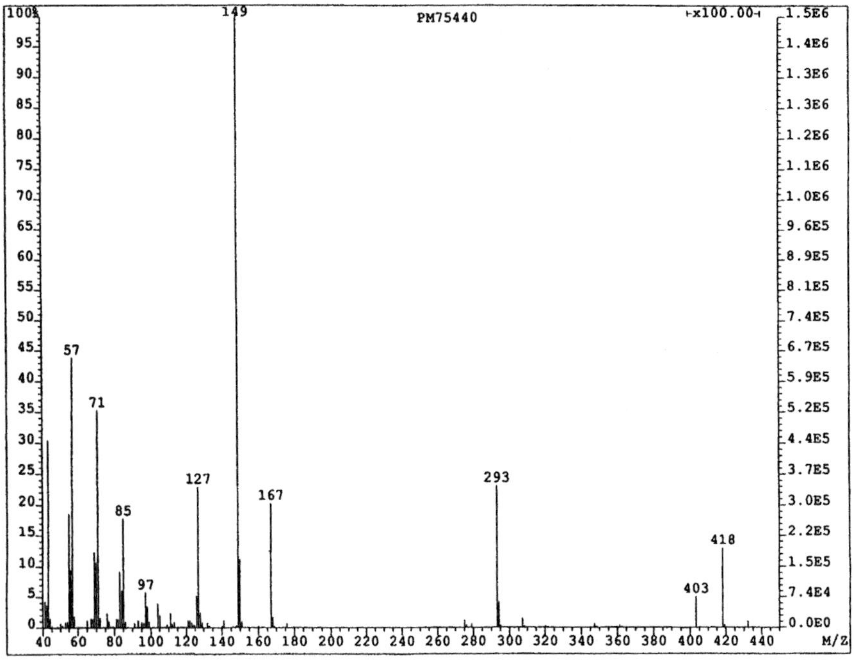

M/Z	Ion Intensity (%)	M/Z	Ion Intensity (%)
43	30.5	85	17.8
55	18.6	97	5.8
56	9.5	126	5.2
57	43.9	127	22.9
69	12.4	149	100.0
70	10.7	150	11.2
71	35.4	167	20.3
83	9.2	279	23.2
84	6.1		

Spectrometer : VG Autospec
Inletsystem : Flow injection particle beam MS
Source Temperature : 250°C
Electron Energy : 70 eV
Scan Range : 40-1000

PM n°	CAS n°	Name
75440	28533-12-0	PHTHALIC ACID, DIISONONYL ESTER

Solvent	$CDCL_3$
SF	250.13
SI	16384
SW	3125
NS	42

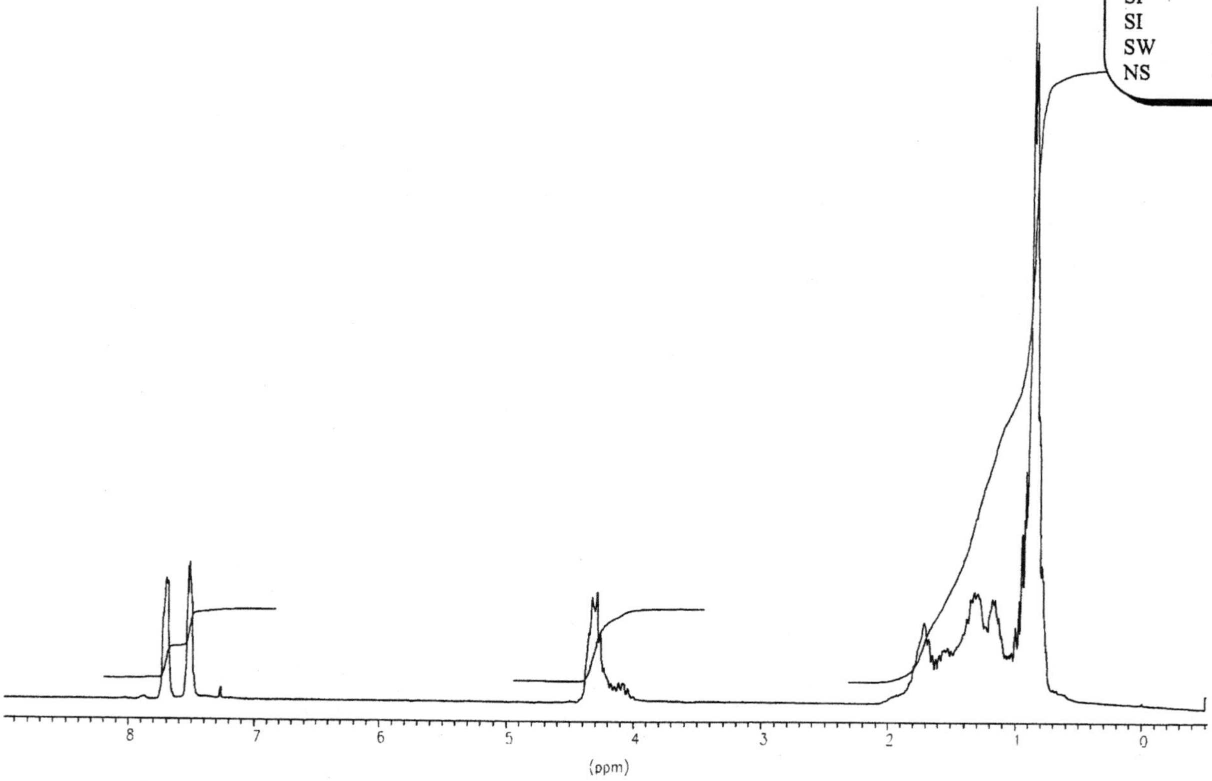

Phthalic acid, diisooctyl ester

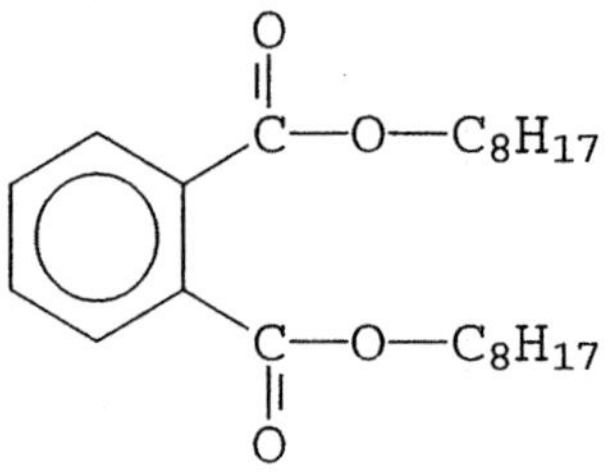

CAS No.	- 27554-26-3
PM Ref. No.	- 75520
Restrictions	- SML = 3.0 mg/kg
Formula	- $C_{24}H_{38}O_4$
Molecular weight	- 390

Alternative names	- Diisooctyl phthalate
Physical Characteristics	- Colourless liquid, mp -43°C, bp 370°C. - Soluble in most organic solvents.
Handling	- Store at room temperature.
Safety	-
Availability	- Standard sample was supplied.
Current uses	- Plasticizer.
Applications	- PVC.
GC Retention time	- Cluster 45.6 - 50.2 min.

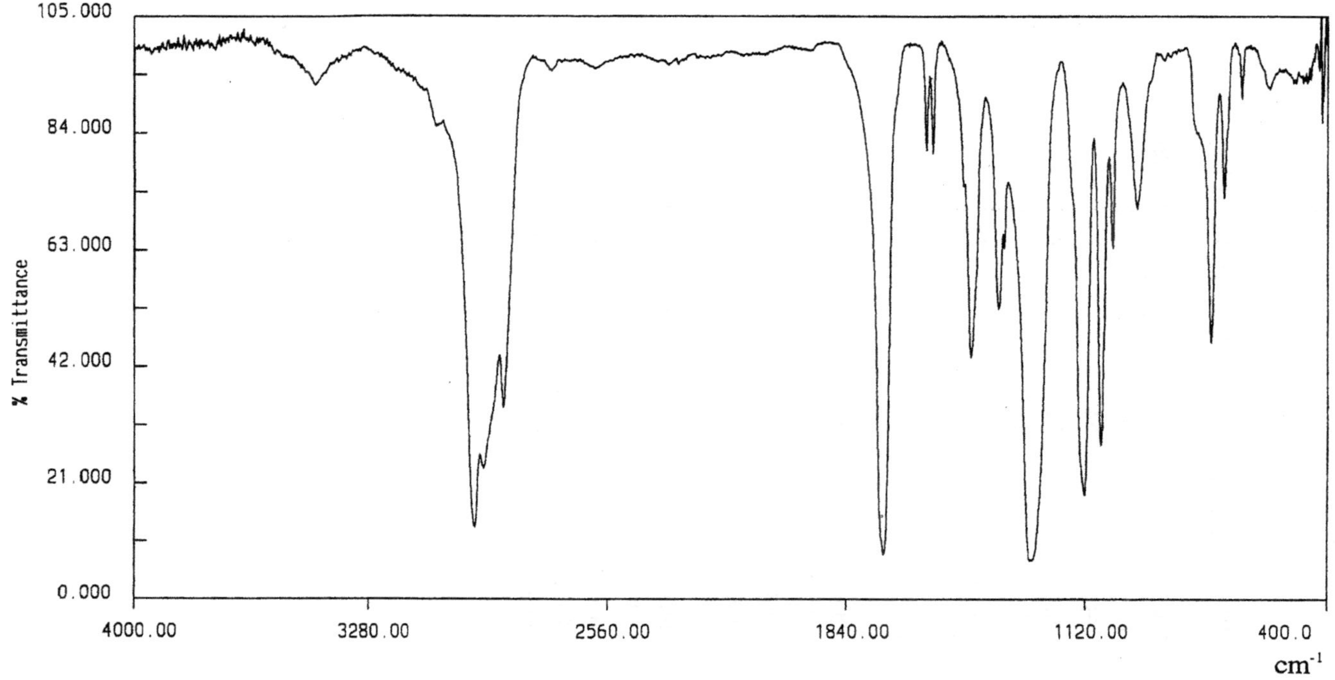

Name : PHTHALIC ACID, DIISOOCTYL ESTER

PM Nr : 75520 CAS Nr : 27554-26-3

KBr pellet

CAS No. 27554-26-3 PM Ref. No. 75520

PHTHALIC ACID, DIISOOCTYL ESTER

Peak wavenumber	% T	Relative intensity
414.8	86.1	14.9
459.1	93.1	7.4
706.0	72.4	29.7
742.7	46.5	57.5
964.5	70.5	31.6
1039.8	63.4	39.3
1074.5	27.9	77.4
1124.6	19.0	87.0
1286.7	6.9	100.0
1367.7	63.5	39.3
1385.1	52.5	51.0
1466.1	43.8	60.4
1581.8	80.3	21.1
1601.1	81.1	20.4
1730.4	8.2	98.6
2874.3	34.7	70.2
2932.2	23.8	81.9
2959.2	13.03	93.4

Phthalic acid, diisooctyl ester

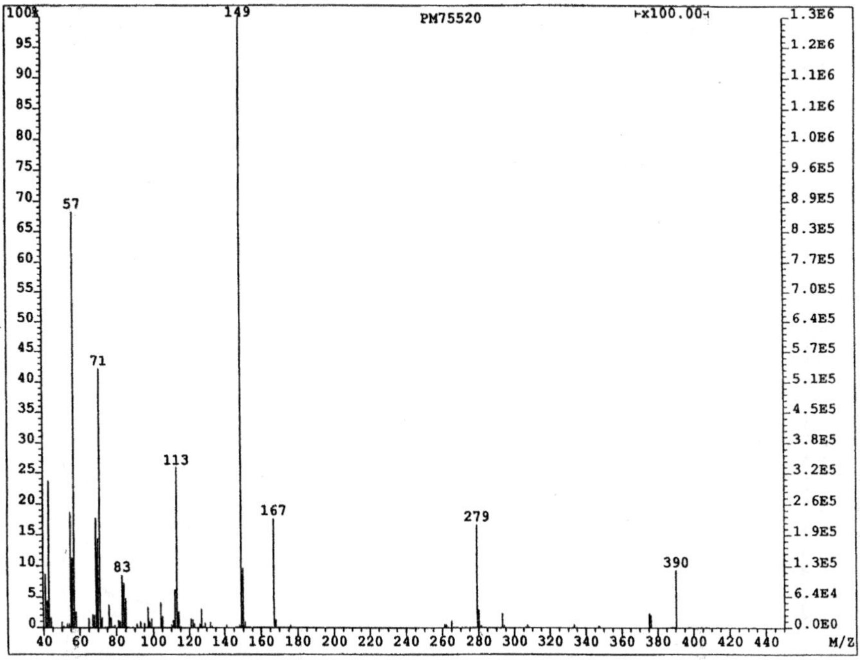

M/Z	Ion Intensity (%)	M/Z	Ion Intensity (%)
41	8.81	83	8.9
43	23.7	84	7.3
55	18.7	112	6.3
56	11.4	113	25.9
57	68.3	149	100.0
69	17.8	150	9.7
70	14.5	167	17.7
71	42.3	279	16.7

Spectrometer	: VG Autospec
Inletsystem	: Flow injection particle beam MS
Source Temperature	: 250°C
Electron Energy	: 70 eV
Scan Range	: 40-1000

PM n°	CAS n°	Name
75520	27554-26-3	PHTHALIC ACID, DIISOOCTYL ESTER

Solvent	$CDCL_3$
SF	250.13
SI	16384
SW	3125
NS	36

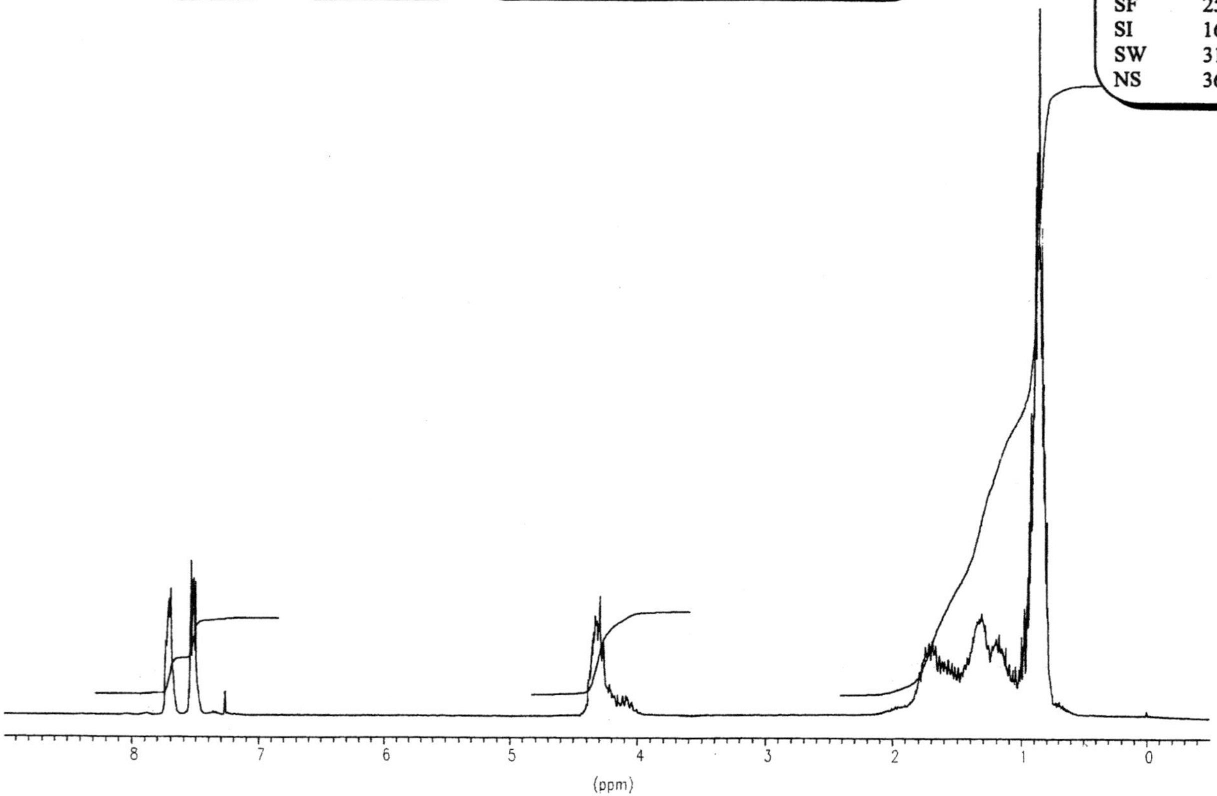

Phthalic acid, dimethyl ester

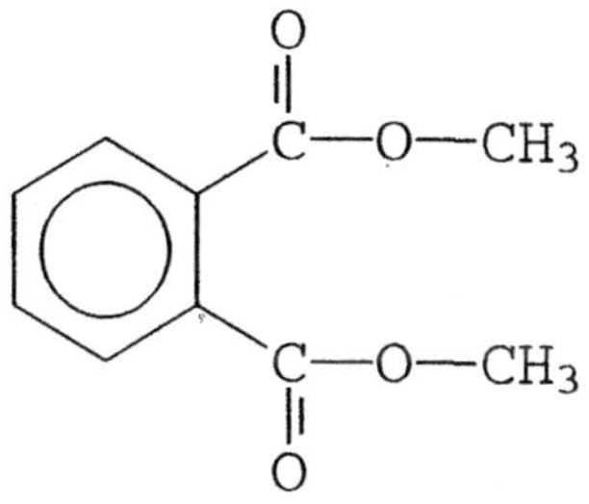

CAS No.	- 00131-11-3
PM Ref. No.	- 75600
Restrictions	- SML = 3.0 mg/kg
Formula	- $C_{10}H_{10}O_4$
Molecular weight	- 194

Alternative names	- Dimethyl phthalate
Physical Characteristics	- Oily liquid, mp 2°C, bp 282°C. - Soluble in most organic solvents.
Handling	- Store at room temperature.
Safety	- Irritant.
Availability	- Standard sample was supplied.
Current uses	- Plasticizer, carrier.
Applications	- PVC.
GC Retention time	- 25.7 min.

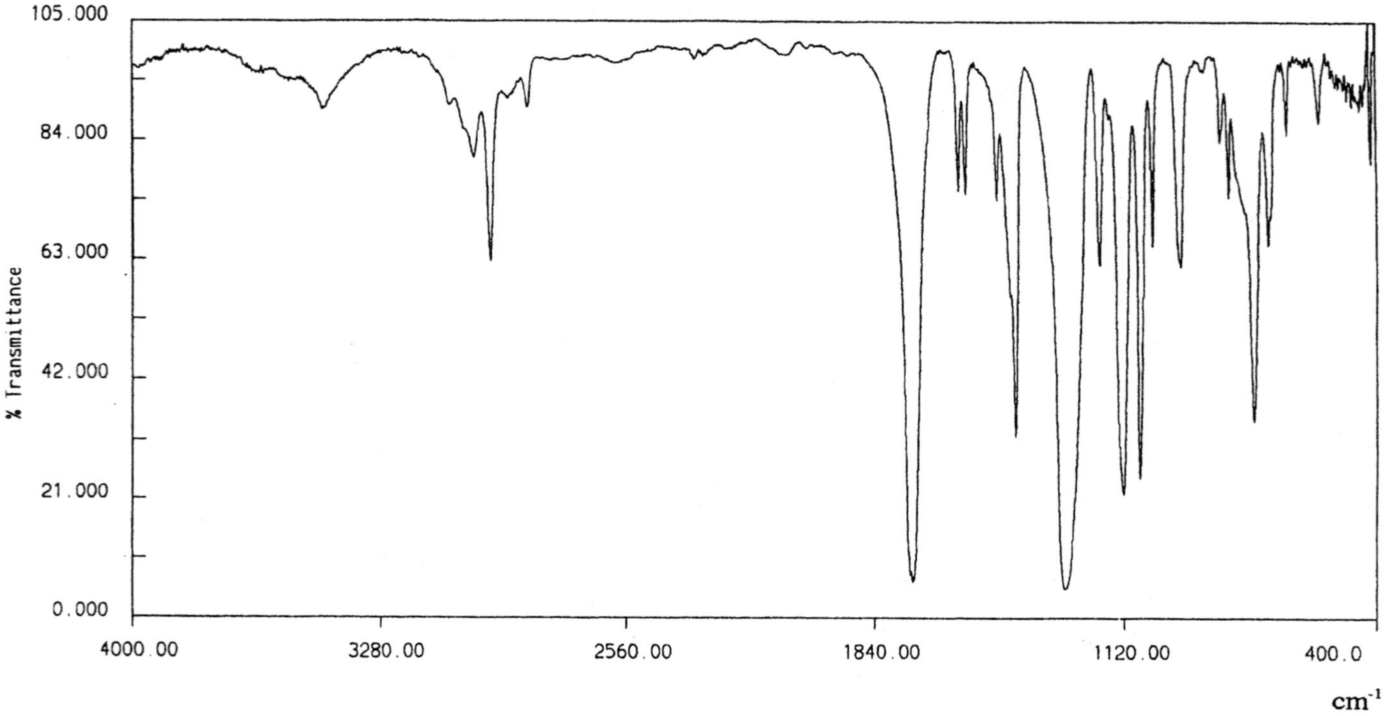

Name : PHTHALIC ACID, DIMETHYL ESTER

PM Nr : 75600 CAS Nr : 00131-11-3

KBr pellet

CAS No. 00131-11-3 PM Ref. No. 75600

PHTHALIC ACID, DIMETHYL ESTER

Peak wavenumber	% T	Relative intensity
412.8	80.1	21.0
434.0	90.4	10.1
445.6	89.4	11.2
466.8	90.0	10.5
480.3	91.2	9.3
559.4	87.3	13.4
652.0	85.2	15.6
706.0	65.7	36.1
746.5	34.9	68.6
817.9	73.9	27.5
843.0	83.8	17.1
958.7	61.8	40.2
1039.8	65.5	36.3
1076.4	24.8	79.2
1124.6	21.8	82.3
1192.2	62.0	40.0
1292.5	5.1	99.9
1435.2	32.1	71.5
1489.2	73.4	28.0
1579.9	74.6	26.7
1601.1	75.1	26.2
1730.4	6.4	98.6
2903.2	91.2	9.3
2955.3	62.6	39.4
3003.5	80.9	20.1

Phthalic acid, dimethyl ester

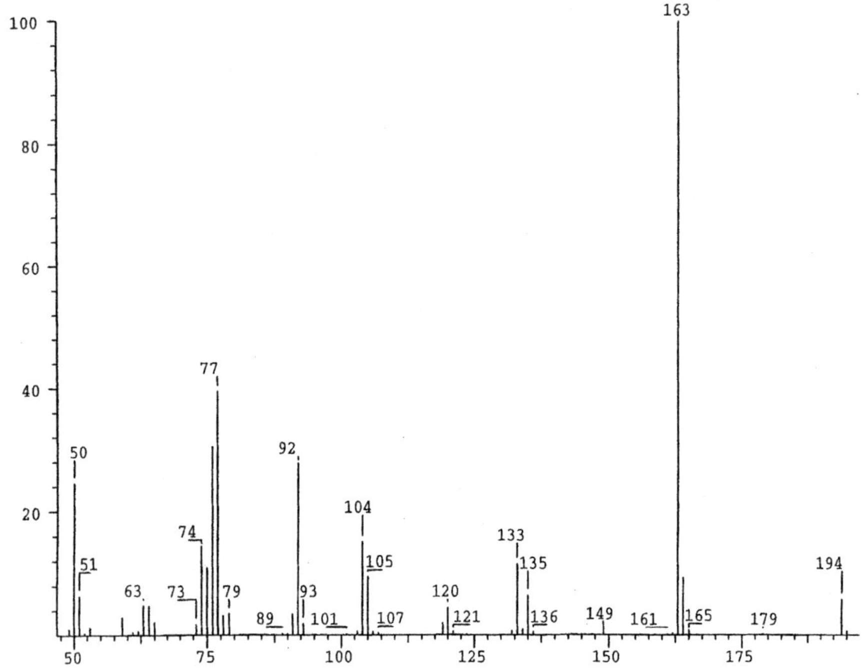

M/Z	Ion Intensity (%)	M/Z	Ion Intensity (%)
50	24.6	92	27.8
51	6.4	104	15.0
63	4.9	105	9.5
64	4.8	120	4.4
74	14.3	133	11.5
75	10.9	135	6.5
76	30.4	163	100.0
77	39.4	164	9.3
78	3.1	194	5.8
91	3.4	195	0.7

Spectrometer	: Finnigan Mat SSQ 700
Inletsystem	: Capillary GC/MS
Source Temperature	: 180°C
Electron Energy	: 70 eV
Scan Range	: 40-500

PM n° 75600

CAS n° 00131-11-3

Name PHTHALIC ACID, DIMETHYL ESTER

Solvent	$CDCl_3$
SF	250.13
SI	16384
SW	3125
NS	32

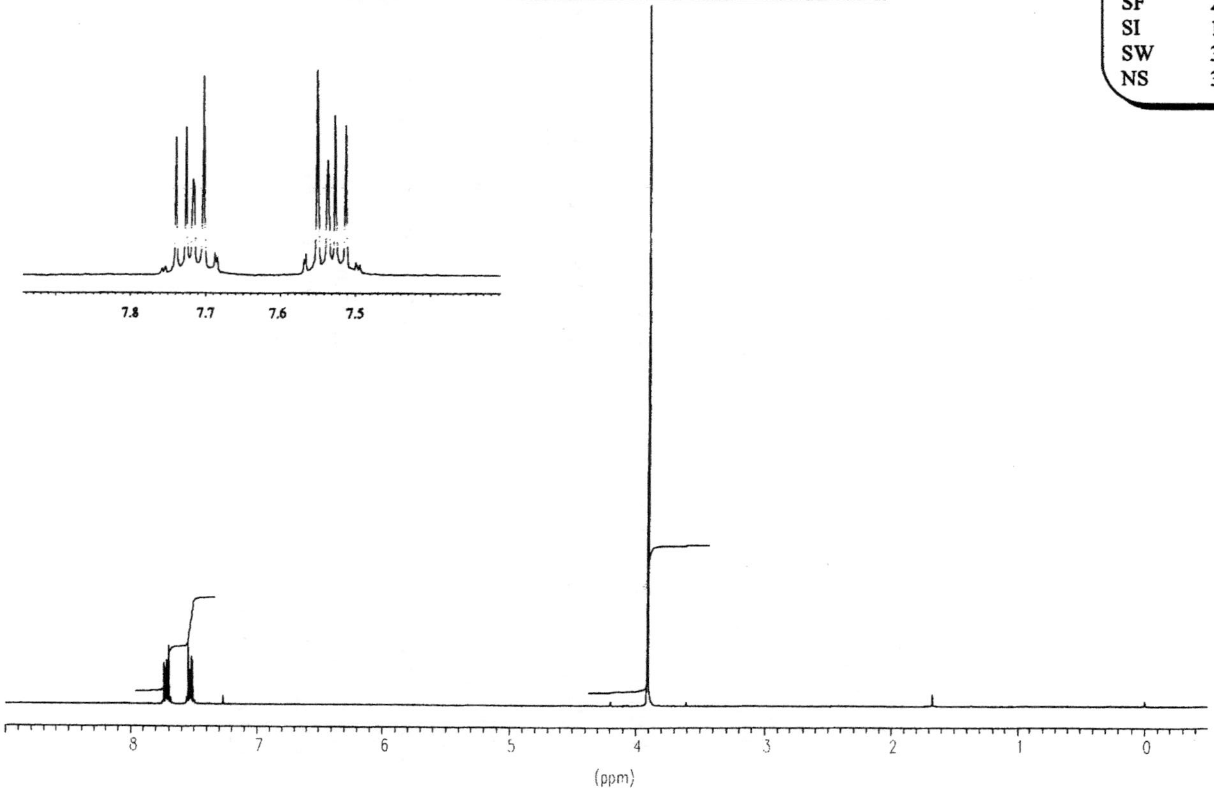

Polydimethylsiloxane

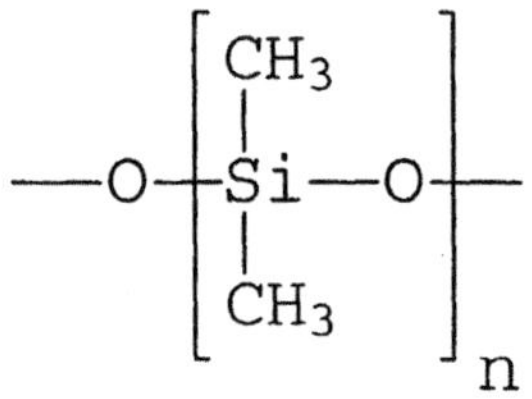

CAS No.	- 09016-00-6 and 61348-62-9
PM Ref. No.	- 76720
Restrictions	- None
Formula	-
Molecular weight	- Approx. 12500

Alternative names	- Dimethicone
Physical Characteristics	- Colourless viscous liquid, mp approx. -34°C. - Soluble in toluene, chloroform, ether.
Handling	- Store at room temperature.
Safety	-
Availability	- Standard sample was supplied.
Current uses	- Lubricant.
Applications	- PS, PVC, PE, PP.
GC Retention time	-

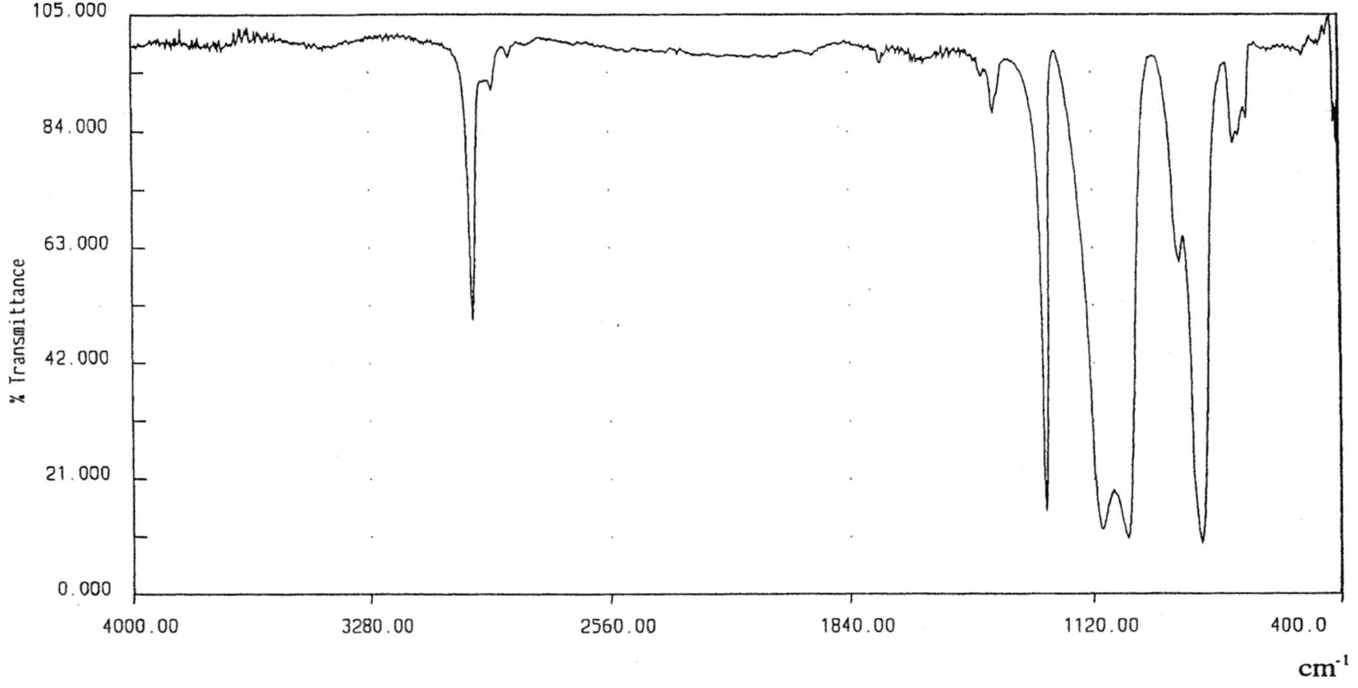

Name : POLYDIMETHYLSILOXANE

PM Nr : 76720 CAS Nr : 09016-00-6
61348-62-9

KBr pellet

CAS No. 09016-00-6 and 61348-62-9 PM Ref. No. 76720

POLYDIMETHYLSILOXANE

Peak wavenumber	% T	Relative intensity
661.7	86.3	15.1
686.7	83.3	18.4
800.6	9.4	100.0
864.2	60.3	43.8
1020.5	10.2	99.1
1097.6	11.9	97.2
2965.0	49.7	55.5

Polydimethylsiloxane

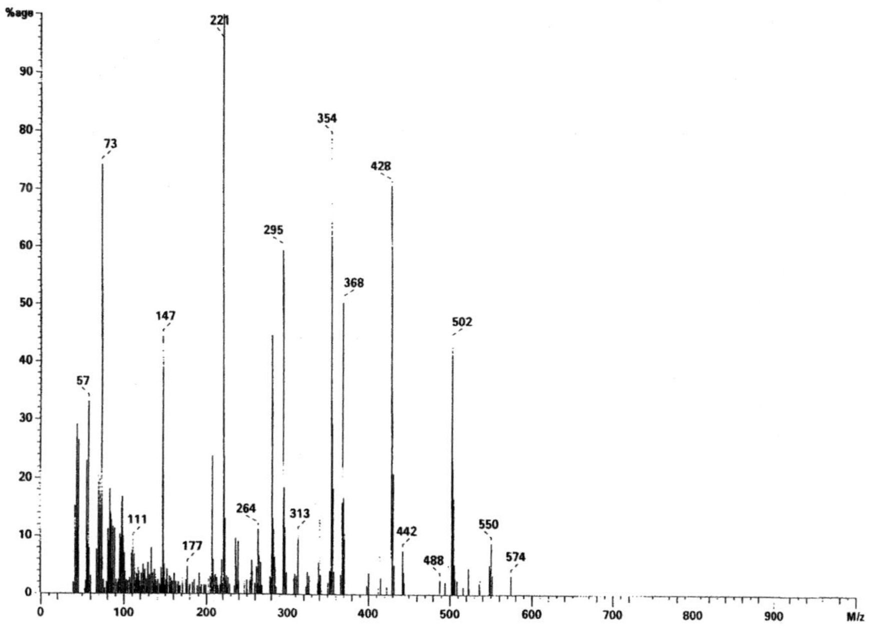

M/Z	Ion Intensity (%)	M/Z	Ion Intensity (%)
43	29.3	281	44.0
45	26.7	295	58.7
55	23.3	354	78.0
57	20.0	355	29.3
70	21.3	368	50.0
73	73.3	428	70.0
147	44.7	429	31.3
207	24.0	430	20.7
221	100.0	502	43.3
222	24.0	503	22.0

Spectrometer	: VG Autospec
Inletsystem	: Probe inlet
Source Temperature	: 350-400°C
Electron Energy	: 70 eV
Scan Range	: 40-1000

PM n°
76720

CAS n°
09016-00-6
63148-62-9

Name
POLYDIMETHYLSILOXANE

Solvent	$CDCL_3$
SF	250.13
SI	16384
SW	6494
NS	16

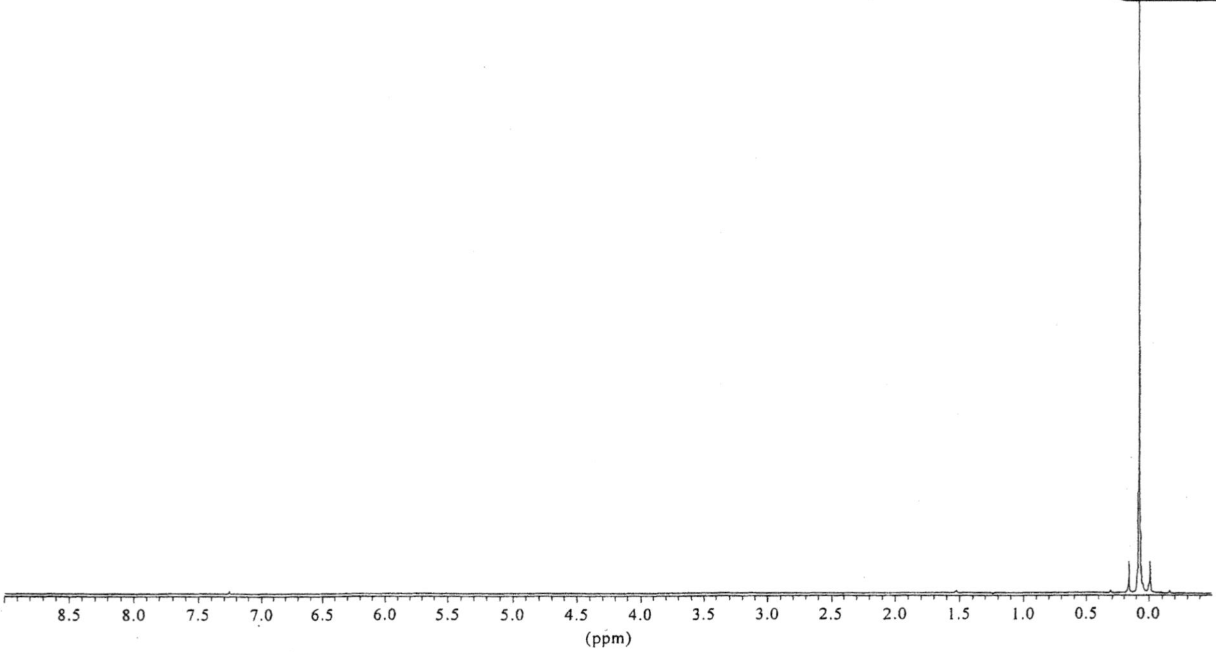

Polyethyleneglycol ether of tallow fatty alcohol

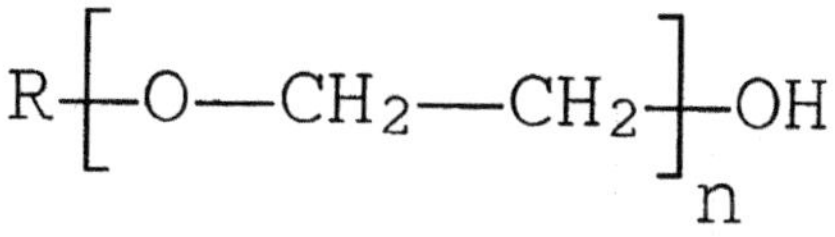

CAS No.	- 61791-28-4
PM Ref. No.	- 77760
Restrictions	- None
Formula	-
Molecular weight	- 10 units ethylene oxide (n=10)
Alternative names	-
Physical Characteristics	- Off white past, mp >30°C. - Soluble in most organic solvents.
Handling	- Store at room temperature.
Safety	- Harmful.
Availability	- Standard sample was supplied.
Current uses	- Lubricant, emulsifier, antistatic agent.
Applications	- PVA, PTFE, polyolefins.
GC Retention time	-

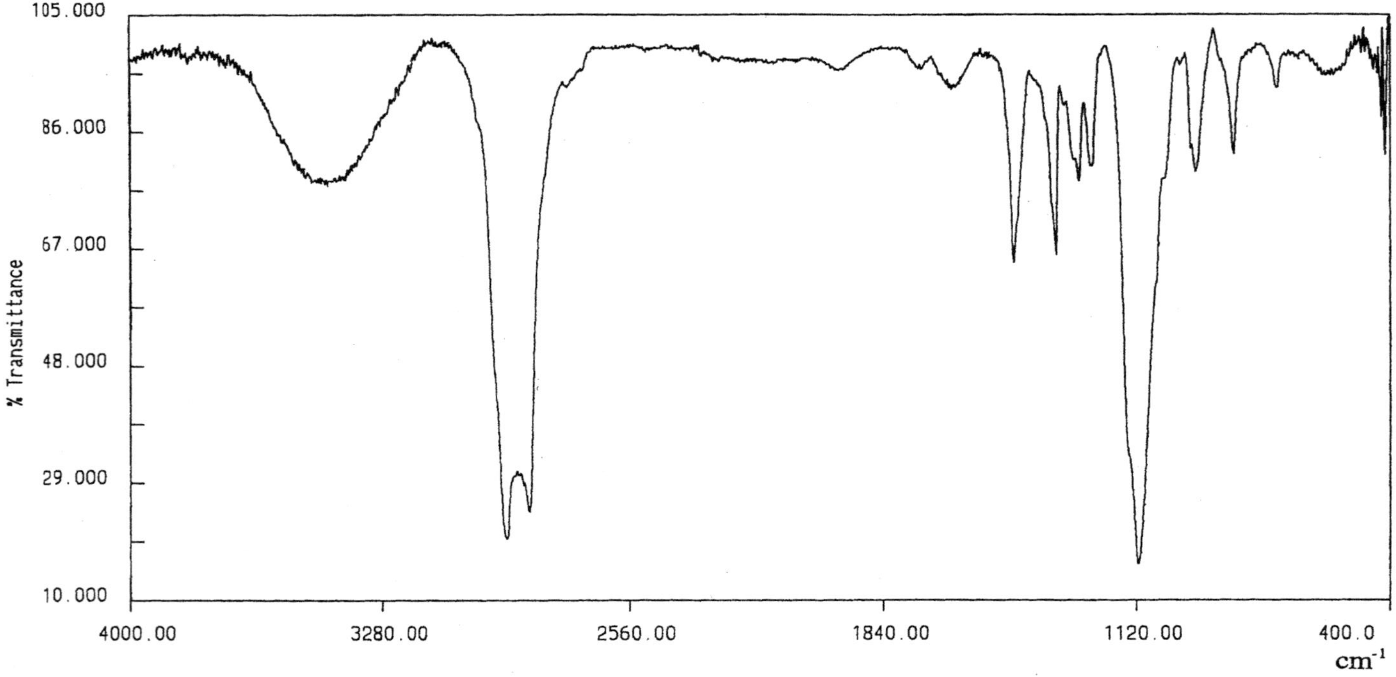

Name : POLYETHYLENEGLYCOL ETHER OF TALLOW FATTY ALCOHOL

PM Nr : 77760 CAS Nr : 61791-28-4

KBr pellet

CAS No. 61791-28-4 PM Ref. No. 77760

POLYETHYLENEGLYCOL ETHER OF TALLOW FATTY ALCOHOL

Peak wavenumber	% T	Relative intensity
412.8	82.2	21.1
424.4	88.2	14.0
580.6	94.9	6.1
723.4	92.9	8.5
843.0	82.3	21.0
951.0	79.3	24.5
1047.5	78.0	26.0
1115.0	15.6	100.0
1282.8	77.7	26.5
1346.5	66.0	40.3
1468.0	64.6	41.9
2855.0	24.2	89.9
2920.6	19.7	95.2
3285.2	86.9	15.6
3343.0	81.4	22.1
3377.8	78.9	25.0

Polyethyleneglycol ether of tallow fatty alcohol

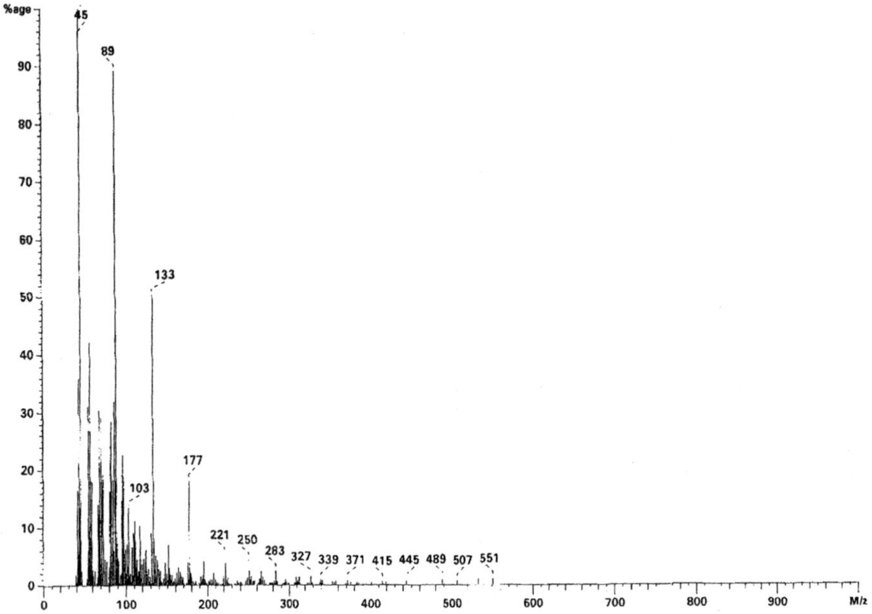

M/Z	Ion Intensity (%)	M/Z	Ion Intensity (%)
41	16.7	82	25.3
43	36.0	83	28.7
45	100.0	85	18.7
55	31.3	87	32.0
57	42.0	88	18.0
59	18.0	89	90.0
69	30.7	96	19.3
71	29.3	97	22.7
72	17.3	133	51.3
73	21.3	177	18.0

Spectrometer	: VG Autospec
Inletsystem	: Probe inlet
Source Temperature	: 350-400°C
Electron Energy	: 70 eV
Scan Range	: 40-1000

PM n°
77760

CAS n°
61791-28-4

Name

POLYETHYLENEGLYCOL ETHER OF TALLOW FATTY ALCOHOL

Solvent	$CDCL_3$
SF	250.13
SI	16384
SW	3125
NS	32

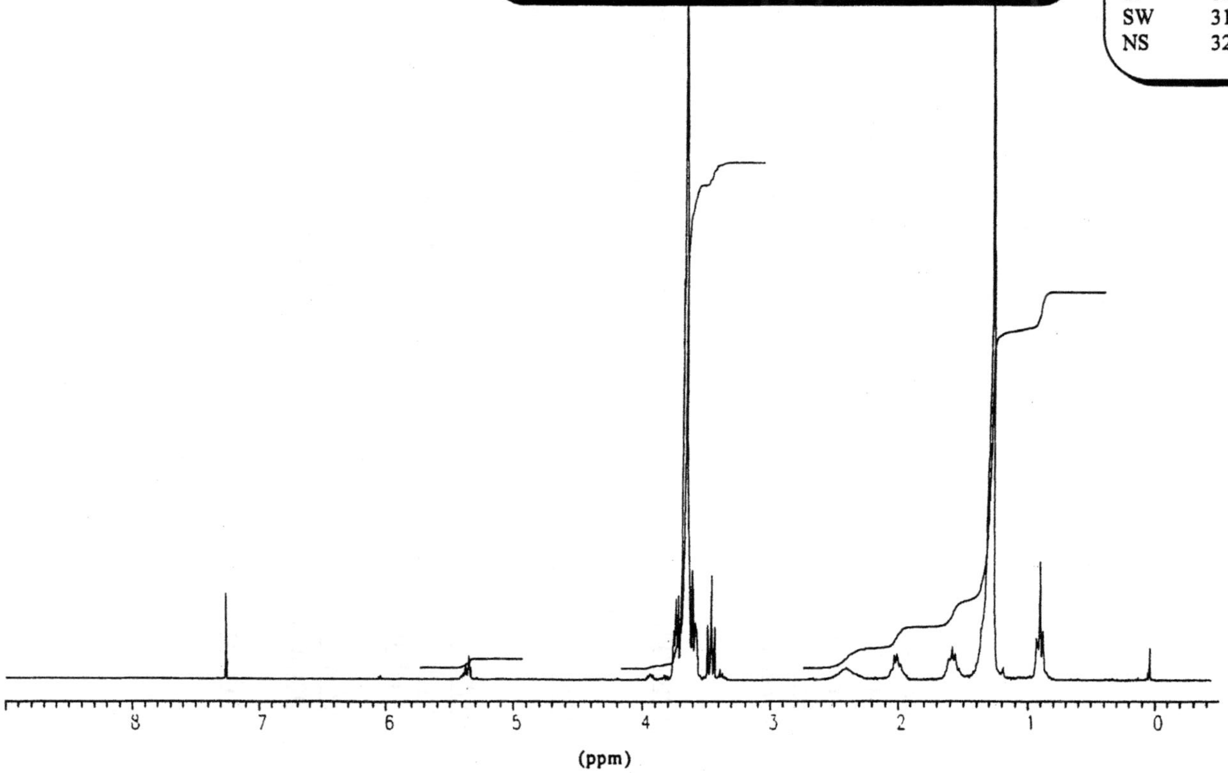

Poly(ethylene propylene)glycol

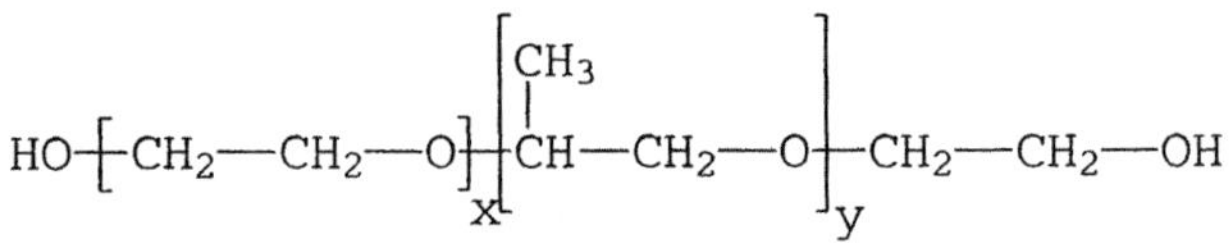

CAS No.	- 9003-11-6
PM Ref. No.	- 79930
Restrictions	- None
Formula	-
Molecular weight	- 8000 - 11000

Alternative names	- Poly(oxyethylene)poly(oxypropylene) block copolymer
Physical Characteristics	- White flakes, mp 55 - 60°C. - Soluble in most organic solvents.
Handling	- Store at room temperature.
Safety	- Low toxicity.
Availability	- Standard sample was supplied.
Current uses	- Lubricant.
Applications	- PS, PVC, PE, PP.
GC Retention time	-

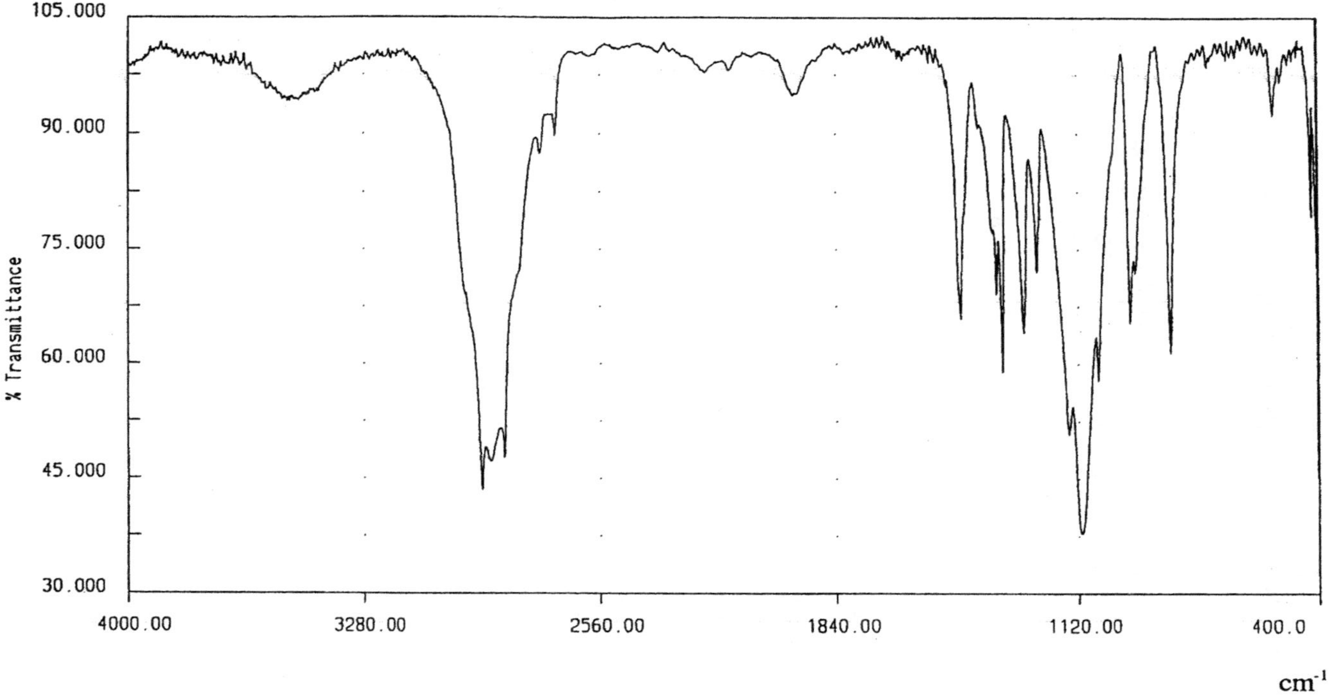

Name : POLY(ETHYLENE PROPYLENE)GLYCOL

PM Nr : 79930 CAS Nr : 9003-11-6

KBr pellet

CAS No. 9003-11-6 PM Ref. No. 79930

Poly(ethylene propylene)glycol

Peak wavenumber	% T	Relative intensity
418.6	79.2	33.3
843.0	61.3	62.0
447.2	71.8	45.2
962.6	65.2	55.8
1061.0	57.6	67.9
1113.1	37.6	100.0
1149.7	50.7	79.0
1242.3	71.8	45.2
1280.9	63.9	57.9
1344.6	58.8	66.0
1360.0	69.0	49.7
1468.0	65.8	54.8
2743.1	87.2	20.5
2851.1	47.6	84.0
2891.7	47.0	84.9
2916.7	43.4	90.7

Poly(ethylene propylene)glycol

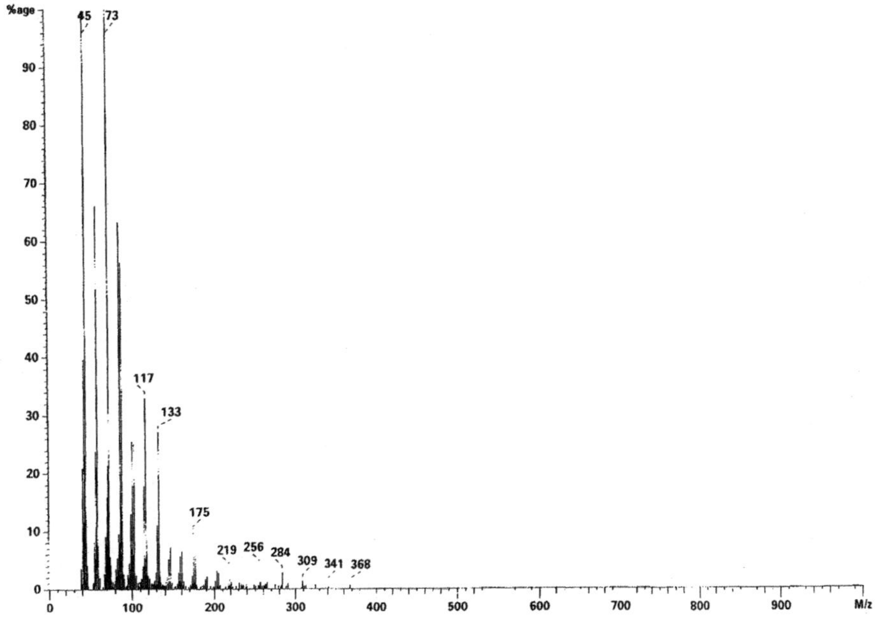

M/Z	Ion Intensity (%)	M/Z	Ion Intensity (%)
41	20.0	73	100.0
42	11.3	87	63.3
43	40.0	88	14.0
44	23.3	89	56.7
45	99.3	99	12.7
57	24.0	101	26.0
58	21.3	103	25.3
59	66.0	115	18.0
71	16.7	117	33.3
72	21.3	133	26.7

Spectrometer	: VG Autospec
Inletsystem	: Probe inlet
Source Temperature	: 350-400°C
Electron Energy	: 70 eV
Scan Range	: 40-1000

PM n°	CAS n°	Name
79930	9003-11-6	POLY(ETHYLENE PROPYLENEGLYCOL)

Solvent	$CDCL_3$
SF	250.13
SI	16384
SW	6494
NS	24

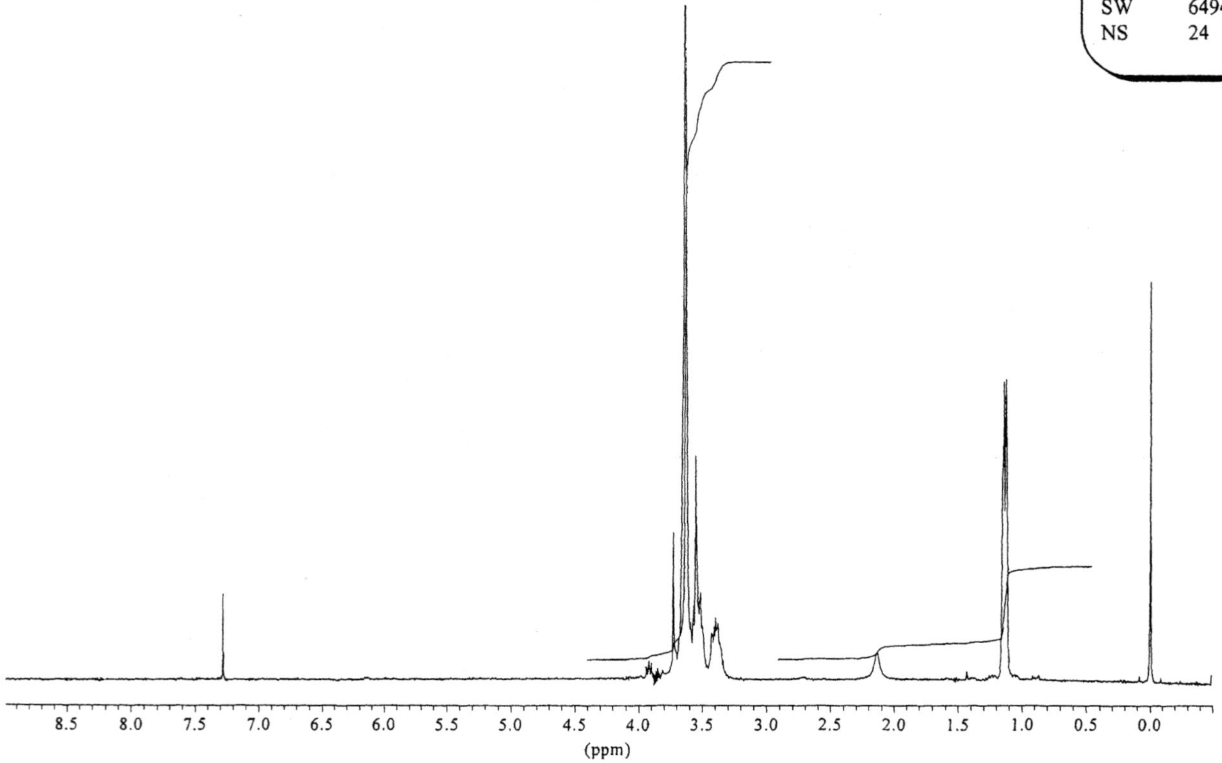

Polyolefins

$C_nH_{(n+2)}$

CAS No.	- 68037-12-7
PM Ref. No.	- 80550
Restrictions	- None
Formula	- $C_nH_{(n+2)}$
Molecular weight	- 500 - 600

Alternative names - Poly(decene-1), hydrogenated

Physical Characteristics - Colourless liquid, mp approx -68°C.

Handling - Store at room temperature.
Safety -
Availability - Standard sample was supplied.

Current uses - Release agent.
Applications - PS, PP, PE.

GC Retention time -

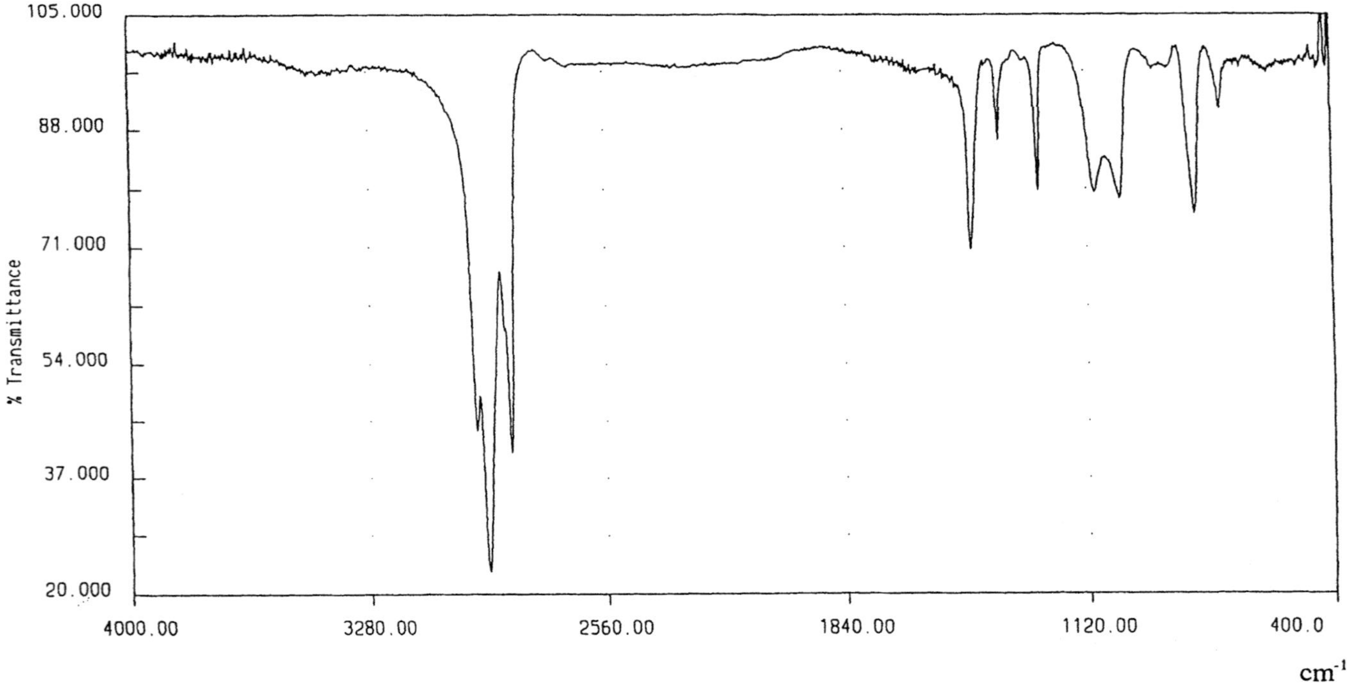

Name : POLY-α-OLEFINS

PM Nr : 80550 CAS Nr : 68037-12-7

KBr pellet

CAS No. 68037-12-7

Pm Ref. No. 80550

POLY-α-OLEFINS

Peak wavenumber	% T	Relative intensity
723.4	90.9	11.8
802.5	75.8	31.5
1022.4	77.9	28.8
1095.7	78.8	27.6
1261.6	79.0	27.3
1466.1	70.6	38.3
2855.0	40.6	77.3
2924.4	23.2	100.0
2959.2	43.9	73.0

Polyolefins

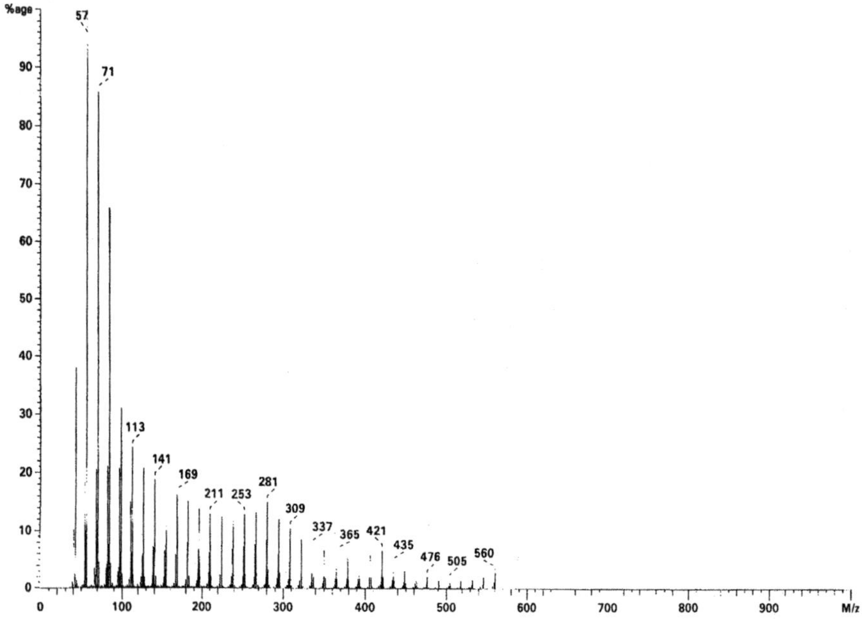

M/Z	Ion Intensity (%)	M/Z	Ion Intensity (%)
43	38.0	113	24.7
55	18.0	127	18.0
57	100.0	141	16.7
69	20.6	155	16.0
71	86.0	169	15.3
83	21.3	183	14.0
85	66.0	197	13.3
97	20.7	211	12.7
99	30.7	225	12.7
111	14.7	239	12.7

Spectrometer	: VG Autospec
Inletsystem	: Probe inlet
Source Temperature	: 350-400°C
Electron Energy	: 70 eV
Scan Range	: 40-1000

PM n°	CAS n°	Name
80550	68037-12-7	POLY-α-OLEFINS

Solvent	$CDCL_3$
SF	250.13
SI	16384
SW	6494
NS	16

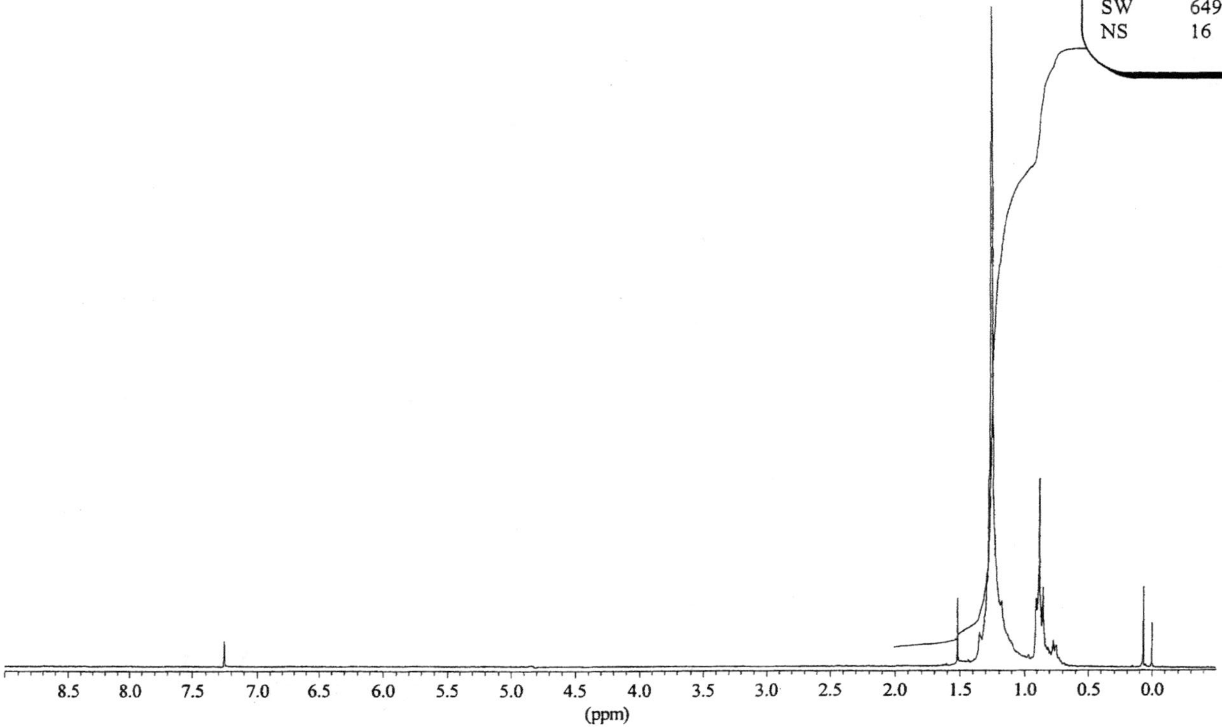

Poly[6-[(1,1,3,3-tetramethylbutyl)amino]-1,3,5-triazine-2,4-diyl]-[(2,2,6,6-tetramethyl-4-piperidyl)-imino]hexamethylene[(2,2,6,6-tetramethyl-4-piperidyl)imino]

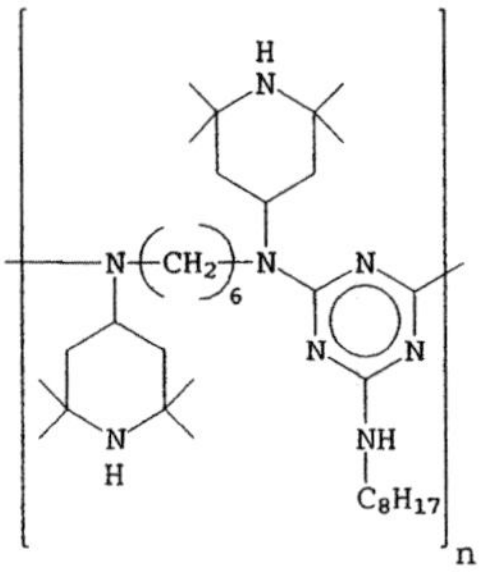

CAS No.	- 71878-19-8
PM Ref. No.	- 81200
Restrictions	- SML = 3.0 mg/kg
Formula	- $C_{35}H_{68}N_8$ when n= 1
Molecular weight	- 600 when n= 1

Alternative names -

Physical Characteristics
- Slightly yellow granules, mp 100 - 135°C.
- Soluble in most organic solvents.

Handling - Store at room temperature.
Safety -
Availability - Standard sample was supplied.

Current uses - Light stabilizer.
Applications - General use.

GC Retention time -

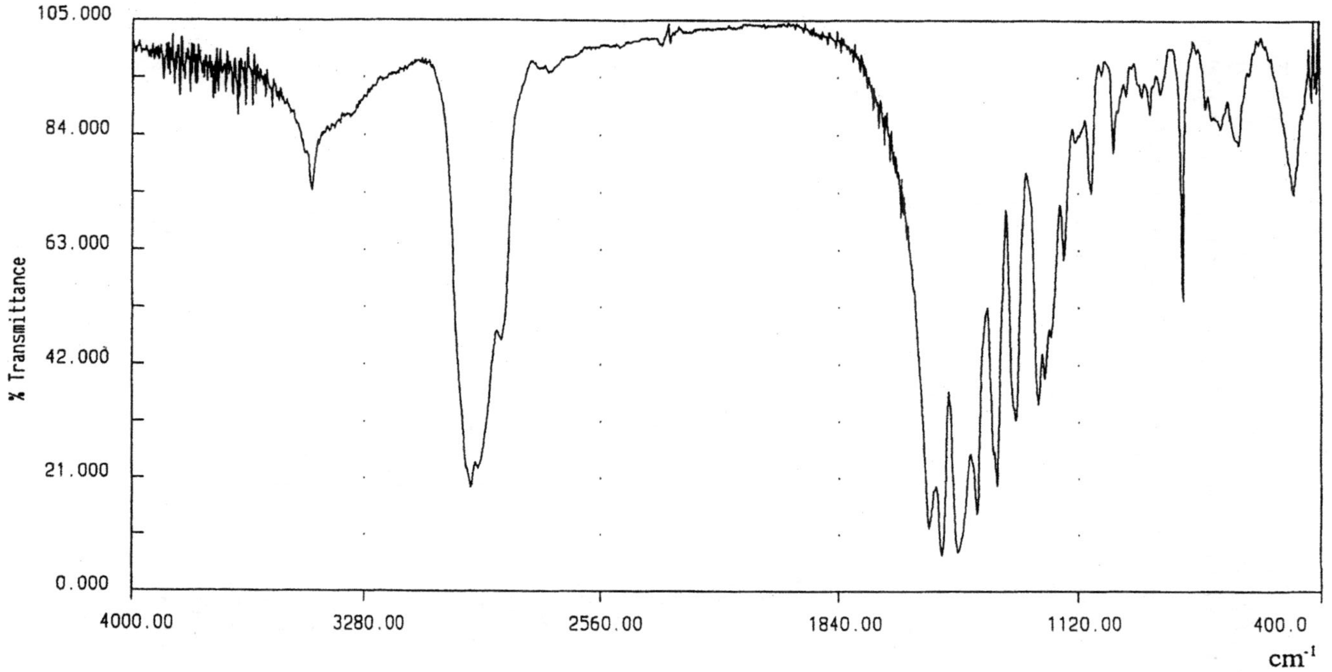

Name : POLY[6-(1,1,3,3,TETRAMETHYLBUTYL)AMINO]-1,3,5-TRIAZINE-2,4-DIYL]-[(2,2,6,6-TETRAMETHYL-4-PIPERIDYL)IMINO]HEXAMETHYLENE[(2,2,6,6-TETRAMETHYL-4-PIPERIDYL) IMINO]

PM Nr : 81200 CAS Nr : 71878-19-8

KBr pellet

CAS No. 71878-19-8 PM Ref. No. 81200

POLY[6-[(1,1,3,3TETRAMETHYLBUTYL)AMINO]-1,3,5-TRIAZINE-2,4-DIYL]-[(2,2,6,6-TETRAMETHYL-4-PIPERIDYL)IMINO] HEXAMETHYLENE[(2,2,6,6-TETRAMETYL-4-PIPERIDYL)IMINO]

Peak wavenumber	% T	Relative intensity
424.4	90.0	10.7
478.4	73.2	28.7
611.5	94.8	5.6
644.3	82.1	19.2
696.4	85.1	16.0
810.2	53.9	49.4
804.7	87.9	13.0
1014.7	80.7	20.7
1082.2	73.5	28.4
1130.4	82.6	18.6
1165.2	61.2	41.6
1203.7	47.0	56.8
1221.1	39.5	64.8
1240.4	34.7	70.0
1307.9	31.7	73.2
1365.8	19.6	86.2
1423.6	14.4	91.7
1483.4	7.20	99.5
1531.7	6.70	100.0
1570.3	11.6	94.7
1555.1	68.9	33.3
1701.4	82.2	19.1
2862.7	46.5	57.3
2932.2	22.7	82.9
2955.3	19.3	86.5
2325.7	87.0	13.9
3443.4	73.7	28.2
3856.2	91.1	9.5

Poly[6-[(1,1,3,3-tetramethylbutyl)amino]-1,3,5-triazine-2,4-diyl]-[(2,2,6,6-tetramethyl-4-piperidyl)-imino]hexamethylene[(2,2,6,6-tetramethyl-4-piperidyl)imino]

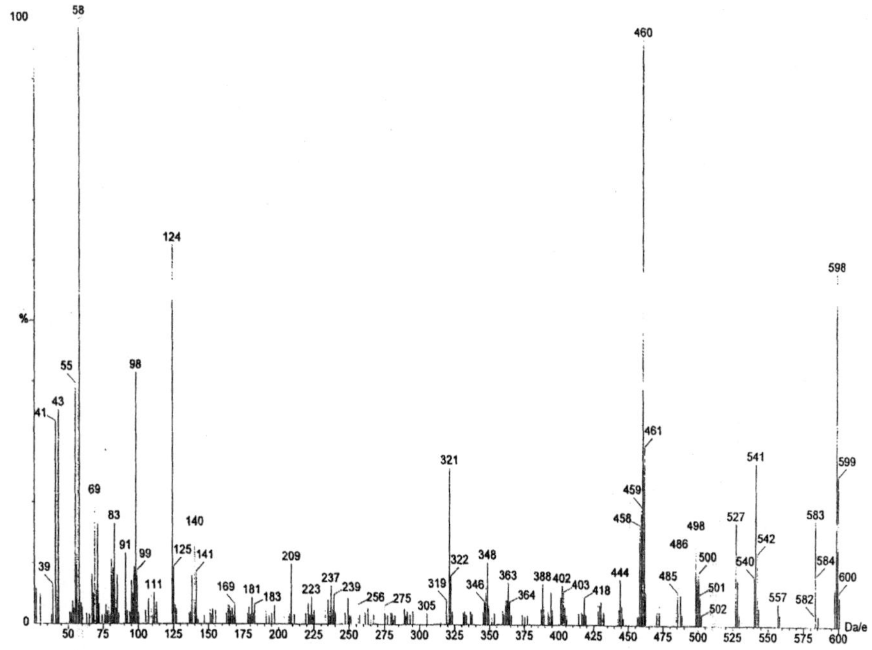

M/Z	Ion Intensity (%)	M/Z	Ion Intensity (%)
41	33.4	321	25.8
42	27.1	458	16.5
43	35.3	459	19.6
55	39.7	460	96.8
57	39.6	461	29.4
58	100.0	527	16.9
69	20.8	541	26.7
83	16.6	583	17.3
98	41.5	598	58.3
124	62.6	599	24.6

Spectrometer	: VG Autospec
Inletsystem	: Probe inlet
Source Temperature	: 350-400°C
Electron Energy	: 70 eV
Scan Range	: 25-1000

PM n°	CAS n°
81200	71878-19-8

Name
POLY[6-[(1,1,3,3-TETRAMETHYLBUTYL)AMINO]-1,3,5-TRIAZINE-2,4-DIYL]-[(2,2,6,6-TETRAMETHYL-4-PIPERIDYL)-IMINO]HEXAMETHYLENE[(2,2,6,6-TETRAMETHYL-4-PIPERIDYL)-IMINO]

Solvent	$CDCL_3$
SF	250.13
SI	16384
SW	3125
NS	36

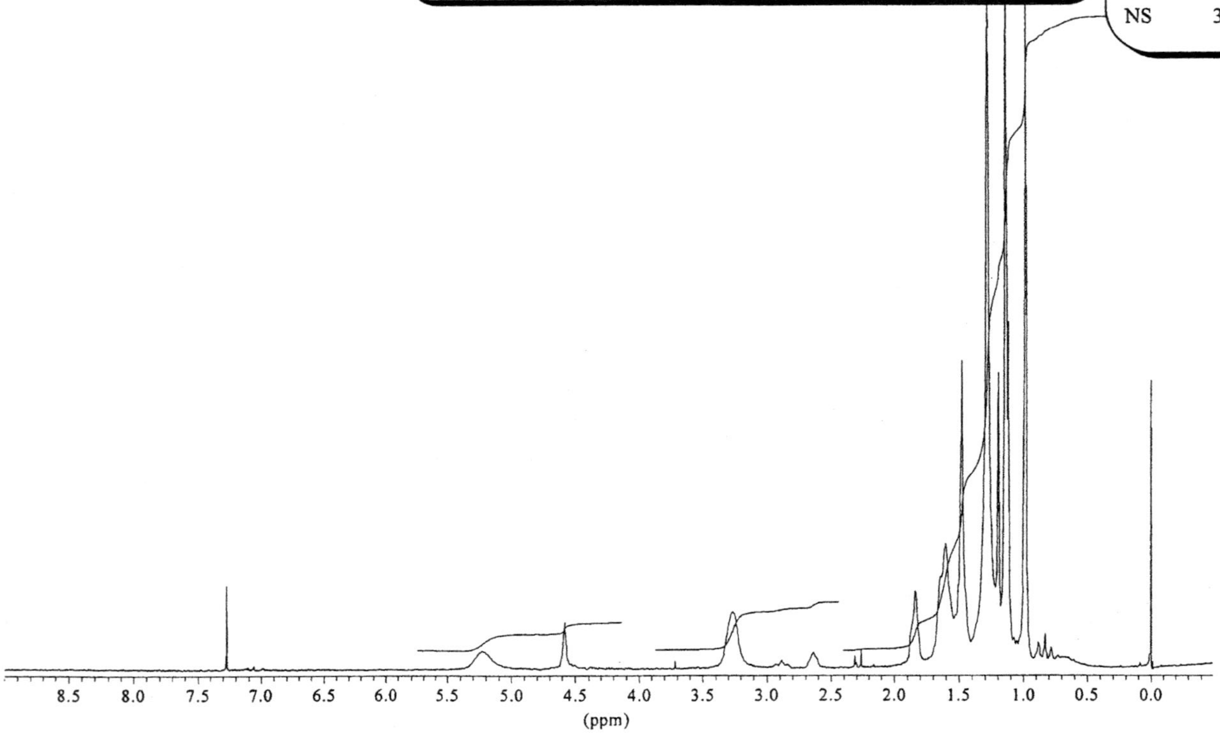

Sebacic acid, bis(2-ethylhexyl) ester

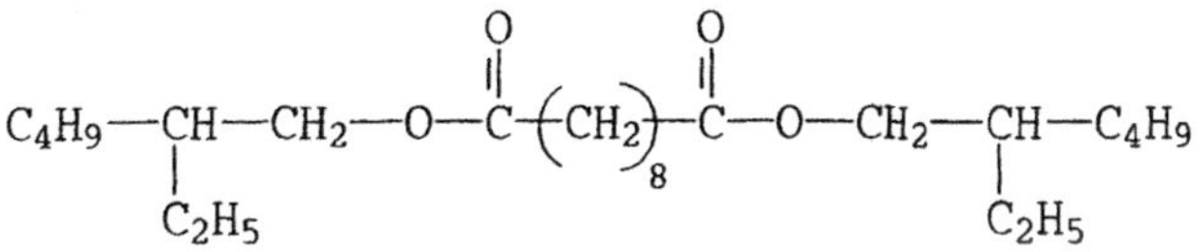

CAS No. - 00122-62-3
PM Ref. No. - 85120
Restrictions - SML = 3.0 mg/kg
Formula - $C_{26}H_{50}O_4$
Molecular weight - 426

Alternative names - Decanedioic acid, bis(2-ethylhexyl) ester
Di(2-ethylhexyl) sebacate

Physical Characteristics - Colourless liquid, mp -48°C,
bp 212°C (1 mm Hg).
- Soluble in alcohol, acetone, benzene, ether.

Handling - Store at room temperature.
Safety -
Availability - Standard sample was supplied.

Current uses - Plasticizer.
Applications - PVC, PVDC, coatings.

GC Retention time - 51.1 min.

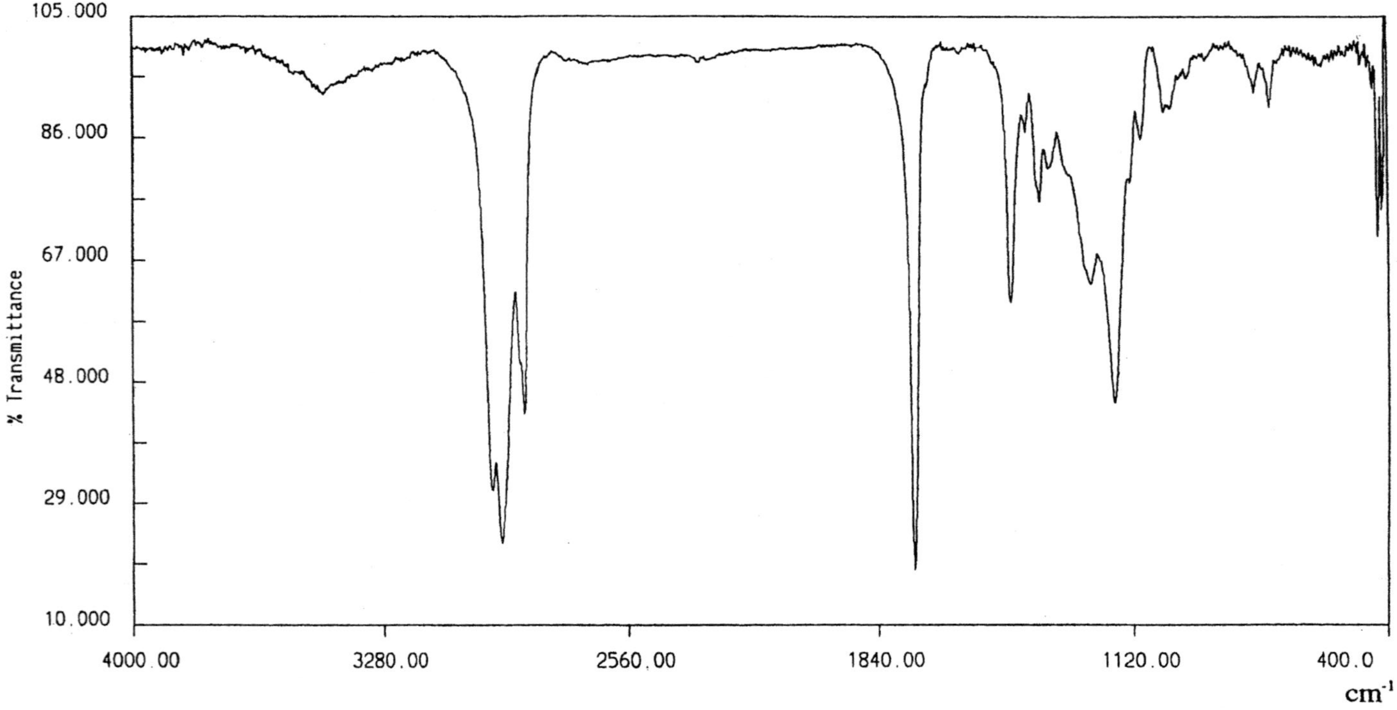

Name : SEBACIC ACID, BIS(2-ETHYLHEXYL) ESTER

PM Nr : 85120 CAS Nr : 00122-62-3

KBr pellet

CAS No. 00122-62-3 PM Ref. No. 85120

SEBACIC ACID, BIS(2-ETHYL-HEXYL) ESTER

Peak wavenumber	% T	Relative intensity
416,7	74,5	31,3
428,3	70,5	36,2
1097,6	85,7	17,5
1130,4	78,8	26,0
1174,8	44,7	67,8
1240,4	63,2	45,1
1358,1	81,0	23,3
1381,2	75,8	29,7
1419,8	86,7	16,3
1464,2	60,2	48,8
1738,1	18,4	100,0
2860,8	42,9	70,0
2930,2	22,5	95,0
2959,2	30,8	84,8

Sebacic acid, bis(2-ethylhexyl) ester

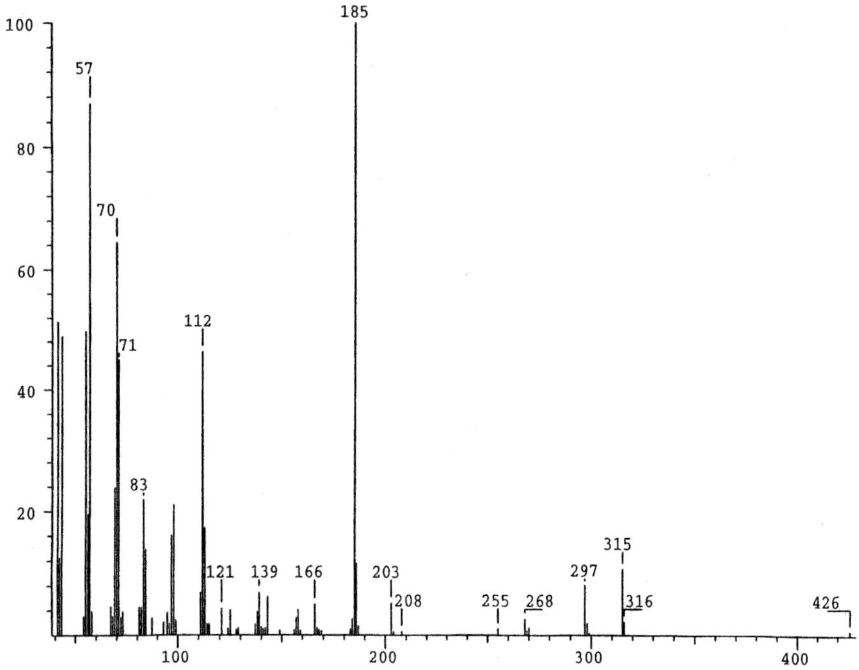

M/Z	Ion Intensity (%)	M/Z	Ion Intensity (%)
43	48.9	98	21.0
55	49.7	112	46.3
56	19.5	113	17.3
57	87.0	185	100.0
69	23.8	186	11.6
70	64.5	203	5.2
71	45.0	297	8.1
83	21.9	315	10.7
84	13.9	426	0.7
97	16.1		

Spectrometer	: Finnigan Mat SSQ 700
Inletsystem	: Capillary GC/MS
Source Temperature	: 180°C
Electron Energy	: 70 eV
Scan Range	: 40-500

PM n°	CAS n°	Name
85120	00122-62-3	SEBACIC ACID, BIS (2-ETHYL-HEXYL) ESTER

Solvent	$CDCL_3$
SF	250.13
SI	16384
SW	3125
NS	32

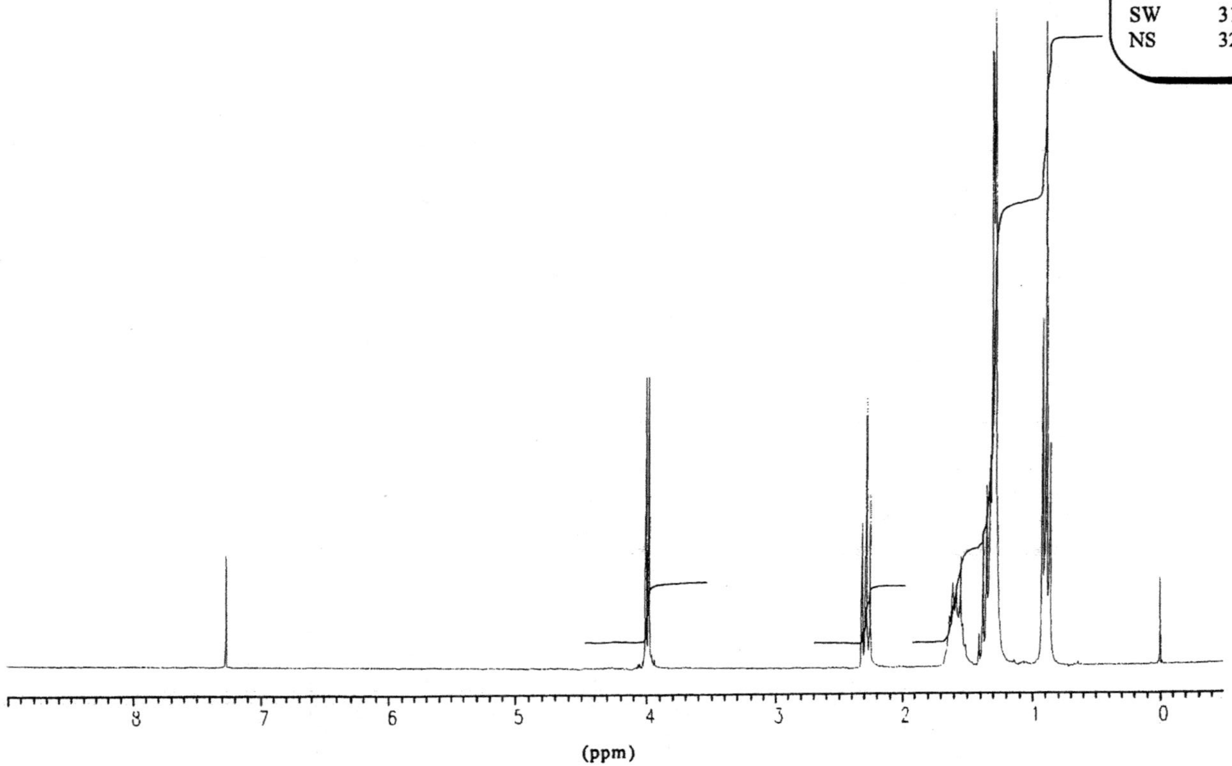

Sebacic acid, dibutyl ester

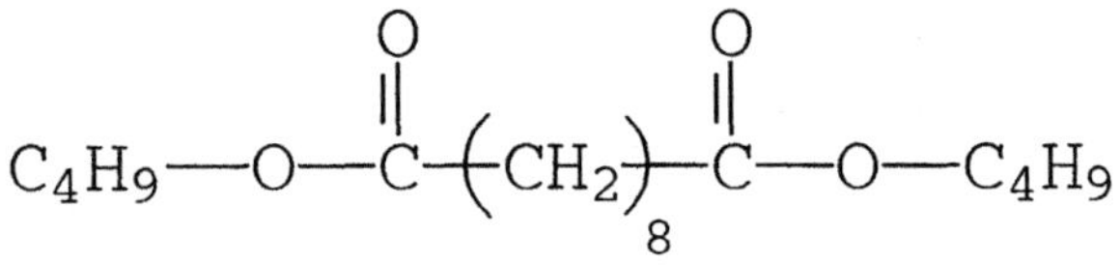

CAS No.	- 00109-43-3
PM Ref. No.	- 85360
Restrictions	- SML = 3.0 mg/kg
Formula	- $C_{18}H_{34}O_4$
Molecular weight	- 314

Alternative names	- Dibutyl sebacate, Decanedioic acid dibutyl ester
Physical Characteristics	- Colourless liquid, mp -10°C, bp 344 - 345°C. - Soluble in ether, alcohol, acetone, benzene.
Handling	- Store at room temperature.
Safety	-
Availability	- Standard sample was supplied.
Current uses	- Plasticizer.
Applications	- PVC, PVDC.
GC Retention time	- 50.0 min.

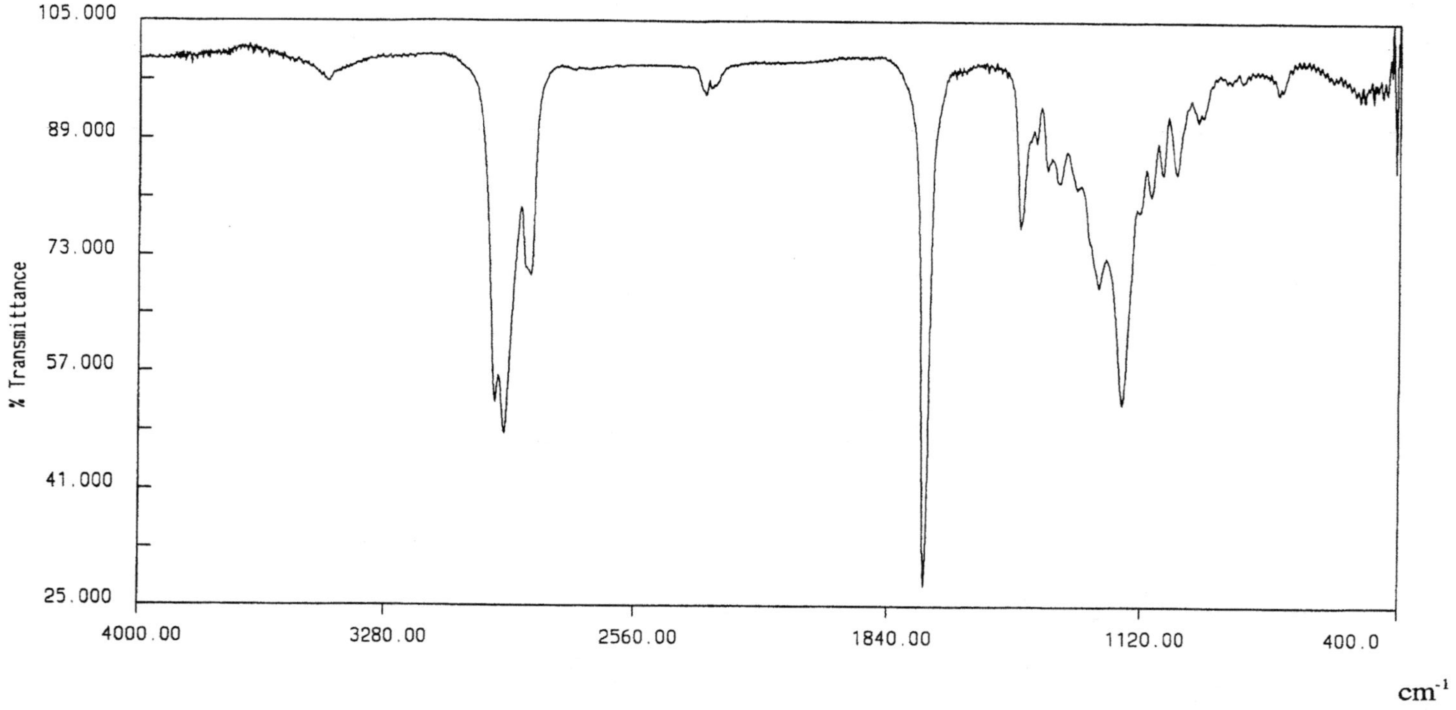

Name : SEBACIC ACID, DIBUTYL ESTER

PM Nr : 85360 CAS Nr : 00109-43-3

KBr pellet

CAS No. 00109-43-3 PM Ref. No. 85360

SEBACIC ACID, DIBUTYL ESTER

Peak wavenumber	% T	Relative intensity
409.0	84.7	21.1
474.6	94.3	7.9
499.6	94.3	7.9
511.2	94.3	7.9
951.0	92.1	11.0
1024.3	84.3	21.8
1064.8	84.2	21.9
1097.6	81.2	26.0
1130.4	79.0	29.0
1174.8	52.6	65.6
1242.3	68.6	43.4
1356.1	83.0	23.5
1419.8	88.5	15.9
1466.1	76.9	32.0
1738.1	27.8	100.0
2858.9	70.2	41.2
2934.1	48.6	71.1
2959.2	52.9	65.2

Sebacic acid, dibutyl ester

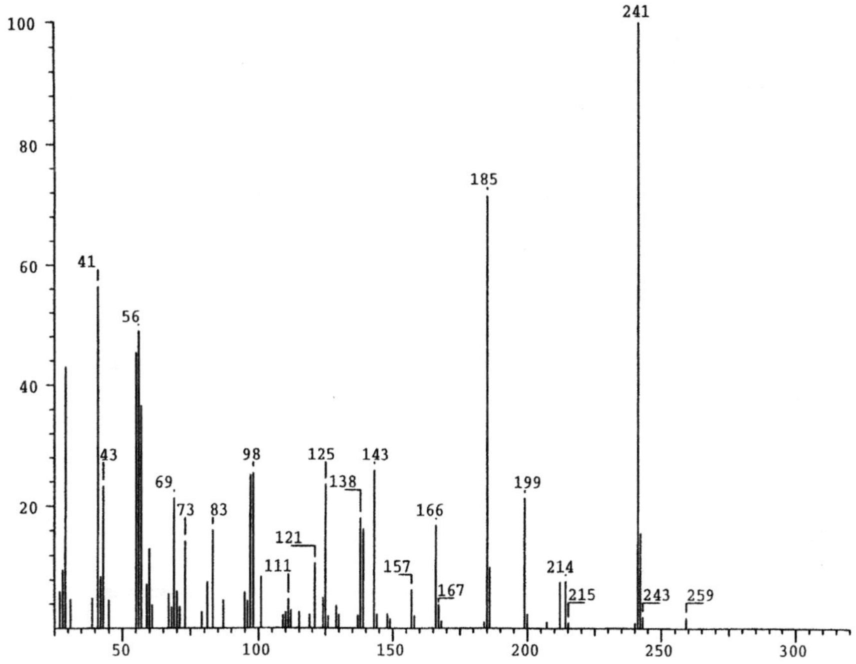

M/Z	Ion Intensity (%)	M/Z	Ion Intensity (%)
41	56.4	125	23.5
55	45.3	143	25.9
56	49.0	166	16.9
57	36.6	185	71.6
69	21.3	186	10.0
83	16.1	199	21.3
97	25.1	241	100.0
98	25.4	242	15.6
121	10.7		

Spectrometer	: Finnigan Mat SSQ 700
Inletsystem	: Capillary GC/MS
Source Temperature	: 180°C
Electron Energy	: 70 eV
Scan Range	: 25-500

PM n°	CAS n°	Name
85360	00109-44-3	SEBACIC ACID, DIBUTYL ESTER

Solvent	$CDCL_3$
SF	250.13
SI	16384
SW	3125
NS	32

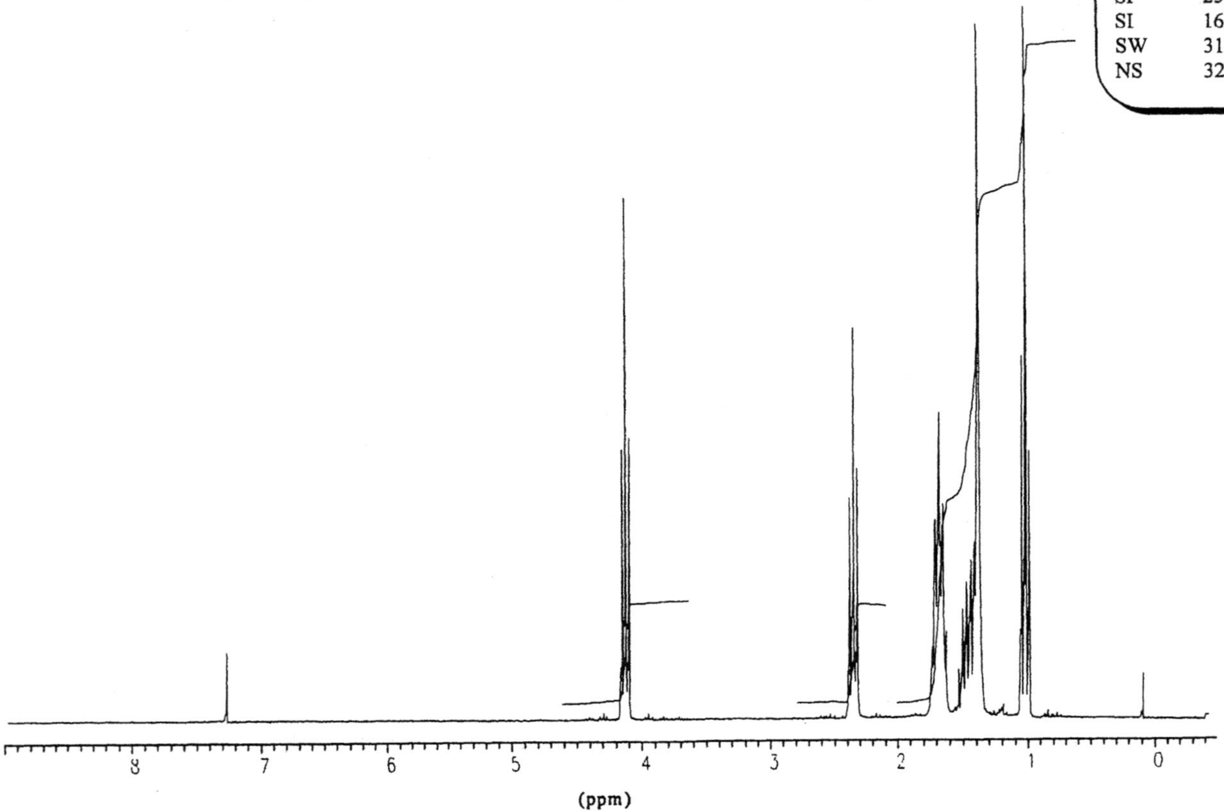

Sorbitan monolaurate

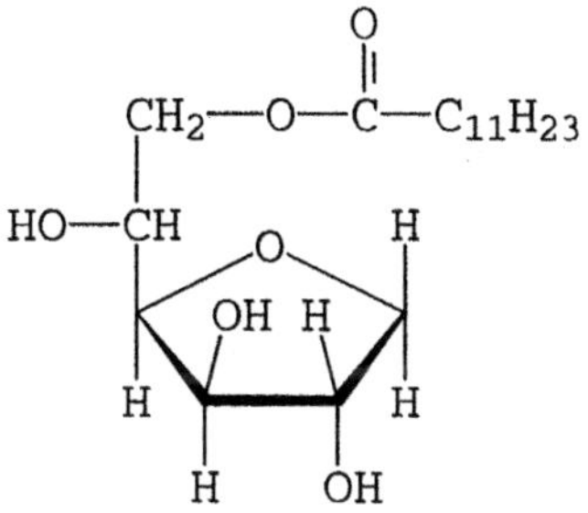

CAS No.	- 01338-39-2
PM Ref. No.	- 87600
Restrictions	- None
Formula	- $C_{18}H_{34}O_6$
Molecular weight	- 346

Alternative names	- Sorbitan monododecanoate
Physical Characteristics	- Yellow liquid, solidification point approx. 19°C. - Soluble in 2-ethoxy ethanol, methanol, ethanol, ether.
Handling	- Store at room temperature.
Safety	-
Availability	- Standard sample supplied.
Current uses	- Emulsifier.
Applications	- General use.
GC Retention time	-

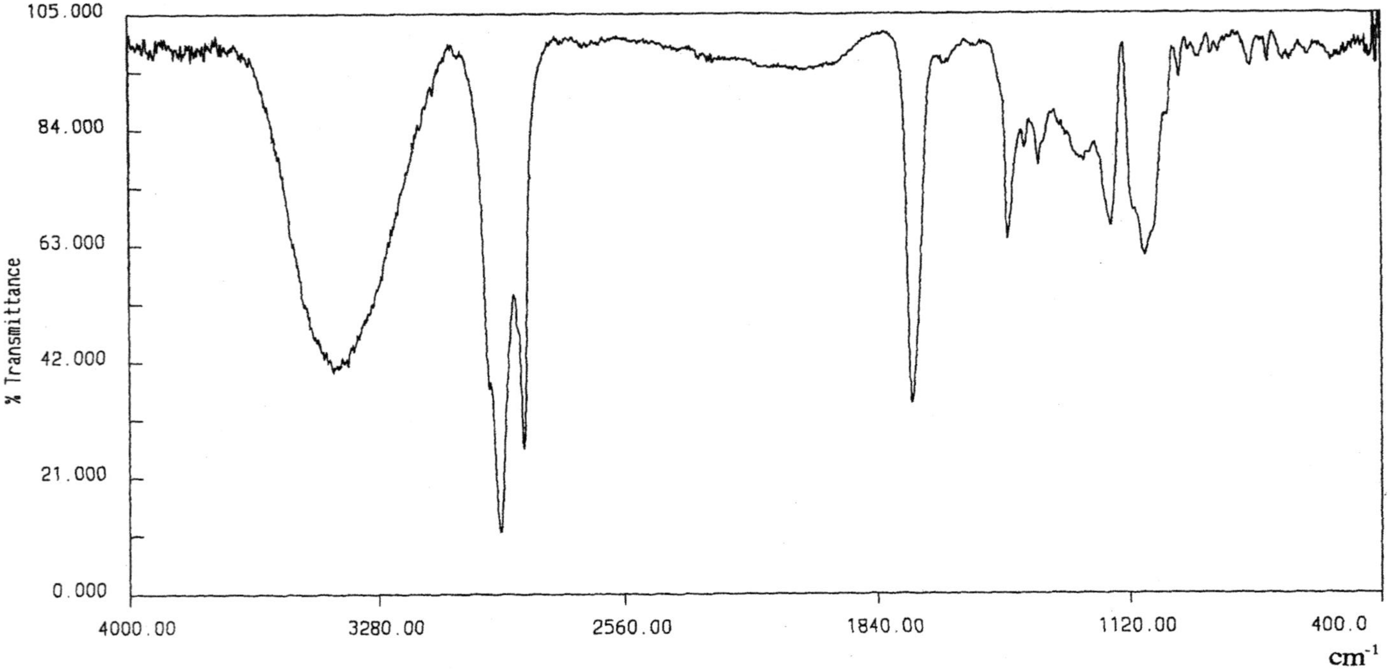

Name : SORBITAN MONOLAURATE

PM Nr : 87600 CAS Nr : 01338-39-2

KBr pellet

CAS No. 01338-39-2 PM Ref. No. 87600

SORBITAN MONOLAURATE

Peak wavenumber	% T	Relative intensity
978.0	93.3	7.5
1012.8	86.3	15.4
1076.4	61.1	43.8
1174.8	66.4	37.8
1379.3	77.5	25.3
1417.9	80.6	21.8
1468.0	64.2	40.3
1741.9	34.5	73.7
2855.0	26.3	82.9
2924.4	11.1	100.0
3159.8	84.0	18.0
3296.8	51.8	54.2
3339.2	46.2	60.5

Sorbitan monolaurate

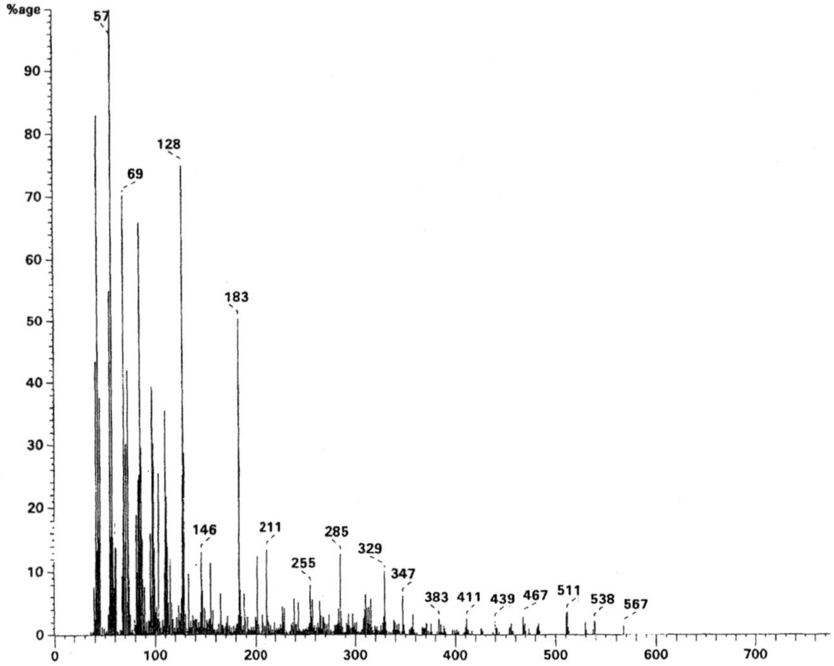

M/Z	Ion Intensity (%)	M/Z	Ion Intensity (%)
41	43.3	85	66.0
43	83.3	86	29.3
45	37.3	97	39.3
55	54.7	98	30.0
57	100.0	103	25.3
69	70.7	110	35.3
71	30.0	111	21.3
73	42.0	128	75.3
83	90.7	129	28.7
84	91.3	183	50.7

Spectrometer	: VG Autospec
Inletsystem	: Probe inlet
Source Temperature	: 350-400°C
Electron Energy	: 70 eV
Scan Range	: 40-1000

PM n°	CAS n°	Name
87600	01338-39-2	SORBITAN MONOLAURATE

Solvent	$CDCL_3$
SF	250.13
SI	16384
SW	2380
NS	32

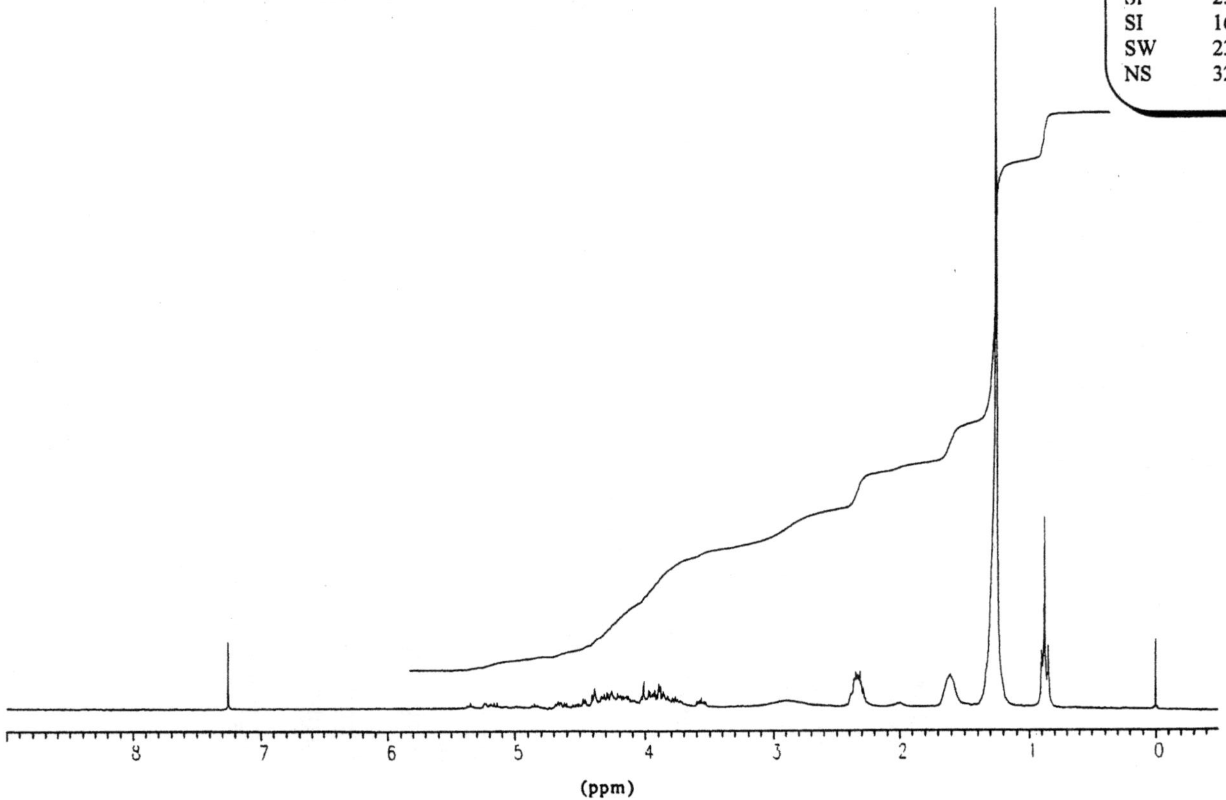

Sorbitan monooleate

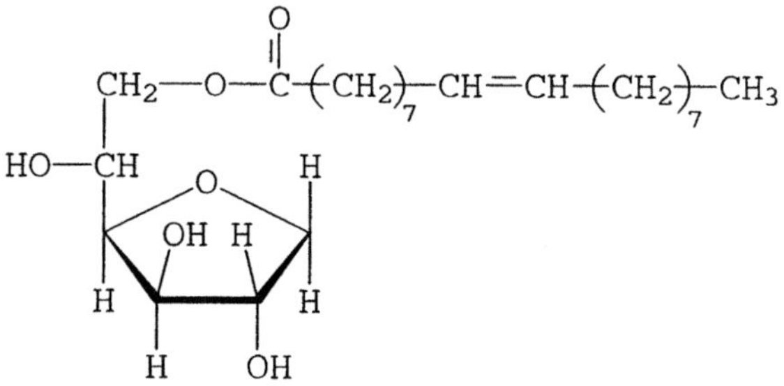

CAS No.	- 01338-43-8
PM Ref. No.	- 87680
Restrictions	- None
Formula	- $C_{24}H_{44}O_6$
Molecular weight	- 428

Alternative names	- Sorbitan mono-9-octadenenoate
Physical Characteristics	- Yellow liquid. - Soluble in 2-ethoxy ethanol, methanol, ethanol, ether.
Handling	- Store at room temperature.
Safety	-
Availability	- Standard sample supplied.
Current uses	- Emulsifier
Applications	- General use.
GC Retention time	-

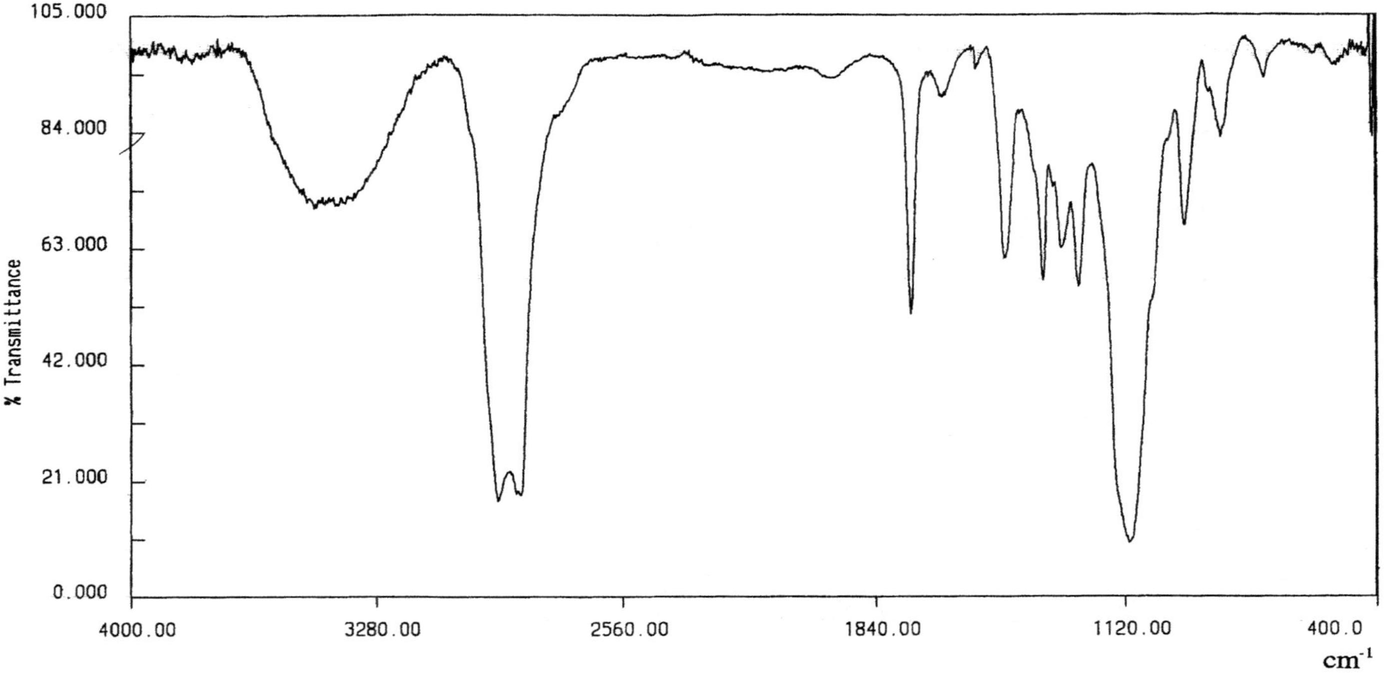

Name : SORBITAN MONOOLEATE

PM Nr : 87680 CAS Nr : 01338-43-8

KBr pellet

CAS No. 01338-43-8 Pm Ref. No. 87680

SORBITAN MONOOLEATE

Peak wavenumber	% T	Relative intensity
410.9	83.2	18.6
723.4	93.6	7.1
846.9	83.1	18.7
949.1	67.0	36.5
1109.2	9.7	100.0
1250.0	56.0	48.8
1300.2	62.9	41.1
1350.3	57.1	47.6
1462.2	61.1	43.1
1738.1	51.0	54.3
2754.7	86.5	14.9
2860.8	18.4	90.4
2924.4	17.4	91.5
3238.9	83.6	18.2
3329.5	73.2	29.7

Sorbitan monooleate

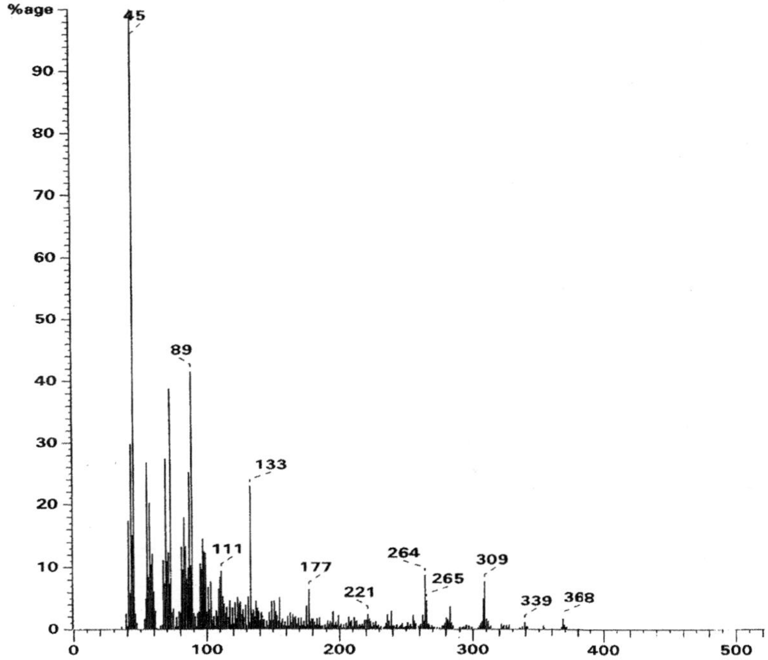

M/Z	Ion Intensity (%)	M/Z	Ion Intensity (%)
41	17.3	73	38.7
43	30.0	81	13.3
45	100.0	83	18.0
50	10.0	84	13.3
55	26.7	87	25.3
57	20.0	89	41.3
59	12.0	97	14.7
69	27.3	98	12.7
70	10.7	99	12.0
71	12.0	133	22.7

Spectrometer	: VG Autospec
Inletsystem	: Probe inlet
Source Temperature	: 350-400°C
Electron Energy	: 70 eV
Scan Range	: 40-1000

PM n°
87680

CAS n°
01338-43-8

Name
SORBITAN MONOOLEATE

Solvent	$CDCL_3$
SF	250.13
SI	16384
SW	3125
NS	32

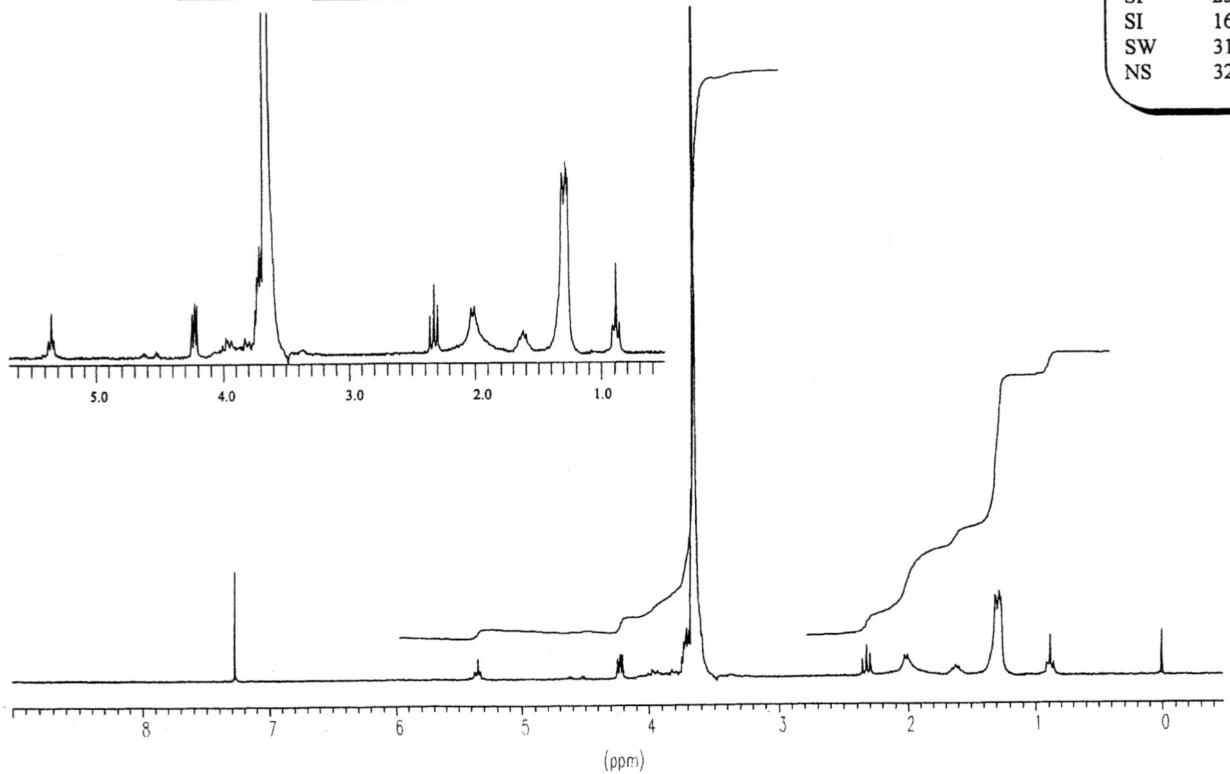

Sorbitan monostearate

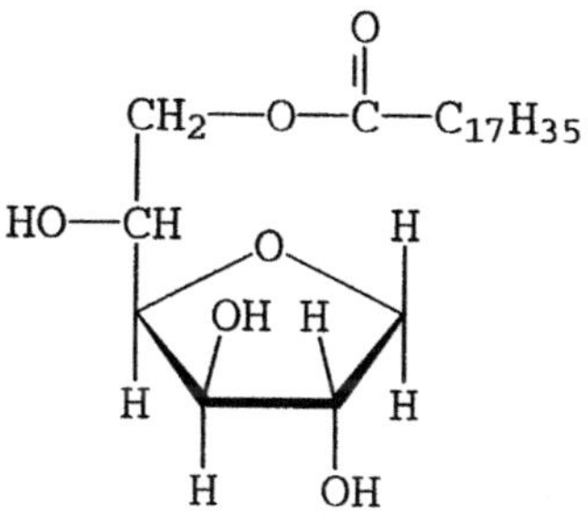

CAS No.	- 01338-41-6
PM Ref. No.	- 87840
Restrictions	- None
Formula	- $C_{24}H_{46}O_6$
Molecular weight	- 430

Alternative names	- Sorbitan monooctadecanoate
Physical Characteristics	- Pale waxy beads, mp 56 - 58°C. - Soluble in 2-ethoxy ethanol, methanol, ethanol, ether.
Handling	- Store at room temperature.
Safety	- Low toxicity.
Availability	- Standard sample supplied.
Current uses	- Emulsifier.
Applications	- General use.
GC Retention time	-

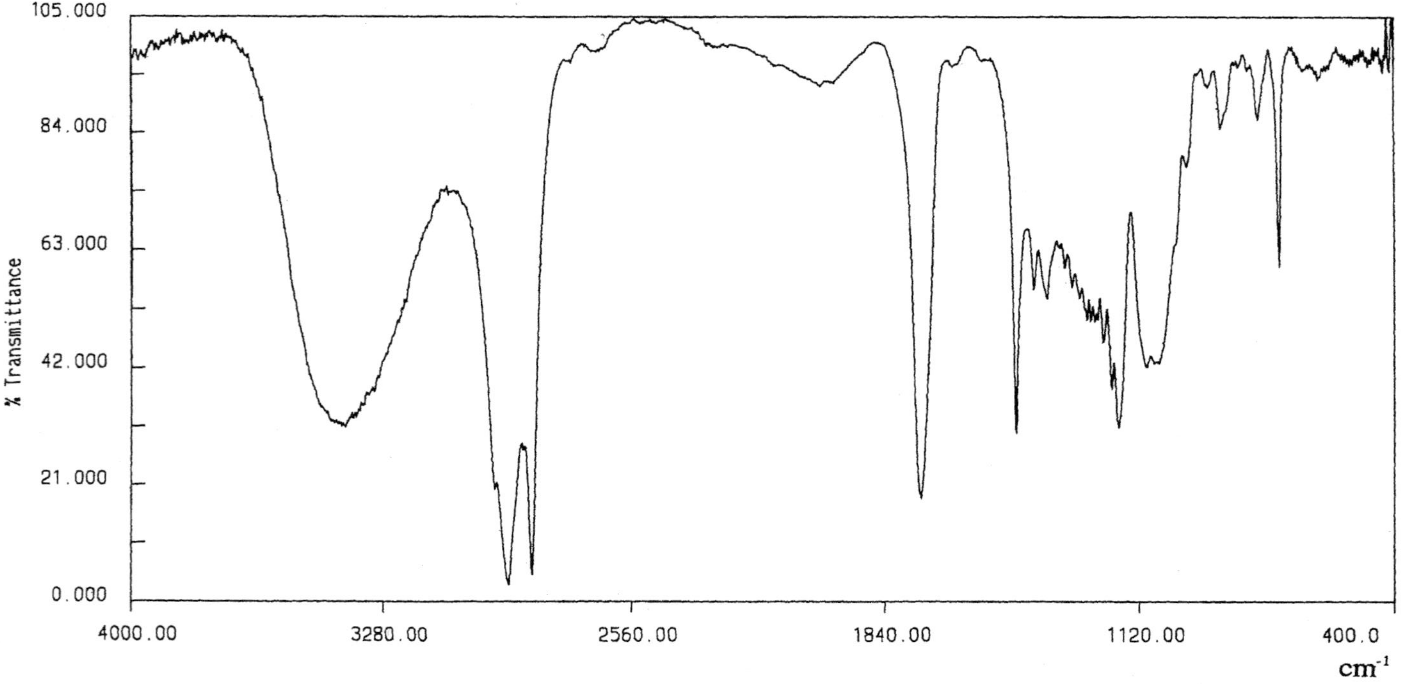

Name : SORBITAN MONOSTEARATE

PM Nr : 87840 CAS Nr : 01338-41-6

KBr pellet

CAS No. 01338-41-6 PM Ref. No. 87840

SORBITAN MONOSTEARATE

Peak wavenumber	% T	Relative intensity
611.5	93.4	6.8
721.5	60.2	41.0
781.3	86.4	14.0
887.4	84.9	15.6
922.1	92.1	8.1
983.8	77.8	22.9
1064.8	42.9	58.9
1178.7	31.4	70.7
1197.9	38.2	63.7
1221.1	46.7	55.0
1244.2	50.3	51.2
1379.3	54.4	47.0
1417.9	56.1	45.2
1468.0	30.3	71.8
1738.1	18.5	83.9
2851.1	4.8	98.1
2916.7	3.0	100.0
3308.3	37.67	64.2

Sorbitan monostearate

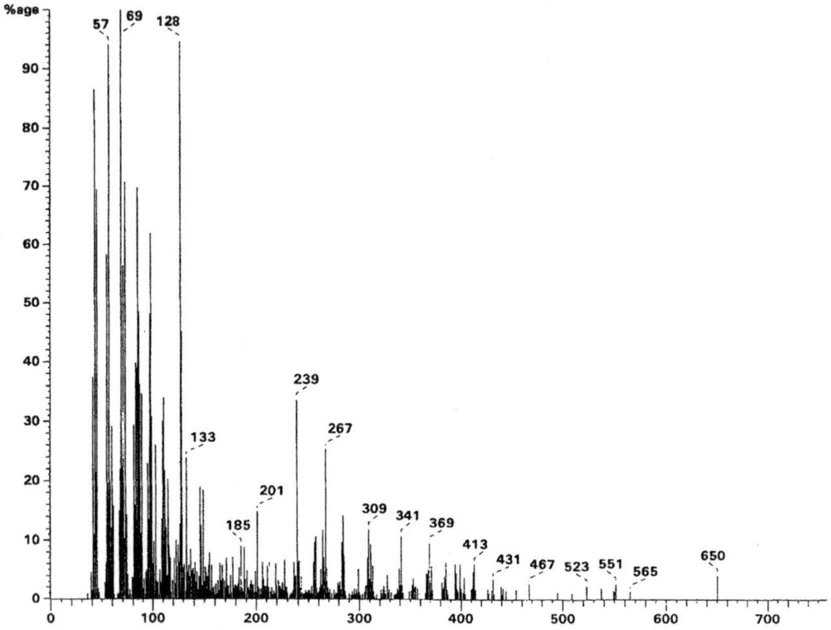

M/Z	Ion Intensity (%)	M/Z	Ion Intensity (%)
41	33.3	85	70.0
43	86.7	87	36.0
45	70.0	89	34.7
55	58.0	97	48.0
57	94.0	98	62.0
69	100.0	99	30.7
71	56.0	111	34.0
73	70.7	128	95.3
83	40.0	129	45.3
84	25.3	239	33.3

Spectrometer	: VG Autospec
Inletsystem	: Probe inlet
Source Temperature	: 350-400°C
Electron Energy	: 70 eV
Scan Range	: 40-1000

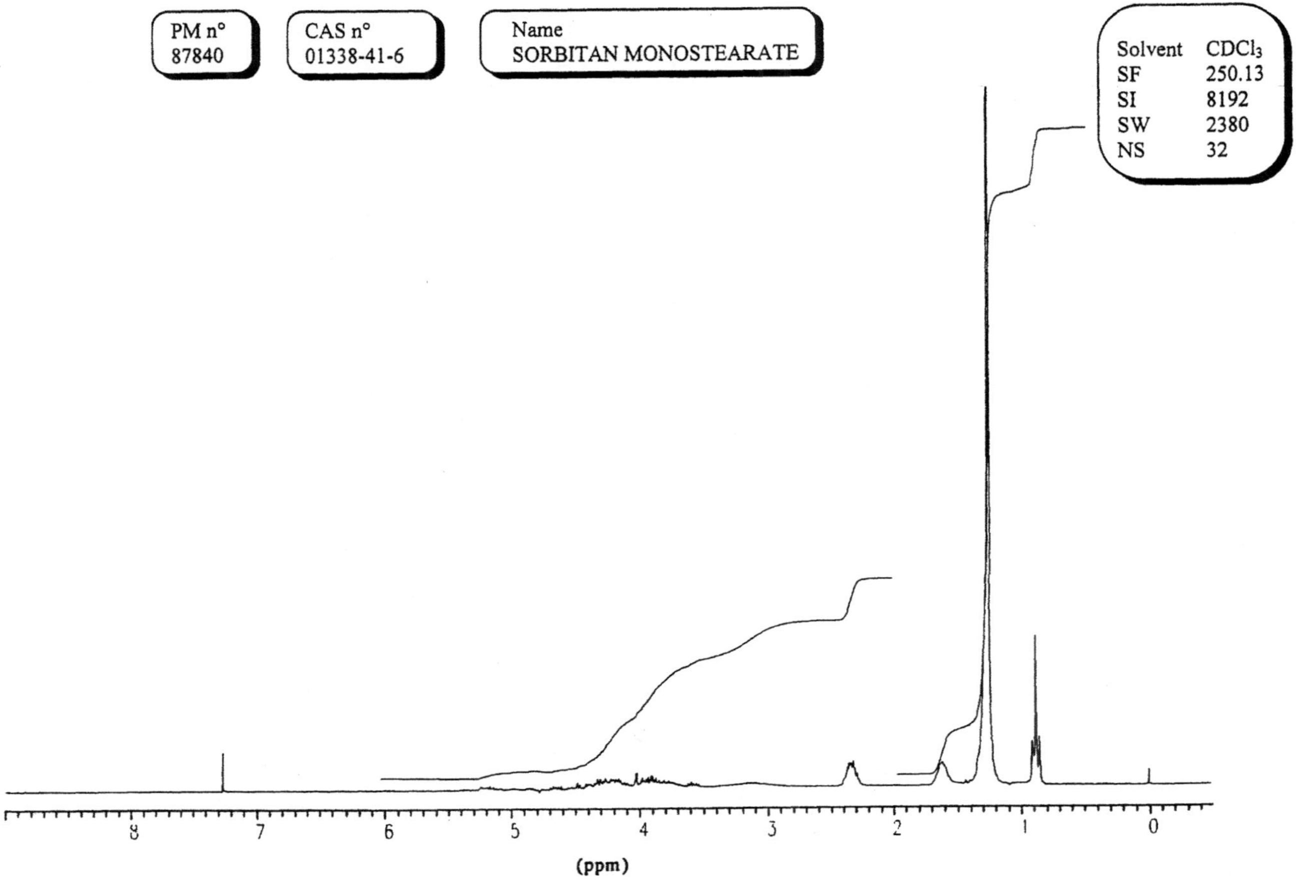
PM n°
87840
CAS n°
01338-41-6
Name
SORBITAN MONOSTEARATE
Solvent CDCl3
SF 250.13
SI 8192
SW 2380
NS 32
8
7
6
5
4
3
2
1
0
(ppm)

Stearamide

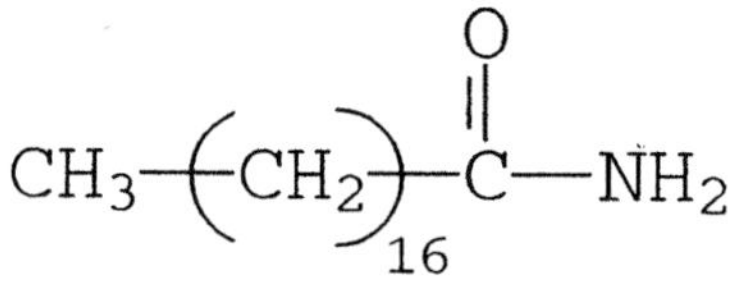

CAS No.	- 00124-26-5
PM Ref. No.	- 88960
Restrictions	- None
Formula	- $C_{18}H_{37}NO$
Molecular weight	- 283

Alternative names - Octadecanoic acid, amide

Physical Characteristics - Slightly yellow cristalline powder, mp 109°C, bp 250 - 251°C.
- Soluble in most organic solvents.

Handling - Store at room temperature.

Safety -

Availability - Standard sample was supplied.

Current uses - Antistatic agent, anti- slipping and blocking agent.

Applications - Polyolefins.

GC Retention time - 44.5 min.

Note - Stearamide contains palmitamide.

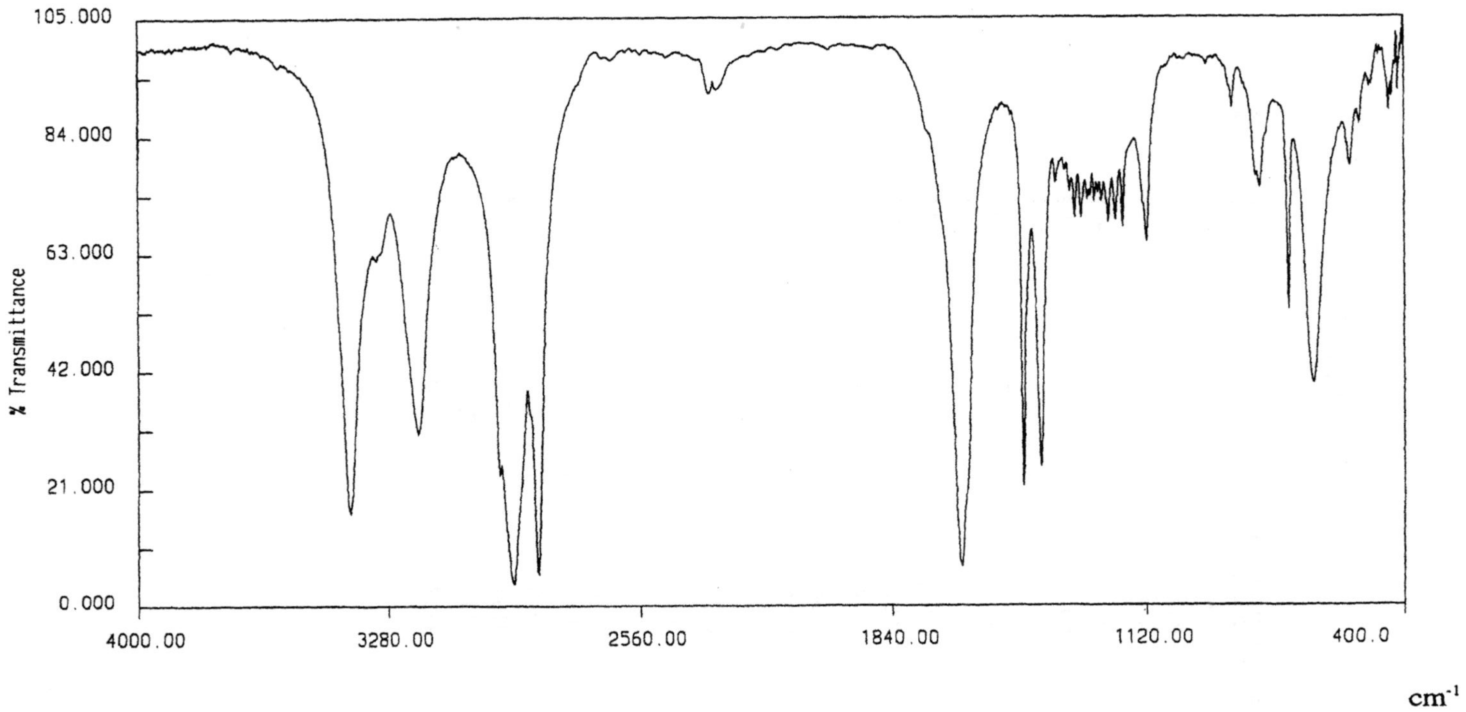

Name : STEARAMIDE

PM Nr : 88960 CAS Nr : 00124-26-5

KBr pellet

CAS No. 00124-26-5 PM Ref. No. 88960

STEARAMIDE

Peak wavenumber	% T	Relative intensity
434.0	90.8	9.6
441.8	88.4	12.1
524.7	85.9	14.7
549.8	78.7	22.2
650.1	39.7	62.9
719.5	52.9	49.1
800.6	74.7	26.3
1120.8	65.1	36.4
1188.3	67.7	33.6
1207.6	68.9	32.4
1228.8	68.9	32.4
1286.7	72.5	28.7
1306.0	69.2	32.1
1323.3	69.4	31.9
1377.3	75.6	25.4
1421.7	25.0	78.2
1469.9	21.5	81.9
1643.6	7.0	96.9
2849.2	5.8	98.2
2918.7	4.0	100.0
3194.5	30.9	72.0
3316.0	62.0	39.6
3393.2	16.8	86.7

Stearamide

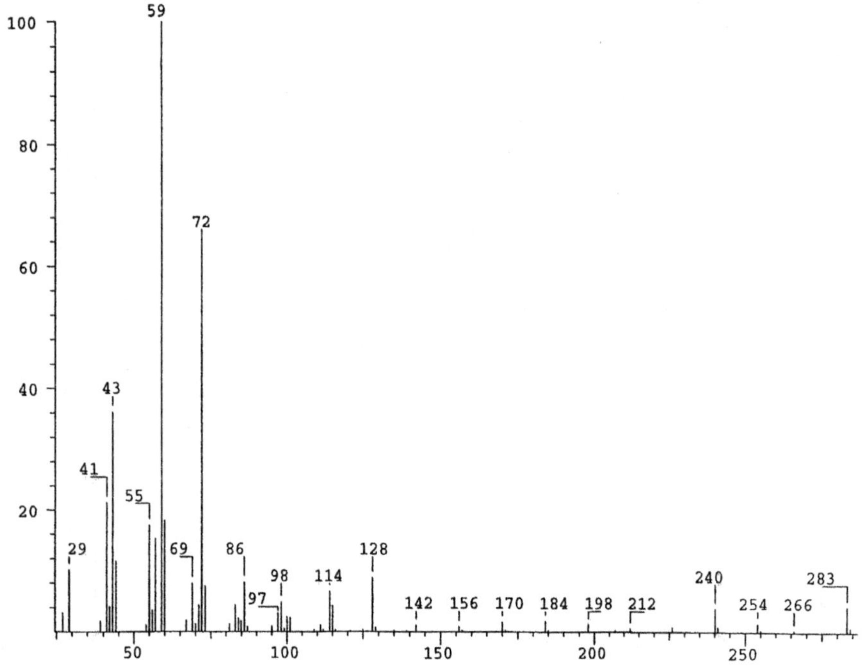

M/Z	Ion Intensity (%)	M/Z	Ion Intensity (%)
29	10.2	60	18.3
41	21.3	72	64.9
43	36.2	128	8.9
44	11.6	240	3.9
55	17.5	283	4.3
57	15.3	284	0.7
59	100.0		

Spectrometer	: Finnigan Mat SSQ 700
Inletsystem	: Capillary GC/MS
Source Temperature	: 180°C
Electron Energy	: 70 eV
Scan Range	: 25-500

PM n°
88960

CAS n°
00124-26-5

Name
STEARAMIDE

Solvent	$CDCL_3$
SF	250.13
SI	16384
SW	3125
NS	32

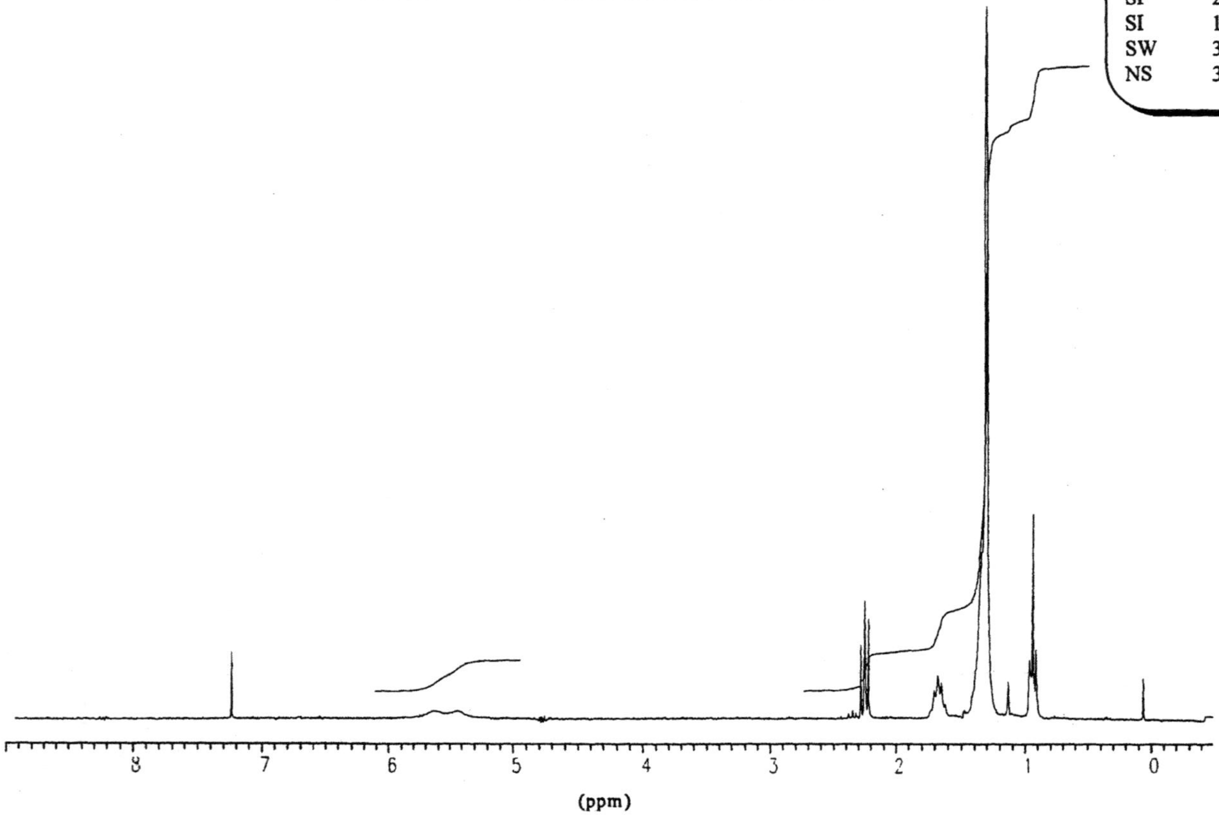

Stearic acid, butyl ester

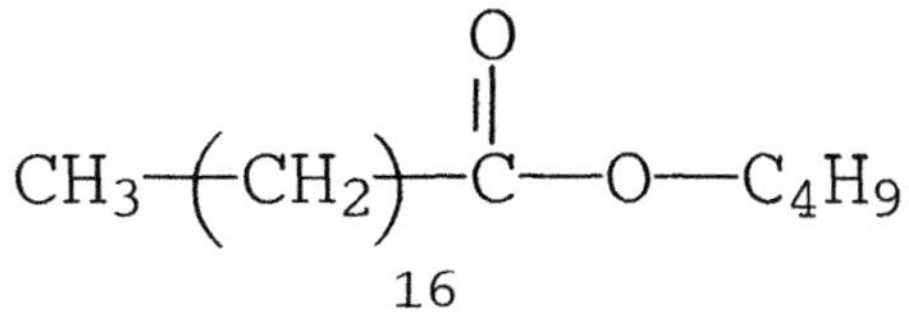

CAS No.	- 00123-95-5
PM Ref. No.	- 89120
Restrictions	- None
Formula	- $C_{22}H_{44}O_2$
Molecular weight	- 340

Alternative names	- Butyl stearate, Octadecanoic acid, butyl ester
Physical Characteristics	- Colourless liquid, mp 16°C, bp 223°C. - Soluble in alcohol, ether, acetone.
Handling	- Store at room temperature.
Safety	-
Availability	- Standard sample was supplied.
Current uses	- Lubricant.
Applications	- General use.
GC Retention time	- 44.9 min.
Note	- Contains Palmatic acid, butyl ester.

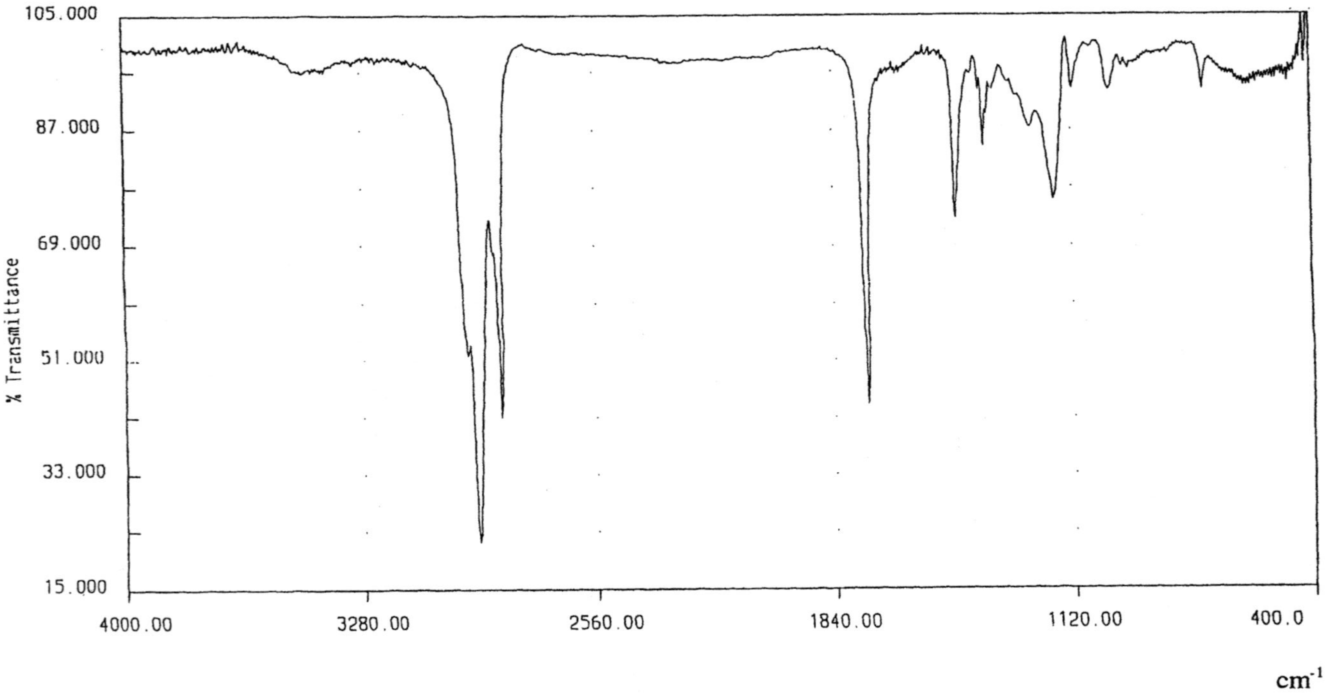

Name : STEARIC ACID, BUTYL ESTER

PM Nr : 89120 CAS Nr : 00123-95-5

KBr pellet

CAS No. 00123-95-5 PM Ref. No. 89120

STEARIC ACID,BUTYL ESTER

Peak wavenumber	% T	Relative intensity
474.6	94.8	6.7
509.3	94.6	6.9
526.6	94.7	6.8
534.4	94.4	7.2
544.0	94.4	7.2
551.7	94.6	6.9
561.4	94.1	7.6
571.0	94.2	7.5
1172.9	75.8	31.1
1379.3	84.3	20.2
1468.0	73.1	34.6
1741.9	43.9	72.1
2855.0	41.8	74.8
2926.4	22.2	100.0

Stearic acid, butyl ester

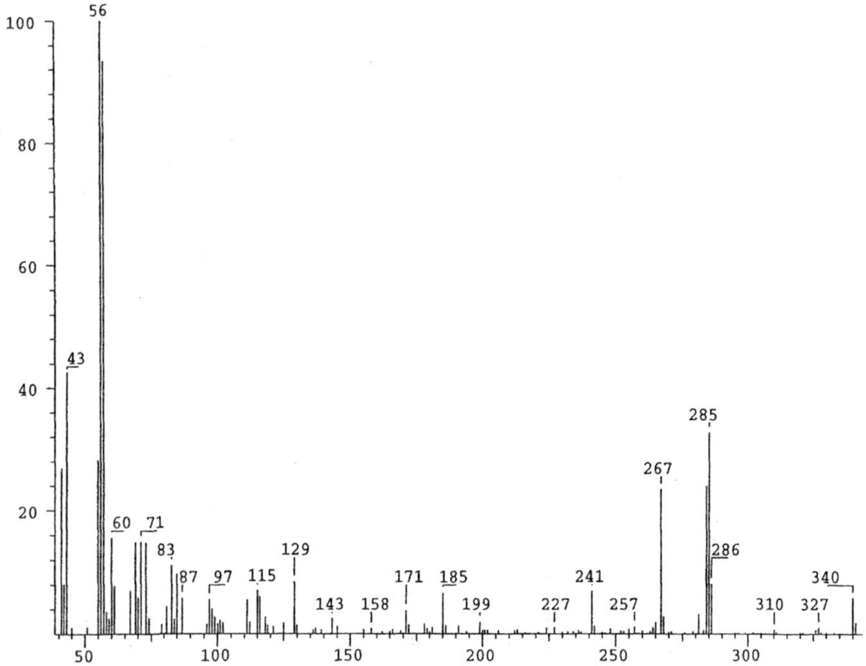

M/Z	Ion Intensity (%)	M/Z	Ion Intensity (%)
41	26.9	171	3.8
42	8.1	185	6.6
43	42.6	199	1.9
55	28.2	241	6.9
56	100.0	267	23.4
57	93.4	284	24.0
70	5.8	285	32.7
71	14.9	286	8.0
115	7.1	340	5.8
129	8.4	341	1.7

Spectrometer	: Finnigan Mat SSQ 700
Inletsystem	: Capillary GC/MS
Source Temperature	: 180°C
Electron Energy	: 70 eV
Scan Range	: 40-500

PM n°
89120

CAS n°
00123-95-5

Name
STEARIC ACID, BUTYL ESTER

Solvent	$CDCL_3$
SF	250.13
SI	16384
SW	6494
NS	16

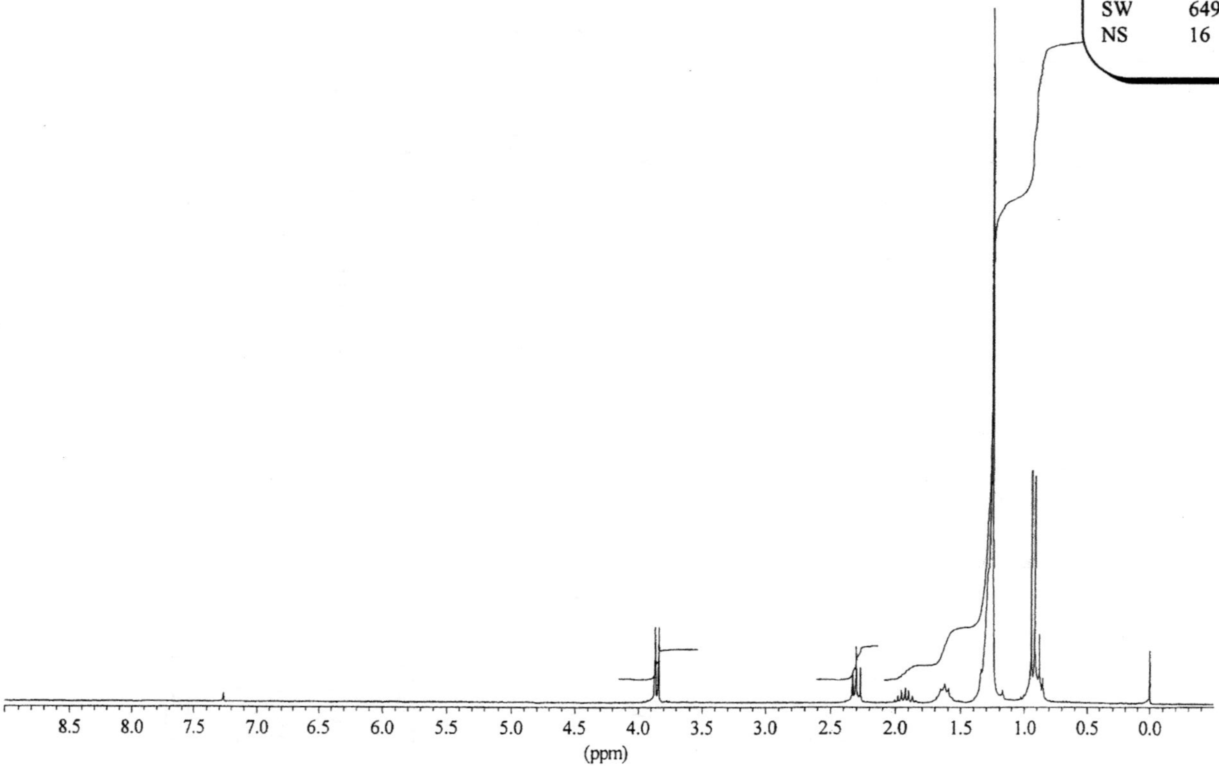

Stearic acid, esters with pentaerythritol

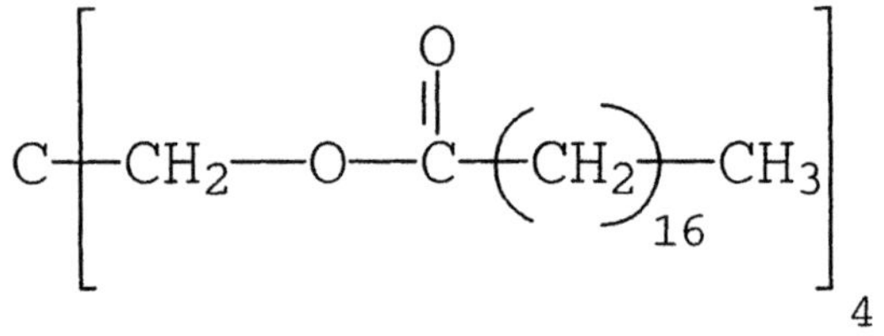

CAS No.	- 08045-34-9 and 00115-83-3
PM Ref. No.	- 89520
Restrictions	- None
Formula	- $C_{77}H_{148}O_8$
Molecular weight	- 1202

Alternative names - Pentaerythritolstearate

Physical Characteristics
- White crystalline powder, mp approx. 60 °C, bp approx. 280°C.
- Soluble in dichloromethane, chloroform, ether, toluene, hexane.

Handling - Store at room temperature.
Safety -
Availability - Standard sample was supplied.

Current uses - Release agent.
Applications - General use.

GC Retention time -

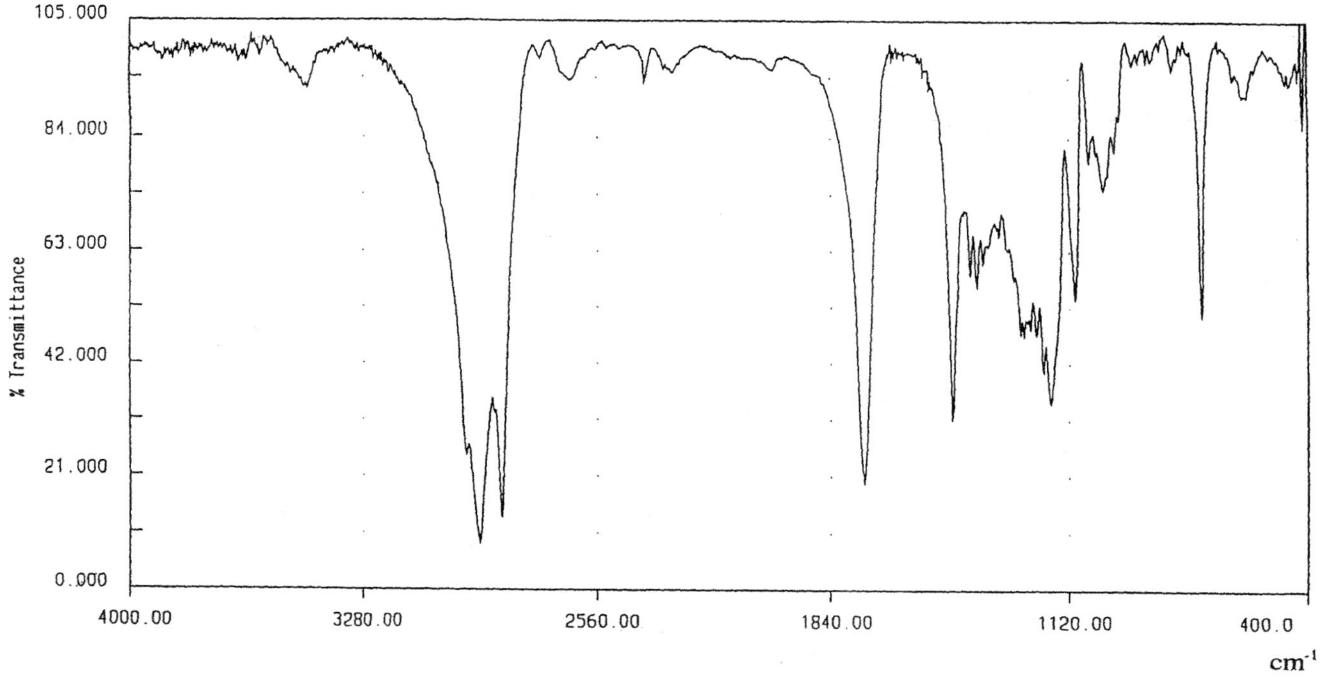

Name : STEARIC ACID, ESTERS WITH PENTAERYTHRITOL

PM Nr : 89520 CAS Nr : 08045-34-9

KBr pellet

CAS No. 08045-34-9 and 00115-83-3 PM Ref. No. 89520

STEARIC ACID, ESTERS WITH PENTAERYTHRITOL

Peak wavenumber	% T	Relative intensity
557.5	94.4	6.1
580.6	89.5	11.5
719.5	49.6	55.0
877.7	86.5	14.7
889.3	83.3	18.2
985.7	68.8	34.1
1018.5	60.6	43.0
1062.9	62.2	41.3
1105.4	42.2	63.1
1174.8	28.1	78.5
1197.9	33.2	72.9
1219.2	39.4	66.2
1236.5	40.7	64.7
1361.9	47.4	57.4
1396.6	51.6	52.8
1415.9	54.1	50.1
1471.9	30.0	76.4
1738.1	19.8	87.6
2851.1	13.0	95.0
2916.7	8.4	100.0

PM n°	CAS n°	Name
89520	08045-34-9	STEARIC ACID, ESTERS WITH PENTAERYTRITOL

Solvent	$CDCL_3$
SF	250.13
SI	16384
SW	6494
NS	50

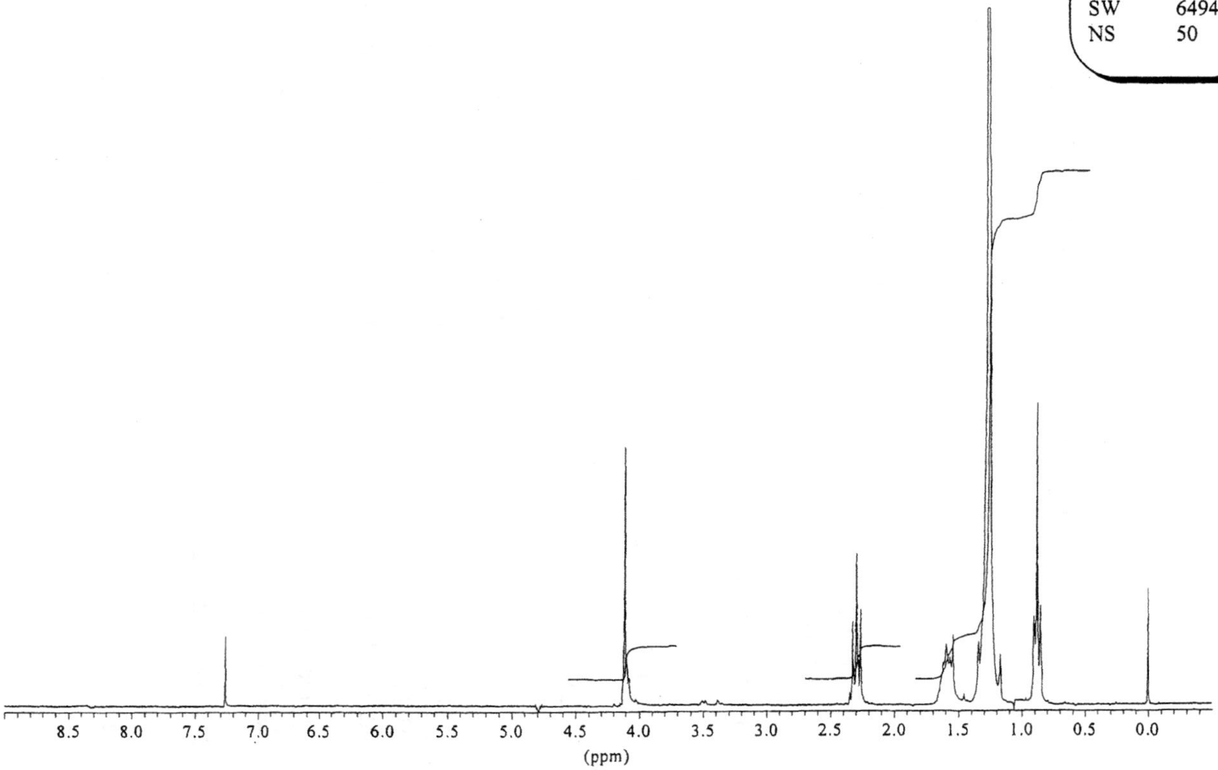

Stearoylbenzoylmethane

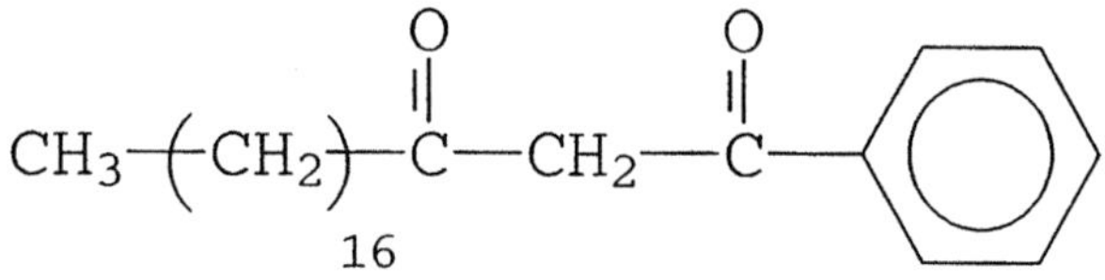

CAS No.	- 58446-52-9
PM Ref. No.	- 90720
Restrictions	- None
Formula	- $C_{26}H_{42}O_2$
Molecular weight	- 386

Alternative names	- 1-Benzoyl-2-keto-nonadecane
Physical Characteristics	- White powder, mp 56 - 61°C, bp 300°C (21 mm Hg). - Soluble in most organic solvents.
Handling	- Store at room temperature.
Safety	-
Availability	- Standard sample was supplied.
Current uses	- Stabilizer.
Applications	- PVC, PVDC.
GC Retention time	- 54.7 min.
Note	- Contains also laurylbenzoylmethane.

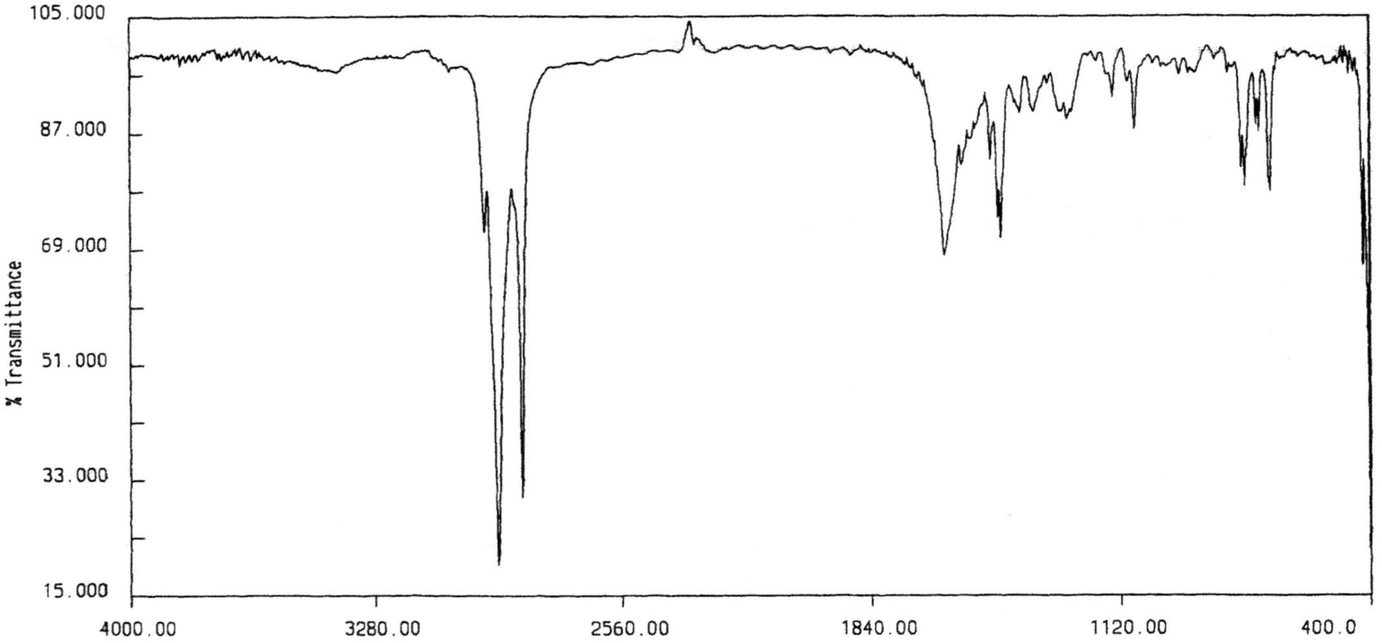

cm^{-1}

Name : STEAROYLBENZOYLMETHANE

PM Nr : 90720 CAS Nr : 58446-52-9

KBr pellet

CAS No. 58446-52-9 PM Ref. No. 90720

STEAROYLBENZOYLMETHANE

Peak wavenumber	% T	Relative intensity
420.5	66.5	41.8
729.2	88.2	14.6
762.0	78.8	26.4
771.6	81.7	22.8
1464.2	70.6	36.6
1473.8	73.9	32.6
1493.1	82.8	21.4
1547.1	86.0	17.5
1576.0	82.0	22.4
1626.2	67.9	40.1
2849.2	30.2	87.0
2916.7	19.8	100.0
2955.3	71.6	35.4

Stearoylbenzoylmethane

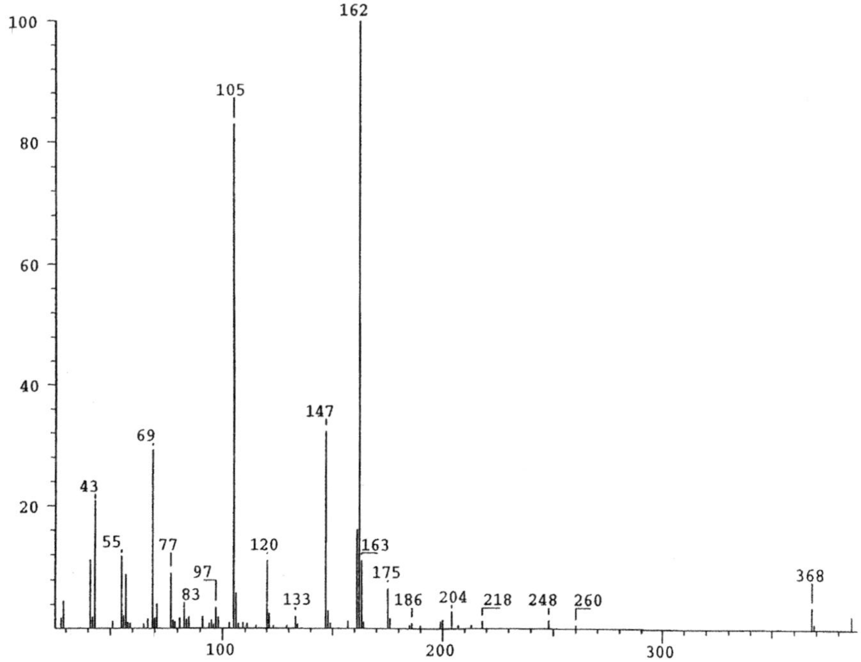

M/Z	Ion Intensity (%)	M/Z	Ion Intensity (%)
41	11.2	147	32.3
43	20.9	161	16.1
55	11.8	162	100.0
69	29.2	163	11.0
77	9.15	175	6.6
105	83.0	368	3.7
120	11.1	386	2.2

Spectrometer	: Finnigan Mat SSQ 700
Inletsystem	: Capillary GC/MS
Source Temperature	: 180°C
Electron Energy	: 70 eV
Scan Range	: 25-500

PM n°	CAS n°	Name
90720	58446-52-9	STEAROYLBENZOYLMETHANE

Solvent	$CDCL_3$
SF	250.13
SI	16384
SW	6494
NS	72

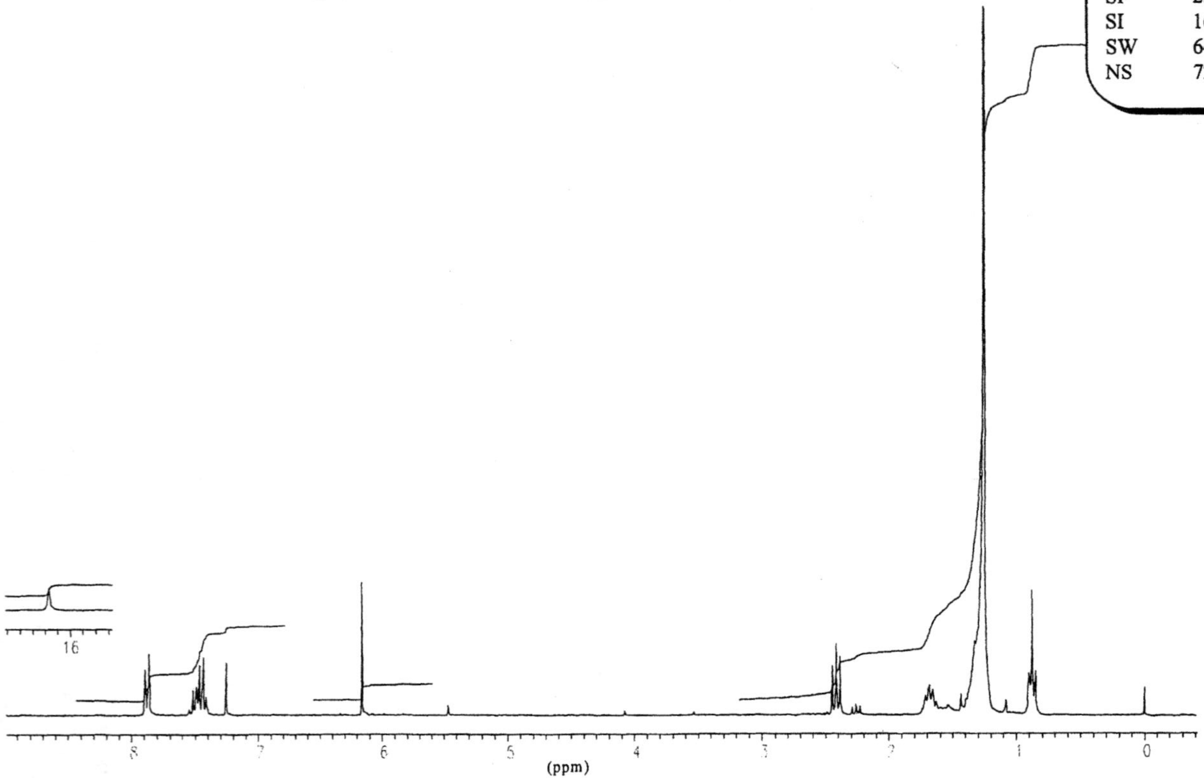

Sulphosuccinic acid, bis(2-ethylhexyl)ester, sodium salt

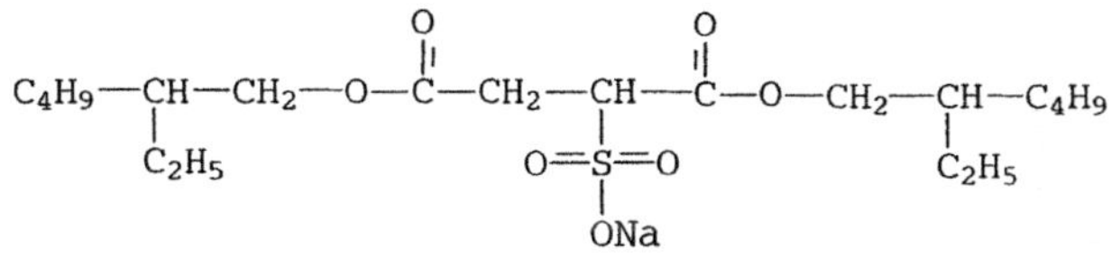

CAS No.	- 00577-11-7
PM Ref. No.	- 91572
Restrictions	- None
Formula	- $C_{20}H_{37}NaO_7S$
Molecular weight	- 444

Alternative names	- Docusate sodium
Physical Characteristics	- Yellow liquid, mp < 0°C, bp 360°C. - Soluble in most organic solvents.
Handling	- Store at room temperature.
Safety	- Irritant.
Availability	- Preparation in alcoholic solution.
Current uses	- Emulsifier, antistatic agent.
Applications	- Styrene co-polymers, coatings.
GC Retention time	-

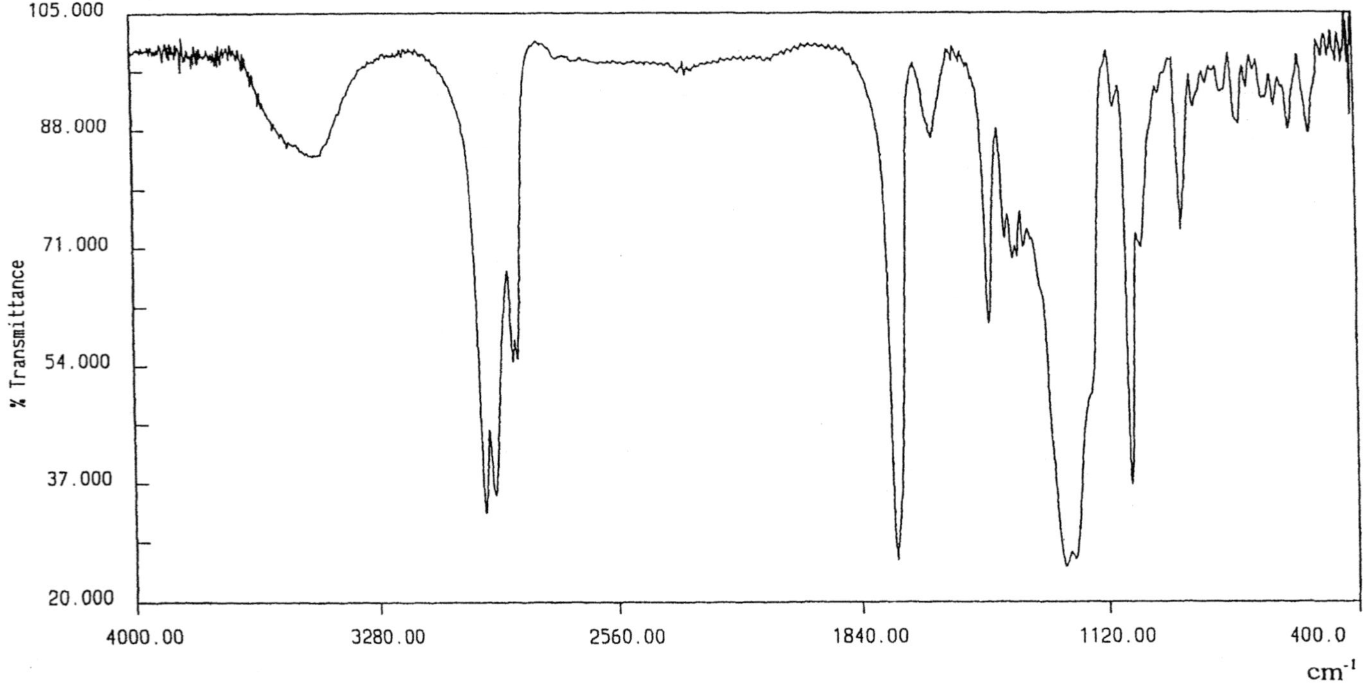

Name : SULPHOSUCCINIC ACID, BIS(2-ETHYLHEXYL) ESTER, SODIUM SALT

PM Nr : 91572 CAS Nr : 00577-11-7

KBr pellet

CAS No. 00577-11-7 PM Ref. No. 91572

SULPHOSUCCINIC ACID, BIS(2-ETHYLHEXYL)ESTER

Peak wavenumber	% T	Relative intensity
409.0	90.3	13.1
528.6	87.5	16.9
586.4	88.1	16.1
729.2	88.9	15.0
900.9	73.7	35.5
1018.5	71.1	39.0
1051.3	36.9	85.2
1219.2	26.0	99.9
1381.2	69.8	40.8
1466.1	60.1	53.9
1738.1	25.9	100.0
2862.7	54.9	60.9
2930.2	35.2	87.5
2961.1	32.6	91.0

Sulphosuccinic acid, bis(2-ethylhexyl)ester, sodium salt

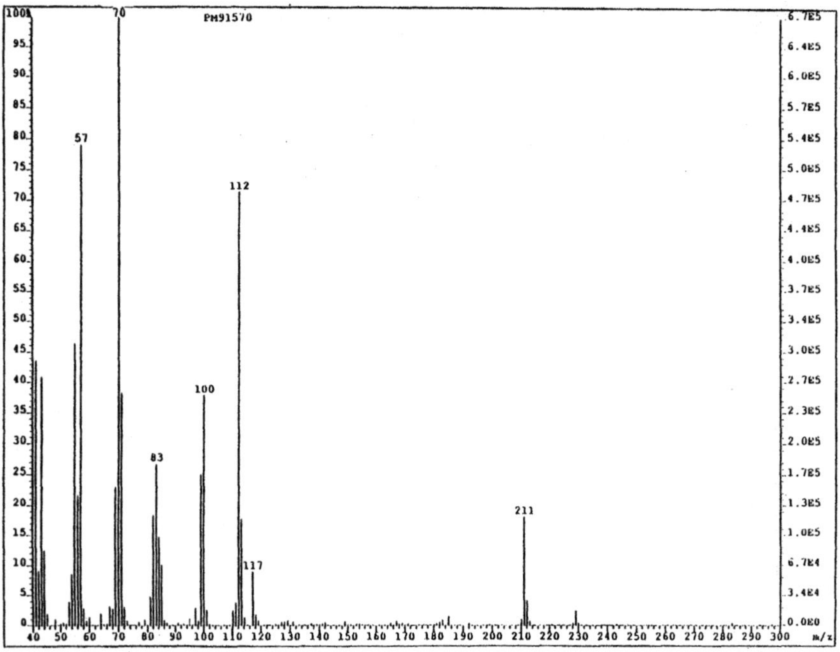

M/Z	Ion Intensity (%)	M/Z	Ion Intensity (%)
41	43.7	82	18.5
42	9.2	83	26.8
43	41.0	84	14.9
44	12.5	85	10.2
55	46.5	99	25.2
56	21.8	100	37.9
57	79.2	112	71.4
69	23.2	113	17.9
70	100.0	117	9.0
71	38.4	211	18.2

Spectrometer	: VG Autospec
Inletsystem	: Flow injection particle beam MS
Source Temperature	: 250°C
Electron Energy	: 70 eV
Scan Range	: 40-1000

PM n°
91572

CAS n°
00577-11-7

Name
SULPHOSUCCINIC ACID, BIS(2-ETHYLHEXYL) ESTER, SODIUM SALT

Solvent	CD_3OD
SF	250.13
SI	16384
SW	3125
NS	64

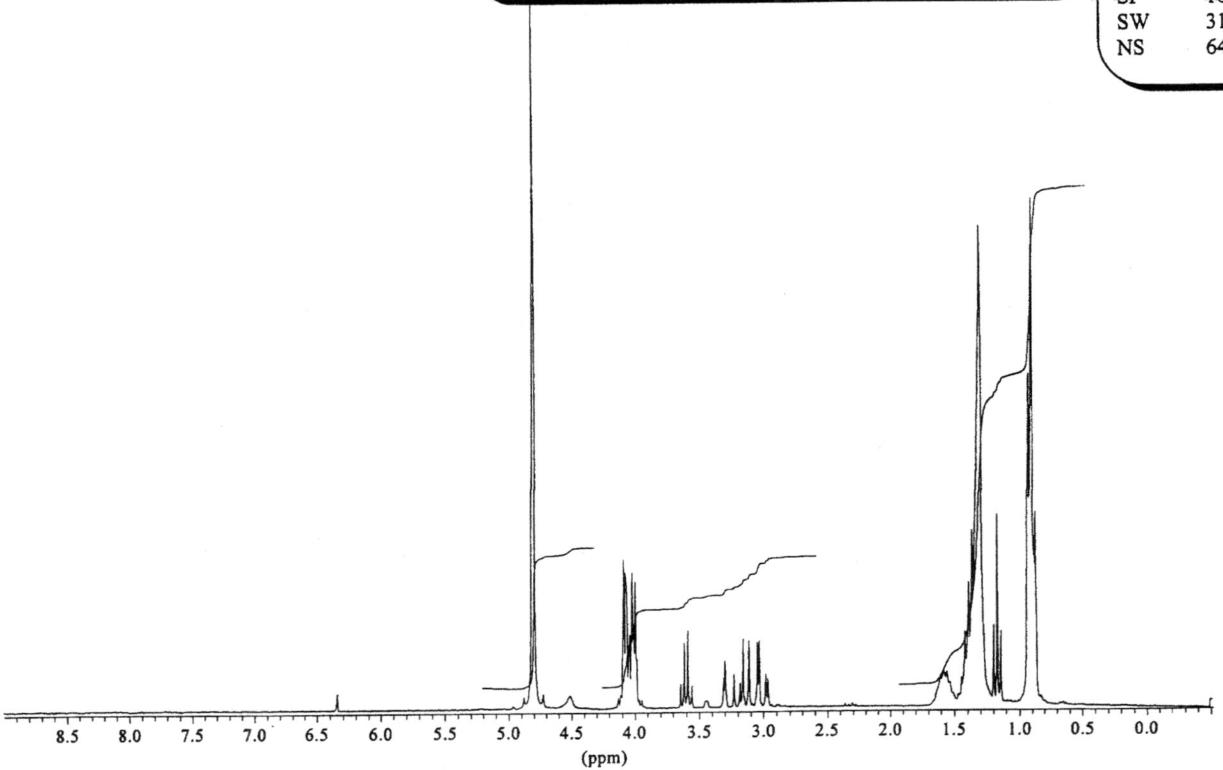

Tetrakis(2,4-di-tert-butylphenyl)-4,4'-biphenylylene diphosphonite

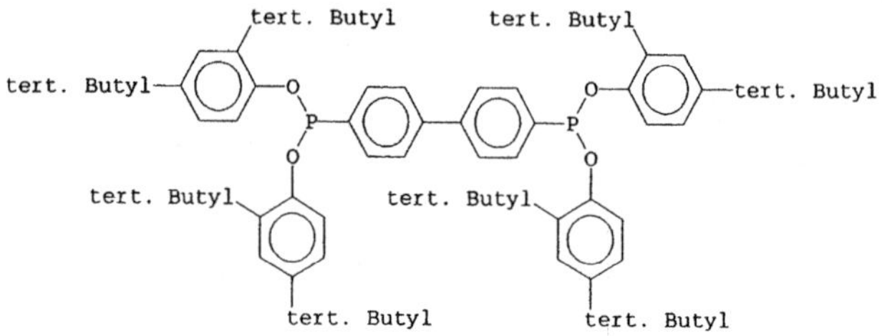

CAS No.	- 38613-77-3
PM Ref. No.	- 92560
Restrictions	- SML = 18 mg/kg
Formula	- $C_{68}H_{92}O_4P_2$
Molecular weight	- 1035

Alternative names -

Physical Characteristics
- Slightly yellow pastilles, mp 75 - 95°C.
- Soluble in most organic solvents.

Handling - Store at room temperature.

Safety -

Availability - Standard sample was supplied.

Current uses - Plastic stabilizer.

Applications - Polyolefins, styrene co-polymers, PC.

GC Retention time -

Note
- Also known as CAS No. 118578-01-1 and 119345-01-6
- Contains PM Ref. No. 74240

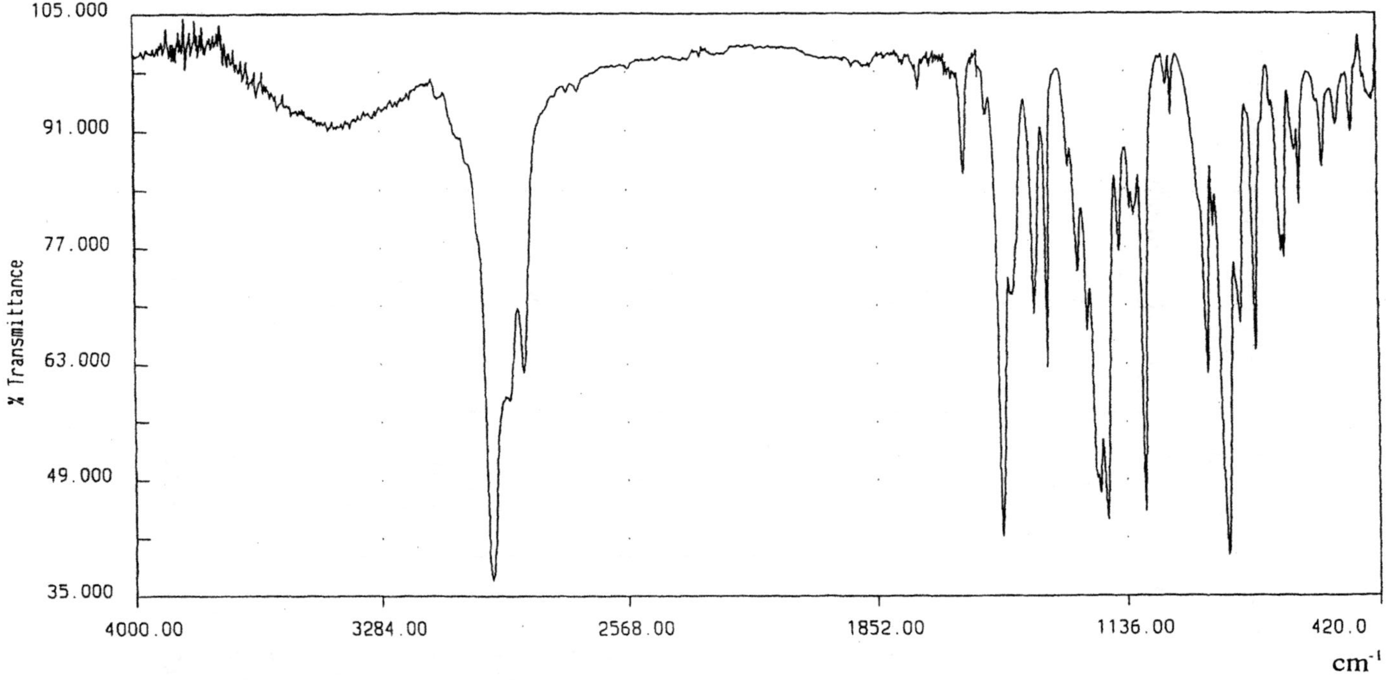

Name : TETRAKIS (2,4-DI-TERT-BUTYLPHENYL)-4,4'-BIPHENYLYLENE DIPHOSPHONITE

PM Nr : 92560 CAS Nr : 38613-77-3

KBr pellet

CAS No. 38613-77-3 PM Ref. No. 92560

TETRAKIS (2,4-DI-TERT-BUTYLPHENYL)-4,4'-BIPHENYLYLENE DIPHOSPHONITE

Peak wavenumber	% T	Relative intensity
451.4	94.7	8.8
463.0	94.4	9.2
495.8	86.4	22.6
538.2	91.3	14.4
576.8	87.8	20.2
644.3	82.2	29.5
688.7	75.1	41.3
771.6	64.2	59.4
814.1	67.0	54.7
850.7	43.8	93.1
889.3	80.1	33.0
906.7	63.7	60.1
1084.1	47.8	86.5
1124.6	75.7	40.3
1155.5	75.6	40.4
1192.2	48.0	86.3
1211.4	51.2	80.8
1248.1	66.63	55.3
1275.1	66.38	55.7
1361.9	64.96	58.1
1400.5	70.49	48.9
1462.2	70.41	49.0
1493.1	46.57	88.5
1601.1	85.27	24.4
1728.4	84.57	25.6
2868.5	62.80	61.6
2907.1	59.82	66.6
2959.2	39.66	100.0

Tetrakis(2,4-di-tert-butylphenyl)-4,4'-biphenylylene diphosphonite

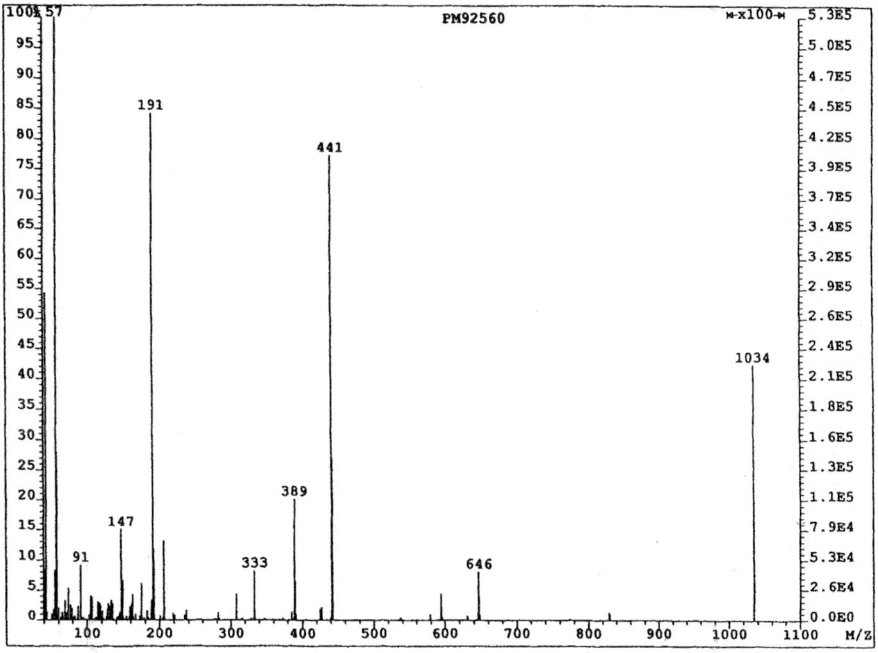

M/Z	Ion Intensity (%)	M/Z	Ion Intensity (%)
41	8.4	191	84.3
43	54.3	192	11.9
55	8.3	206	13.2
57	100.0	333	8.2
58	28.1	389	20.1
74	5.4	390	6.6
91	9.2	441	77.3
147	15.1	442	27.0
149	6.7	646	8.1
175	6.1		

Spectrometer	: VG Autospec
Inletsystem	: Flow injection particle beam MS
Source Temperature	: 250°C
Electron Energy	: 70 eV
Scan Range	: 40-1000

PM n°	CAS n°	Name
92560	38613-77-3	TETRAKIS(2,4-DI-tert-BUTYL-PHENYL)-4,4'-BIPHENYLYLENE DIPHOSPHONITE

Solvent	$CDCL_3$
SF	250.13
SI	16384
SW	3125
NS	32

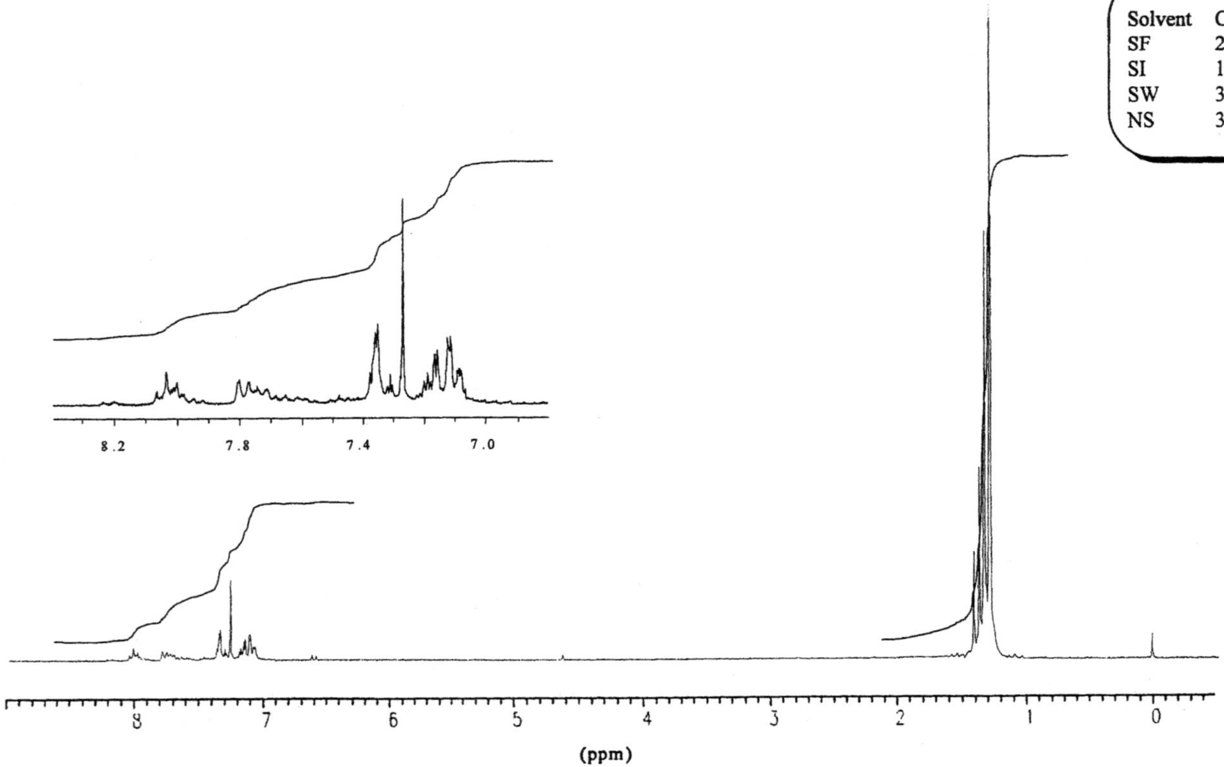

4,4'-Thiobis(6-tert-butyl-3-methylphenol)

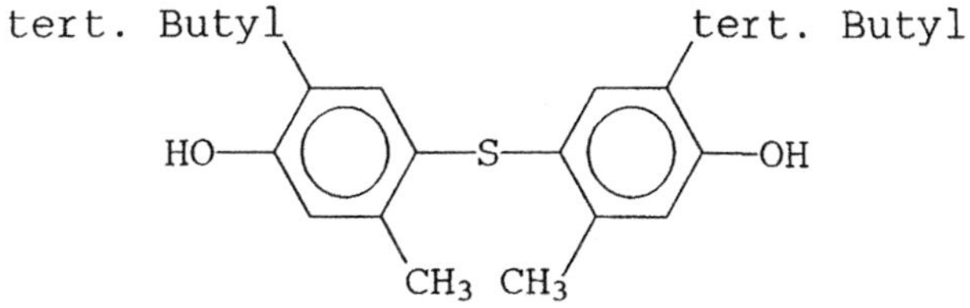

CAS No.	- 00096-69-5
PM Ref. No.	- 92800
Restrictions	- SML = 0.48 mg/kg
Formula	- $C_{22}H_{30}O_2S$
Molecular weight	- 358

Alternative names - 4,4'-Thiobis(6-tert-butyl-m-cresol), Di(2-methyl-4-hydroxy-5-tert-butylphenyl) thioether

Physical Characteristics - White to light yellow crystalline powder, mp > 158°C.
- Soluble in most organic solvents.

Handling - Store at room temperature.

Safety -

Availability - Standard sample was supplied.

Current uses - Antioxidant.

Applications - Polyolefins, styrene co-polymers.

GC Retention time - 50.3 min.

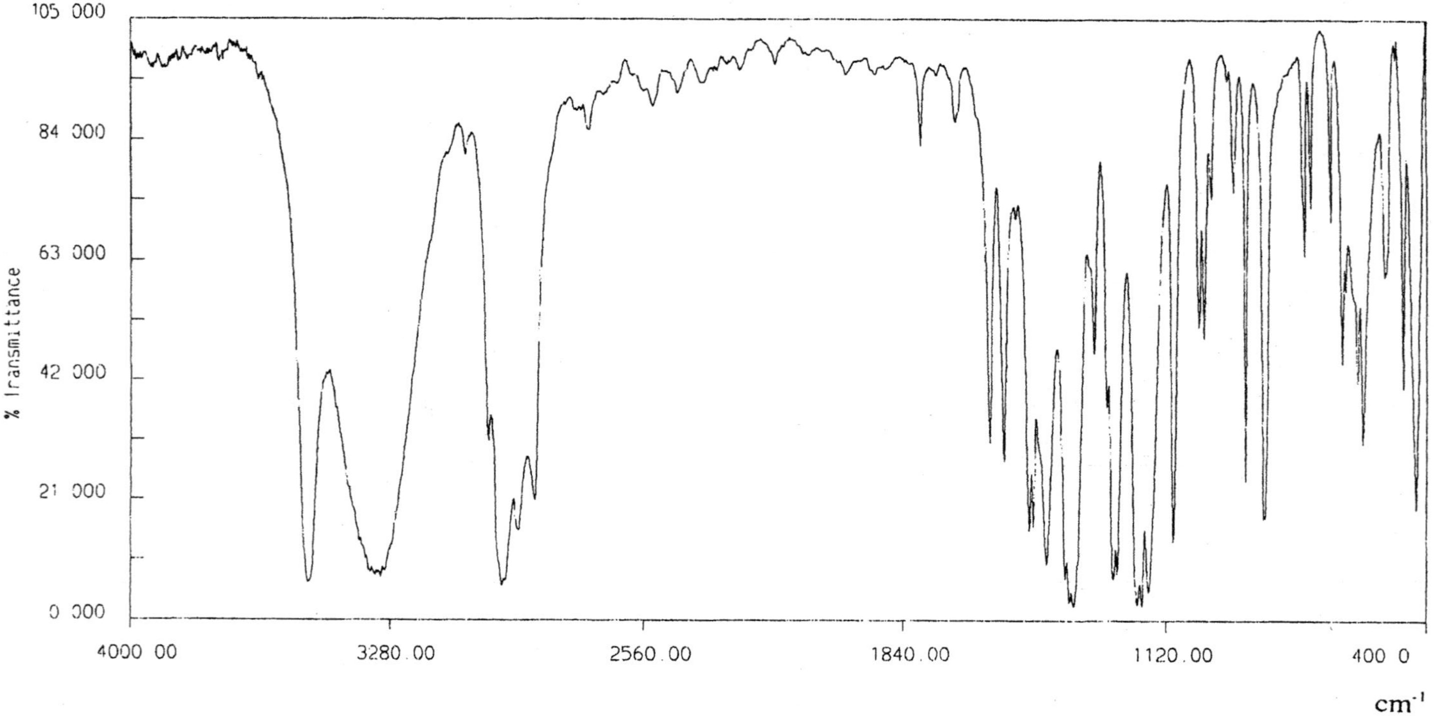

Name : 4,4'-THIOBIS(6-Tert-BUTYL-3-METHYLPHENOL)

PM Nr : 92800 CAS Nr : 00096-69-5

KBr pellet

CAS No. 00096-69-5 PM Ref. No. 92800

4,4'-THIOBIS(6-tert-BUTYL-3-METHYLPHENOL)

Peak wavenumber	% T	Relative intensity
426,3	19,3	82,8
461,0	40,7	60,8
511,2	60,3	40,7
571,0	30,8	71,0
586,4	41,5	60,0
628,9	45,0	56,4
735,0	63,9	37,0
848,8	17,6	84,5
898,9	24,4	77,6
933,7	74,9	25,8
995,4	73,8	26,9
1014,7	49,5	51,8
1028,2	51,4	49,9
1099,6	13,7	88,5
1169,0	4,9	97,6
1186,4	2,5	100,0
1199,9	2,6	99,9
1255,8	8,1	94,3
1265,5	7,1	95,3
1315,6	46,9	54,5
1371,6	2,5	100,0
1446,8	9,6	92,8
1483,4	16,1	86,1
1495,0	15,5	86,7
1562,5	27,8	74,1
1601,1	31,0	70,8
2708,4	85,6	14,8
2868,5	20,8	81,3
2916,7	15,5	86,7
2961,1	6,0	96,4
3001,6	31,3	70,5
3065,3	81,4	19,1
3117,4	81,4	19,1
3296,8	8,6	93,8
3510,9	6,6	95,8

4,4'-Thiobis(6-tert-butyl-3-methylphenol)

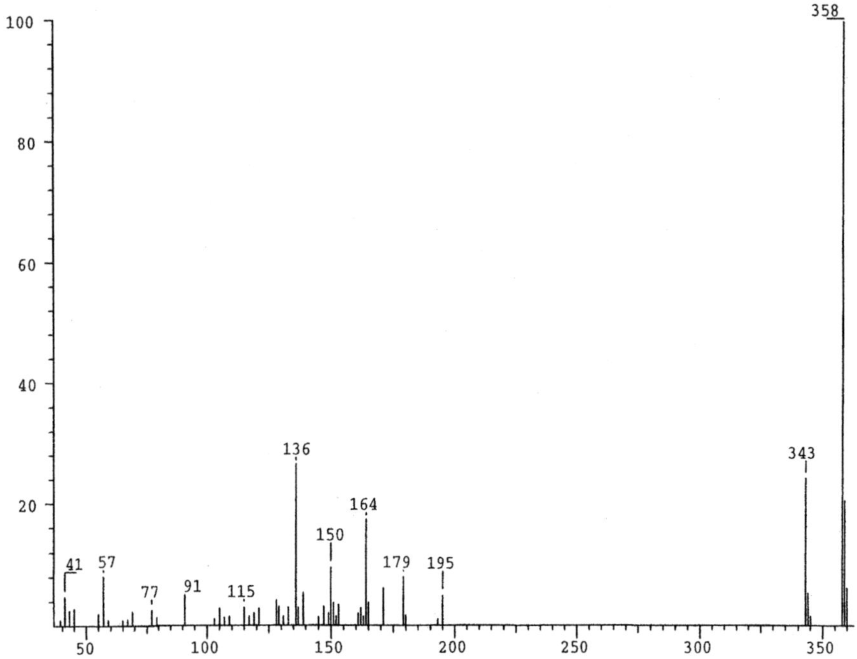

M/Z	Ion Intensity (%)	M/Z	Ion Intensity (%)
41	4.8	195	5.0
57	8.1	343	24.3
91	4.1	344	5.4
136	26.6	358	100.0
150	9.5	359	20.6
164	17.4	360	6.3
179	8.0		

Spectrometer	: Finnigan Mat SSQ 700
Inletsystem	: Capillary GC/MS
Source Temperature	: 180°C
Electron Energy	: 70 eV
Scan Range	: 35-500

PM n°	CAS n°	Name
92800	00096-69-5	4,4'-THIOBIS(6-TERT-BUTYL-3-METHYLPHENOL)

Solvent	$CDCL_3$
SF	250.13
SI	16384
SW	3125
NS	64

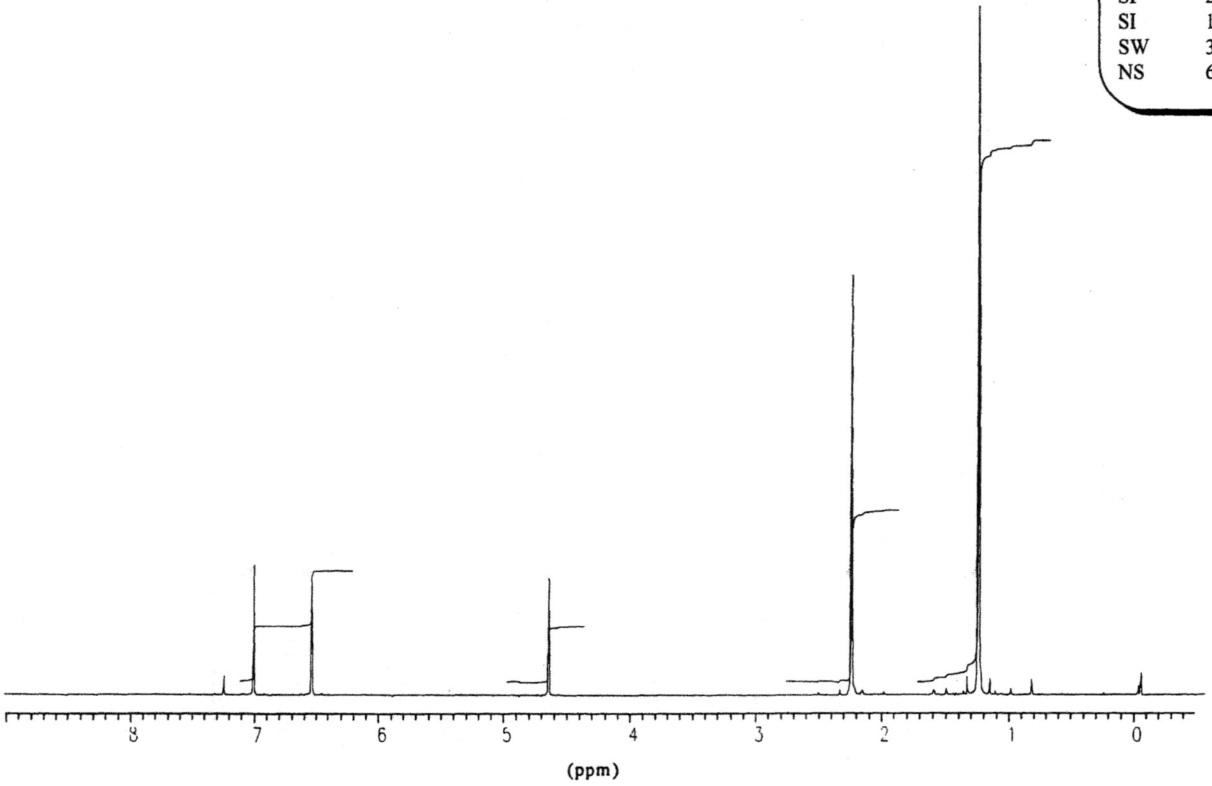

Thiodiethanol bis(3-(3,5-di-tert-butyl-4-hydroxyphenyl) propionate)

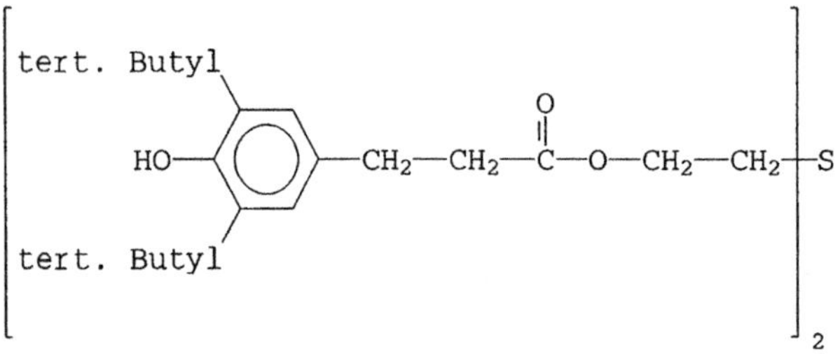

CAS No.	- 41484-35-9
PM Ref. No.	- 92880
Restrictions	- SML = 2.4 mg/kg
Formula	- $C_{38}H_{58}O_6S$
Molecular weight	- 642

Alternative names	- Benzenepropanoic acid, 3,5(1,1-dimethylethyl)-4-hydroxy-,thiodi-2,1-ethanediyl ester
Physical Characteristics	- White cristalline powder, mp 63 - 67°C. - Soluble in most organic solvents.
Handling	- Store at room temperature.
Safety	- Toxic.
Availability	- Standard sample was supplied.
Current uses	- Antioxidant.
Applications	- Styrene co-polymers, polyolefins.
GC Retention time	- 75.7 min.

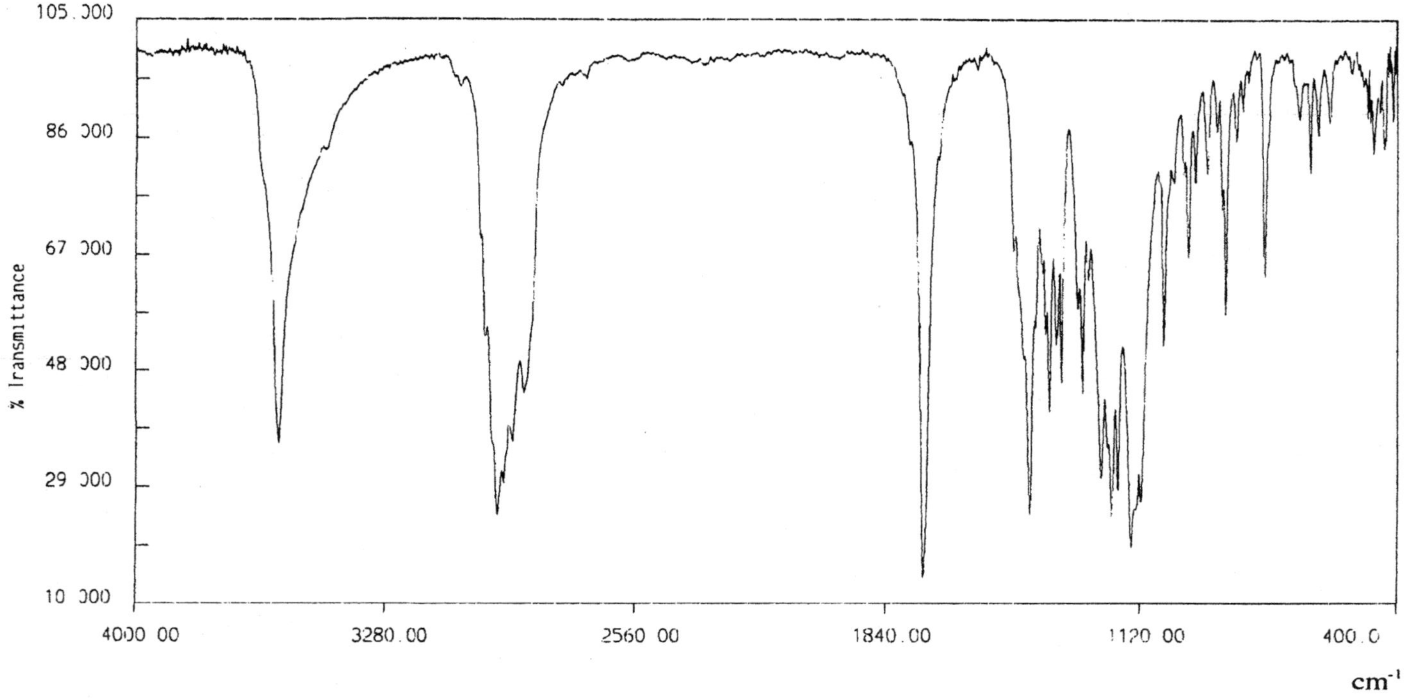

Name : THIODIETHANOL BIS(3-(3,5-DI-Tert-BUTYL-4-HYDROXYPHENYL)PROPIONATE)

PM Nr : 92880 CAS Nr : 41484-35-9

KBr pellet

CAS No. 41484-35-9 PM Ref. No. 92880

THIODIETHANOL BIS(3-(3,5-DI-tert-BUTYL-4-HYDROXYPHENYL)PROPIONATE

Peak wavenumber	% T	Relative intensity
410,9	88,7	13,2
436,0	84,3	18,3
464,9	83,6	19,1
619,2	86,6	15,6
642,4	80,4	22,9
671,3	89,1	12,7
767,8	63,5	42,6
829,5	90,4	11,2
848,8	85,4	17,0
879,7	57,3	49,8
904,7	86,9	15,3
931,7	80,2	23,1
966,5	78,9	24,6
985,7	66,6	38,9
1026,3	78,6	25,0
1055,2	52,0	56,0
1120,8	26,6	85,6
1147,8	19,1	94,3
1186,4	28,4	83,5
1203,7	24,1	88,5
1236,5	30,5	81,0
1273,2	62,5	43,7
1286,7	44,2	65,1
1300,2	58,1	48,9
1346,5	45,9	63,1
1361,9	52,2	55,7
1379,3	41,2	68,6
1435,2	24,6	87,9
1738,1	14,3	99,9
2882,0	44,1	65,2
2914,8	36,1	74,5
2937,9	29,7	82,0
2955,3	24,6	87,9
3454,9	83,9	18,8
3591,9	35,9	74,7

Thiodiethanol bis(3-(3,5-di-tert-butyl-4-hydroxyphenyl) propionate)

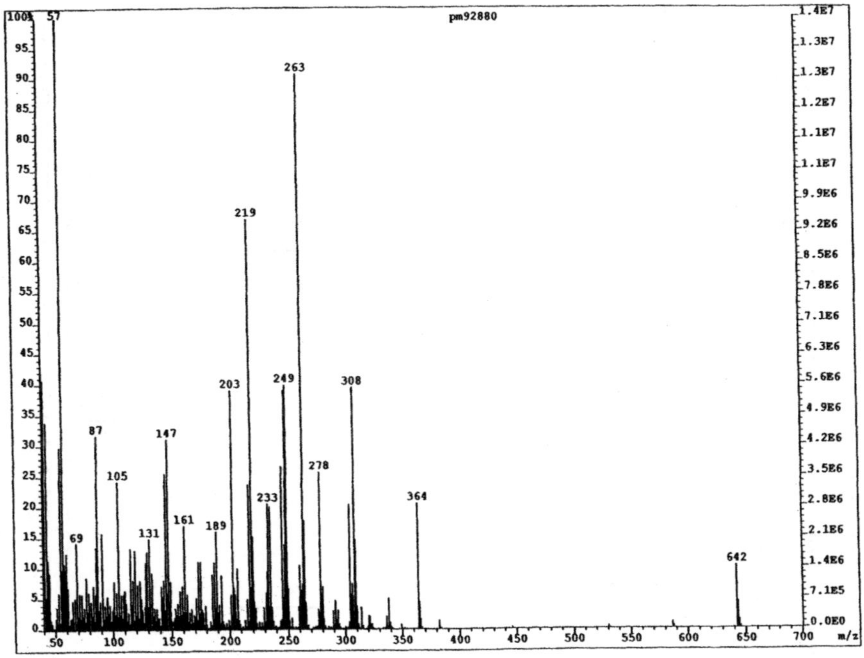

M/Z	Ion Intensity (%)	M/Z	Ion Intensity (%)
41	41.1	219	67.5
43	34.4	233	20.9
55	30.1	249	39.9
57	100.0	263	91.4
87	32.5	278	25.8
105	24.5	304	20.9
145	25.2	308	39.9
147	31.3	364	20.9
203	39.3	642	13.5
217	23.9	643	4.8

Spectrometer	: VG Autospec
Inletsystem	: Flow injection particle beam MS
Source Temperature	: 250°C
Electron Energy	: 70 eV
Scan Range	: 40-1000

PM n°
92880

CAS n°
41484-35-9

Name
THIODIETHANOL BIS(3-(3,5-DI-TERT-BUTYL-4-HYDROXYPHENYL) PROPIONATE)

Solvent	$CDCL_3$
SF	250.13
SI	16384
SW	3125
NS	44

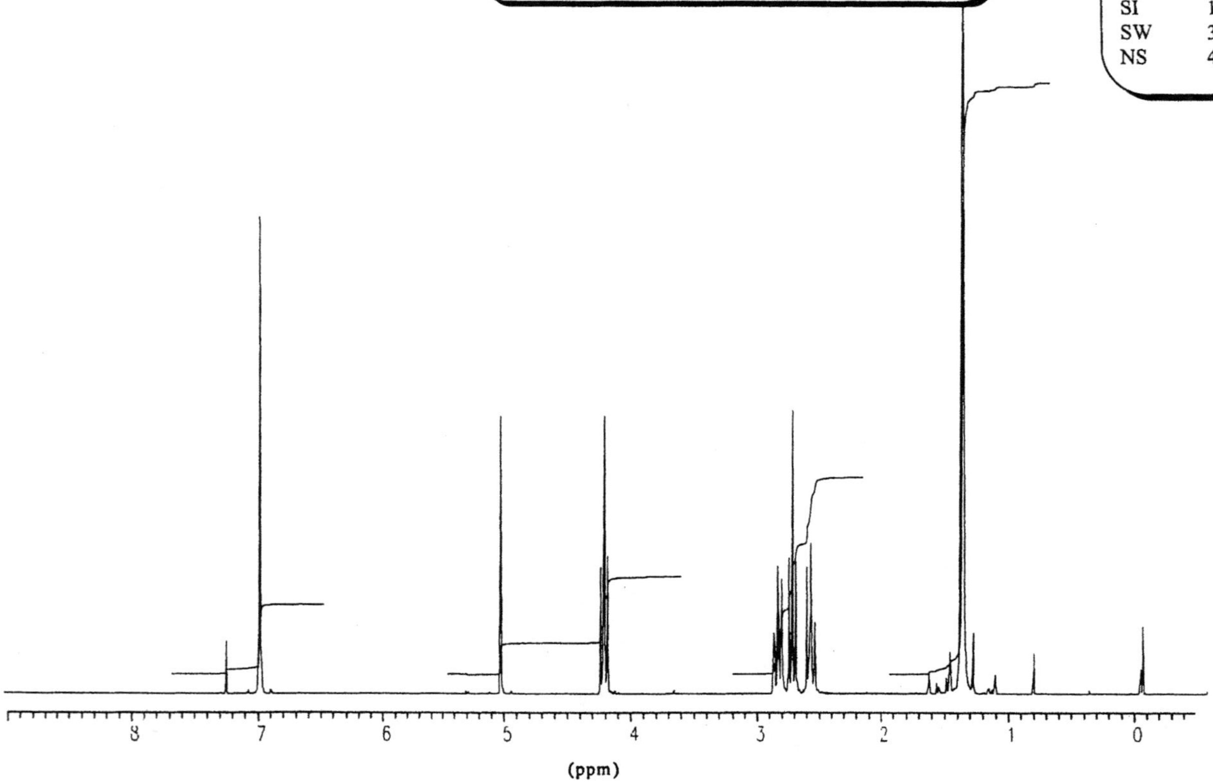

Thiodipropionic acid, didodecyl ester

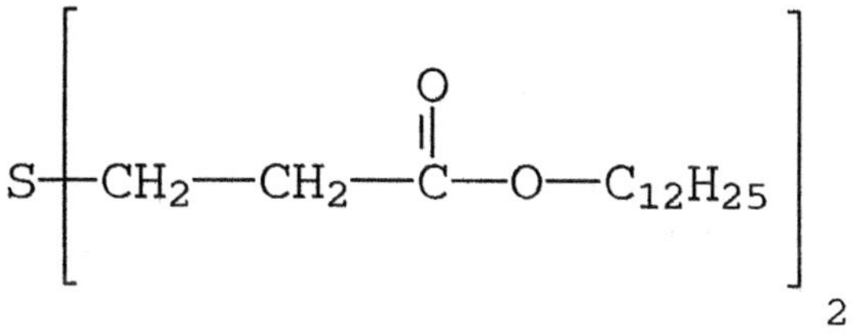

CAS No.	- 00123-28-4
PM Ref. No.	- 93120
Restrictions	- SML = 3.0 mg/kg
Formula	- $C_{30}H_{58}O_4S$
Molecular weight	- 515

<u>Alternative names</u> - Propanoic acid, 3,3'-thiobis-, didodecyl ester

<u>Physical Characteristics</u> - White powder, mp 38 - 40°C.
- Soluble in most organic solvents.

<u>Handling</u> - Store at room temperature.
<u>Safety</u> -
<u>Availability</u> - Standard sample was supplied.

<u>Current uses</u> - Co-stabilizer.
<u>Applications</u> - General use.

<u>GC Retention time</u> - 70.0 min.

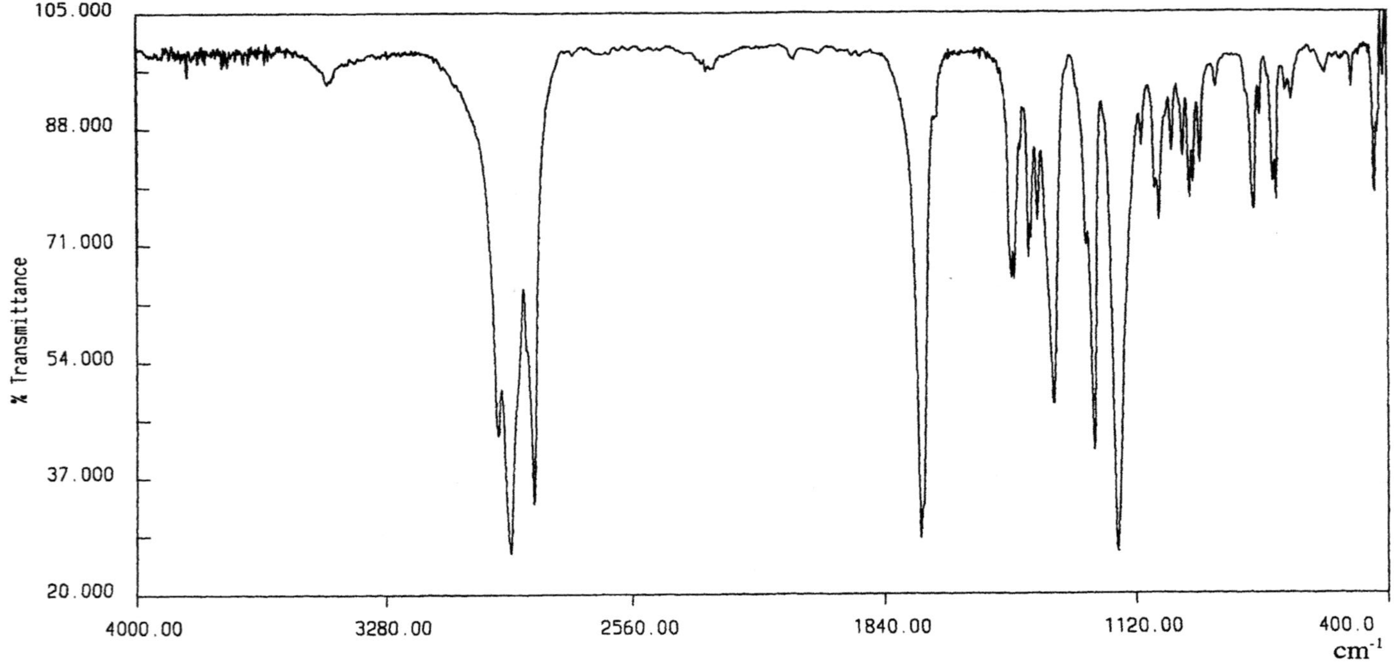

Name : THIODIPROPIONIC ACID, DIDODECYL ESTER

PM Nr : 93120 CAS Nr : 00123-28-4

KBr pellet

CAS No. 00123-28-4 PM Ref. No. 93120

THIODIPROPIONIC ACID, DIDODECYL ESTER

Peak wavenumber	% T	Relative intensity
436	78,2	29,4
501,6	93,7	8,5
675,2	92,0	10,8
719,5	77,2	30,8
785,1	75,9	32,5
933,7	82,8	23,2
954,9	79,8	27,3
964,5	77,5	30,4
983,8	83,8	21,9
1016,6	84,6	20,8
1051,3	74,4	34,6
1101,5	85,4	19,7
1172,9	26,0	99,9
1238,5	40,8	79,9
1352,3	47,8	70,5
1396,6	74,4	34,6
1423,6	69,0	41,9
1466,1	65,7	46,3
1738,1	27,9	97,3
2851,1	33,1	90,3
2918,7	26,0	100,0
2955,3	43,0	77,0

Thiodipropionic acid, didodecyl ester

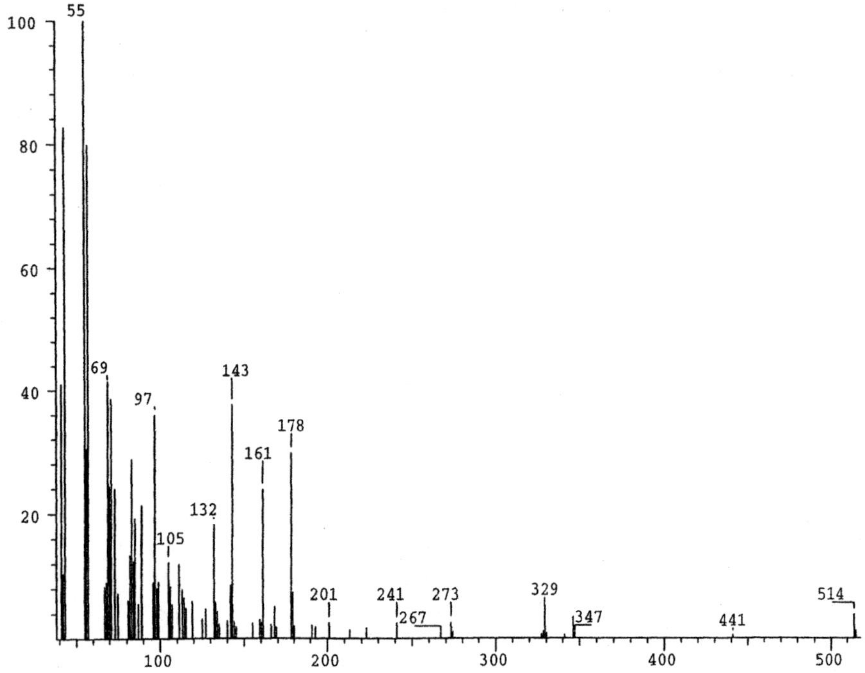

M/Z	Ion Intensity (%)	M/Z	Ion Intensity (%)
43	82.7	97	35.9
55	100.0	105	12.2
56	30.5	111	11.9
57	79.9	132	18.3
69	41.5	143	37.6
71	38.6	161	23.8
73	24.0	178	29.7
83	28.8	329	5.5
85	19.2	514	3.8
89	21.4	515	1.2

Spectrometer	: Finnigan Mat SSQ 700
Inletsystem	: Capillary GC/MS
Source Temperature	: 180°C
Electron Energy	: 70 eV
Scan Range	: 35-600

PM n°	CAS n°	Name
93120	00123-28-4	THIODIPROPIONIC ACID, DIDODECYL ESTER

Solvent	$CDCl_3$
SF	250.13
SI	16384
SW	3125
NS	24

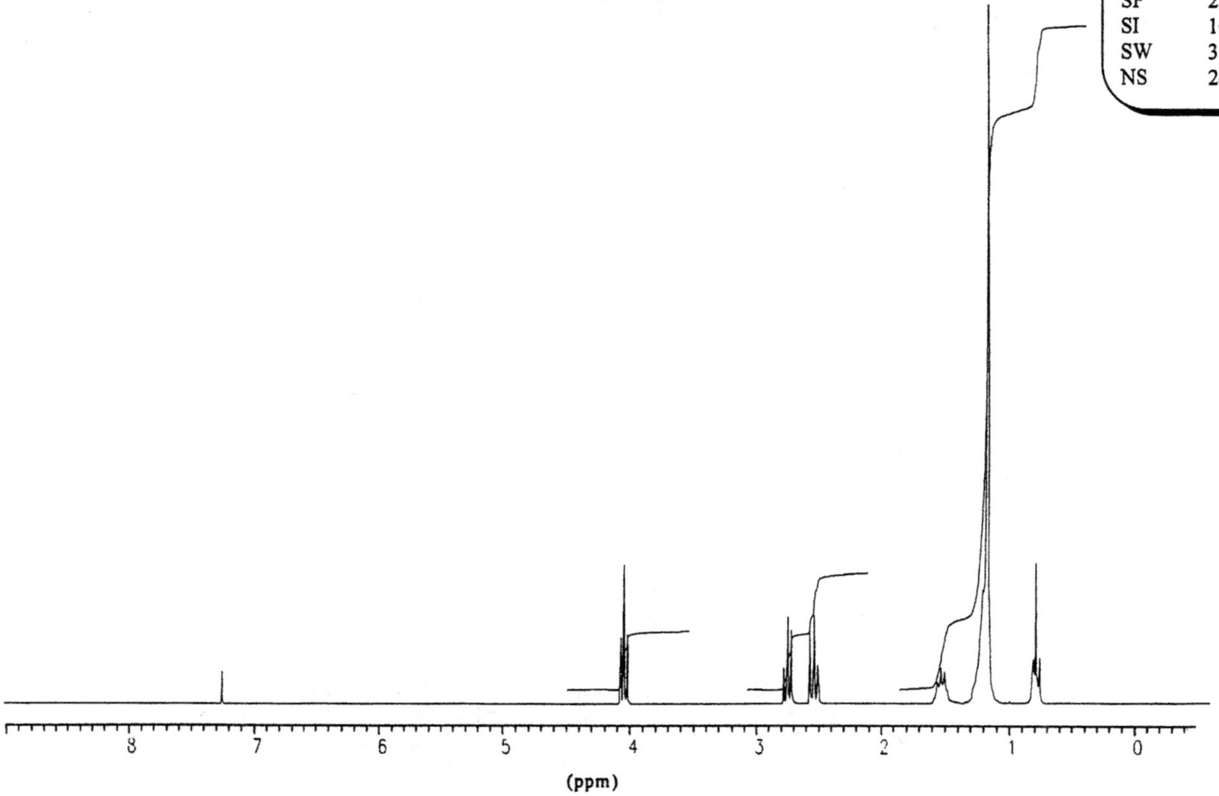

Thiodipropionic acid, ditetradecyl ester

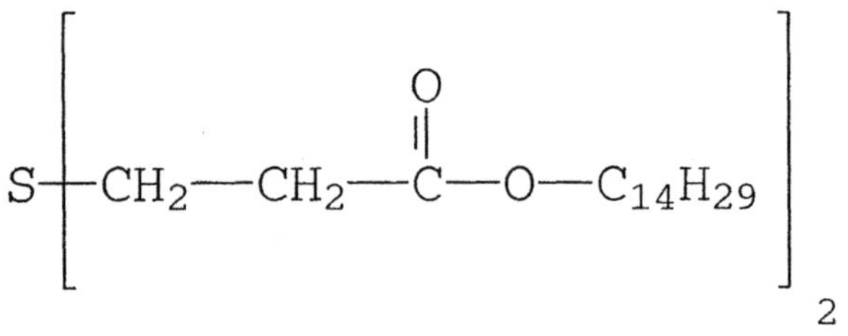

CAS No.	- 16545-54-3
PM Ref. No.	- 93360
Restrictions	- SML = 3.0 mg/kg
Formula	- $C_{34}H_{66}O_4S$
Molecular weight	- 570

Alternative names	- Propanoic acid, 3,3'-thiobis-, ditetradecyl ester
Physical Characteristics	- White crystalline powder, mp 50°C. - Soluble in most organic solvents.
Handling	- Store at room temperature.
Safety	-
Availability	- Standard sample was supplied.
Current uses	- Stabilizer.
Applications	- General use.
GC Retention time	- 67.9 min.

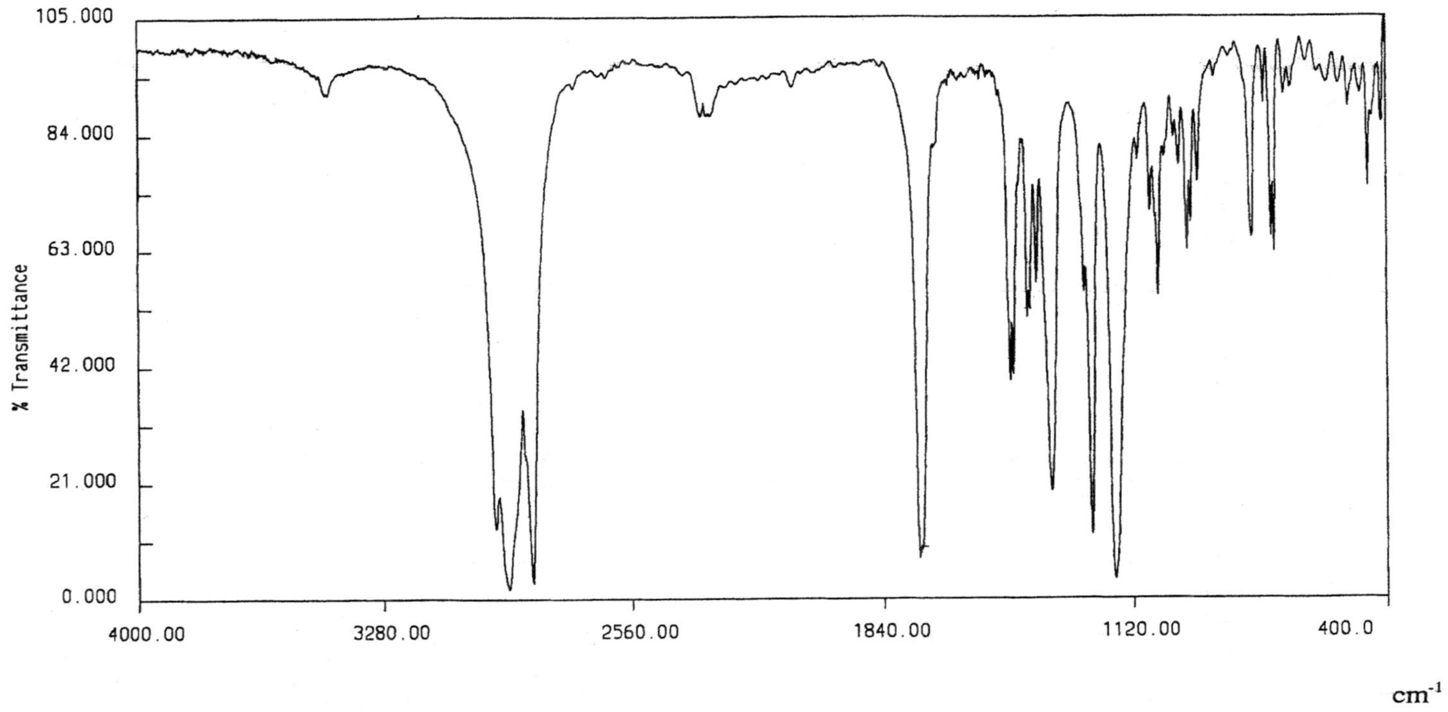

Name : THIODIPROPIONIC ACID, DITETRADECYL ESTER

PM Nr : 93360 CAS Nr : 16545-54-3

KBr pellet

CAS No. 16545-54-3 PM Ref. No. 93360

THIODIPROPIONIC ACID, DITETRADECYL ESTER

Peak wavenumber	% T	Relative intensity
412.8	86.2	14.0
439.8	87.1	13.1
451.4	74.5	26.0
719.5	62.6	38.0
729.2	65.5	35.1
785.1	65.4	35.2
954.9	68.0	32.5
964.5	63.0	37.7
989.6	78.3	22.1
1032.0	79.9	20.5
1049.4	54.6	46.2
1072.6	70.1	30.4
1107.3	79.3	21.1
1172.9	3.6	98.0
1238.5	11.6	89.9
1259.7	55.5	45.2
1352.3	19.5	81.9
1396.6	57.0	43.7
1423.6	50.8	50.0
1464.2	40.5	60.5
1473.8	39.3	61.7
1738.1	7.3	94.2
2330.3	87.3	13.0
2849.2	3.0	98.6
2918.7	1.7	100.0
2955.3	12.8	88.7
3441.4	91.2	8.9

Thiodipropionic acid, ditetradecyl ester

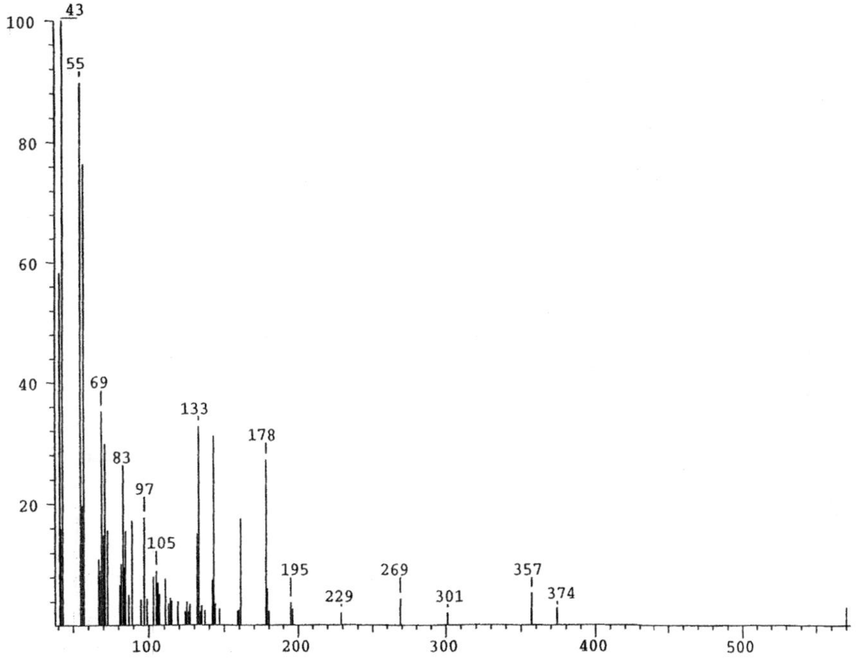

M/Z	Ion Intensity (%)	M/Z	Ion Intensity (%)
43	100.0	97	17.7
55	89.8	105	9.0
57	76.4	133	32.7
69	35.2	143	31.1
71	29.8	161	17.5
73	25.7	178	27.2
82	10.1	269	4.5
83	25.3	357	5.4
85	15.5	570	3.0
89	17.2		

Spectrometer : Finnigan Mat SSQ 700
Inletsystem : Capillary GC/MS
Source Temperature : 180°C
Electron Energy : 70 eV
Scan Range : 40-700

PM n°	CAS n°	Name
93360	16545-54-3	THIODIPROPIONIC ACID, DITETRADECYL ESTER

Solvent	$CDCL_3$
SF	250.13
SI	16384
SW	3125
NS	32

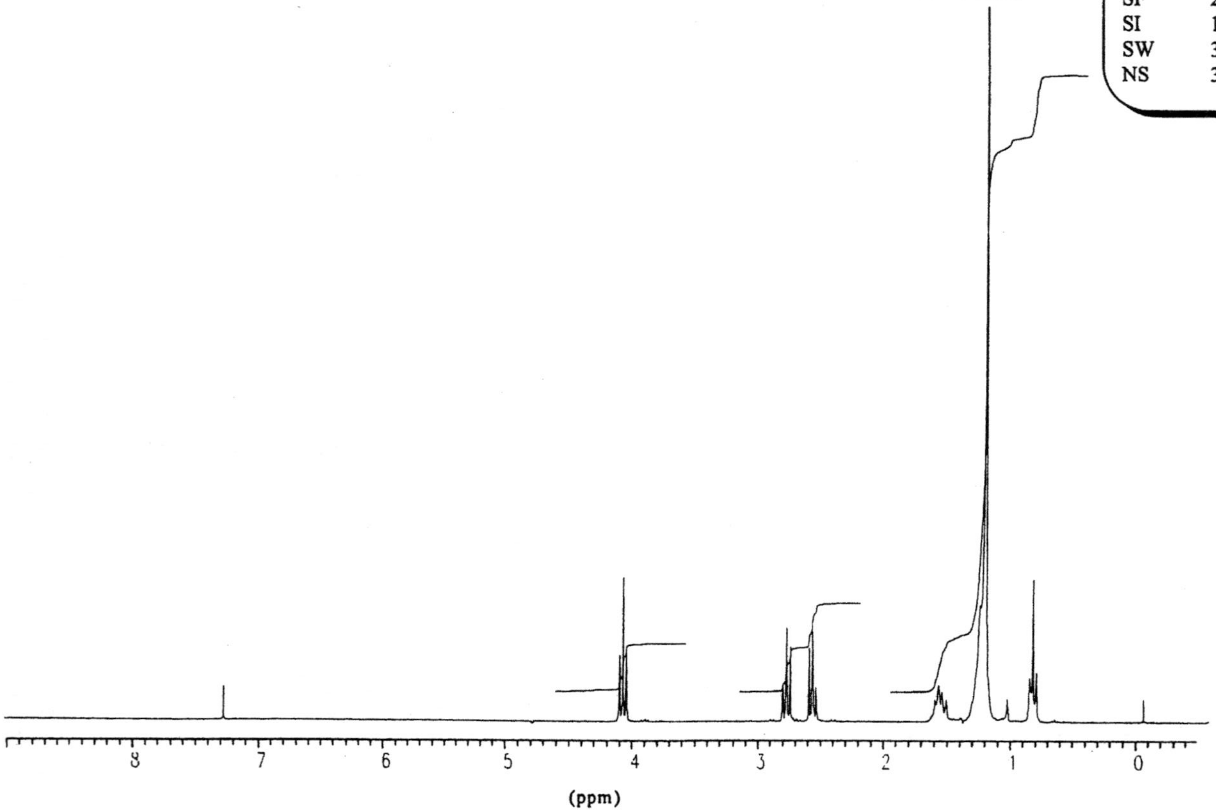

Triethyl acetylcitrate

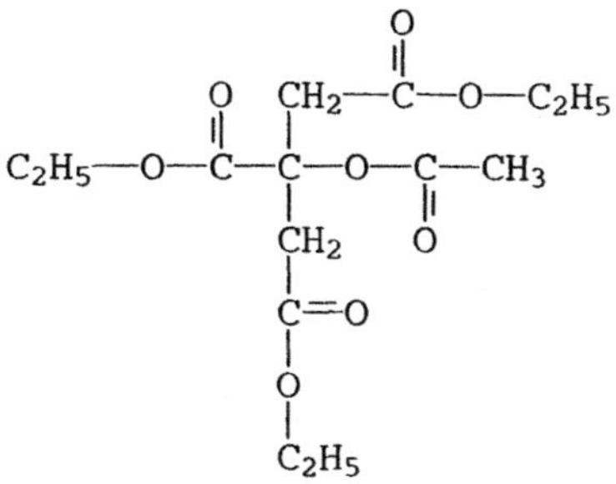

CAS No.	- 00077-89-4
PM Ref. No.	- 94240
Restrictions	- None
Formula	- $C_{14}H_{22}O_8$
Molecular weight	- 318

Alternative names	- 1,2,3-Propanetricarboxylic acid, 2-(acetyloxy)-, triethyl ester
Physical Characteristics	- Colourless liquid. bp 228 - 229°C (100 mm Hg). - Soluble in most organic solvents.
Handling	- Store at room temperature.
Safety	-
Availability	- Standard sample was supplied.
Current uses	- Plasticizer.
Applications	- PVC, PVDC.
GC Retention time	- 32.7 min.

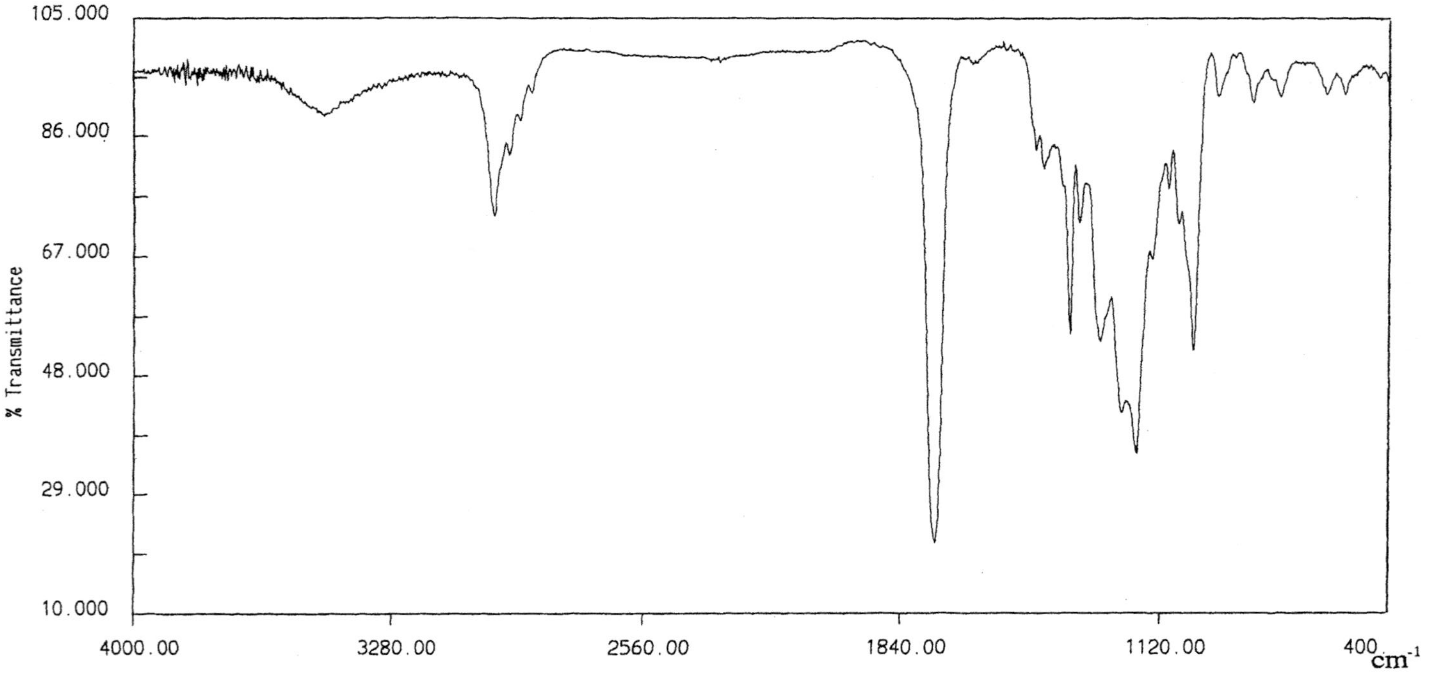

Name : TRIETHYL ACETYL CITRATE

PM Nr : 94240 CAS Nr : 000077-89-4

KBr pellet

CAS No. 00077-89-4 PM Ref. No. 94240

TRIETHYL ACETYL CITRATE

Peak wavenumber	% T	Relative intensity
414.8	87.4	16.0
428.3	89.8	12.9
488.1	94.7	6.7
603.8	92.6	9.4
655.9	92.7	9.3
783.2	92.1	10.0
860.4	91.3	11.0
1028.2	52.0	60.9
1097.6	77.5	28.6
1142.0	66.4	42.6
1186.4	35.5	81.8
1286.7	53.4	59.0
1346.5	72.2	35.3
1371.6	54.5	57.7
1446.8	80.5	24.7
1743.9	21.1	100.0
2910.9	88.3	14.9
2941.8	82.8	21.8
2984.2	73.2	34.0

Triethyl acetylcitrate

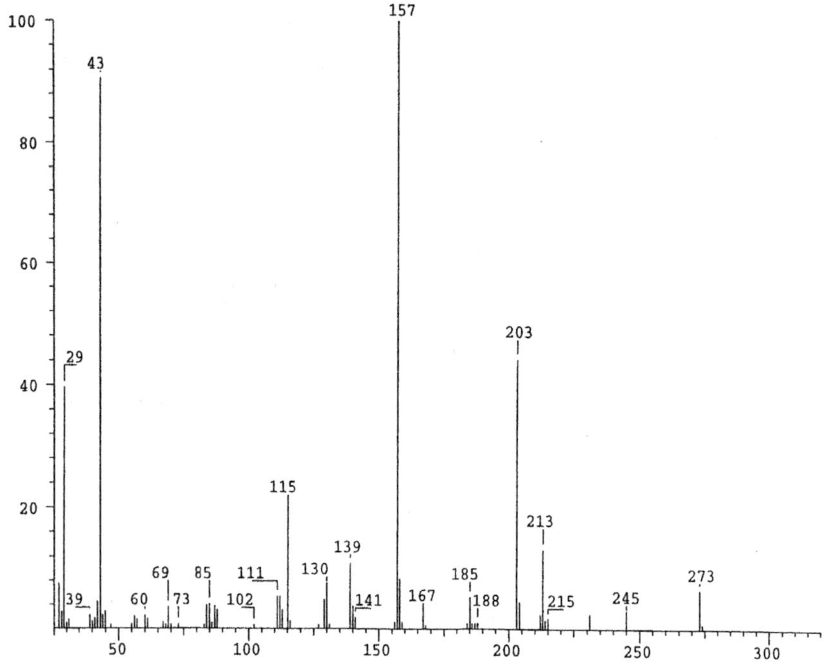

M/Z	Ion Intensity (%)	M/Z	Ion Intensity (%)
29	39.6	157	100.0
43	90.6	158	8.4
115	21.0	203	44.3
130	7.6	213	13.2
139	10.8	273	6.5

Spectrometer	: Finnigan Mat SSQ 700
Inletsystem	: Capillary GC/MS
Source Temperature	: 180°C
Electron Energy	: 70 eV
Scan Range	: 25-350

PM n°
94240

CAS n°
00077-89-4

Name
TRIETHYL ACETHYL CITRATE

Solvent	$CDCL_3$
SF	250.13
SI	16384
SW	3125
NS	32

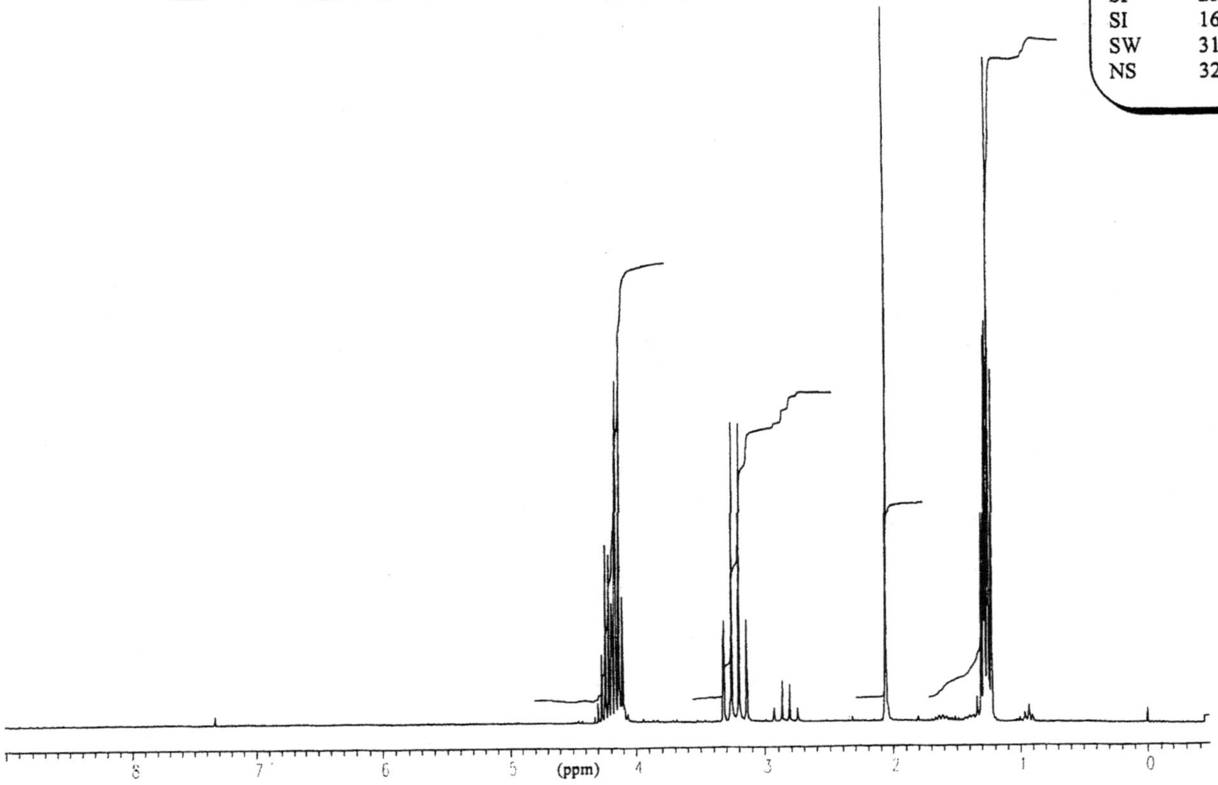

Triethyleneglycol bis(3-(3-tert-butyl-4-hydroxy-5-methylphenyl) propionate)

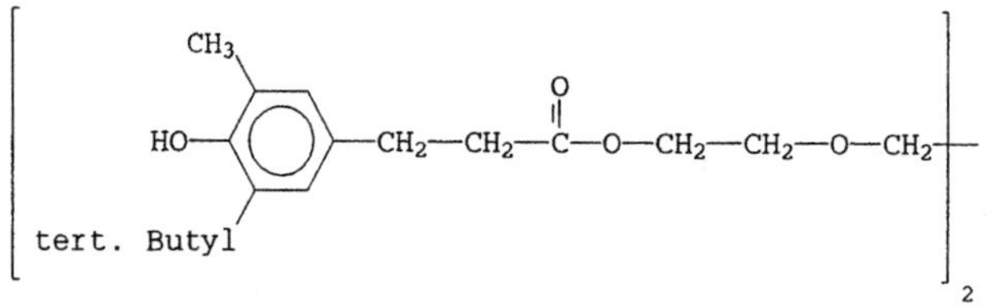

CAS No.	- 36443-68-2
PM Ref. No.	- 94400
Restrictions	- SML = 3.0 mg/kg
Formula	- $C_{34}H_{50}O_8$
Molecular weight	- 586

Alternative names
- Ethylene bis(oxyethylene)-bis(3-tert-butyl-4-hydroxy-5-methylhydrocinnamate)

Physical Characteristics
- Slightly yellow granules, mp 76 - 79°C.
- Soluble in acetone, benzene, methanol, chloroform.

Handling
- Store at room temperature.

Safety
- Harmful.

Availability
- Standard sample was supplied.

Current uses
- Antioxidant.

Applications
- PVC, PVDC, PA, styrene co-polymers, polyesters, polyacetals.

GC Retention time
- 71.9 min.

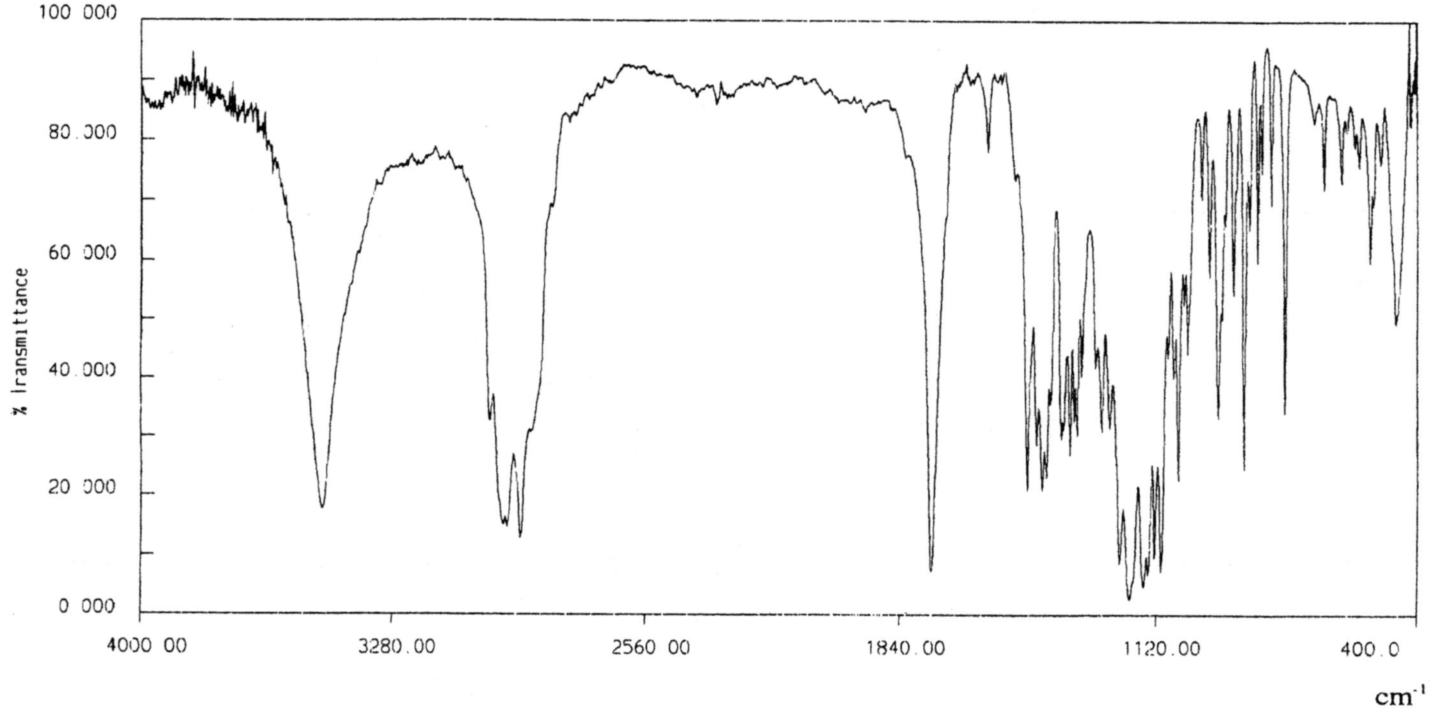

Name : TRIETHYLENEGLYCOL BIS-3-(3-Tert-BUTYL-4-HYDROXY-5-METHYLPHENYL)PROPIONATE

PM Nr : 94400 CAS Nr : 36443-68-2

KBr pellet

CAS No. 36443-68-2 PM Ref. No. 94400

TRIETHYLENEGLYCOL BIS-3-(3-tert-BUTYL-4-HYDROXY-5-METHYLPHENYL)PROPIONATE

Peak wavenumber	% T	Relative intensity
414,8	82,4	18,1
455,3	49,3	52,0
495,8	76,2	24,4
524,7	59,8	41,2
553,6	75,7	24,9
586,4	81,4	19,1
599,9	72,8	27,9
648,2	71,9	28,8
673,2	83,2	17,2
756,2	33,9	67,8
794,8	69,3	31,5
821,8	74,6	26,1
833,4	59,7	41,3
856,5	65,4	35,5
870,0	24,2	77,8
900,9	54,3	46,9
943,3	33,1	68,6
968,4	57,4	43,7
989,6	70,0	30,8
1030,1	44,1	57,4
1041,7	52,5	48,7
1057,1	22,7	79,3
1070,6	40,0	61,6
1086,1	43,4	58,1
1107,3	7,2	95,2
1126,6	9,6	92,8
1143,9	6,7	95,7
1155,5	4,7	97,8
1196,0	2,5	100,0
1225,0	8,6	93,8
1252,0	31,4	70,4
1275,1	30,8	71,0
1292,5	41,8	59,7
1331,0	40,3	61,3
1344,6	30,2	71,6
1365,8	26,7	75,2
1388,9	29,6	72,2
1419,8	35,4	66,3
1431,4	23,0	79,0
1442,9	20,9	81,2
1458,4	28,4	73,5
1485,4	20,9	81,2
1597,3	78,1	22,5
1682,1	88,3	12,0
1753,5	7,2	95,2
2359,2	85,9	14,5
2660,2	89,4	10,9
2822,2	68,8	32,0
2882,0	30,7	71,1
2912,9	12,9	89,4
2949,5	14,5	87,7
2995,8	32,6	69,2
3375,8	61,0	40,0
3485,8	17,7	84,4
3750,1	83,5	16,9
3800,2	85,8	14,6
3852,3	85,1	15,3
3869,7	87,1	13,2

Triethyleneglycol bis(3-(3-tert-butyl-4-hydroxy-5-methylphenyl) propionate)

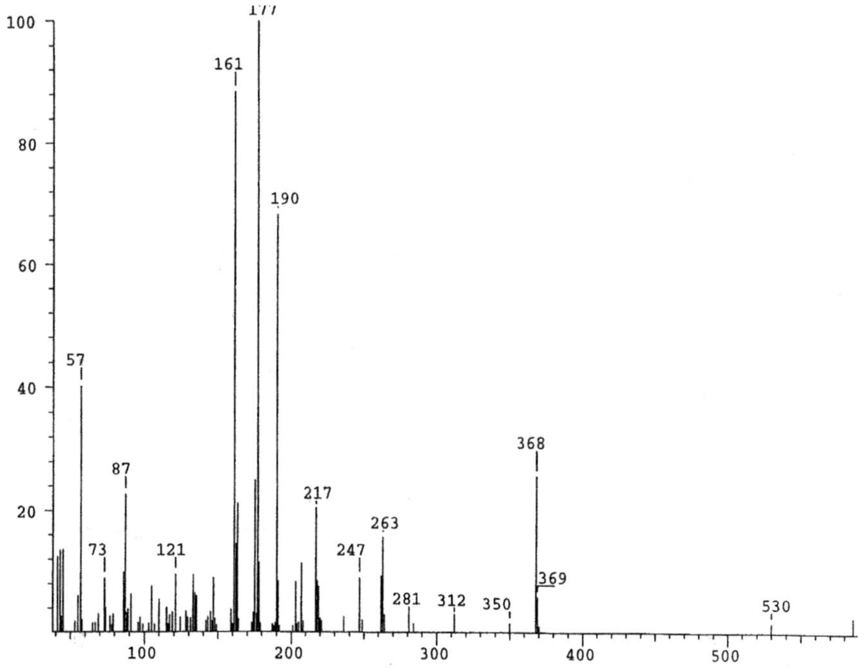

M/Z	Ion Intensity (%)	M/Z	Ion Intensity (%)
41	12.5	178	11.6
43	13.5	190	68.3
45	13.7	207	11.4
57	40.2	217	20.5
87	22.7	262	9.4
161	88.5	263	15.8
162	14.6	368	25.7
163	21.2	530	1.6
175	25.0	586	2.6
177	100.0		

Spectrometer	: Finnigan Mat SSQ 700
Inletsystem	: Capillary GC/MS
Source Temperature	: 180°C
Electron Energy	: 70 eV
Scan Range	: 40-700

PM n°
94400

CAS n°
36443-68-2

Name
TRIETHYLENEGLYCOL BIS-3-(3-TERT-BUTYL-4-HYDROXY-5-METHYLPHENYL)PROPIONATE

Solvent	$CDCL_3$
SF	250.13
SI	16384
SW	3125
NS	76

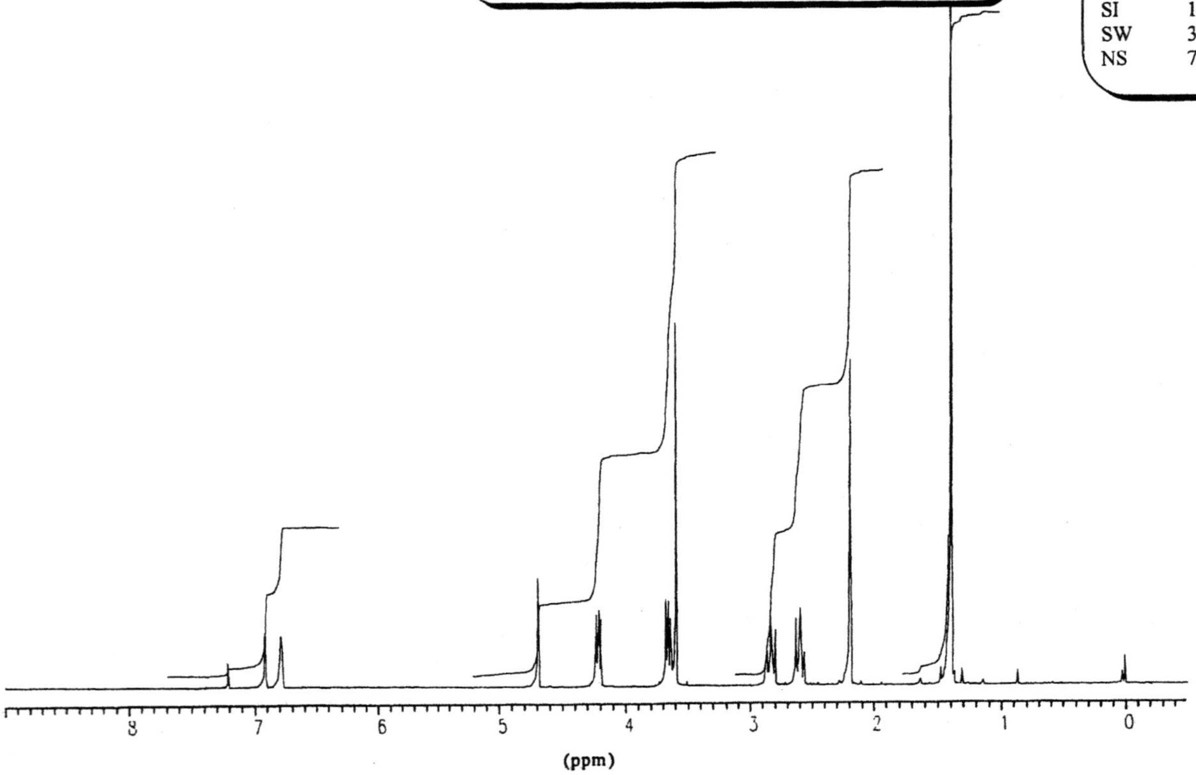

1,1,1-Trimethylolpropane

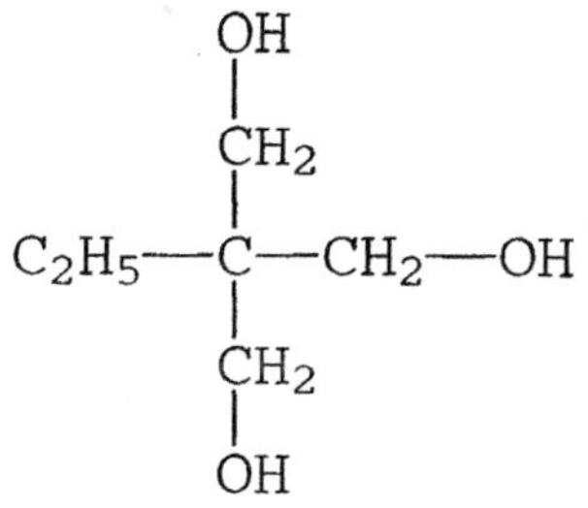

CAS No.	- 00077-99-6
PM Ref. No.	- 94960
Restrictions	- SML = 6.0 mg/kg
Formula	- $C_6H_{14}O_3$
Molecular weight	- 134

Alternative names	- 1,1,1(Trihydroxymethyl)propane 2,2(Dihydroxymethyl) butanol
Physical Characteristics	- White flaked solid, mp 59°C, bp 297°C. - Soluble in water, alcohol, acetone.
Handling	- Store at room temperature.
Safety	-
Availability	- Standard sample was supplied.
Current uses	- Stabilizer.
Applications	- PVC, PVDC, PU, Polyesters.
GC Retention time	- 20.4 min.

Name : TRIMETHYLOLPROPANE

PM Nr : 94960 CAS Nr : 00077-99-6

KBr pellet

CAS No. 00077-99-6 PM Ref. No. 94960

1,1,1-TRIMETHYLOLPROPANE

Peak wavenumber	% T	Relative intensity
409	65,3	42,9
422,5	73,2	33,1
538,2	71,4	35,3
576,8	59,7	49,8
638,5	49,8	62,0
684,8	46,4	66,2
767,8	49,8	62,0
783,2	45,5	67,3
866,1	86,4	16,8
960,7	53,9	56,9
989,6	30,9	85,3
1016,6	19,0	100,0
1064,8	25,4	92,1
1176,7	71,8	34,8
1196,0	67,1	40,6
1292,5	61,1	48,0
1346,5	43,5	69,8
1439,1	58,1	51,7
1460,3	52,0	59,3
1475,7	47,8	64,5
1489,2	49,8	62,0
2675,6	79,7	25,1
2762,4	72,3	34,2
2858,9	46,2	66,4
2883,9	45,3	67,6
2934,1	34,5	80,9
2968,8	40,8	73,1
3119,3	50,7	60,9
3152,1	42,3	71,3
3217,7	29,6	86,9
3233,1	25,8	91,6
3271,7	23,5	94,5

1,1,1-Trimethylolpropane

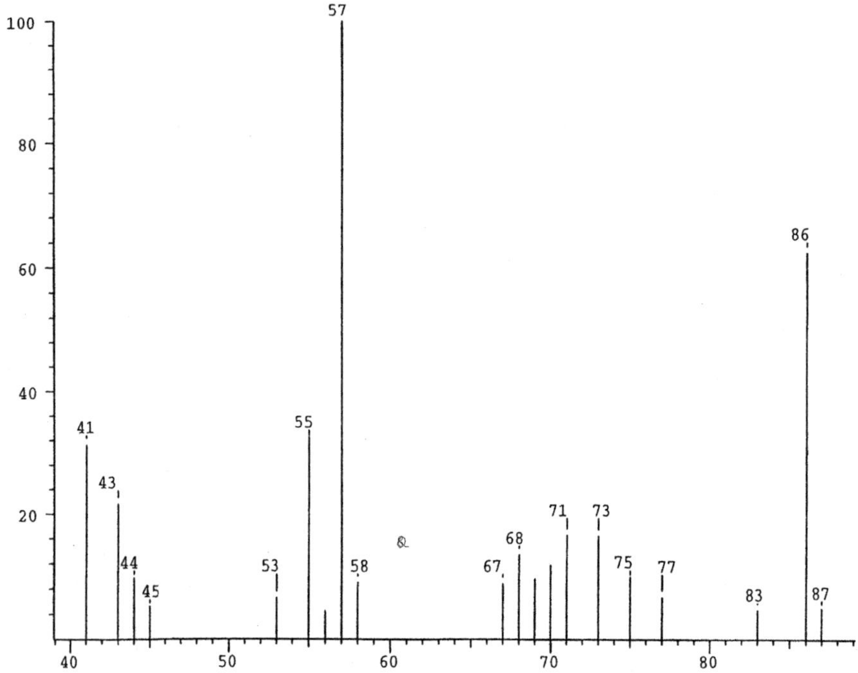

M/Z	Ion Intensity (%)	M/Z	Ion Intensity (%)
41	31.3	68	13.4
43	21.6	69	9.6
44	9.8	70	11.8
45	5.3	71	16.6
53	6.6	73	16.4
55	32.6	75	9.9
56	4.6	77	6.8
57	100.0	83	4.7
58	9.0	86	62.7
67	8.8	87	5.1

Spectrometer	: Finnigan Mat SSQ 700
Inletsystem	: Capillary GC/MS
Source Temperature	: 180°C
Electron Energy	: 70 eV
Scan Range	: 40-500

PM n°	CAS n°	Name
94960	00077-99-6	1,1,1-TRIMETHYLOLPROPANE

Solvent	$CDCl_3$
SF	250.13
SI	16384
SW	3125
NS	116

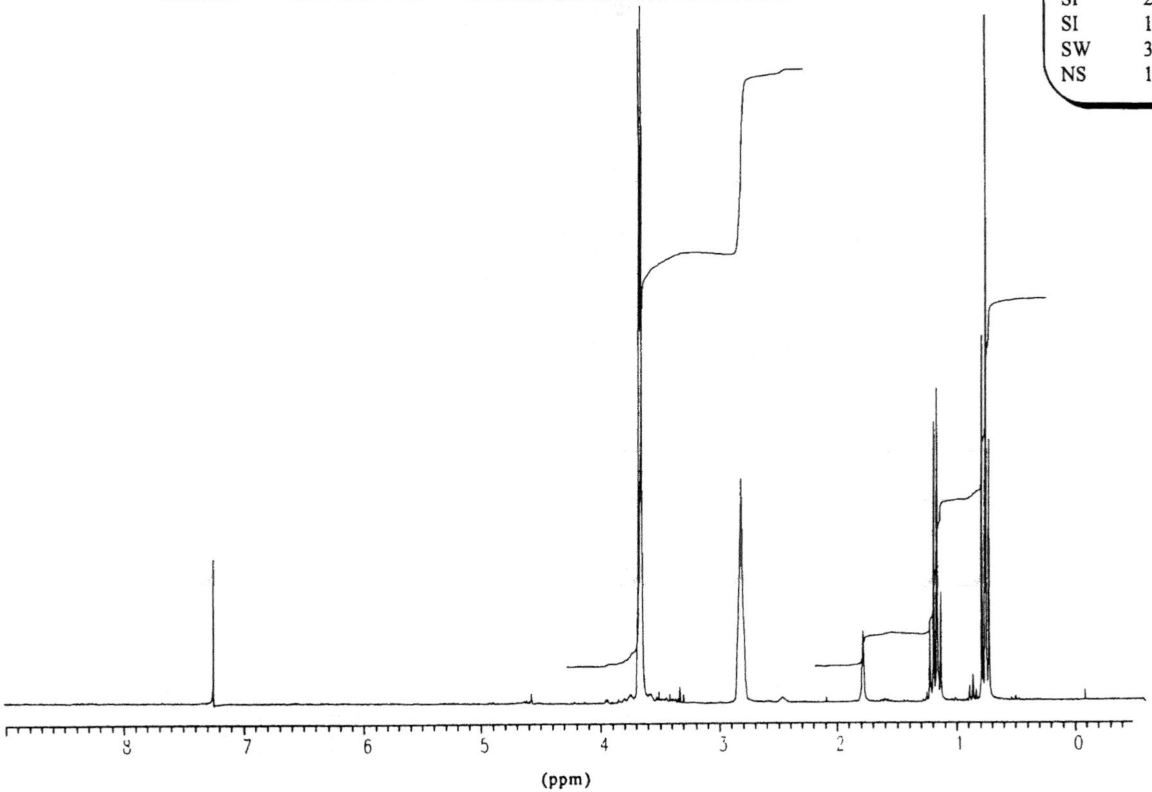

1,3,5-Trimethyl-2,4,6-tris(3,5-di-tert-butyl-4-hydroxybenzyl)benzene

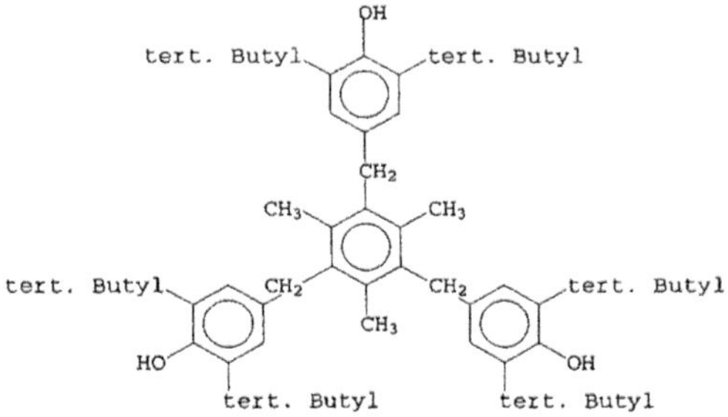

CAS No.	- 01709-70-2
PM Ref. No.	- 95200
Restrictions	- None
Formula	- $C_{54}H_{78}O_3$
Molecular weight	- 774

Alternative names - Phenol, 4,4',4"-((2,4,6-trimethyl-1,3,5-benzenetriyl)-tris(methylene)) tris(2,6-bis(1,1-dimethylethyl))-

Physical Characteristics - White powder, mp 240 - 245°C.
- Soluble in most organic solvents.

Handling - Store at room temperature.
Safety -
Availability - Standard sample was supplied.

Current uses - Antioxidant.
Applications - General use.

GC Retention time -

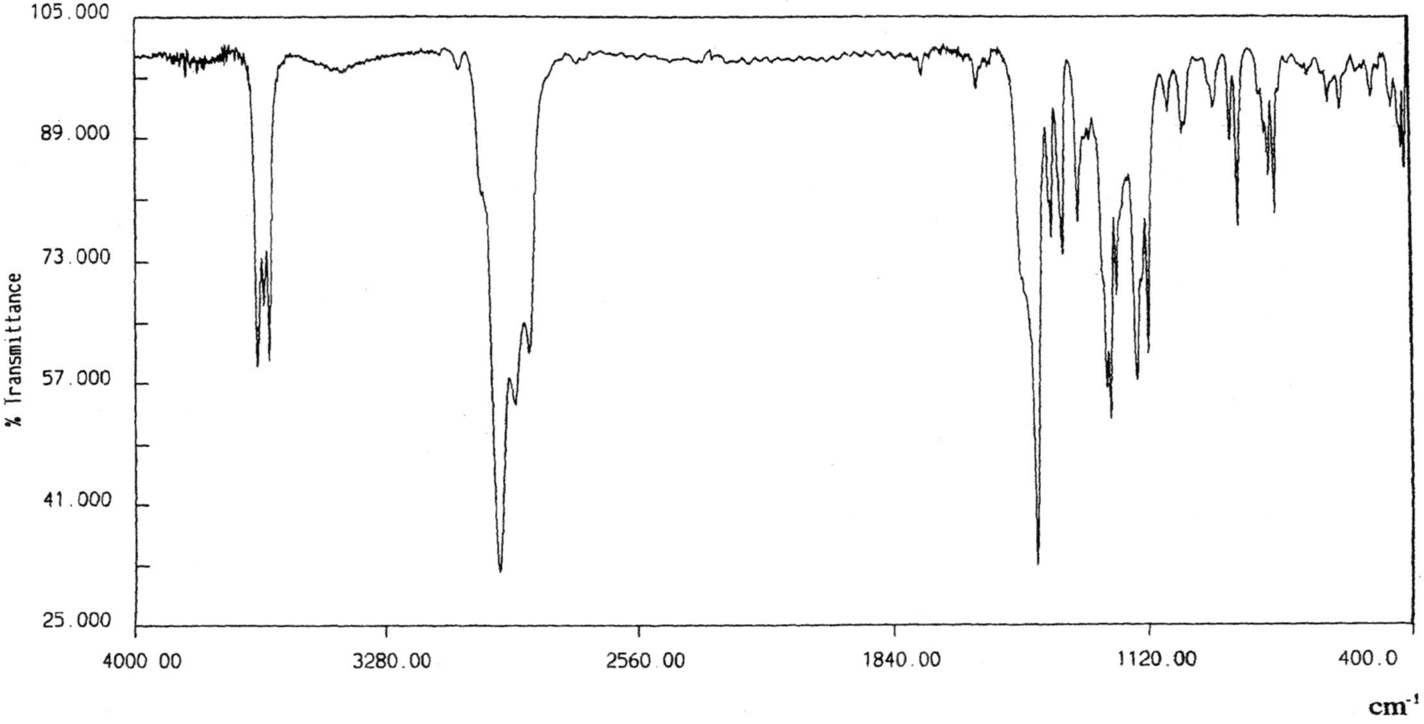

Name : 1,3,5-TRIMETHYL-2,4,6-TRIS(3,5-DI-tert-BUTYL-4-HYDROXYBENZYL)BENZENE

PM Nr : 95200 CAS Nr : 01709-70-2

KBr pellet

CAS No. 01709-70-2 PM Ref. No. 95200

1,3,5-TRIMETHYL-2,4,6-TRIS(3,5-DI-tert.BUTYL-4-HYDROXYBENZYL)BENZENE

Peak wavenumber	% T	Relative intensity
414,8	85,0	22,1
424,4	87,7	18,1
449,5	93,0	10,3
505,4	94,4	8,2
588,4	92,7	10,8
621,2	93,6	9,4
769,7	79,2	30,6
787,1	84,0	23,6
870,0	77,6	33,0
889,3	88,6	16,8
1024,3	89,5	15,5
1120,8	61,0	57,4
1153,6	57,5	62,6
1209,5	68,7	46,1
1226,9	52,3	70,3
1236,5	56,4	64,2
1317,5	78,1	32,3
1360,0	74,0	38,3
1390,9	76,1	35,2
1435,2	33,0	98,7
2872,4	61,0	57,4
2912,9	54,3	67,3
2957,2	32,1	100,0
3630,5	67,3	48,2
3649,8	59,2	60,1

1,3,5-Trimethyl-2,4,6-tris(3,5-di-tert-butyl-4-hydroxybenzyl)benzene

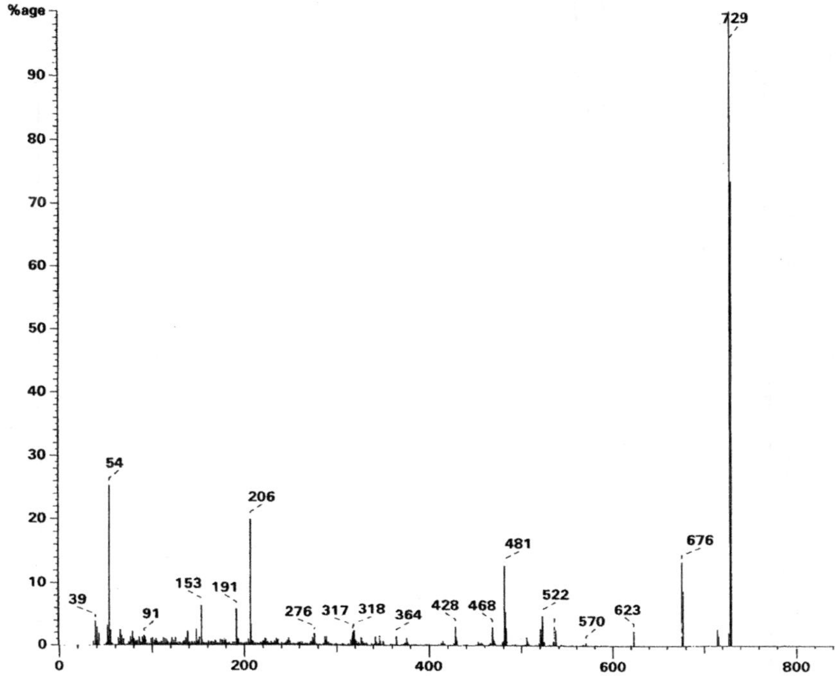

M/Z	Ion Intensity (%)	M/Z	Ion Intensity (%)
54	25.3	482	5.3
153	6.0	676	13.3
191	5.3	677	8.7
206	20.0	729	100.0
481	12.7	730	73.3

Spectrometer	: VG Autospec
Inletsystem	: Probe inlet
Source Temperature	: 350-400°C
Electron Energy	: 70 eV
Scan Range	: 40-1000

PM n°	CAS n°	Name
95200	01709-70-2	1,3,5-TRIMETHYL-2,4,6-TRIS(3,5-di-tert-BUTYL-4-HYDROXYBENZYL)BENZENE

Solvent	$CDCl_3$
SF	250.13
SI	16384
SW	3125
NS	42

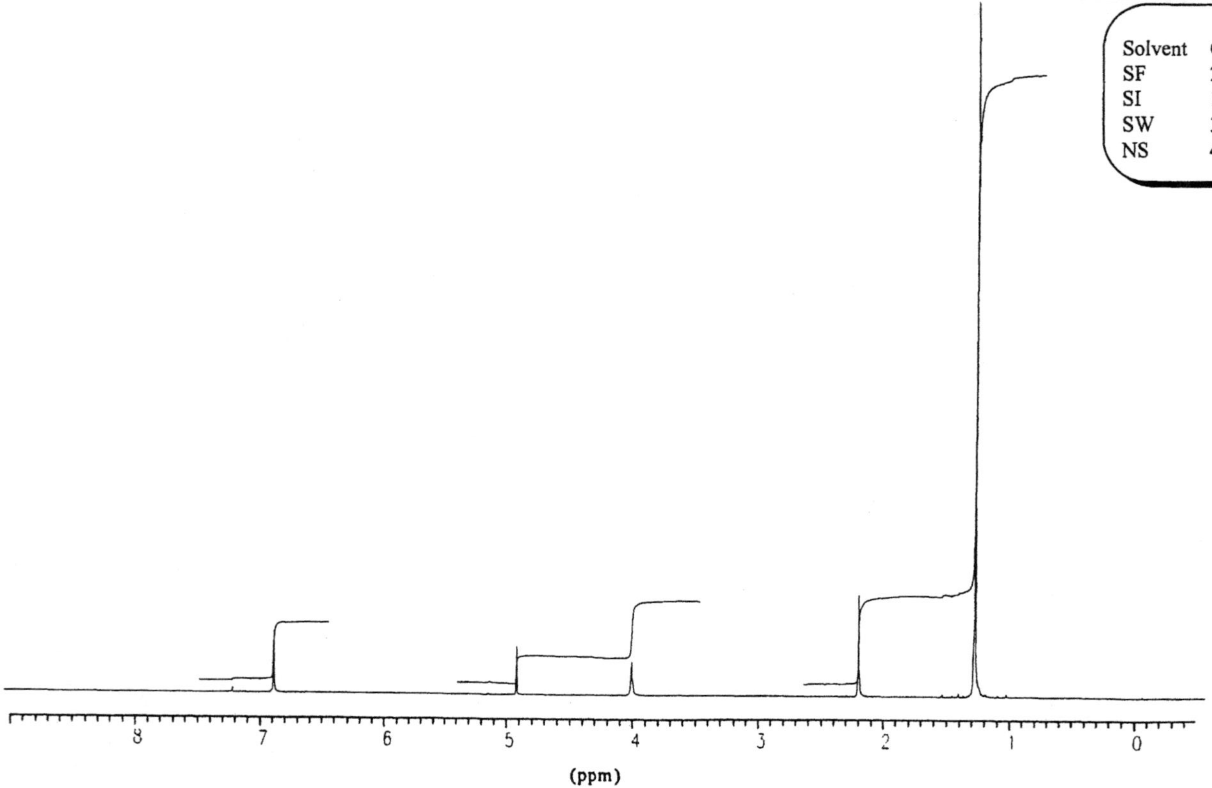

1,3,5-Tris(3,5-di-tert-butyl-4-hydroxybenzyl)-1,3,5-triazine-2,4,6(1H,3H,5H)trione

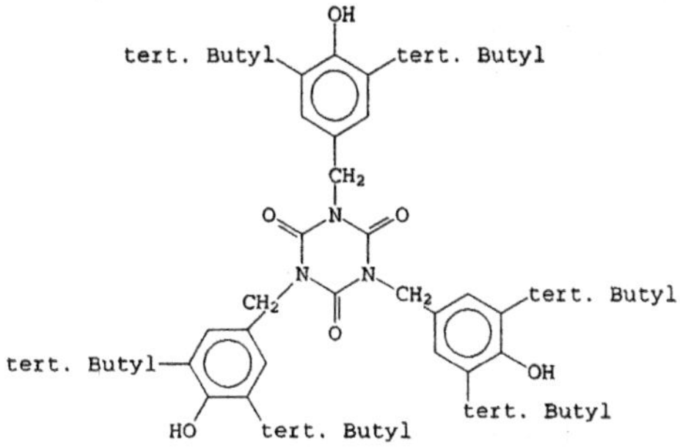

CAS No.	- 27676-62-6
PM Ref. No.	- 95360
Restrictions	- SML = 5.0 mg/kg
Formula	- $C_{48}H_{69}N_3O_6$
Molecular weight	- 784

Alternative names -

Physical Characteristics
- White powder, mp 219.5 - 225.5°C.
- Soluble in acetone, benzene, chloroform.

Handling - Store at room temperature.

Safety -

Availability - Standard sample was supplied.

Current uses - Antioxidant.

Applications - General use.

GC Retention time -

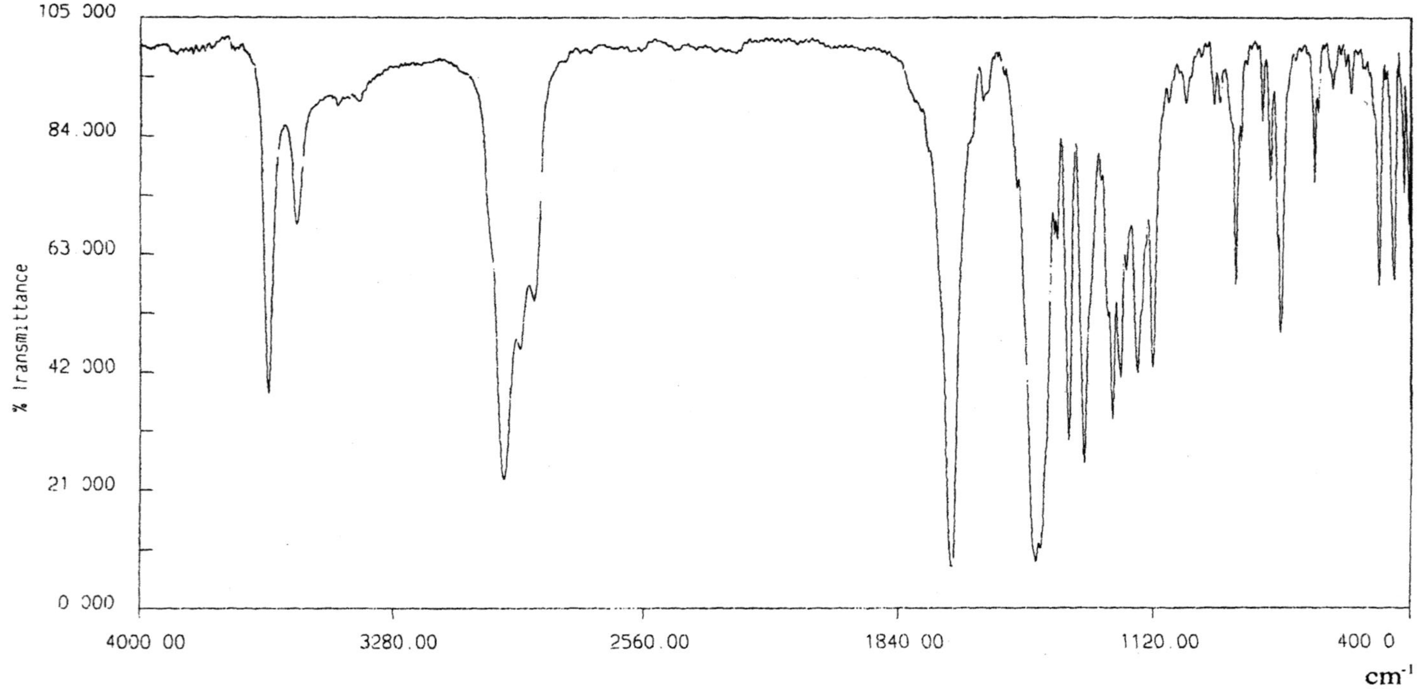

Name : 1,3,5-TRIS(3,5-DI-Tert-BUTYL-4-HYDROXYBENZYL)-1,3,5-TRIAZINE-2,4,6(1H,3H,5H)-TRIONE

PM Nr : 95360 CAS Nr : 27676-62-6

KBr pellet

CAS No. 27676-62-6 PM Ref. No. 95360

1,3,5-TRIS(3,5-DI-tert-BUTYL-4-HYDROXYBENZYL)-1,3,5-TRIAZINE-2,4,6(1H,3H,5H)-TRIONE

Peak wavenumber	% T	Relative intensity
420,5	73,9	28,2
451,4	57,8	45,5
490,0	57,6	45,7
657,8	88,1	12,8
669,4	75,7	26,2
763,9	49,1	54,9
792,8	75,0	27,0
814,1	86,7	14,4
889,3	57,5	45,9
1122,7	42,8	61,7
1167,1	41,9	62,7
1197,9	60,0	43,2
1213,4	40,9	63,8
1236,5	33,6	71,6
1315,6	25,6	80,3
1360,0	29,7	75,9
1390,9	65,6	37,1
1439,1	10,7	96,4
1452,6	8,2	99,1
1603,0	90,2	10,6
1691,8	7,3	100,0
2876,2	54,6	49,0
2916,7	46,1	58,2
2961,1	22,9	83,2
3377,8	90,0	10,8
3557,2	68,2	34,3
3638,2	38,2	66,7

1,3,5-Tris(3,5-di-tert-butyl-4-hydroxybenzyl)-1,3,5-triazine-2,4,6(1H,3H,5H)trione

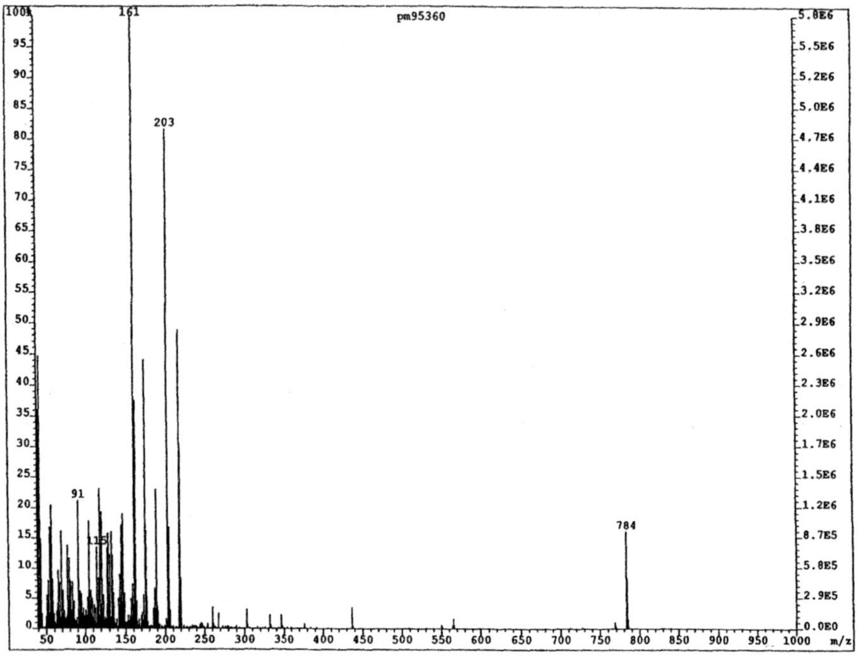

M/Z	Ion Intensity (%)	M/Z	Ion Intensity (%)
41	36.2	161	100.0
43	44.8	162	18.4
55	17.2	175	44.2
57	20.9	190	23.3
91	21.5	203	81.6
105	17.8	205	17.2
119	23.3	219	49.1
133	17.2	220	30.7
145	17.2	784	16.0
147	19.0	785	8.6

Spectrometer	: VG Autospec
Inletsystem	: Flow injection particle beam MS
Source Temperature	: 250°C
Electron Energy	: 70 eV
Scan Range	: 40-1000

PM n°	CAS n°	Name
95360	27676-62-6	1,3,5-TRIS(3,5-DI-TERT-BUTYL-4-HYDROXYBENZYL)-1,3,5-TRIAZINE-2,4,6(lH,3H,5H)-TRIONE

Solvent	$CDCL_3$
SF	250.13
SI	16384
SW	3125
NS	80

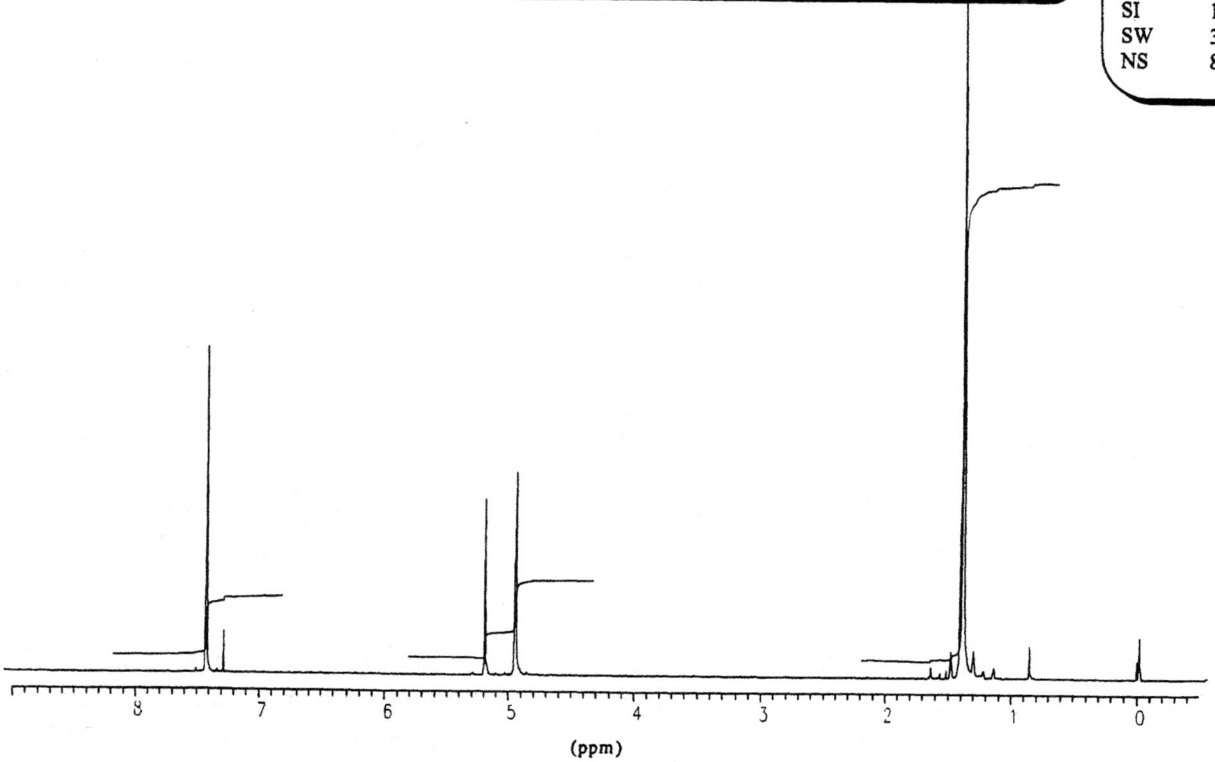

CONTENTS IN CAS NUMBER ORDER.

POLYMER ABBREVIATIONS

- CA	Cellulose acetate
- EVA	Ethylene vinyl acetate
- PA	Polyamide
- PBT	Polybutadiene terephthalate
- PC	Polycarbonate
- PE	Polyethylene
- PET	Polyethylene terephthalate
- PMMA	Polymethylmethacrylate
- PP	Polypropylene
- PS	Polystyrene
- PSO	Polysulfone
- PTFE	Polytetrafluoroethylene
- PU	Polyurethane
- PVA	Polyvinyl alcohol
- PVC	Polyvinyl chloride
- PVDC	Polyvinylidene chloride

GPSR Compliance

The European Union's (EU) General Product Safety Regulation (GPSR) is a set of rules that requires consumer products to be safe and our obligations to ensure this.

If you have any concerns about our products, you can contact us on ProductSafety@springernature.com

In case Publisher is established outside the EU, the EU authorized representative is:

Springer Nature Customer Service Center GmbH
Europaplatz 3
69115 Heidelberg, Germany

Batch number: 10755248

Printed by Printforce, the Netherlands